Molecular Panbiogeography
of the Tropics

SPECIES AND SYSTEMATICS

www.ucpress.edu/go/spsy

The Species and Systematics series will investigate fundamental and practical aspects of systematics and taxonomy in a series of comprehensive volumes aimed at students and researchers in systematic biology and in the history and philosophy of biology. The book series will examine the role of descriptive taxonomy, its fusion with cyber-infrastructure, its future within biodiversity studies, and its importance as an empirical science. The philosophical consequences of classification, as well as its history, will be among the themes explored by this series, including systematic methods, empirical studies of taxonomic groups, the history of homology, and its significance in molecular systematics.

Molecular Panbiogeography of the Tropics

Michael Heads

UNIVERSITY OF CALIFORNIA PRESS
Berkeley · Los Angeles · London

University of California Press, one of the most
distinguished university presses in the United States,
enriches lives around the world by advancing
scholarship in the humanities, social sciences, and
natural sciences. Its activities are supported by the
UC Press Foundation and by philanthropic
contributions from individuals and institutions.
For more information, visit www.ucpress.edu.

Species and Systematics, Vol. 4
For online version, see www.ucpress.edu.

University of California Press
Berkeley and Los Angeles, California

University of California Press, Ltd.
London, England

Library of Congress Cataloging-in-Publication Data

Heads, Michael J.
    Molecular panbiogeography of the tropics / Michael
Heads.
        p.   cm.—(Species and systematics; v. 4)
    Includes bibliographical references and index.
    ISBN 978-0-520-27196-8 (cloth : alk. paper)
    1. Biogeography—Tropics.   2. Biology—
Classification—Molecular aspects.   3. Variation
(Biology)—Tropics.   I. Title.
QH84.5.H43 2012
578.01'2—dc23   2011016690

19   18   17   16   15   14   13   12
10   9   8   7   6   5   4   3   2   1

The paper used in this publication meets the minimum
requirements of ANSI/NISO Z39.48-1992 (R 1997)
(*Permanence of Paper*).∞

Cover photograph: Mist rising from the Borneo
rainforest. Photo by Rhett A. Butler, mongabay.com.

Contents

Preface

The theme of this book is the distribution of plants and animals and how it developed. The subject is approached using the methods of pan-biogeography, a synthesis of plant geography, animal geography, and geology (Craw et al., 1999). The methodology is based on the idea that distribution is not due to chance dispersal; instead, range expansion (dispersal) and allopatric differentiation are both mediated by geological and climatic change. The biogeographic patterns discussed below mainly concern spatial variation in DNA, and so the subject can be termed "molecular panbiogeography." The book focuses on molecular variation in plants and animals as this shows such clear geographic structure. Molecular analysis has revealed an intricate, orderly, geographic pattern in most groups examined, even in those that are apparently well dispersed, such as birds and marine taxa. This molecular/geographic structure has often been described as "surprising," as, for example, by Worth et al. (2010), reporting on bird-dispersed trees in Winteraceae, and it is certainly impressive. The discovery of this structure has been one of the most exciting developments in molecular biology, and it has intriguing, far-reaching implications for evolutionary studies in general. This book analyzes and integrates inherited information at the largest scale—the geographic distributions—and at the smallest scale—the molecular variation.

Molecular research has had a revolutionary impact on all aspects of biology and has led to revised ideas on the evolution and classification of many groups. Yet molecular variation is just morphology on a small scale, and there is no real conflict between the traditional morphological data and the new molecular data. Traditional taxonomic groups that were well supported in morphological studies are often corroborated in molecular work, and many of the radical realignments suggested by molecular studies are in groups and areas that morphologists have acknowledged as difficult. With respect to biogeography, most patterns shown in molecular clades were already documented in earlier systematic studies of some group or other.

In order to assess the reliability and importance of a proposed phylogeny, it is necessary to know many details of the particular study. These include the sample size, the part or parts of the genome sequenced, the methods of establishing sequences, the methods of analyzing them in order to produce a phylogeny, and the statistical support of the groups. These are not provided here because the book is not about these parameters. In the same way, accounts of biogeography using morphological taxonomy do not cite the morphological characters that were used to construct the taxonomies. The distributional and phylogenetic data cited herein are introduced as "facts" for discussion, that, hopefully, the reader will accept. This may not always be the case, but most of the studies referred to are exemplary accounts and most of the clades mentioned have good statistical support.

The first two chapters in this book deal with general aspects of interpreting evolution in space and time. The next eight chapters comprise a biogeographic "transect" around the tropics, from America to Africa, Asia, the Pacific, and back to America. The book does not give a systematic, area-by-area treatment, and only selected localities are covered in any detail. Australasia is covered in a separate volume. The main aim in this book is to provide worked examples and to illustrate principles using a new method of analysis. The groups that are discussed were chosen because their distributions are reasonably well known and they have been the subject of recent, detailed molecular study.

Acknowledgments

I am very grateful for the help and encouragement I've received from friends and colleagues, especially Lynne Parenti (Washington, D.C.), John Grehan (Buffalo), Isolda Luna-Vega and Juan Morrone (Mexico City), Jürg de Marmels (Maracay), Mauro Cavalcanti (Rio de Janeiro), Guilherme Ribeiro (São Paulo), Jorge Crisci (Buenos Aires), Andres Moreira-Muñoz (Santiago), Pierre Jolivet (Paris), Alan Myers (Cork), Robin Bruce and David Mabberley (London), Gareth Nelson and Pauline Ladiges (Melbourne), Malte Ebach (Sydney), Rhys Gardner (Auckland), Frank Climo and Karin Mahlfeld (Wellington), Bastow Wilson and Robin Craw (Dunedin), and Brian Patrick (Alexandra).

1

Evolution in Space

Many different ways of analyzing *spatial* variation in biological diversity—the biogeographic patterns—have been employed by different authors, and some of the assumptions in these methods are discussed here. The *chronological* aspect of evolution is discussed in the next chapter.

Every kind of plant or animal has its own particular distribution and ecology, and this was already well understood in ancient times. Yet portraying a distribution is not straightforward. New collections are always being made and ideas on the delimitation of taxonomic groups change. Outline maps are generalized simplifications only but are useful for comparative purposes. Although dot maps showing sample localities give more detail, they are always incomplete, the accuracy of the dot locations can often be questioned, and the entities that the dots represent—the populations or individuals—are constantly changing position due to birth, death, and movement. A distribution is dynamic and so a distribution map represents an approximation, a probability cloud, not an actual distribution. Nevertheless, the fact that so many distribution maps have been made reflects the high value that biologists and many others have put on them.

This chapter incorporates material previously published in the *Biological Journal of the Linnean Society* (Heads, 2009b), reprinted here with permission from John Wiley and Sons.

Knowledge of organic distribution is useful for simple survival and economic development, as the plants, animals, and microorganisms of a particular place are often among its most distinctive and valuable features, and also its most poisonous and dangerous. Many groups have particular, idiosyncratic distributions; the details of these are known by local people and broader-scale distributions are documented in the literature.

Organisms are distributed spatially in three dimensions and while the questions treated in this book mainly involve differentiation in the horizontal plane, in latitude and longitude, the altitudinal component of a clade's distribution must also be considered. While the elevation of a group is sometimes assumed to reflect its ecological preference, in some cases there is an ecological lag and historical effects are important. For example, an area may be uplifted along with its biota, and some of the biota will likely survive to become montane taxa. Depending on where it is located, a population may be uplifted or not during an episode of mountain building, and so biogeography can determine ecology, rather than the reverse.

THE METHOD OF MULTIPLE WORKING HYPOTHESES

The focus in this book is on distribution patterns and their interpretation in terms of evolutionary processes. Most biogeographic interpretation over the last 2,000 years has been based on a single paradigm, the center of origin/dispersal model of historical development. But having only a single working hypothesis to explain a set of phenomena can lead to problems, and over time it becomes easy to accept that the single hypothesis is the truth.

Although much modern work in biogeography stresses supposed consensus, in science and philosophy, as in art and literature, a diversity of views and approaches can be a good thing. Puritans of all sorts (whether Oliver Cromwell or Louis XIV) cannot stand anyone having a view different from their own. The inflexible schemes of these great simplifiers, levelers, and systematizers can hold up progress for decades. In contrast, geologists (Chamberlin, 1890, reprinted 1965) and now molecular biologists (Hickerson et al., 2010) cite the method of "multiple working hypotheses," which proposes that it is never desirable to have just one working hypothesis to explain a given phenomenon. Accepting a single interpretation as definitive can be counterproductive and lead to the decline of a subject.

It is unfortunate that the interpretations of the data currently given in most molecular studies are all based on the same fundamental concepts. This "plug-and-play" biogeography involves the following steps: Assume that the study group has a center of origin and use a suitable program to find one; accept that fossil-calibrated clock dates give the maximum age of the group; describe possible dispersal routes from the center of origin. The axioms that are assumed here can be questioned, though, and a Socratic approach may be useful. Canetti (1935, reprinted 1962) wrote that "A scholar's strength consists in concentrating all doubt onto his special subject," and a healthy scepticism is one of the pillars of science, both in history and in everyday practice. When identifying unfamiliar plants and animals on the reef or in the rainforest, it is tempting, but often dangerous, to jump to conclusions before considering a wide range of possibilities, and the same is true for biogeographic interpretation.

The case studies of different groups discussed below adopt certain assumptions and concepts, and some of these are outlined next.

PHYLOGENIES, CLASSIFICATIONS, AND NESTED SETS: HIERARCHICAL SUMMARIES OF CHARACTER DISTRIBUTIONS

A related group of organisms forms a branch or clade in a phylogeny or evolutionary family tree. A clade may or may not be be formally named as a taxon (plural: taxa). The closest relative of a group is termed its sister group. In most published phylogenies, the clades in a group are shown in a strictly hierarchical system of nested clades. Phylogeny is the general process of the evolution or genesis of clades, and "a phylogeny" is also a term for a branching diagram or a tree, a symbolic arrangement of hierarchical, nested sets of clades. Nested sets of groups are depicted in traditional dichotomous keys, nomenclatural systems, cladograms, phylogenies, trees, and so on; all represent the same thing, an Aristotelian classification. This is only one way of representing variation; another is ordination, a method which shows trends rather than groups and which is often used in ecology.

Ideas on the evolutionary process still reflect the Aristotelian, classificatory approach in many ways. This sometimes leads to a misplaced emphasis on the clades rather than on the morphological and molecular characters that underlie them. The usual units of analysis in this book are indeed clades, as presented in molecular phylogenies, but these should not be taken too literally. Biogeographic areas may be problematic, and biological groups—"monophyletic clades"—may also be complex. Most groups have characters/genes that show phylogenetic and geographic variation within the group that is "incongruent" with those of other characters/genes, and this will be discussed below. Ultimately, in a hypocladistic approach, the focus is on the evolution of the underlying characters rather than on particular combinations of characters, including the clades.

IS THE SPECIES SPECIAL? THE DARWINIAN SPECIES CONCEPT

Evolution results in a continuum of differentiation. Entities may differ by a smaller or greater amount and, depending on this level, may be recognized as barely distinct populations, subspecies, species, genera, families, and so on. The focus here is on the process of differentiation rather than any of its particular products, and the species is seen here simply as a point on a trajectory between subspecies and genus; it has no "special" value. This is the species concept used by Darwin (1859) and Croizat (1964) (see also Ereshefsky, 2010). Neither the species nor any of the other taxonomic categories have any absolute value, and a species or genus in one group cannot necessarily be compared with species- or genus-level differentiation in another group.

In contrast with the Darwinian species, the species in the neo-Darwinian synthesis are very special indeed, as they have a reality that subspecies, genus, and the other categories do not. In this return to medieval nominalism, subspecies, genera, and other "universals" are seen as really just names and not things. Only species are real things ("individuals"). This distinction is not accepted in the Darwinian approach used here, in which clades (monophyletic groups) of any rank and their characters replace the species as the basic units of analysis. The most detailed information available on geographic differentation happens to concern "monophyletic" clades in morphological and molecular phylogenies, although geographic variation in any single character would be just as useful.

DEGREE OF DIFFERENCE

The interpretation of degree of difference (branch length) between groups is discussed in the next chapter. The particular degree of difference of a group and its taxonomic rank are not necessarily related to time; instead, they may reflect aspects of prior genome architecture in

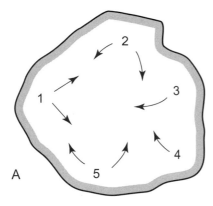

FIGURE 1-1. The distribution of hypothetical group A. Its center of origin might occur in the area with a highest diversity in the group (1), in the region of the oldest fossil (2), in the area of the most "advanced" form (3), in the area of the most "primitive" form (4), or in the area of the basal group (5).

the ancestor. The focus here is on spatial differentiation in any clades, whatever rank or branch length is involved.

SPATIAL ANALYSIS OF GROUPS: THE CENTER OF ORIGIN/DISPERSAL MODEL AND THE VICARIANCE MODEL

The center of origin/dispersal model and the vicariance model are often contrasted in biogeographic studies, and their applications have caused a great deal of debate.

Center of Origin/Dispersal Model

How did the distribution of a plant or animal develop? Consider a hypothetical distribution (Fig. 1-1). One theory is that such a pattern originated by a plant or animal evolving at a point somewhere within its current area and spreading out from there to the limits of its present range. Researchers attempt to locate the "center of origin" or "ancestral area" by studying the distribution and phylogeny in the group itself and by using different criteria. The center of origin has been thought to occur in the area that shows one or more of the following:

- the highest diversity of forms within the group ("1" in Fig. 1-1),
- the oldest fossil ("2" in Fig. 1-1),
- the most "advanced" form (cf. Darwin, 1859; Briggs, 2003) ("3" in Fig. 1-1),
- the most "primitive" form (cf. Mayr, 1942; Hennig, 1966) ("4" in Fig. 1-1), and

- the "basal" clade or grade of the group (most modern studies) ("5" in Fig. 1-1).

Several computer programs designed to find the center of origin of a group are now available, for example, DIVA (Ronquist, 1997) and Lagrange (Ree and Smith, 2008).

Many other criteria for locating a group's center of origin have been proposed in addition to those listed above, and the confusion that this implies was pointed out by Cain (1943). This paper led to the modern critique of the center of origin that has been developed in panbiogeography (Craw et al., 1999) and paleontology (López-Martínez, 2003, 2009; Cecca, 2008).

Cecca (2008) characterized two models of evolutionary biogeography: center of origin/dispersal theory as developed by Darwin (1859) and Wallace (1876), and vicariance, as developed by Sclater (1864) and Croizat (1964). In discussions of these models, the phrase "center of origin" does not simply mean a center where a group has originated (all taxa originate somewhere), but refers to a specific concept used in the dispersal model. In this model, a group's ancestor evolves as a monomorphic, homogeneous entity in a restricted area (the center of origin) following a chance dispersal event, and the group attains its distribution by physical movement out of this center.

Vicariance Model

Finding the center of origin of a group is a fundamental aim of many studies, and groups may be analyzed in ever-increasing detail in order to locate the center. The center of origin of a group is often located by examining the group itself. An alternative approach considers a group not on its own, but in relationship to its closest relative or sister group (Fig. 1-2). It may be difficult to understand the origin of a group by studying the group itself, especially if groups come into existence together with at least one other, by vicariance.

In many cases two sister groups have neatly allopatric (vicariant) distributions, with one group replacing or representing the other in a second area, often nearby or even adjacent to the first. Each of the two groups may have arisen not by spreading out from a point, but by geographic (allopatric) differentiation in its respective area from a widespread ancestor. In this process ("vicariance"), there is no physical movement, only differentiation, with populations in area A evolving

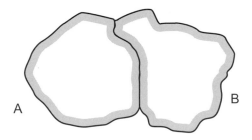

FIGURE 1-2. Group A and its sister group, B, two allopatric clades.

into one form and populations in area B into another. This vicariance theory is a basic component of panbiogeography. In a vicariance event, the distribution of a group comes into existence with the group itself. A group's "center of origin" may be more or less the same as its distribution, especially if it is part of an allopatric series. (A distribution range may expand or contract after its initial formation, leading to secondary overlap; this is discussed below.)

Despite the development of the vicariance model, the center of origin/dispersal model of the evolutionary process is still widely assumed by paleontologists (Eldredge et al., 2005), ecologists (Levin, 2000; Gaston, 2003: 81), and some biogeographers. For example, Cox and Moore (2010: 204) wrote: "Let us imagine that a species has recently evolved. It is likely, to begin with, to expand its area of distribution or *range* until it meets barriers of one kind or another." But this does not necessarily happen in a vicariance event, as the new species already abut their relatives. If a globally widespread form evolves by breaking down into, say, two allopatric species, one in the northern hemisphere and one in the southern hemisphere, neither one may expand its range. In a vicariance model, a new clade is just as likely to contract its range as to expand it.

In the vicariance approach, the focus is on tracing the originary breaks *between* groups, not on locating a point center of origin *within* a group. In a dispersal analysis, the first question is: Where is the center of origin? In a vicariance analysis, the first question is: Where is the sister group? The focus is not on the group itself or on details of its internal geographic/phylogenetic structure, but on its geographic and ecological relationship with its sister group and other relatives.

In this model, a group originates by the breakdown of a widespread ancestor, not by evolving at a point and spreading out from there. Analysis of any group can start either with a point center of origin or, alternatively, with a widespread ancestor. In the latter model, a group evolves on a broad front over the region it occupies, by "fracturing" with its sisters (vicariance) at phylogenetic and biogeographic breaks or nodes. A node is not a center of origin or an ancestor; it is a break where the distributions of two or more groups meet.

In one example, Chakrabarty (2004) supported a vicariance history for the freshwater fish family Cichlidae. He compared the process with a mirror being struck several times with a hammer. There is no movement of the individual shards, which are all neatly vicariant. The sequence of the hammer blows (i.e., phases of differentiation) is seen in the phylogeny or cladogram. The process also resembles the development of vascular tissue in a young organ out of ground tissue. There is no physical movement and the veins do not grow by pushing their way through tissue, but by differentiating *in situ*, in accordance with the genetic program.

Modes of Speciation: Dispersal and Vicariance

The mode of differentiation of groups in general (phylogenesis) and of species in particular (speciation) is problematic, and the interpretation of even the simplest cases is debated. As with clades in general, two allopatric species can be explained as the result of either *vicariance* in a widespread ancestor (dichopatric speciation) or *founder dispersal* from a center of origin (peripatric speciation).

Origin of the Ancestor

Two descendant groups may have originated by vicariance, but what about the ancestor of the two? Surely the ancestor must have dispersed to achieve its wide range? In fact, this is not necessary, as the ancestor of the two groups A and B, in areas A and B, may itself have originated as an allopatric member of a broader complex, the ancestor of A + B + C, that also occurred in area C. This in turn may have differentiated from the ancestor of A + B + C + D, as indicated by the four clades in Figure 1-3. Here there is no center of origin. In the center of origin/dispersal theory, each of the four allopatric groups in Figure 1-3 would have a separate center of origin, and their *distributions* are not directly related to their *origins*—the groups formed first and the distributions were established later. The boundaries of the four groups are secondary and the distributions only met after the four groups spread out from their respective centers of origin. Instead, in panbiogeography

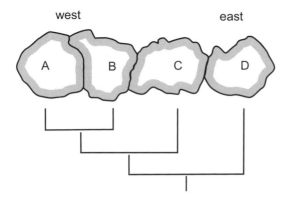

FIGURE 1-3. Four groups A–D with allopatric distributions in areas A–D. The phylogeny of the four groups is also shown.

the mutual boundaries are interpreted as phylogenetic and geographic breaks or nodes. These recur at the same localities in many different groups with different ecology, and so a chance explanation is unlikely.

Overlap in Distribution

In many cases, the allopatry between close relatives is not perfect and the groups show marginal overlap or interdigitation. This can represent local, secondary overlap by range expansion following the original, allopatric differentiation.

Sometimes a group shows extensive secondary overlap with its sister; this occurs mainly at higher taxonomic levels such as families and orders. A worldwide group and its worldwide sister may have each occupied half the Earth before the overlap developed. A vicariance analysis of the groups' history does not involve finding each group's center of origin, but tracing possible original breaks between the two groups. For example, one of the worldwide sister groups may be fundamentally northern, the other southern.

A few species, more genera, and many higher-level taxa, such as birds and flowering plants, are worldwide and overlap with each other everywhere. This pattern probably reflects the older age of the higher categories. The overlap of groups shows that vicariance cannot be the only biogeographic process. If it were, there would be pure allopatry—each point on Earth would only have one, locally or regionally endemic, life form. As it is, most places have biotas that include many kinds of plants and animals, indicating the overlap of clades. The overlap may be due to early phases of large-scale range expansion by whole communities.

If the affinities of any group are traced far enough, these will be found to make up a worldwide complex. Beyond this stage, if not before, there must be overlap with relatives. The widespread overlap among, for example, families and orders of birds indicates phases of range expansion. In primates, for example, the main branches—Old World monkeys, New World monkeys, lemurs, tarsiers—are notable for their high degree of allopatry. On the other hand, primates as a whole (together with their close relatives) show wide overlap with their sister group, the rodents and lagomorphs. At some stage (probably between the origin of primates and the origin of the main primate subgroups) there was a phase of overlap between the proto-primate complex and the proto-rodent complex.

Phases of vicariance affecting whole communities are often attributed to geological changes in the past, such as the opening of the Atlantic. In the same way, phases of population mobilism, with range expansion and colonization, probably occurred during and following the great geological revolutions. For example, the last, great phase of marine transgressions, in the Mesozoic, produced dramatically extended coastlines and associated habitats. Marine transgressions occurred on all the continents and, at the same time, rifting and continental breakup also produced new seaways. During this phase of mobilism, groups with suitable coastal, marginal ecology colonized vast areas of new habitat and became widespread globally. This was followed by a phase of immobilism through the Cenozoic, during which local differentiation predominated.

GEOLOGY AND VICARIANCE

It is often suggested that the main factor distinguishing evolution by vicariance and by dispersal is the time of appearance of the "barrier"—before evolution in dispersal and during evolution in vicariance. A more general point is that in a vicariance model, the Earth and its life evolve together, whereas in the dispersal model they do not. In dispersal theory, every taxon has its own unique history caused by one-off, chance events, and there are no community-wide biogeographic patterns with single causes. (Molecular clock studies based on dispersal theory also support this notion; this will be discussed in Chapter 2.) Yet many biologists would be reluctant to abandon the idea that geology can cause community-wide vicariance and generate both large- and small-scale community patterns. The idea that Earth and life evolve together is seen in the geographic concordance of many aspects of biogeography. This agreement even occurs among groups with completely different ecology, such as intertidal marine groups and montane groups. To cite just one example, the Hawaiian Islands and the Marquesas Islands form a center of endemism that is unexpected, given the direction of the currents, and yet it is defined by reef fishes, insects, and montane plants (see Chapter 7). Geographic congruence among different groups and also a general congruence between biogeography and geology have been known for a long time. For example, "The primary geographical divisions in the global mammal fauna clearly coincide with geology and plate boundaries" (Kreft and Jetz, 2010: 19). For example, in mammals, major phylogenetic/geographic breaks (nodes) occur between South America and Africa, and between Madagascar and Africa.

THE FOUR PROCESSES PROPOSED IN BIOGEOGRAPHY AND THE TWO THAT ARE ACCEPTED HERE

Four key processes have been proposed in biogeography. As discussed above, differentiation (e.g., speciation) can be due to *vicariance* of a widespread ancestor or to *founder dispersal* from a center of origin. In addition, two overlapping sister clades can be explained as the result of range expansion (by *normal ecological dispersal*, simple physical movement) or by *sympatric differentiation.*

Of the four processes just cited, *vicariance* and *normal ecological dispersal* are accepted as important by all authors. They are the two processes that are accepted in this book as explaining distributions. Normal ecological dispersal can involve movement within the distribution area or outside it, and this may lead to range expansion. Range expansion explains overlap; it does not explain allopatry.

The third process, *sympatric differentiation*, was controversial, although it is now accepted in some cases (Schluter, 2001; Friesen et al., 2007; Bolnick and Fitzpatrick, 2007). If it does occur, it is probably quite rare and many cases of supposed sympatry between sister groups prove, on closer examination, to involve only partial geographic overlap and significant allopatry. Other apparent cases of sympatry may involve allopatry at a small scale. As noted above, low-level clades are often allopatric with their sisters, whereas higher-level clades show more overlap, so overlap can generally be regarded as a secondary process that has developed over time from original allopatry, rather than by sympatric evolution.

The fourth process, differentiation by *founder dispersal*, is controversial and may not exist.

"DISPERSAL": ONE WORD, SEVERAL CONCEPTS

Three quite different processes have all been termed "dispersal," and they can be contrasted as follows.

Normal Ecological Dispersal. This is the normal physical movement seen in plants and animals. It includes daily and annual migrations, along with the dispersal of juveniles. The movement is made possible by the well-known mechanisms observed in different groups. Normal ecological dispersal occurs every day and does not lead to differentiation (speciation, etc.). It may take place over long distances—for example, in sea-birds—or over much shorter distances, depending on the organism. Following their origin by reproduction, all individual organisms have dispersed to where they are by this process. Normal ecological dispersal is seen in the weeds that soon colonize a disturbed area, whether this is a newly dug garden, an area of burnt vegetation, the area in front of a retreating glacier, a landslide, or a volcanic island such as Krakatau in Indonesia that has been devastated by a recent explosive eruption.

Despite appearances, this process of simple movement does not necessarily explain the distribution area occupied by a taxon—in particular, any allopatry with related taxa—as it does not account for evolutionary differentiation, and this can, by itself, produce a distribution. Thornton and New (2007) titled their book *Island Colonization: The Origin and Development of Island Communities*, yet studies on the colonization of Krakatau, for example, only concern the ecological origin and development of communities, not their evolutionary origin. The community on Krakatau is a subset of the weedy community that already existed on the islands in the region, and its evolutionary origin dates to long before the last eruption on Krakatau.

Range Expansion. Following a "normal dispersal" event, an organism's new position may lie within the former range of the taxon or it may lie outside it and represent a range expansion. Range expansion is seen in historical times in the anthropogenic spread of weeds and at other times in geological and evolutionary history. Range expansion, when it does occur, may be very rapid and a more or less local plant or animal may become worldwide in hundreds rather than millions of years. This takes place by normal ecological dispersal using the normal means of dispersal in the group, not the rarely used or unknown means sometimes cited to explain the more spectacular events of founder dispersal.

A global ancestor may have achieved its range during Mesozoic range expansion. This mobilism eventually stabilized and was replaced with a phase of immobilism through the Cenozoic. This was a period of *in situ* evolution that produced local differentiation, mainly at species and subspecies levels. Phases of mobilism may alternate with phases of immobilism in which allopatric evolution (vicariance) takes place. Earlier phases of population mobilism would have occurred as new landscapes emerged from the devastation of the Permo-Carboniferous ice ages, centered in the southern hemisphere and much more severe and long-lasting than the Pleistocene ice ages. These cycles of biogeographic mobilism and immobilism may take tens of millions of years to complete, as with the geological cycles of mountain uplift, erosion, deposition, and further uplift.

Naturally, it is far more difficult to analyze the biogeography of a time prior to the one in which the "modern," extant patterns developed, and many aspects of the premodern patterns will never be known. The modern centers of endemism, in their turn, will not last forever. A new geological or major climatic catastrophe will eventually lead to massive extinction and renewed mobilism, with weedy taxa taking over before they settle and establish new regional blocks of endemic taxa.

"Long-distance Dispersal"/"Speciation by Founder Dispersal." The defining feature of this process is not so much the long distance but the fact that it involves a unique, extraordinary dispersal event by a founder *across a barrier*. This leads to isolation and speciation (or at least some differentiation). As Clark et al. (2008) emphasized, there is an important distinction between dispersal as normal individual movement and range expansion on one hand, processes that are seen every day, and long-distance dispersal involving founder speciation on the other. The latter (termed "dispersal-mediated allopatry" in Clark et al., 2008) is a theoretical construction. Normal ecological movement and range expansion, along with other kinds of "dispersal" such as daily and annual migrations, are accepted here; long-distance dispersal/founder speciation is not.

As noted, every individual plant and animal moves as part of its normal means of survival, at least during one stage of its life cycle. With the exception of some colonial taxa, all individual organisms have reached their present position by dispersing there. This normal ecological movement should not be confused with "long-distance" or "founder" dispersal, which leads to new lineages. Dispersalists argue that "When lineages arrive in new habitats they will usually diverge and sometimes speciate" (Renner, 2005). But any patch of newly cleared garden will soon be colonized by "weedy" flora and fauna, later by less weedy taxa, and none of these will speciate there. Again, founder dispersal is quite distinct from normal dispersal. Authors supporting the center of origin/founder dispersal view for one or other group have often concentrated on proving that ecological dispersal, or ordinary movement, does occur, but this may not be relevant to the issue of founder dispersal.

For differentiation or speciation to occur, a fundamental change in the population ecology from a state of mobilism to one of relative immobilism has to occur. Dispersal on its own might explain how primates came to be in America if the American primates were the same as those of Africa or Asia. But physical movement on its own cannot explain why the American primates are different and form their own group. One main problem with founder dispersal is explaining how movement between populations could be occurring at one time, but then at some point stop or at least decrease (leading to differentiation). What is the reason for the crucial change from high rates of dispersal to low rates? In theory, this might be due to changing behavioral patterns in animals or means of dispersal in plants, but this cannot explain repeated patterns in unrelated animals and plants. Geological or climatic change is one obvious possibility, and this is the basis of vicariance. In center of origin theory, dispersal and speciation are instead determined by chance—the change from movement to no (or less) movement is created by a "barrier" which is permeable to a chance crossing by a single founder, but is then, somehow, impermeable to all others. In this view, the evolutionary biogeography of a group is due to chance, and so there is no need to examine any details of distributions that cannot be attributed to local ecology.

Dispersalists have sometimes suggested that island endemic taxa, for example, had much more effective means of dispersal in the past than they do now, and that these were lost with evolution, "trapping" taxa on an island (Carlquist, 1966a, 1966b). This is an ingenious and logical solution to the general problem that is often not mentioned—what causes the change from a phase of dispersal to a phase of no dispersal? Unfortunately, the idea of "loss of means" is probably wrong, as most endemics in most places have not lost their means of dispersal. But the fact that the idea was proposed at all indicates there is a problem that cannot be solved simply by citing "chance." Profound geological and ecological change is a more likely reason for cycles of immobilism–mobilism–immobilism.

Dispersal theory accepts that normal ecological dispersal and founder dispersal both occur in nature, whereas vicariance theory only accepts the first, but in any case it is important to distinguish between the two processes. The fact that a distinction is not made between contiguous range expansion (by normal dispersal) and across-barrier, founder dispersal is a serious drawback with programs such as DIVA (Kodandaramaiah, 2010). This conflation of the two different processes is a defining feature of dispersal biogeography (Matthew, 1915) and is also the basis of the confusing criticism that panbiogeography denies "dispersal."

To summarize: Most modern biogeographers follow Mayr (1982, 1997) in accepting that allopatry can be caused either by vicariance (dichopatry) or by founder dispersal (peripatry), but only vicariance is accepted here. Allopatry is accounted for by immobilism and vicariance, while overlap among groups can be attributed to range expansion and population mobilism ('dispersal').

Dispersal: "Any and All Changes in Position"

Many birds, primates, and other groups show daily and annual migrations that involve significant distances and are repeated through the millennia. These need to be accounted for in biogeography and ecology, although they are usually dealt with separately. Clements and Shelford (1939) realized the problem and introduced the highly generalized, rigorously geometric concept of "any and all changes in position." This would include changes in position due to physical movement or to evolution. The authors' suggestion that this concept be termed "dispersal" or "migration" was elegant but confusing and never caught on. This does not detract from the value of the concept. As with global phylogenies and the evolution of major groups, daily migrations of animals, even at a local scale, may reflect either current ecological conditions or past features such as former streams, rivers, or coastlines.

"BASAL" GROUPS

A phylogeny often has its main division, its basal break, between a small group and a more diverse sister group containing several clades. The smaller group is termed "basal," although strictly speaking only the nodes or breaks between groups are basal; no group is more or less basal than its sister group. The term "basal group" is thus potentially misleading, as a basal group is no more primitive than its sister, and is not ancestral to it (Krell and Cranston, 2004; Crisp and Cook, 2005; Santos, 2007; Omland et al., 2008). Nevertheless, by now the term "basal group" is widely used and understood in its purely topological sense, and it is a useful term for a smaller sister group. It should probably always be used in quote marks, to indicate the problem, but then terms such as "clade," "monophyletic," "dispersal," "center," "gene," and so on would have to be treated in the same way.

The phrases "sister to the rest of" and "basal in" are used here more or less interchangeably. The difference between the two is arbitrary and mainly nomenclatural—a basal group is considered to be part of the sister group and has the same name; a sister is a separate group and has a different name.

Although an ancestor would be basal in a phylogeny, a basal group is not necessarily ancestral or structurally primitive. For example, *Amborella* is likely to be the basal angiosperm, sister to all the rest, but Pennisi (2009: 28) went one step further and suggested: "Given that placement, *Amborella*'s tiny flowers may hint at what early blossoms were like." In fact, there is no reason why one (*Amborella*) or the other (all the other flowering plants) of the two sister branches should have a flower that is more primitive.

In the same way, the basal clade in a group is often interpreted as occupying the center of origin for the group, although this cannot be justified (Crisp and Cook, 2005). Likewise, morphological analysis may show that the oldest fossil clade in a group is phylogenetically basal to the rest, but it cannot be assumed to be ancestral to the others; it may simply be an extinct sister group.

Thus the idea that basal groups in a phylogeny are ancestral can be rejected as a generalization. Basal groups are simply less diverse sister groups, and their distribution boundaries may represent centers of differentiation in what were already widespread ancestors, not centers of origin for the whole group (Heads, 2009a).

Despite these arguments, modern phylogeographic studies often assume that a "basal" clade is primitive, ancestral, and located near the group's original center of origin, while advanced members of a clade have migrated away (Avise, 2000). This idea is derived from Mayr (1942) and Hennig (1966), who proposed that the primitive member of a group occurs at the center of origin. This is in contrast with the

Darwinian model, which assumes that an advanced form would out-compete the older forms and force them to migrate away (Darwin, 1859; Matthew, 1915; Darlington, 1966; Frey, 1993; Briggs, 2003). In this view, the center of origin is occupied by derived forms. The conflict between the Darwinians and the Mayr/Hennig/phylogeography school over the center of origin is irrelevant in the vicariance model, where there is no center of origin to begin with.

To summarize, a "basal" group is not ancestral; it is simply the smaller of two sister groups. Both will have the same age and neither one is derived from the other, or more advanced or primitive than the other.

BASAL GROUPS AND CENTERS OF ORIGIN

Good examples of basal group/center of origin analyses are seen in the extensive literature proposing dispersal into and out of the Caribbean. In the mockingbirds, Mimidae, a Yucatán endemic is basal to a largely Caribbean clade. Lovette and Rubenstein (2007: 1045) argued that this "is suggestive of a pathway of colonization into the Antilles from central America via Cuba." Conversely, in butterflies, the Greater Antilles genus *Antillea* is basal to a widespread clade (Phyciodina) of North, South, and Central America. So from a center of origin in the Antilles, "The ancestral Phyciodina colonized the [Antilles–Venezuela] landspan and spread south to the Guyanan Shield and then quickly to the Brazilian Shield" (Wahlberg and Freitas, 2007: 1265). (The subsequent scenario involved a convoluted history of transcontinental dispersals and back-dispersals, although the authors described these butterflies as "well-known to be relatively sedentary.")

In fact, no migration into or out of the Caribbean is required for the mockingbirds or the butterflies. In both groups, the location of the basal clade in the Yucatán/Greater Antilles region and the distribution of the rest of the group elsewhere can be explained by simple vicariance somewhere around Yucatán/Greater Antilles in an already widespread ancestor. The basal node represents an early center of differentiation in already widespread groups, not a center of origin. In a similar example, Sturge et al. (2009) wrote that molecular phylogeny "confirms" that the New World oriole *Icterus* (Icteridae) colonized South America from the Antilles, but this was only because South American species were nested in an otherwise Antillean clade and a widespread ancestor is a more parsimonious solution.

PHYLOGENIES CAN REPRESENT SEQUENCES OF DISPERSAL EVENTS
OR SEQUENCES OF DIFFERENTIATION EVENTS IN A WIDESPREAD
ANCESTOR

Consider the group of four taxa A–D shown in Figure 1-3 that are found in allopatric areas A–D and have a phylogeny (D (C (B + A))). In modern dispersal theory, the sequence of nodes in a phylogeny is read as a sequence of dispersal events, with taxa invading a new region, differentiating there, and then invading another region. The center of origin is occupied by the basal population in the basal group, D, and the phylogenetic sequence reflects a series of dispersal events from area D to C, B, and A. Each of the four taxa has its own individual center of origin somewhere within its range.

The model has been criticized by vicariance biogeographers because the simple allopatry among the four clades might not be due to dispersal but to a sequence of *in situ* differentiation events in an ancestor that was already widespread in A–D. The phylogeny (D (C (B + A))) would then reflect a sequence of breaks among the areas: D versus A + B + C, C versus A + B, A versus B. In this case, the sequence of differentiation shows a simple progression from east to west in Figure 1-3, and there is no physical movement.

In other cases, the phylogeny does not follow a simple geographic progression. Figure 1-4 shows a pattern in which the two sequential basal clades in a group, A and B, are not adjacent geographically and are separated by other groups, C and D. Dispersal theory would attribute this to jump dispersal. In vicariance theory it indicates that a widespread ancestor differentiated first at two basal nodes (between A and the rest, then between B and C + D) and finally at a node geographically between A and B.

Thus a phylogeny may convey the impression of a center of origin at the locality of the basal clade and "dispersal" from there, but if there was a widespread ancestor this is not necessary. Major disjunctions of tens of thousands of kilometers often occur between taxa at consecutive nodes on a phylogeny, even in groups in which long-distance, colonizing dispersal is improbable. Here it is especially likely that a phylogeny reflects a sequence of vicariance events. In many groups, differentiation has taken place repeatedly and more or less simultaneously around just a few globally significant nodes, such as the southwest Pacific basin and southwest Indian Ocean basin. Consider a phylogeny in five groups: Australia (Madagascar (Australia (Madagascar (Australia)))).

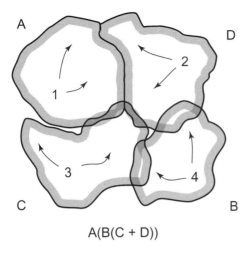

A(B(C + D))

FIGURE 1-4. Four groups A–D with allopatric distributions in areas A–D. The phylogeny of the four groups is also shown.

In a dispersal interpretation, this would require repeated long-distance dispersal events backward and forward between the two centers. A vicariance interpretation of the same pattern proposes repeated differentiation at the same nodes in a widespread ancestor, perhaps caused by reactivation of tectonic features.

Vicariance interpretations of phylogenetic sequences are given in some recent literature. For example, in the New World snake *Bothriechis*, earlier studies deduced a process of dispersal from Costa Rica to southern Mexico. Instead, Castoe et al. (2009: 98) interpreted the phylogeny as indicating "a more simplistic northward *progression of cladogenesis* that requires no inference of dispersal" (italics added). A similar pattern is also seen in other snakes of the area, "suggesting vicariance as the primary driving force underlying speciation." The Great American Biotic Interchange theory proposes dispersal across Central America, so the idea that evolution in the region may not have involved physical movement is of special interest. In another example, Doan (2003) interpreted a phylogeny of Andean lizards as reflecting a northward sequence of speciation in a widespread ancestor, rather than northward dispersal. In a botanical study, the phylogeny of *Rhododendron* (Ericaceae) in Malesia was interpreted as a geographical progression of cladogenesis (Brown et al., 2006). The same method of interpreting phylogeny used in these papers is adopted here.

Summing up, phylogenies of extant clades can indicate a sequence of divisions (nodes) between sister groups, rather than a sequence of ancestors and descendants. Dispersalists have adopted the second option, but this has only led to long-lasting, unresolved debates about the center of origin in particular groups, about how to locate the center of origin in the first place, and what the means of dispersal could be. In the model of evolution proposed here, there is no center of origin (other than the point of break, which is a margin rather than a center), there is no founder dispersal speciation, and there is no "radiation" from a center. If the ancestor is already widespread geographically (and probably also ecologically) before the differentiation of the descendant groups, the issue is no longer about how the modern groups "reached" a certain area, but how they evolved there—in other words, where the breaks occurred that led to their differentiation. Once the spatial context is clarified, the question of timing should be more straightforward.

DISPERSAL-VICARIANCE ANALYSIS (DIVA)

In the "dispersal-vicariance analysis" of Ronquist (1997), inferences of dispersal events are minimized as they attract a "cost." Extinction also attracts a cost, but vicariance does not. It was not explained why this approach should be taken and, as suggested above, it is based on a confusion of the two different concepts of "dispersal." Dispersal in the sense of ordinary movement should not attract any cost in any model. Jump or founder dispersal would attract no cost in a traditional dispersalist model, although in a vicariance model of speciation or evolution it is rejected *a priori*.

In most modern studies, the spatial analysis of phylogeny has been based on the idea of a center of origin, and so authors employ programs, such as DIVA, that will often find one. Authors looking for a particular center of origin sometimes complain that DIVA will find a widespread ancestor if, for example, all the extant groups are allopatric. But even when they are not, a widespread ancestor can still be proposed, as original allopatry may have been obscured by subsequent range expansion or extinction. There is no logical need to interpret a phylogeny as a series of dispersal events.

GROUPS THAT ARE RECIPROCALLY MONOPHYLETIC IN TWO AREAS

If one group occurs on a mainland, a, and its sister group occurs on a much smaller island, b, (Fig. 1-5), the island group is often assumed to

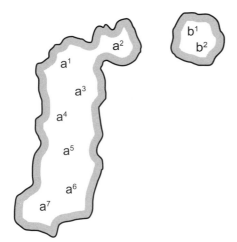

FIGURE 1-5. Two sister groups, a and b, on a mainland and an island. The phylogeny is: $(a^1 + a^2 + \cdots a^7)$ $(b^1 + b^2)$, and the mainland clade and the island clade are reciprocally monophyletic.

have been derived from the mainland group by dispersal. The island forms are predicted to be related to particular populations in their large sister group. Yet well-sampled molecular studies now show that in many of these cases the phylogeny has the pattern: $(a^1, a^2, a^3 \ldots)$ $(b^1, b^2, b^3 \ldots)$, where the superscripts indicate different areas within a and b. The groups in the two areas are reciprocally monophyletic and the group in b is not related to any one population in a. In this type of pattern, dispersal can still be salvaged as an explanation, but only if it occurred prior to any other differentiation in the groups, and this is often unlikely. On the other hand, reciprocal monophyly is the standard signature of simple vicariance of a widespread ancestor at a break between a and b. Even if groups in the two areas a and b are not reciprocally monophyletic, vicariance is still possible, and this is discussed next.

GROUPS WITH A BASAL GRADE IN ONE REGION OR HABITAT TYPE

A monophyletic clade includes all the branches derived from a single node. A paraphyletic group or *grade* comprises several sequential branches of a phylogeny, but does not include all the branches derived from a node (e.g., in Fig. 1-3, the clades B, C, and D, but not A). Many groups comprise a *basal grade* located in one area, *A*, and a disjunct population or clade in a second area, *B*. The pattern is usually explained as the result of dispersal of the clade from *A* to *B*. Instead, a grade located in a single area may represent a phase of differentiation

area *A*   area *B*

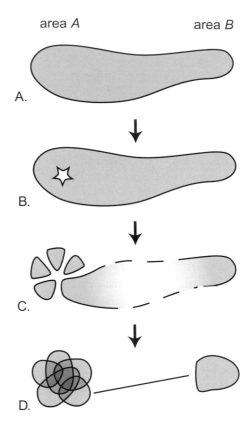

A.

FIGURE 1-6. A hypothetical example of distribution in a taxon currently found in two areas. A. A widespread ancestor. B. This begins to differentiate around a node (star) associated with the formation of a mountain range or inland sea, for example. C. The ancestor has differentiated into five allopatric clades, four with a narrow range and one widespread. Their ranges begin to overlap while some of the populations of the widespread clade suffer extinction (broken line). D. The clades now overlap but the ranges still show traces of their original allopatry. Following extinction of populations between areas *A* and *B*, the outlier in *B* may appear to be a secondary feature and the result of long-distance dispersal.

there, not a center of origin (Heads, 2009b). For example, within a widespread ancestor already in *A* and *B* (Fig. 1-6), allopatric evolution may occur around a node at *A*. If this is followed by secondary overlap at *A* and extinction of populations between *A* and *B*, this will produce a basal grade in area *A* (Fig. 1-6D). In actual cases, the overlapping clades in area *A* often show slight but significant differences in their distribution (as indicated in Fig. 1-6D), and these may represent traces of the earlier phase of allopatry. Many biogeographic analyses treat all the species in an area such as *A* as having the same distribution, overlooking any allopatry, and this can confuse analysis. To summarize: A dispersal analysis interprets the pattern described here, with a basal grade in *A*, by long-distance dispersal from *A* to *B* "across a barrier." Instead, a vicariance analysis infers differentiation in a widespread ancestor followed by local overlap within *A* by normal means of dispersal.

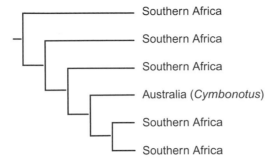

FIGURE 1-7. The phylogeny and distribution of clades in the Arctotidinae (from Funk et al., 2007).

Examples of Groups with Basal Grades in One Area

The Arctotidinae (Asteraceae) are a good example of the basal grade pattern. The group comprises a basal grade of three southern African clades as well as *Cymbonotus* of southern Australia, which is sister to two other southern African genera (Fig. 1-7; Funk et al., 2007). *Cymbonotus* is sister to a southern African clade and embedded in an Australian–southern African clade; it is not embedded in a southern African clade. It is likely that the southern African clades show significant differences in their distributions within the region.

In another example, *Protea* (Proteaceae) is diverse in the Cape region of South Africa and also has a few species widespread throughout tropical Africa. Valente et al. (2010) found that "non-Cape species are nested within a wider radiation of Cape lineages and all except two of them belong to a single clade." "Therefore," the authors suggested, "most extant lineages outside the Cape originated by in situ diversification from a single ancestor that arrived there from the Cape" (p. 746). This logic is not accepted here (cf. Fig. 1-6). The authors only interpreted the phylogeny in terms of a dispersal model because of the programs they used; an alternative vicariance scenario was not considered. Nevertheless, the sister genus of *Protea* is *Faurea*, centered in tropical Africa (where *Protea* has low diversity) and Madagascar (where *Protea* is absent); a vicariance analysis would be simple and of considerable interest.

In a third example, Grandcolas et al. (2008: 3311) argued that "within certain New Caledonian groups, multiple species are nested within larger clades with taxa from Australia, New Zealand or New Guinea, calling for explanations in terms of recent dispersal [to New Caledonia]." Thus the phylogeny: (Australia (Australia (Australia (New Caledonia)))) is taken to reflect a center of origin in Australia, where the "basal grade" occurs. An alternative explanation for the pattern would be differentiation in a widespread Australian–New Caledonian ancestor, as in Figure 1-6.

Center of Origin/Basal Grade Theory and Ancestral Habitat Reconstruction

As discussed, a basal grade does not necessarily indicate a geographic center of origin, and this argument also applies to ecology, as a basal grade is often thought to occupy the ancestral habitat. For centropagid crustaceans, Adamowicz et al. (2010) concluded that "Species occupying saline lakes are nested within freshwater clades, indicating invasion of these habitats via fresh waters rather than directly from the ocean or from epicontinental seas" (p. 418). But, using the same argument given above for geography, this ecological phylogeny: (freshwater (freshwater (freshwater (saline lakes)))), does not necessarily mean that freshwater habitat was the center of origin for the saline lake clade. The ancestor of the whole group may have occupied both freshwater and saline lakes and itself be derived from a marine ancestor, for example, following marine transgression and regression.

In other crustaceans, the spiny lobsters, Palinuridae, are widespread in warmer seas and especially common around Australasia. Tsang et al. (2009) reasoned that the three genera restricted to the southern high latitudes (*Jasus*, *Projasus*, and *Sagmariasus*) are the basal lineages in the family, "suggesting a Southern Hemisphere origin for the group." In the same way, the authors assumed that the basal groups indicated the ecological center of origin. For one clade, they wrote, "the shallow-water genus *Panulirus* is the basal taxon in Stridentes, while the deep-sea genera *Puerulus* and *Linuparus* are found to be derived. This indicates that the spiny lobsters invaded deep-sea habitats from the shallower water rocky reefs and then radiated." Again, the habitat of the basal taxon is not necessarily a center of origin for the other groups; the ancestor may have already been widespread in both deep and shallow water before it differentiated into the modern genera.

GROUPS WITH A BASAL GRADE IN ONE REGION AND WIDESPREAD DISTAL CLADES

Many groups have a basal grade in one region, as in the last pattern, and also have a widespread distal clade.

Cichorieae

The dandelion tribe Cichorieae (= Lactuceae) of Asteraceae is cosmo-politan and has its basal clades and also its sister group (Gundelieae), that is, a basal grade, in the Mediterranean region (Funk et al., 2005). As usual, this does not mean the Mediterranean basin was a center of origin from which the tribe itself spread. Instead, the region may have been an early center of differentiation in an already global ancestral Cichorieae.

Asteraceae

The family Asteraceae as a whole, the largest plant family, with ~24,000 species, is another example of a group with a basal grade in one region and a widespread distal clade. Its basal group, subfam. Barnadesioideae, is in South America. The sister groups of the family are centered on the eastern Pacific margin (Calyceraceae of South America) and western Pacific (Goodeniaceae, mainly in Australia; Stevens, 2010). Bremer (1994) concluded that "the geographic origin of the Asteraceae most probably involved South America and the Pacific"—a vast area—whereas Stuessy et al. (1996) instead suggested a much smaller area (about 200 miles across) in southern Argentina. The conflict reflects different approaches to the center of origin concept and to fossils: Stuessy et al. (1996) wrote that a vicariance model "seems unlikely because available fossils provide no evidence...," whereas Bremer (1994) accepted that "groups may be much older than their fossil record." Stuessy et al. (1996) suggested that Barnadesioideae are not only the sister group to all the other Asteraceae, they actually "gave rise" to the rest of the family. This is unnecessary and unlikely. The "sister as ancestor" theory is often adopted, as it is compat-ible with a localized center of origin model, but Bremer (1994) took a broader perspective in suggesting that the origin of the Asteraceae could have been linked to the history of the Pacific area.

A recent phylogeny for the Asteraceae (Fig. 1-8, from Funk et al., 2005, and Panero and Funk, 2008) is: (Barnadesioideae (Mutisioideae (Stifftioideae (Wunderlichioideae (Gochnatioideae + seven remaining subfamilies))))). The sister group of the whole family, Calyceraceae, and the five basal branches are all small groups mainly found in South America, although some have outliers elsewhere. (Mutisioideae have a few representatives in Africa and North America, and one genus, *Liebnitzia*, is in China and Mexico. Wunderlichioideae have two gen-era in Southeast Asia. Gochnatioideae also occur in Central America, North America, and Asia.)

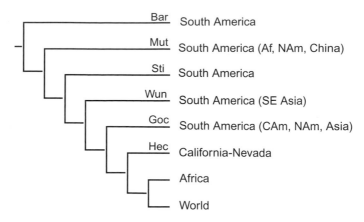

FIGURE 1-8. Phylogeny of Asteraceae (from Funk et al., 2005; Panero and Funk, 2008). Areas listed in brackets have lower levels of diversity (Af = Africa, CAm = Central America, NAm = North America). See main text for subfamily names.

Most authors accept that this phylogeny indicates a center of origin in South America. Instead, it may reflect evolution of a worldwide ancestor in which the modern groups differentiated at breaks in or around what is now South America. The process is the same as that shown in Figure 1-6, with South America equivalent to the area on the left, although in Asteraceae the distal clade has subsequently expanded its range to include South America. Funk et al. (2005) wrote, "it appears incontrovertible" that the family itself had a center of origin in southern South America. But there is no reason why the site of the initial *splits* in the modern group should indicate the *distribution* of the ancestor, and the authors did not consider the possibility of a globally widespread ancestor.

Some of the oldest known fossil flowers of Asteraceae are from the Eocene of Patagonia. These show affinities with subfamily Mutisioideae and also the distal clade Carduoideae. Barreda et al. (2010) regarded the fossils as support for the hypothesis of a South American origin of Asteraceae and an Eocene age of divergence. Yet the fossils are also compatible with a much broader area of origin for the family and much earlier divergence within the family. The authors suggested that "an ancestral stock of Asteraceae may have formed part of a geoflora developed in southern Gondwana." Another possibility is that the group was already more or less worldwide when it originated by splitting from its sister group, Calyceraceae, in or around its range in southern South America.

After the five basal South American groups diverged from the remaining Asteraceae, the monotypic *Hecastocleis* of California and Nevada (mountains in the Mojave Desert) is sister to the rest, within which a large African group is sister to the remaining, cosmopolitan clade. Funk et al. (2005) wrote that the South American differentiation followed by the African phase "might suggest a Gondwanan origin for the family." Again, the observed phylogeny could instead have developed from an ancestor that was already cosmopolitan. Differentiation in the global ancestor occurred first around phylogenetic/biogeographic nodes or breaks in or near South America, then California (part of the Pacific, not Gondwana), and finally Africa, and this could reflect a sequence of differentiation events, not a route of dispersal.

Funk et al. (2005) wrote that "the few data from pollen records and geology seem to indicate a more recent [Cenozoic] origin for the family" and relied on this date, along with the phylogeny, in deducing intercontinental dispersal. However, this is not necessary if the fossil pollen dates and calibrations are only minimum, not absolute dates and migration leading to overlap occurred over the very different geography of the Mesozoic. As Funk et al. stressed, the general perception of ecology in Asteraceae as simply "weedy" is incorrect. Some species are indeed cosmopolitan or pantropical weeds, yet the great majority are restricted-range endemics. The main clades are notable for their conspicuous geographic centers of diversity in different areas—for example, Stifftioideae in eastern Brazil, Liabeae in Peru, Heliantheae s.lat. from Mexico to the northern Andes, Calenduleae in South Africa, and Gundelieae around the Mediterranean. The extensive regional endemism means that while the family is very large, there are relatively few global clades (several of the larger tribes and subtribes, a few large genera). Thus in the history of the family there have only been a small number of widespread ancestors (e.g., groups such as Senecioneae and Astereae each require their own global ancestor). These few ancestors may have undergone a phase of active mobilism in the Mesozoic during which they occupied much of Earth's land surface, before settling down into a Cenozoic phase of immobilism and speciation. Although many modern species of Asteraceae are narrow endemics, even these often retain a weedy ecology within their local centers of endemism, occupying unstable sites such as cliffs and rocky outcrops on steep mountain slopes.

OTHER GROUPS WITH BASAL CLADES OR GRADES IN SOUTH AMERICA

The following groups resemble Asteraceae in having their basal nodes in or around South America and provide further illustrations of the principles discussed here.

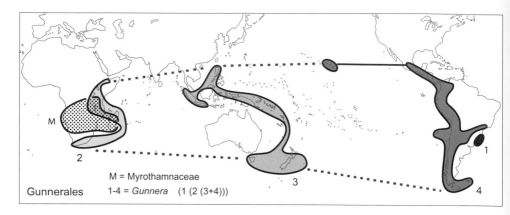

FIGURE 1-9. Gunnerales: *Gunnera* (Gunneraceae) and Myrothamnaceae (distribution from van der Meijden, 1975, phylogeny from Wanntorp et al., 2002).

Gunnera *(Gunneraceae)*

The distribution of the plant *Gunnera*, the only genus in Gunneraceae, is shown in Figure 1-9 along with its sister, the Myrothamnaceae. The molecular phylogeny of *Gunnera* (Wanntorp et al., 2002) matches the morphological classification into subgenera and sections (van der Meijden, 1975). The basal clade is a very small plant of Uruguay and southern Brazil (Rio Grande do Sul, Santa Catarina) that forms dense mats on seepages in coastal sand dunes. The rest of the genus occurs in the mountains of the tropics and the south temperate zone, although it is absent from eastern tropical America and western tropical Africa. Fossil pollen from the Early Cretaceous is recorded in the southern continents and also from North America (Stevens, 2010). The Myrothamnaceae replaces *Gunnera* in large areas of south-central Africa.

Reading the phylogeny of *Gunnera* from the bottom to the top, the first phylogenetic break occurred around what is now southeastern Brazil, isolating the basal clade from adjacent populations in South America and Africa. The second break occurred in the Indian Ocean (between clades 2 and 3), the third in the Pacific (between 3 and 4). The phylogeny follows a geographic sequence from South America eastward, and this would usually be interpreted as a sequence of dispersal events. Instead, starting with a widespread ancestor, the sequence of phylogeny could represent a series of vicariance events, a wave of evolution passing around the Earth through the population.

In a dispersal analysis, the focus is on the question: How did *Gunnera* cross the Atlantic, Indian, and Pacific Oceans? In a vicariance analysis, the focus is instead on the phylogenetic and geographic breaks, their exact locations, and their possible geological causes. If *Gunnera* evolved by vicariance with Myrothamnaceae (perhaps at a node between western and eastern Africa), it may have already been widespread in the southern continents at the time of its origin.

Verbenaceae, Solanaceae, and Bignoniaceae

Marx et al. (2010) proposed that these plant families all originated in South America and colonized the Old World on multiple occasions, but this was only because the basal clades in each family are in South America. A widespread ancestor/vicariance interpretation is an alternative possibility.

Grasses (Poaceae)

The basal clade in the grass family is Anomochlooideae (Bouchenak-Khelladi et al., 2010), which comprises two rainforest genera:

Streptochaeta: throughout mainland tropical America,

Anomochloa: coastal Brazil (Bahia).

The pair highlights the significance of a break at or near coastal Brazil.

Genlisea (Lentibulariaceae)

Genlisea occurs in tropical America (mainly in the east; absent in the Andes) and in Africa. It comprises two main clades (Fleischmann et al., 2010):

1. Subg. *Tayloria*: southeastern Brazil
2. Subg. *Genlisea*: Africa + South America, including southeastern Brazil.

Fleischmann et al. (2010) adopted a center of origin model and because of the basal grade in South America inferred dispersal from South America to Africa (and then back again to South America). In an alternative, widespread ancestor/vicariance model, the primary split is between Africa + America (except southeastern Brazil) on one hand,

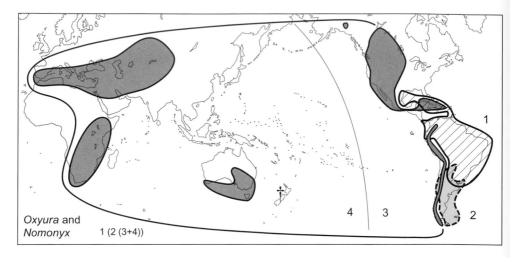

FIGURE 1-10. Distribution of the ducks *Oxyura* (two main clades in light gray and dark gray) and *Nomonyx* (diagonal lines). Breeding ranges only shown. *O. jamaicensis* of the western United States is a seasonal migrant to the eastern United States, where there are also rare breeding records, but these are not shown here (cf. maps at www.massaudobon.org and www.natureserve.org). The New Zealand fossil record of *Oxyura* (dagger) is from Worthy (2005). The nested sequence of numbers indicates the phylogeny (from McCracken and Sorenson, 2005).

and southeastern Brazil on the other. The only dispersal required is a range expansion of clade 2 into southeastern Brazil to account for the overlap with clade 1. The split in clade 2 at the Atlantic occurred after the split between clades 1 and 2 around southeastern Brazil.

A Group with a Basal Grade in South and Central America: *Oxyura* and *Nomonyx* Ducks

Allocation of clades to *a priori* geographic areas, such as the continents, in the initial stages of biogeographic analysis has often involved incorrect assumptions of sympatry. This in turn has led to the idea that the "areas of sympatry" were centers of origin. Biogeographic analysis does not require the use of any area other than those defined by the taxa themselves. For example, the ducks *Oxyura* and *Nomonyx* (Fig. 1-10) form a globally widespread clade with a phylogeny that could be presented as: (America (America (America + Old World))). Although this is accurate descriptively, it would be misleading, as the basal clades are presented as sympatric and "America" could be misinterpreted as an area and as a center of origin. In fact, the groups are largely allopatric, and the phylogeny is better summarized as: (eastern America (southern America (western America + Old World))). Although there are clades in different parts of America, no clade defines South America or America, and these geographic entities may not have been important for the biogeographic history.

For the *Oxyura–Nomonyx* group, McCracken and Sorenson (2005) suggested a center of origin in lowland South America, as the two basal members occur there. Subsequently, a complex series of intercontinental dispersal events led to the modern distribution. Yet this scenario does not explain the fact that the three main clades are largely allopatric, with secondary overlap restricted to the Greater Antilles, Colombia, and northern Argentina. There is an overall east/west split (with, for example, *Nomonyx* in eastern Mexico, *Oxyura* in western Mexico), and the three main clades meet at a node in northwestern Argentina. This may be near the original break. The three Old World clades are also allopatric (in Eurasia, Africa, and Australia), and so the simplest scenario is a worldwide ancestor breaking up into six vicariant clades, with some local overlap. Testing this idea would require examining the patterns in detail at intercontinental and also local, ecological scales and seeing if they are shared with other groups.

If we break "America" down into its components, the phylogeny for the group of ducks is: (Brazil + Greater Antilles (Argentina (western North America + Greater Antilles + Andes) (Old World))). The first division is between Brazil–Greater Antilles (*Nomonyx*) and the rest of the world (*Oxyura*), with breaks around the Greater Antilles, Colombia/Ecuador, and northern Argentina. America does not appear as a monophyletic area, but as a composite of eastern and western sectors, each with endemic taxa that are not sisters. This is compatible with the tectonic division of the Americas into eastern (cratonic) and western (orogenic/accreted terrane) provinces. The second division in the phylogeny involves breaks in western Argentina, perhaps around the basins that were later incorporated in the uplift of the Andes. The third division implies differentiation of the western North America–Antilles–Andes group and its Old World sister. Ducks are known from Cretaceous fossil material, and this provides a useful minimum age for the group (Clarke et al., 2005).

Populations of *Oxyura* and *Nomonyx* inhabit lakes, swamps, and sometimes brackish water (Carboneras, 1992; Kear, 2005). *Nomonyx* is recorded in mangrove. In North America, *O. jamaicensis* breeds inland in the northern prairies and south into the intermontane basins and valleys of the western United States, reaching the coast in southern California. In the northern Andes, *O. j. andina* occurs at 2,500–4,000 m. These populations are now montane and even alpine, but they may have been derived from ancestral complexes which lived in the mangrove swamps and lakes of the pre-Andean Cretaceous basins. Uplift in the Andean region began in the Cretaceous, and the birds here would have been lifted up with the land, a process suggested for Andean parrots by Ribas et al. (2007). In the ducks, most of the populations that differentiated in Brazil as *Nomonyx* lay too far east to be caught up in the main Andean orogeny and remained in the lowlands.

CASE STUDIES: BASAL NODES AROUND THE PHILIPPINES AND MADAGASCAR

The same methods used above can be applied to distribution centered in and around the Indian Ocean basin. This is an important biogeographic pattern, and the three examples cited next illustrate sequences of differentiation that are typical for groups in the region.

Old World Rats and Mice

All four subfamilies of the family Muridae are restricted to the Old World. Strangely, despite the weedy ecology of some members, the group was completely absent from the Americas before human introductions. In contrast, all five subfamilies of the related Cricetidae are indigenous in America. The two families overlap in Eurasia.

One of the murid subfamilies, the Murinae, includes typical rats and mice and occurs throughout the Old World. In Murinae, one of the clades in the Philippines, the pair *Batomys* and *Phloeomys*, is sister to all the rest (Steppan et al., 2005; Rowe et al., 2008). Nevertheless, "[w]hether the Philippine Old Endemics represent a relictual distribution from the periphery or the core of the ancestral range of the Murinae cannot be determined" (Steppan et al., 2005: 382). In its geography and its phylogeny the Philippines group is peripheral to the main bulk of the Old World group and might be thought to have "budded off" its more widespread sister. But this would have to have occurred before any other differentiation in the group; instead, the Philippines group could represent the sole remains of a group that was formerly more widespread in the Pacific region.

As with America, the Philippines archipelago is a geological composite, made up of many fault-bounded blocks of crust with independent histories, or terranes (see the Glossary for this and other geological terms). Some Philippines terranes have an Asian origin, while others formed further east in the central Pacific (Metcalfe, 2006). It is possible that the two main clades of Murinae—the small basal clade in Philippines and the rest—are, respectively, west Pacific and Indian Ocean in origin. Secondary juxtaposition of the two followed the collision of many Pacific terranes with New Guinea and the Philippines.

Terrane accretion in the west Pacific also explains the strong biogeographic connection between the Philippines and northern New Guinea, including its offshore islands. In the main group of Murinae, beyond the basal genera, Steppan et al. (2005) and Rowe et al. (2008) found a close relationship between another Philippines clade (*Apomys*, *Archboldomys*, etc.) and a large clade termed the "Old endemic genera" of New Guinea and Australia. In this group, three of the four tribes are mainly in New Guinea. Steppan et al. (2005) described this Philippines–New Guinea/Australia connection as "perhaps [their] most surprising finding." In geology, the sector New Guinea–northern Moluccan Islands–eastern Philippines is recognized as a belt of deformation; Pubellier et al. (2003, 2004) suggested that the portion of the Philippine Sea plate carrying the Taiwan–Philippine arc may have originated closer to Papua New Guinea. Biogeographic connections between the Philippines and New Guinea may be explained in this way.

Steppan et al. (2005) suggested that the murines "appear to have originated in Southeast Asia and then rapidly expanded across all of the Old World," but this was only because "three of the four basal branches . . . include taxa almost entirely restricted to South east Asia." In fact, the four clades have quite different distributions (Rowe et al., 2008), as follows:

- the Philippines (the basal group),
- Southeast Asia,
- Southeast Asia and west to Eurasia and Africa,
- Southeast Asia and areas to the east and south (Philippines and New Guinea/Australia).

In a simple vicariance model, a widespread Old World/Pacific ancestor has undergone early differentiation around the Philippines/Southeast Asia. Subsequently, the clades have been juxtaposed and there has been range overlap. Rowe et al. (2008: 97) suggested that "The pattern that emerges from these phylogenies [of Murinae] is of rapid and probably adaptive radiations after colonization of landmasses previously unoccupied by muroid-like rodents." Yet there is no special reason to assume that the phylogeny reflects a sequence of dispersal events rather than differentiation in an already widespread ancestor.

Otus (strigidae)

The main group in the owl genus *Otus* comprises three allopatric clades, structured as follows (Fuchs et al., 2008):

1. Africa, Pemba Island (Tanzania), Mediterranean to central Asia (*O. scops*, etc.)
2. Philippines (*O. mirus* and *O. longicornis*)
3. Madagascar, the Comoros, the Seychelles, Sri Lanka, and eastern mainland Asia to northern China (*O. sunia*, *O. rutilus*, etc.)

The phylogeny is: 1 (2 + 3). The first break is in Mozambique Channel; the next is around the Philippines. As usual, the two breaks may represent successive divergence events in an already widespread ancestor, rather than dispersal across the Indian Ocean and then back again. It is easy to see how even a simple phylogeny of a widespread group with allopatric clades can generate a convoluted dispersal scenario if every differentiation event requires physical movement.

Pachychilidae

The freshwater gastropod family Pachychilidae (Köhler and Glaubrecht, 2007) has a widespread Indo-Pacific distribution, with the clades distributed as follows:

1. Philippines
2. Madagascar
3. India to southern China and Java
4. Sulawesi, Torres Strait, Central America

The phylogeny is: 1 (2 (3 + 4)). As in *Otus*, the sequence of differentiation "jumps" across the Indian Ocean, in this case from around the Philippines to around Madagascar. Clade 4 is a typical trans-tropical Pacific affinity; this pattern is discussed further below (see Chapter 6).

Magpie Robins

The magpie robins, *Copsychus* s.s. (Muscicapidae), comprise three main clades (Lim et al., 2010):

1. Philippines
2. Madagascar
3. Seychelles
4. India to China and Borneo

The phylogeny is 1 (2 (3 + 4)), giving a sequence of differentiation events similar to that of the pachychilids, despite the very different ecology and means of dispersal in the two clades.

CHARACTER INCONGRUENCE IN MODERN CLADES AND POLYMORPHISM IN ANCESTRAL COMPLEXES

The center of origin/dispersal model of biogeography is based on the idea that the ancestor of a clade (either a single parent pair or at most a uniform species) is monomorphic and that the clade had a small, single center of origin. The new taxon is separated from the ancestor by a chance, one-off dispersal event. In this model, each character in the ancestor has only one state, and so all characters in the new taxon are either primitive (resembling the ancestor) or new. The new taxon is not simply a recombination of ancestral polymorphism and does not "emerge" over a broad region.

In contrast, vicariance theory stresses that a hierarchical tree diagram does not represent all aspects of phylogeny. A phylogeny is a summary diagram only and cannot portray all aspects of differentiation, such as characters showing variation that is "incongruent" with the phylogeny. For example, in three genera with a phylogeny a (b + c), a character may occur in a and b, but not c. Although hierarchical classifications are often useful, they are only summaries and have limitations. The many structural features, both morphological and molecular, that underlie the clades can be distributed in different ways in the phylogeny and may show different geographic patterns. The characters and their variation within a clade may date back to before the origin of the clade as such, for example, if evolution has proceeded by hybridism or incomplete lineage sorting. In the latter case, character incongruence in a phylogeny reflects polymorphism that was already present in the ancestor before it began to differentiate into the modern taxa. In this

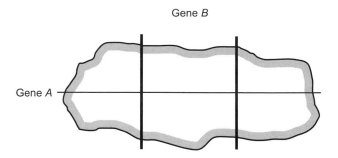

FIGURE 1-11. Distribution of a hypothetical clade (gray) in which geographic patterns of variation in two parts of the genome lie at right angles to each other.

way, as in hybridism, new clades may emerge over a broad range as new recombinations of ancestral characters.

Many examples of groups are now known in which the variation and geography of one gene is incongruent with that of another, and this cannot be depicted in a single tree. Workers on some groups do not take the tree metaphor as seriously as others. For example, in prokaryotes, horizontal gene transfer is pervasive and "very few gene trees are fully consistent, making the original tree of life concept obsolete" (Puigbó, 2009). Botanists often feel the same way, and there is increasing appreciation of the role of hybridism in animals. In the case of incomplete lineage sorting, the ancestral polymorphism may be topologically incongruent with the phylogeny of the modern descendants and also geographically incongruent in its geographic distribution.

Figure 1-11 shows a hypothetical clade in which patterns of variation in two parts of the genome show geographic trends lying at right angles to each other. One of the patterns may represent variation in the ancestor. Figure 1-12 shows an actual example from the fish *Nemadactylus*, in which the pattern of variation in cytochrome *b* is phylogenetically and geographically incongruent with clades shown in D-loop sequences (Burridge, 1997). Burridge discussed whether this pattern was due to hybridism or to incomplete lineage sorting and concluded in favor of the latter. In any case, patterns of "incongruent" variation have long been known in morphological studies and are now well documented in many molecular accounts. It seems that lineages do not necessarily evolve in a linear or hierarchical way and that phylogeny may develop through the recombination of ancestral characters, rather than the evolution of any new, uniquely derived characters.

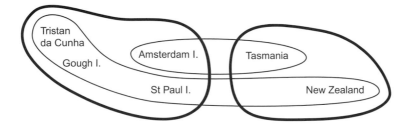

FIGURE 1-12. Patterns of variation in the fish *Nemadactylus* (incl. *Acantholatris*) (Cheilodactylidae) (Burridge, 1997). Thick line = clades defined by cytochrome *b* sequences, fine line = clades defined by D-loop sequences.

A typical example of incongruence occurs in *Bystropogon* (Lamiaceae), endemic to the Canary Islands and Madeira. Chloroplast DNA supports a relationship of *Bystropogon* with New World groups, while nuclear DNA shows connections with Old World groups (Trusty et al., 2004). The authors concluded that due to the "apparent conflict . . . we are not certain of the true biogeographic relationship of *Bystropogon*" (p. 2004). In a vicariance framework, both sets of affinities are accepted as valid and there is no real conflict. The New World, Macaronesian, and Old World genera each evolved *in situ* out of a widespread ancestor by recombination of ancestral characters.

EXTINCTION, FOSSILS, AND CENTERS OF DIVERSITY: *NOTHOFAGUS* (NOTHOFAGACEAE), THE SOUTHERN BEECHES

All clades go extinct, and so extinction is clearly a major factor in distribution. How does it affect analysis of larger clades? Other things being equal, extinction may occur first in areas of lower diversity, and this would be most likely in large, widespread clades. This idea was tested in the tree genus *Nothofagus*, the only member of Nothofagaceae (Heads, 2006a). This genus was studied as it is a diverse, widespread group with a fossil record that is exceptionally good relative to most other groups.

Nothofagus trees dominate forests in many parts of Australasia and southern South America. In a dispersal paradigm, this transoceanic distribution presents two problems: Where is the center of origin? And how exactly did *Nothofagus* move from Australasia to South America (or vice versa)? Instead, if the problem is seen in a broader phylogenetic and geographic context, the first question is how to integrate the distribution of Nothofagaceae with that of its relatives. Nothofagaceae are the basal group in the Fagales (oaks and beeches, etc.), a group that is distributed all around the world. The different families of Fagales have their respective centers of diversity in different places: Nothofagaceae in Australasia and southern South America, Casuarinaceae in Australia, Rhoipteleaceae in China and Southeast Asia, Fagaceae in North America and Eurasia, Ticodendraceae in Central America, and so on. This suggests that Nothofagaceae and the other families arose by vicariance in a widespread ancestor that was already present in Australasia and South America before Nothofagaceae evolved.

There was probably dispersal (range expansion) in the early Fagales, before they attained their worldwide distribution and differentiated into families. In contrast, the families themselves show a large degree of allopatry, and so there may be no need to infer any great range expansion of *Nothofagus* itself, for example, across the South Pacific.

After the allopatric evolution of the families, overlap has developed around their margins. For example, Fagaceae extend south from the northern hemisphere to New Guinea and overlap there with *Nothofagus* (which is usually at higher elevation). Casuarinaceae have a clear center of diversity in Australia, but a few widespread species occur in New Guinea, and a few fossil species are known from New Zealand and South America. In New Caledonia, both families are diverse. Thus, although the distribution of Casuarinaceae overlaps that of *Nothofagus* completely, with one exception this only occurs in areas where the levels of diversity in Casuarinaceae are low. The exceptional overlap in New Caledonia, geologically a composite island, is a frequent pattern.

Vicariance with some secondary overlap occurs between *Nothofagus* and its relatives, and also occurs within *Nothofagus*. Counting extant species, the four subgenera have vicariant centers of diversity as follows:

Subg. *Fuscospora* New Zealand,

Subg. *Brassospora* New Guinea and New Caledonia,

Subg. *Lophozonia* Chile north of Chiloé Island,

Subg. *Nothofagus* Chile south of Valdivia.

All four subgenera are known from Upper Cretaceous fossils. Thus, the process of differentiation from within the order to within the genus may have been caused by vicariance and there is no need for any center of origin or long-distance, transoceanic dispersal. The only dispersal required is that needed to explain the secondary overlap of *Nothofagus* with Fagaceae (in New Guinea) and Casuarinaceae (especially in New Caledonia) and local overlap among the *Nothofagus* subgenera.

What has been the affect of extinction on the biogeography of *Nothofagus*? *Nothofagus* is an abundant, wind-pollinated tree with an extensive and diverse fossil record from the Cretaceous on. Many fossil species have been described, and if these are added to the extant ranges, the total distributions of the subgenera are all enlarged and overlap. Nevertheless, the centers of subgeneric diversity do not change; they are the same as those cited above for the extant species alone (Heads, 2006a). This is compatible with the idea that, in general, total extinction occurs first outside a group's center of diversity rather than in the center of diversity itself. If the areas of overlap among the *Nothofagus* subgenera do represent secondary range expansions, the original differentiation of the subgenera may have been purely allopatric.

Most clades go extinct eventually, but the main centers of diversity, as in the families of Fagales and the subgenera of *Nothofagus*, may survive extensive and long-lasting phases of "erosion" and local extinction. A group with 100 species in South America, one in Hawaii, and one in Taiwan may suffer extinction, but in terms of biogeography it is likely to be the outlying single species that disappear, not the center of diversity in South America.

It is sometimes suggested that a local or regionally endemic clade may have formerly occurred everywhere and that extinction alone has produced its distribution. Fossils show that extinction has reduced the range of many taxa (for example, the *Nothofagus* subgenera) and wiped out others completely. But biogeography has to explain the data available, both fossil and extant records, rather than invoking possible extinction as a reason to give up analysis and, in any case, extinction may tend to occur in areas that were only occupied secondarily to begin with. As Allwood et al. (2010: 676) concluded: "[W]ithout fossils, hypotheses on extinction are ad hoc, unable to be tested and can be used to explain almost any biogeographic pattern . . . while acknowledging the potentially confounding effects of extinction, we have no evidence that this process has misled our biogeographic interpretations."

DISPERSAL AND TWO CONCEPTS OF CHANCE

A key factor often cited in chance dispersal is the "long distance" involved. On the other hand, chance or jump dispersal is also used to explain differentiation over short distances, for example, across a river. Thus the defining feature of chance dispersal is not so much distance as the fact that the physical movement by the founder is a single, freak event. The process is "chance," not in the sense of being analyzable statistically, as normal ecological dispersal is often analyzed, but in the sense of not being analyzable at all. In this model, plants and invertebrates that are restricted to dense rainforest and move only meters in their lifetime are proposed to have moved—just once—thousands of kilometers across open ocean to attain a distribution in, say, New Guinea and Colombia. No further analysis is given, even if there are no obvious means of dispersal. Chance dispersal is justified because "anything can happen given enough geological time," but the process is unfalsifiable and explains all distributions and none at the same time.

The new, revolutionary idea of chance as a calculated probability began with Pascal and Fermat in the 17th century and became a founding principle of modern science. For example, a probability-cloud diagram of seed dispersal depicts how "chance" (in the new, modern sense) determines aspects of normal ecological dispersal. On the other hand, "chance dispersal" invokes chance in the old sense of "factors beyond our understanding" or "given enough time anything can happen"; "chance" here is a very different concept and is not a real explanation for anything. If the process leading to biogeographic distribution is due to "chance" in this old sense, every clade would have its own individual biogeographic history independent of all others. There would be no real biogeographic patterns, only pseudopatterns including pseudovicariance, that is, vicariance caused by chance geographic coincidence of chance dispersal events at different times. Community-wide patterns of allopatric differentiation caused by a new mountain range or a new channel of the sea, for example, would not occur. In contrast, molecular work has shown high levels of precise vicariance in most taxa, including widespread marine species and even bacteria (Fenchel, 2003). In protozoa, the amoeba *Nebela vas* was found to have a typical Gondwana plus South Pacific islands distribution (Smith and Wilkinson, 2007), contradicting the usual paradigm of microbial distribution in which "everything is everywhere." Foissner (2006), reviewing microbial biogeography, also stressed endemism in different regions, including Laurasia, Gondwana, and west Gondwana (eastern South America, Africa). Foissner compared the patterns with those of spore plants, "many of which occupy distinct areas, in spite of their minute and abundant means of dispersal" (p. 111).

The profound geographic structure evident in communities is not explained simply by appealing to chance. Authors who conclude that long-distance dispersal has occurred often admit that the evident, normal means of dispersal in the studied organism do not appear sufficient for the postulated dispersal events. Sometimes authors suggest that more study is required to clarify the means of dispersal, but the dispersal events suggested are one-off events, unique in geological time, and it is hard to imagine how they could ever be studied. By definition, they do not conform to any general laws and are completely unpredictable. Many authors who support dispersal wisely avoid discussing the problem.

A good example is the alpine cushion-plant genus *Abrotanella* (Asteraceae). There are three clades, distributed in:

- eastern Tasmania,
- western Tasmania and Australasia, and
- Stewart Island (southern New Zealand) and South America.

The last two clades overlap only on the summit ridge of the highest mountain in Stewart Island, Mount Anglem. Wagstaff et al. (2006) concluded that distributions in *Abrotanella* "undoubtedly" reflect a "convoluted history of dispersal," but this idea (based on a fossil-based chronology) does not explain the precise dovetailing of the clades at intercontinental, regional, and even local scales. In addition, these plants do not have the feathery pappus on the "seed" that is usual in members of Asteraceae such as dandelion and allows effective wind dispersal. Wagstaff et al. (2006) concluded that long-distance dispersal across the South Pacific (10,000 km) must have occurred, but they did not speculate as to how it could have actually happened. Perhaps it never did.

WHAT IS AN "ORIGIN" IN EVOLUTION?

In the vicariance model of evolution, a group does not originate at a point within its range, but over a large area and by articulation at its margins with another group. Instead of the group having an absolute origin at an ideal point, its ancestor was always already widespread and differentiated, and so the modern group "originated" only in the sense that it became recognizable. In the unfolding of organic evolution, entities originate by differentiation of former entities, not by starting at a point and expanding. In contrast, in center of origin/dispersal theory, an ancestor develops at a single point and is homogeneous—there is no differentiation. Its descendant taxa start from a single parent pair and develop into phylogenetic clades. Other biological entities that have an absolute origin include the uniquely derived characters of cladistics and the taxa of special creation. Again, if the ancestor is polymorphic, the clades do not necessarily represent an absolute beginning and may instead be recombinations of features and genes that already existed.

CRITIQUE OF FOUNDER SPECIATION

Founder speciation has been criticized in studies of genetics and biogeography, with the critiques based on different data sets, and the concept may not be necessary.

The Critique of Founder Speciation in Population Genetics Studies

In "founder dispersal," unlike normal ecological dispersal, the founder is isolated from its parent population by "dispersing over a barrier"— an apparent contradiction—and the new population then diverges into a new species. In the biogeography of the modern synthesis, this is achieved by a second unlikely process, the "genetic revolution" produced by the founder effect. The founder effect is well established in genetics, but whether it can lead to speciation, via some sort of a "genetic revolution," is much more controversial. Nevertheless, until the recent revival of ecological speciation, dispersal theory was based on it.

Mayr (1954) stressed geography in speciation and did not accept that speciation was simply ecological. He introduced the founder effect/ genetic revolution as a mode of geographic differentiation in order to account for various distribution patterns in New Guinea birds. He developed the idea from field observations, not from genetic studies. He was especially interested in birds endemic to small islands north of New Guinea, separated from relatives on the mainland only by narrow sea passages. The idea he proposed to explain the pattern, a combination of chance dispersal and founder effect/genetic revolution, became almost universally accepted in explanations of island endemism and as a paradigm for geographic evolution in general. One of the most studied localities in this research has been Hawaii, and the classic case there is the fruit fly *Drosophila*. There are about a thousand species of *Drosophila* in Hawaii (O'Grady et al., 2009), many more than would be expected given the area of the islands, and geneticists developed an explanation for this diversity anomaly based on Mayr's theory (Templeton, 1980, 1998; Carson and Templeton, 1984). This work has provided a theoretical basis for modern accounts of founder speciation/chance dispersal.

While many biogeographers have accepted the argument from genetics, geneticists themselves have been less convinced. Tokeshi (1999) argued that the genetic founder effect does not seem to be an effective means of speciation, and Nei (2002) cited "one of the most important findings in evolutionary biology in recent years: that speciation by the founder principle may not be very common after all." Orr (2005) wrote that despite the early popularity of the idea, "it is difficult to point to unambiguous evidence for founder effect speciation, and the idea has grown controversial." The experiments of Moya et al. (1995) failed to corroborate predictions of founder effect speciation, and subsequent studies have also found no evidence for it (Rundle et al., 1998; Mooers et al., 1999; McKinnon and Rundle, 2002; Rundle, 2003). Crow (2008) called the idea of genetic revolution "vague and misguided" (see also Crow, 2009). Even in birds, founder effects 'may be unnecessary' (Grant, 2001; cf. Walsh et al., 2005). The passerine *Zosterops* is often cited as the classic case of a taxon that has evolved by founder speciation (Mayr and Diamond, 2001), yet a detailed study of clades in the southwest Pacific concluded that the focus on founder effects in this group "has been overemphasized" (Clegg et al., 2002).

Florin (2001) described how "The vicariance model of allopatric speciation has been repeatedly confirmed empirically, while peripatric [founder effect] speciation has suffered severe criticism for being both implausible and empirically unsupported." In her own studies on flies she found "no support for speciation through founder effects." In recent years the debate has heated up, and advocates of dispersal theory have found it necessary to publish an article stressing "The reality and importance of founder speciation in evolution" (Templeton, 2008). This was a reply to Coyne and Orr's (2004: 401) conclusion that "there is little evidence for founder effect speciation." Coyne (1994) wrote that the idea that the theory "has infected evolutionary biology with a plague of problematic work."

Templeton (1998) had earlier introduced a new method of analysis—nested clade analysis—that incorporated founder effect speciation (as "range expansion") and promised to distinguish between the results of this process and those due to the fragmentation of populations, or

vicariance. Apart from its problematic use of founder effect speciation (and confusion with range expansion), this method has been criticized on other grounds (Knowles and Maddison, 2002; Beaumont and
Panchal, 2008; Petit, 2008). Neigel (2002) wrote that in nested clade
analysis "it is predicted that gene genealogies will exhibit a 'star phylogeny' . . . in a population that has been expanding in size. . . . These
descriptive patterns are useful as heuristics and help us understand how
we can use population genetic data to reconstruct population histories
(Templeton 1998). However it is not clear that they provide reliable
evidence of particular kinds of population histories. Often the same
pattern may be created by any of several different population histories.
For example, a star phylogeny could also result if a once continuous
population became fragmented. . . ."

In another critique of nested clade analysis, Knowles (2008) stressed
that simulation studies have found consistently high error rates using
the method. Knowles noted that the method had been cited 1,700 times
and questioned why it is so popular despite having such obvious problems. Nested clade analysis is a logical development of Templeton's
earlier work supporting founder effect speciation and dispersal biogeography. As a formalization of earlier work by Mayr (1954) and
Carson and Templeton (1984), nested clade analysis incorporates the
core concepts of orthodox dispersal theory and the modern synthesis,
and so it may remain popular largely for historical reasons.

Mayr's Critique of Ecological Speciation

Although geneticists such as Coyne and Orr (2004) dispute speciation
by the genetic founder *effect*, this does not mean that they reject center
of origin/dispersal and founder *events*. Nevertheless, founder effect speciation was one of the pillars of modern dispersal theory, and rejecting
it has important consequences. In particular, if it is rejected but dispersal itself is retained, other modes of differentiation must be invoked and
ecological speciation is often inferred. Yet, although many sister species have different ecologies, ecological differentiation seems unlikely
to be the initial cause of many variants. Mayr (1954: 168) recognized
this and discussed the issue when he proposed his ideas on "genetic
revolution."

Mayr had an extensive field knowledge of New Guinea and he
developed his founder theory to account for a striking fact. This was
the marked morphological difference between bird taxa on the New

Guinea mainland and their representatives on small, offshore islands close to the mainland (Mayr, 1942, 1954, 1992). Biological and physical differences between the two nearby environments, both with lowland rainforest, seemed relatively minor (Mayr, 1954: 158, 168). Of course, any two areas show ecological differences, but in this case the environmental variation seemed much less than that within the vast areas of rainforest through mainland New Guinea, where there was often less differentiation in the sister group. As Mayr later emphasized: "The crucial process in speciation is not selection, which is always present, even when there is no speciation, but isolation" (Mayr, 1999: xxv). In the case of the islands, any appeal to ecology seemed ad hoc and with no supporting evidence; instead, the obvious factor seemed to be the "water barrier" isolating the island and mainland forms. Although Mayr's ideas on the particular genetic mechanism were probably incorrect, as geneticists now suggest, they were attempts to explain a genuine, fundamental problem in biogeography, ecology, and evolution, exemplified by the New Guinea birds.

If chance dispersal with genetic revolution and founder effect speciation is untenable and ecological speciation is unlikely, what are the alternatives? The answer may involve isolation, as Mayr thought, but mediated by tectonics, not dispersal (Heads, 2001, 2002). Authors have recognized that tectonic geology is relevant for evolution and biogeography, but its importance has been underestimated. In addition, knowledge of tectonics in places such as New Guinea has increased greatly since the 1930s when Mayr was formulating his ideas. Many tectonic structures and processes have only been recognized recently by geologists, and some of these are geographically coincident with biogeographic features.

Founder Dispersal and New Ideas on Rift Tectonics

New developments in geology indicate that founder dispersal may not be required, and this adds to the problems with founder effect speciation raised by geneticists. The northern part of New Guinea and its offshore islands are one of the most mobile parts of the Earth's crust and the region is now interpreted as a plate margin or series of margins. It is traversed by major transform faults and strike-slip movement on these has transported terranes, fault slivers, and biota in different directions over hundreds of kilometers. This has produced whole series of centers of endemism, along with dramatic disjunctions and juxtapositions

(Heads, 2001, 2002). The latter include the "strange" and "intriguing" distributions reported in Murinae by Musser et al. (2008). Following Mayr's New Guinea surveys, another important study of New Guinea birds involved a transect across the main tectonic boundary in New Guinea, the craton margin (Diamond, 1972). As in Mayr's studies, Diamond did not refer to structural geology, but instead attributed the many faunistic and phylogenetic differences found along the transect to ecological factors. Some of the geology was only clarified more recently, after these two biologists were writing. Nevertheless, a later, book-length collaboration explaining evolution and biogeography in the region (Mayr and Diamond, 2001) continued to rely on founder dispersal over current or recent geography. Instead, the distributions can be explained by tectonics.

The Critique of Founder Dispersal in Studies of Biogeography

Biogeographers have known for years that similar distributional phenomena occur in different, unrelated taxa, each with different means of dispersal and ecology. Because of this, and because the shared patterns are intricate and precise, many biogeographers have rejected a concept of long-distance or founder dispersal that relies on chance. Instead, they have developed concepts of the orderly evolution of entire communities. The global vicariance biogeography produced by 19th-century systematists such as the zoologist T.H. Huxley and the botanist J.D. Hooker was rejected by the dispersalists of the 20th century as "land-bridge building." But in retrospect these early, global analyses of the main groups were often more or less valid. By the late 19th century, museums had extensive collections from around the world, and the main geographic patterns were well understood. Many biogeographers of the time analyzed the groups without reliance on centers of origin or dispersal, and it was this research tradition, not that of dispersalism, that led to Wegener's (1912, 1915) prescient synthesis of geology and biogeography. Pure "chance dispersalism" denies any true patterns in distribution other than those related to ecology and means of dispersal. Yet two botanists working on the diverse flora of Indonesia and Malaysia (Malesia) concluded in a similar way. Van Steenis (1936) wrote: "On the whole I cannot trace any relation between distribution and what is known of [means of] dispersal," and Kalkman (1979) suggested: "The dispersal method [of dispersal] is of subordinate importance for areogenesis as compared with other factors."

Vicariance biogeography is sometimes portrayed as a new, iconoclastic view that only developed in the 1970s after the rise of plate tectonics. This is not correct, and the idea that distribution must be related to means of dispersal has often been questioned. After studying the species of Gentianaceae and their distributions, T.H. Huxley (1888) wrote that "one conclusion appears to me to be very clear; and that is, that they are not to be accounted for by migration from any 'centre of diffusion.'" Mueller (1892) discussed biogeographic affinities between Australia and Africa and wrote that "the enigma cannot be explained by migration. . . . Floras of certain districts are not homogeneous, but exhibit amongst their warp a woof which has nothing in common with it, but exhibits the stamp of a totally different flora. . . . I do not deny [migrations] when they are opportune, and I know very well that wind and weather, animals and men, are able to distribute species sometimes over large areas; but it is quite a different thing when we have to deal with the spreading of whole floras, sufficient to impress one district with the stamp of another, where all the species are united in an organic association, so that one cannot be understood without the other. This cannot ever have been accomplished by a migration of a mechanical nature" (pp. 432–433).

MOLECULAR STUDIES AND ECOLOGICAL SPECIATION

It is sometimes observed that while Darwin's (1859) book is titled *On the Origin of Species by Means of Natural Selection*, it does not actually cite any examples of this. Does natural selection really lead to speciation? Darwin himself stressed that regions with a similar environment have very different biotas, indicating that ecology does not explain the difference. Later authors such as Mayr (1942) and Croizat (1964) have also accepted geographic factors as the main cause of speciation, with ecology as a secondary factor.

Many clades comprise species that are allopatric but show little or no obvious difference in their ecology. These are often interpreted as nonadaptive "radiations," rather than adaptive radiations caused by ecological differentiation. Examples include the plant *Nigella* in the Aegean Archipelago (Comes et al., 2008) and landsnails in the Azores (Jordaens et al., 2009). Rundell and Price (2009) suggested that sympatric species that are ecologically differentiated could arise if speciation occurs through geographical isolation and nonadaptive radiation and is followed by ecological differentiation and range expansion into sympatry.

The environments of any two allopatric sister groups always differ in some variable or other, but this does not necessarily explain the differentiation. Simple ecological speciation would lead to species with sisters that are sympatric, with more or less exactly the same geographic range, and different ecology, but this unusual. If allopatric evolution is the usual mode of speciation, there should be a positive relationship between the relative divergence time of taxa and their degree of geographic range overlap. Barraclough and Vogler (2000) studied a range of animal groups and found "a general pattern of increasing sympatry with relative node age . . . consistent with a predominantly allopatric mode of speciation." In a similar study, Kamilar et al. (2009) examined biogeography and phylogeny in 19 species of cercopithecid monkeys in Africa and also found a positive relationship between age and overlap. This is good evidence for allopatric speciation being the usual mode of diversification in cercopithecids, and there is no indication that this group has unusual biogeography.

The best evidence for ecological speciation (to eliminate the possibility of geographical factors) is provided by sympatric, ecologically separated sister species. Examples include the reef fishes *Hexagrammos agrammus* and *H. otakii* in the Sea of Japan, with the former species in seaweed beds, the latter at rocky sites (Crow et al., 2010). But in practice, true sympatry of sister species, not just secondary overlap at some point of their range, is uncommon. Usually there is a significant degree of geographic allopatry between the two, and this probably reflects the original, allopatric break (as shown in many examples in this book). Unfortunately, studies of ecological speciation (e.g. Rundle and Nosil, 2005; Funk and Nosil, 2008; Schluter, 2009) often lack geographic analysis or even distribution maps, so it is difficult to assess the geographic component in the speciation. Bolnick and Fitzpatrick (2007) observed that "Biogeographic comparative studies of range overlaps have not been conducted for the two groups most widely thought to exhibit sympatric speciation, phytophagous insects and lacustrine fishes."

Studies of particular clades that have included biogeographic information show that supposed cases of sympatric and ecological speciation have little support. For example, the different races of the apple maggot fly *Rhagoletis pomonella* (Tephritidae) on hawthorn and apple trees are often cited as evidence for sympatric speciation. Sequence data instead suggested a geographic scenario with allopatric differentiation followed by overlap (Feder et al., 2003).

The threespined stickleback *Gasterosteus aculeatus* (Gasterosteidae) is widespread in the north temperate zone. In a very small part of this range, near Vancouver Island, six lakes on three separate islands each contain two sympatric "species" of *G. aculeatus* occupying, respectively, benthic habitat near the lake floor and limnetic habitat higher in the water column. The two forms are morphologically and reproductively isolated from each other. (They are referred to as species by geneticists studying ecological speciation, e.g., Taylor et al., 2006, but not by taxonomists.) The pattern has been interpreted as the result of sympatric, ecological speciation in the six lakes, and this model predicts that the pair in each lake should form a monophyletic clade. In fact, though, molecular studies found that this is not the case (Taylor and McPhail, 2000). The sympatric pairs of sticklebacks are only known from a very small part of the total species range, the Strait of Georgia, suggesting that the history of the region rather than ecological speciation has promoted "species pair" evolution. The area has been submerged by the sea twice, at 12,000 BP and 1500 BP, and Taylor and McPhail (2000) concluded that the presence of two forms in each lake was due to invasion of the coastal lakes by different populations of marine sticklebacks on two separate occasions, rather than ecological speciation (cf. Rundle and Schluter, 2004).

Yeung et al. (2011) studied the population genetics of the royal spoonbill *Platalea regia* in Australasia and its allopatric sister group, the black-faced spoonbill *P. minor* in eastern Asia. Their results "do not support founder effect speciation in *PP. regia*." The authors suggested that the divergence between the two species was probably driven by selection. Gay et al. (2009) found high levels of divergence between species of gulls (*Larus*) and concluded: "Such divergence is unlikely to have arisen randomly and is therefore attributed to spatially varying selection." Yet whether or not all evolution is due to selection is controversial, and there is great confusion about the role of natural selection. Wilson (2009: vii) proposed that "All elements and processes defining living organisms have been generated by evolution through natural selection," yet in the same book, Ruse and Travis (2009: x) wrote that "natural selection is not the only evolutionary force. . . ."

Although ecology and natural selection explain why there are no alpine plants with large, membranous leaves, they do not explain why the particular families, genera and species that are found on a mountain occur there in the first place. Natural selection might explain why the black form of the peppered moth *Biston bistularia* became more common than the mottled form during the industrial revolution, due to increased pollution. But it does not explain why there were black forms there to start with, before the industrial revolution. To summarize: "Natural selection does not design an organism or its features; it merely filters existing variation" (Travis and Resnick, 2009: 114).

Island Biogeography with Metapopulations and without Founders

Many taxa occur on "ecological islands," small areas of suitable habitat found scattered through large regions. Ecological islands are often more or less ephemeral and include puddles, landslides, or areas of fresh lava. Among the classic examples of ecological islands are young, volcanic islands. Here the groups in the rainforest and on the reef exist as metapopulations, populations of populations, constantly dispersing from older islands to younger ones within a region as new islands appear and older ones subside. An island clade endemic to the central Pacific and widespread there can survive more or less indefinitely, in the same region, on islands that are individually ephemeral.

Many botanists have denied that there is a true regional Polynesian flora, perhaps because dispersal theory suggests one should not exist. Instead, Philipson (1970) argued that plants endemic and widespread in Polynesia, such as *Meryta* (Araliaceae), *Tetraplasandra* (Araliaceae), *Fitchia* (Asteraceae), *Sclerotheca* (Campanulaceae), and many others, indicate that "the southern Pacific islands must be credited with a flora specific to this region. . . . Clearly land has been present for long periods in this area of the Pacific because well-marked genera are endemic to it. The flora characteristic of this region could survive provided a few oceanic volcanoes projected above the sea at all times. Such oceanic islands characteristically rise and fall relative to sea-level so that they are precarious footholds for a flora, but collectively they form a secure base." In this view, new individual islands will be colonized by ordinary movement, on the scale seen every day. This is an observable ecological phenomenon (unlike long-distance dispersal/founder effect speciation) that functions using the organisms' ordinary means of survival.

Founder Theory, Island Biogeography and New Ideas on Volcanism

Founder theory and the associated "equilibrium" theory of island biogeography both assume that volcanic islands appear at random with respect to other islands. But volcanism and volcanic islands tend to recur around the same tectonic features (subduction zones, propagating fissures, hotspots, etc.), where terrestrial taxa survive and evolve more or less *in situ* as metapopulations. New ideas on volcanism in the Pacific, for example, emphasize the importance of the large, igneous plateaus emplaced in the Cretaceous. These are mainly submarine but some intercalated sedimentary strata include fossil wood, and the plateaus have many large, flat-topped seamounts that were once high islands. The simple hotspot model is not sufficient to explain many aspects of intraplate volcanism, and in contrast with the traditional mantle-plume hotspots, new tectonic models explain linear volcanic chains by propagating fissures. These could be caused by stress fields induced by normal plate tectonics (Foulger and Jurdy, 2007a).

ECOLOGY AND BIOGEOGRAPHY

Darwin (1859: 346) observed that "In considering the distribution of organic beings over the face of the globe, the first great fact which strikes us is, that neither the similarity nor the dissimilarity of the inhabitants of various regions can be accounted for by their climatal and other physical conditions." Few taxa occur throughout all areas on Earth with similar ecology—most are restricted to particular geographic regions. Areas of rainforest, Mediterranean climate, desert, etc. in different parts of the world have similar climate and vegetation but almost completely different floras and faunas. This indicates the primary importance of historical factors rather than ecological ones in determining global distribution. Of course, *within* a biogeographic region, distribution may reflect ecological factors.

For example, within an area of endemism in tropical hill forest ten or a hundred kilometers across, the taxa are sorted out into the standard ecological zones and habitats. Some will always occur on the valley floors, others on the ridge crests and summits. Outside this biogeographic region, the species pool changes. Thus biogeography involves differentiation in space at a large scale; ecology involves smaller-scale spatial differentiation along ecological gradients within a biogeographic region. Some ecological phenomena are large scale, though, and the three latitudinal belts, northern, tropical, and southern, are among the most obvious phenomena in biogeography. Ecology can explain the differentiation among these, and the smaller-scale spatial differentiation in communities within a biogeographic region, but it cannot explain many large-scale patterns; these are usually attributed to historical factors.

Examples of important phylogenetic/geographic breaks where there are no obvious climatic or topographic breaks include Wallace's line and many others that are less well known. A group's ecology does not usually explain its biogeography, although it explains much of its spatial distribution at a small scale—across meters rather than kilometers—within its biogeographic region.

Ecological Lag

The elevation of a community determines many aspects of its ecology and can change rapidly in geological or even ecological time. As discussed above, ducks and parrots in the Andes may have been uplifted tectonically during orogeny. *Salicornia* (Amaranthaceae) and Frankeniaceae are usually maritime shrubs, but occur at over 4,200 m elevation in the Andes (Ruthsatz, 1978). Rapid uplift in New Guinea may have carried populations of coastal mangrove associates such as *Pandanus* (Pandanaceae) to treeline at ~3,000 m in just 1 million years. In a similar way, populations of plants and animals may be lowered with erosion or subsidence, or stranded in the middle of a continent following marine transgression and regression. Thus the current ecology of a group may have little to do with its original one, and a population can find itself in a new type of habitat without having moved from its initial substrate. There may be a long lag phase while the groups readjust to the new conditions, but in areas of active tectonism, equilibrium may not be reached and ecological anomalies may persist. At this scale of time and space, there is little distinction between ecology and biogeography; factors determining the elevation of a group (uplift, subsidence, erosion) may also result in biogeographic differentiation in latitude and longitude.

The CODA Paradigm and Pan-adaptationism

Modern interpretations of distribution are based on a center of origin and dispersal from there, and also involve the idea of adaptation. A clade moves out from its center of origin by dispersal to a new locality and habitat. Here it faces new extrinsic needs and changes its morphology and physiology in response to these, bringing about adaptation. Lomolino and Brown (2009) referred to this as the CODA model, as it is based on center of origin, dispersal, and adaptation, and this is an accurate characterization. There is no denying the intricate fit that exists between most organisms and their environment, and this is necessary for their survival, but instead of structure being determined by function and extrinsic needs, structure may determine function. In this view, the morphology of the teeth determine many aspects of the diet; the diet does not determine the teeth. If the teeth evolve, the diet changes.

It is usually taken for granted that adaptation is the main factor in structural evolution, although this has been questioned. Scheiner (1999: 145) wrote that

> By the end of the Modern Synthesis much of ecology and related disciplines were in the grip of . . . pan-adaptationism. Most traits were seen as adaptations, and organisms were viewed as being nearly perfectly adapted to their environment. This viewpoint came about through taking the theory of natural selection and attempting to fit the entire world under its banner. . . . Many advocates of the modern synthesis were guilty of pan-adaptationism. . . . It tended to be ecologists, rather than evolutionary biologists, who carried this banner. . . . Natural selection was the focus of many studies so the investigated traits, not surprisingly, were shown to be adaptations. . . . Since then we have spent much more effort on the other processes that contribute to evolution such as drift, mutation, and developmental constraints.

Scheiner (1999) did not accept natural selection as a primary driver in evolution, and it is suggested below that in many groups the biogeographic history, not selection, has determined the current ecology. For example, a group may be in an area of highlands, or desert, or on an atoll because its ancestors were already in the region before the uplift, the desertification, or the subsidence and erosion of a high island. If the group already had structures and physiology suitable for the new environment (*pre*-adaptations), it survived; if not, it went extinct. Adaptation itself may have relatively minor significance.

Metapopulations and Weeds

All clades go extinct in the end, but usually only after many millions of years, and during that time they survive while their environment changes around them. Weeds are plants that can survive in disturbed and marginal areas, but all plants and animals are, or have been, "weedy" to some extent. If not, they would never have survived the climatic and topographic revolutions of geological time, such as volcanism, marine transgression, orogeny, and glaciation. Populations of "fire weeds" can survive burning *in situ*, while montane weeds survive uplift and thrive in highly disturbed areas around glaciers. Both of these occur around the around the Pacific margin, where acitve plate margins have led to orogeny and volcanism. The Pacific "ring of fire" maintains its own endemic biota and this is diverse in areas such as Indonesia and the Andes. Fire weeds found in and around craters, fumaroles, and ash slopes include members of Ericaceae, and the plants survive in these dynamic, disturbed regions as metapopulations (Heads, 2003). A well-studied "volcano weed" in Ericaceae is *Vaccinium membranaceum*, an animal-dispersed shrub of western North America. Yang et al. (2008) examined the genetics of a recently founded population on new volcanic deposits at Mount St. Helens, Washington, 24 years after the 1980 eruption. They found that "While founders were derived from many sources, about half originated from a small number of plants that survived the 1980 eruption in pockets of remnant soil embedded within primary successional areas. We found no evidence of a strong founder effect in the new population; indeed genetic diversity in the newly founded population tended to be higher than in some of the source regions" (p. 731).

Major geological events include phases of extensive, very large-scale volcanism, as in the central Pacific (during the Cretaceous), and also asteroid impacts, as at Chicxulub, Yucatán (Cretaceous/Paleocene boundary), and at Chesapeake Bay, Virginia (Late Eocene). These types of events will cause local, regional, or even global extinction of some clades. Yet they will also provide opportunities for range expansion in many taxa, especially those that are tolerant of extreme disturbance and marginal ecology. In many cases the local endemics of today, even in the least disturbed, oldest forests, may be derived directly from Cretaceous weeds. Following the original range expansion—determined by geology, not chance dispersal—they have settled into a phase of immobilism through the Cenozoic.

THE RETURN OF DISPERSALISM

Throughout history it has been assumed that because animals and plants move, their distributions have been caused by this movement. The center of origin has been a recurrent motif in philosophy and literature for the last two or three thousand years, at least. In this model, each kind of plant and animal originated at one point and attained its distribution by spreading out from there. This is seen in literal readings of biblical accounts, such as the stories of Eden and Ararat, and in many modern evolutionary studies. The only times that the center of origin assumption was widely questioned were in the late 19th century until the first world war, a period when systematic activity peaked, and briefly in the

1970s and 1980s. The current popularity of dispersalism in molecular phylogenetics is sometimes portrayed as an exciting new development, but in reality it is the orthodox view of the last two millennia.

THE ROLE OF GEOGRAPHIC DATA IN EVOLUTIONARY BIOLOGY

Wallis and Trewick (2001) suggested that plate tectonics, Hennigian systematic, and molecular clock dating "have placed biogeographic data at center stage of evolutionary debate." But biogeographic data have been central to evolutionary biology ever since the origin of the subject. In an 1845 letter, Darwin referred to geographic distribution as "that almost keystone of the laws of creation" (Darwin, 1887) and he often used it to frame and test evolutionary hypotheses.

Trewick and Wallis (2001: 2178) suggested that "The tendency to identify and focus upon repeated pattern is an important feature of biological research, but one that has usurped the role of hypothesis testing among some biogeographers." There is more to science than testing hypotheses, though. After all, medieval scholastics rigorously tested their hypotheses by examining scripture, Aristotle, and miracles. The main interest of molecular phylogenies for biogeography is not that they suggest hypotheses, but that they illustrate geographic patterns so clearly.

Authors such as McDowall (2007) have criticized panbiogeography as "a search for general patterns." The criticism proposes that individual taxa have their own evolutionary and biogeographic history and that the correct topic of study is a particular taxon. But many critics of panbiogeography are taxonomic specialists on particular groups, not comparative biogeographers. Many have not investigated general biogeographic patterns or described any new ones, but have devoted their time to studying distributions of particular taxa in their group. Comparative biogeography cannot be concerned with one-off phenomena, such as unusual distribution patterns seen only once in a single group. Instead it focuses on general patterns and laws in distribution and explains particular cases in terms of these. Often what are assumed by specialists to be rare or unique distributions are quite common in other groups. One main conclusion of panbiogeography is that while the details of distribution patterns are virtually infinite and no distribution pattern is exactly like that of any other group, the main phylogenetic and biogeographic breaks or nodes are few. Nothing important in biology only happens once, and multitudinous aspects of differentiation occur in many groups of animals and plants at the same standard nodes.

DEDUCTION AND INDUCTION IN BIOGEOGRAPHY

Cain (1943: 151) concluded his critique of the center of origin idea by writing that geobotany and geozoology "carry a heavy burden of hypothesis and assumptions which has resulted from an over-employment of deductive reasoning." In other words, a theory such as the center of origin model has been taken as fact and used, along with other ideas ingeniously deduced from it, to explain certain observations. Cain argued that geobiologists neglected to test the initial assumption: "What is most needed in these fields is a complete return to inductive reasoning with assumptions reduced to a minimum. . . . In many instances the assumptions rising from deductive reasoning have so thoroughly permeated the science of geography and have so long been part of its warp and woof that students of the field can only with difficulty distinguish fact from fiction." Modern authors have agreed: "[E]volutionary biology is an inductive science, one in which generalities emerge not as the result of theoretical deduction or the conduct of critical experiments, but rather through the summation of many evolutionary case studies" (Losos, 2010: 623; cf. Gavrilets and Losos, 2009). This suggests that induction is at least as important as deduction, and the main section of this book (following Chapter 2) is devoted to particular case studies.

MORPHOLOGICAL WORK AND MOLECULAR PHYLOGENIES

The case studies discussed herein are from molecular accounts. Why concentrate on these when so many morphological phylogenies are available? The main reason is the inherent biogeographic interest of the molecular clades, as they show so much orderly, complex geographic structure. For a biogeographer, molecular systematics is confirmed as a productive research program by the close correlation between phylogenetic groupings and geography. The geographic patterns shown by the molecular clades are not necessarily new. Many of the most interesting ones have already been documented in traditional taxonomic revisions, although the authors of the new studies are often unaware of the earlier work. Nearly all the maps in this book depict molecular phylogenetic arrangements of groups, but the details of the geographic distributions are often based on morphological observations and collections. Working out these concordances supplements the molecular phylogeny with the more extensive samples of traditional work.

While most morphological classifications can be used for biogeographic analysis, there are problems with morphological work. Taxonomists have sometimes cherry-picked certain combinations of characters and left out others in order to achieve a desired end. One author can "demolish" the taxonomy of another simply by selecting an appropriate combination of characters and in this way morphological classifications can be manipulated to fit preconceived notions of evolution and biogeography. Molecular work is not immune to this, but it seems to be less of a problem.

In addition to their impressive biogeography, molecular phylogenies tend to make sense morphologically, although not always with respect to the traditional homologies and categories. Structural concepts such as the "flower" are often based on 18th-century notions that require revision. Molecular phylogeny now places a raspberry (Rosaceae) and a mulberry (Moraceae) together in a novel grouping, a radically changed Rosales (Stevens, 2010). At first sight the fruit look very similar. But using traditional homologies, the first grows from a single flower, the second from an inflorescence or group of flowers, and the "flowers" in both represent opposite poles of floral morphology—they are as different as possible. Taking the hint from the molecular phylogeny, this traditional homology may be incorrect and the flower in one may be equivalent to the inflorescence in the other. In other words, it is not only the morphological classification that is incorrect, but also the morphological concepts underlying it. (Flower–inflorescence intergrades also occur in plants such as *Euphorbia*; Prenner and Rudall, 2007). Thus molecular work may stimulate reexamination of the morphological categories and homologies. It is not simply a question of finding a new, cryptic character of morphology that corresponds to the new molecular grouping.

The main problems with the molecular work include sampling limitations, both phylogenetic and spatial. This is improving rapidly due to technical advances and as workers come to recognize the great value of extensive sampling. Another problem is not so much a technical limitation as a conceptual one: Modern molecular work often interprets its excellent data using old, inefficient biogeographic methods, such as divining the center of origin, dating the phylogeny with fossils, and explaining common patterns by extraordinary events of chance dispersal. Molecular data provide a fabulous new source of information on biogeographic differentiation and evolution, but satisfying interpretations of the patterns have lagged far behind their description. Donoghue and Moore (2003) advocated the standard methodology now employed in hundreds of papers: Use DIVA or a similar program to locate the center of origin and use fossil-based calibrations (minimum ages) to give maximum ages for clades (see next chapter). It is not surprising that the end result duplicates the biogeography of a century ago (Matthew, 1915), as both share the same principles of center of origin, fossil-based chronology, and Cenozoic evolution of the orders by long-distance dispersal. Instead, all these concepts can be abandoned, as the molecular phylogenies and modern tectonics suggest more likely, alternative reconstructions of groups' histories.

Although molecular and morphological phylogenies may differ in individual groups, many of the broader biogeographic patterns indicated by molecular phylogeny were known to the early workers. As Avise (2007) concluded, "traditional non-molecular systematists generally seem to have done an excellent job in identifying and classifying salient historical discontinuities in the biological world." Avise emphasized that intraspecific molecular groups are nearly always allopatric and that their geographic distributions "usually make biogeographic sense," as they "orient well" with known patterns. These comments also apply to many molecular clades above species level. Authors sometimes describe the geographic distribution of a molecular clade as idiosyncratic or peculiar when in fact it is a standard pattern, and this provides good evidence for both the grouping and the distribution pattern. Many molecular clades conform to specific distribution patterns that have been of interest to comparative biogeographers for decades or longer.

To summarize, one of the main themes of molecular studies has been the precise geographic structure evident in molecular clades. Evolution can be interpreted using the methods of molecular biology and panbiogeography, as Croizat (1977) suggested, and the most useful studies on groups include detailed data on both phylogeny and comparative distribution.

In a remarkable phase of the molecular revolution, from about 1990 to 2000, biologists worked out the higher-level molecular phylogenies of the large, worldwide groups such as birds and angiosperms. In a second phase, researchers have produced detailed accounts of lower-level groups that involve regional and local differentiation. The new molecular work shows that clades relate closely to geography, and so researchers have started to annotate their phylogenies with geographic information and to map clades. The molecular work shows classic biogeographic patterns with improved clarity, revealing many details that were previously obscure or unknown, and the patterns are the subject of this book.

2

Evolution in Time

Differentiation in morphology and molecules occurs over space and through time. Spatial variation can be observed directly; establishing the age of a clade is more difficult. Molecular clock dates for clades are reported in the mass media and even the scientific literature as more or less factual, yet in reality there are fundamental problems transforming sequence data—aspects of structure or *form*—into dates, which vary along the *time* axis. Modern molecular clock dates give an overall chronology of evolution in which many large taxa, such as plant families and mammal orders, only develop in the Cenozoic. This same chronology is seen in the fossil record and the similarity results from the fact that the molecular clock dates were fossil-calibrated. The fossil-calibrated molecular dates are often no more than minor varations of fossil-based estimates.

Molecular phylogenies are of great interest because of their clades and biogeographic patterns, but dating is one of the weaker aspects of molecular analysis. The confidence and triumphalism that surround molecular dating can be misleading. One author suggested that "Until recently it has been difficult to distinguish between vicariance and dispersal. . . . Although morphological characters and the fossil record can be used to examine relationships among species it is difficult to estimate the

This chapter incorporates material first published in *Cladistics* (Heads, 2005a) and reproduced here with permission from John Wiley and Sons.

timing of evolutionary events from these data. In contrast, molecular data allow inferences about both pattern and time" (Winkworth, 2009: 93–94). Despite this suggestion, there does not appear to be any fundamental difference between the way macro-scale morphology and molecular structure evolve in space and time. Molecular sequences are features of form in the broad sense, and their spatial and chronological evolution needs to be interpreted—the dates do not just magically appear in the sequences any more than they do in the gross morphology.

Differences between sequences need to be calibrated and a model of how evolution proceeds—for example, whether it is clock-like or not—has to be adopted. Winkworth (2009: 94) wrote: "By calibrating genetic diversity with a known time point we can . . . assign absolute ages [to nodes]." But there is no "known time point" in phylogeny, and the same problems with calibration occur whether using morphological or molecular data for dating.

The real change that led to molecular clock theory was not so much the molecular data as the idea that evolution is clock-like. If this is true, the degree of morphological or molecular difference between clades (branch length) is related to the time since their divergence. But this is controversial, and the rate of evolution may show great variation between lineages (and genes) and within the same lineage at different times.

IS EVOLUTIONARY RATE CONSTANT (CLOCK-LIKE)? THE EVOLUTIONARY CLOCK IN MORPHOLOGICAL AND MOLECULAR BIOLOGY

Many biogeographic studies are based on phylogenetic trees in which the nodes are dated and this information is used to test between vicariance and dispersal. These studies often assume that evolution is clock-like, with lineages showing a constant rate of sequence divergence of, say, 2% per million years. But whether evolutionary mode may in fact deviate from clock-like, and if so, by how much, is debated, and the different statistical tests for clock-like evolution make their own assumptions. The history of the evolutionary clock idea is implicated with the rise of the modern synthesis in 20th-century biology, and this history is worth outlining.

The Biogeography of Hutton and Chapman Was Replaced with the Evolutionary Clock of Matthew, Mayr, and the Modern Synthesis

Based on their broad knowledge of biogeography and geology, 19th-century evolutionists such as Hutton (1872) concluded that "differentiation of form, even in closely allied species, is evidently a very fallacious guide in judging of lapse of time." Nevertheless, the idea that morphological or molecular evolution is roughly clock-like was accepted by the initiators of the modern synthesis. Matthew (1915) and most subsequent authors overlooked the 19th century work and assumed that evolution (morphological or molecular) is indeed clock-like. It follows from this that the groups in a biogeographic pattern must have all evolved at different times, as reflected in their degree of divergence, and so community-wide vicariance can be rejected. This idea became the dominant paradigm and one of the foundations of the modern synthesis. For example, employing the evolutionary clock idea, Matthew (1915) concluded that "the Malagasy mammals point to a number of colonizations of the island by single species of animals at different times," and this remains the standard interpretation (Yoder and Nowak, 2006).

Applying Matthew's (1915) ideas, Mayr (1931) discussed the avifauna of Rennell Island, in the Solomon Islands, and wrote: "The different degree of speciation suggests that the time of immigration has not been the same for all the species" (p. 9). Likewise, for the birds of New Caledonia, Hawaii, and other islands, he wrote that "Strikingly different degrees of differentiation indicate colonization at different ages" (Mayr, 1944: 186).

Other ornithologists of the time did not adopt this argument, but instead followed the line of reasoning advocated earlier by Hutton (1872). One example was Chapman (1931), who discussed the birds of the Pantepui mountains in southern Venezuela (Mount Roraima, etc.). He cited the case of *Oxyruncus* (Tityridae), with a single species and widely disjunct populations in Costa Rica/Panama, the Roraima area, south-central Brazil, and southeastern Brazil. Chapman concluded that the four populations had diverged only to subspecies level "in spite of their complete and obviously prolonged separation—a fact which warns us to be cautious when we attempt to judge of the age of a race by the extent of its differentiation."

Despite this advice, Mayr and Phelps (1967: 290) accepted the clock idea in their own analysis of the Pantepui birds: "We now know that populations of different species and on islands of different size may differ rather widely in the rate at which they diverge morphologically under the influence of the same degree of isolation. *Nevertheless, all other things being equal, degree of differentiation of an isolated population reflects length of isolation.* By a study of the degree of endemicity [i.e., degree of differentiation] that has developed among the 96 upland species of Pantepui, it should be possible to determine whether the colonization took place at a single period of time or whether, on the contrary, it was a continuous process" (italics added). But in the group of 96 Pantepui birds, "all other things" are not equal and at any one time there will be clade-specific differences in the genetic potential of the different taxa of a region to evolve and diverge.

Matthew's (1915) idea of a single rate and an "evolutionary clock," as advocated by Mayr (1931, 1944) and Mayr and Phelps (1967), became part of the modern synthesis and led, for example, to the idea that "given enough time, speciation is an inevitable consequence of populations evolving in allopatry" (Turelli et al., 2001). But if degree of divergence is related to prior genome architecture, not simply time, many populations that become separated will not speciate.

The evolutionary clock idea supports the idea of chance dispersal, with each clade in a pattern having its own age and individual history, and so it will always conflict with the idea of general, community-wide vicariance. Later the evolutionary clock concept was applied to molecular data and termed the "molecular clock." But adopting this method seems illogical in light of earlier observations (Hutton, 1872; Chapman, 1931) and the studies on different rates that Mayr and Phelps (1967: 290) themselves cited at the start of the passage quoted. The simplistic assumption that the clock method relies on is so dubious that it was questioned and rejected over a century ago.

The Components of a Single Pattern Have Different Branch Lengths: dispersal or vicariance?

Later authors in the modern synthesis have continued to accept the evolutionary clock idea. Ehrendorfer and Samuel (2000) discussed the pattern of disjunction between South America and New Zealand seen in many groups. They wrote that, "Judged by their morphological (and molecular) divergence, these disjunctions, which range from the infraspecific (e.g., in *Hebe*), to the specific (e.g., in *Anemone*), sectional, or even subgeneric level (e.g., in *Fuchsia* or *Nothofagus* p.p.), must be of very different ages." This interpretation makes the implicit assumption that rates of evolution, even in different orders, are more or less equal and that the taxonomic rank of a clade supplies its age.

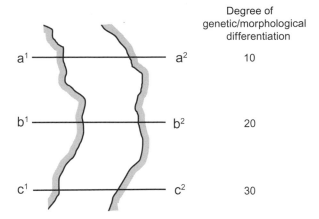

FIGURE 2-1. A biogeographic disjunction across an ocean basin. The components of the pattern, clades a, b, and c, all show different degrees of genetic differentiation across the basin.

Nevertheless, the fact that Ehrendorfer and Samuel (2000) highlight is important—different groups with the same biogeographic pattern— for example, a disjunction between South America and New Zealand, or between South America and Africa—show different degrees of differentiation. Morphological studies show that the components of any important biogeographic pattern include taxa of different taxonomic rank (subspecies, species, subgenera, etc.). The principle can now be seen to apply in molecular evolution, as most component clades of a geographic pattern, where more than one has been sequenced, have different branch lengths (Fig. 2-1). This has been reported in South African plants (Galley and Linder, 2006), western/eastern Pacific fishes (Lessios and Robertson, 2006), southwestern/southeastern Australian groups (Crisp and Cook, 2007: Fig. 2; Morgan et al., 2007), plants disjunct between Eurasia and western North America, and between North America and South America (Wen and Ickert-Bond, 2009), in vertebrates east and west of the Mississippi River (Pyron and Burbrink, 2010), and in birds east and west of Bering Strait (Humphries and Winker, 2011). Based on morphological observations, it can be predicted that molecular studies of any important biogeographic pattern will always find a great range of branch lengths in the taxa. If an evolutionary clock is assumed, this finding implies a great range of values for the ages of the taxa involved, from zero in undifferentiated taxa up to high values in the most differentiated taxa.

To summarize, at all biogeographic breaks, for example, the Atlantic Ocean, different pairs of clades show differing degrees of divergence. There are two possible explanations:

- Evolution is clock-like and the different pairs of clades split at different times.
- Evolution is not clock-like and the different pairs of clades split at the same time. Degree of divergence is related to prior genome architecture, not to time.

If the first option is correct, a general vicariance event cannot explain any biogeographic pattern; distribution patterns must all be pseudopatterns and the result of chance coincidence. In this model, speciation is always peripatric, involving dispersal over a barrier, and never dichopatric, with a new mountain range, seaway, or other feature developing and dividing a population. Arguing against this is the intricate geographic structure now revealed in so many clades, including groups with efficient means of dispersal. As shown above, 19th-century evolutionists had already concluded that degree of differentiation of form, even in close relatives, cannot be used to judge the age of the groups, and so a biogeographic pattern could have developed at the same time in the different groups. A species in one group may have the same age as a genus or even a family in another group.

Avise's Paradox

Degree of differentiation may often not be related to time elapsed and could instead be due to prior aspects of genome architecture that determine the evolutionary "propensity" or "evolvability" of a group. This explanation for branch length accounts for Avise's (1992) paradox: "[C]oncordant phylogeographic patterns among independently evolving species provide evidence of similar vicariant histories. . . . However, the heterogeneity of observed genetic distances and inferred speciation times are difficult to accommodate under a uniform molecular clock" (p. 63).

In other words, all biogeographic patterns are shared by groups that show significant geographic concordance, indicating general vicariance, but all patterns are shared by different taxa showing different degrees of differentiation, which indicates separate histories and thus dispersal. Morphological examples of Avise's paradox include the New Zealand–South American groups cited above or low-level clades in bryophytes and higher-level clades in vascular plants (Mishler, 2001). Molecular examples include the South African plants (Galley and Linder, 2006) and others already mentioned. The usual response to Avise's paradox has been to ignore it and carry on accepting a more or less uniform clock, such as the "2% rule" for birds (Weir and Schluter, 2008). This will imply that speciation events in a pattern occurred at different times, and so the pattern has been caused by separate dispersal events, not vicariance. But this does not account for the concordance in community-wide geographic patterns that Avise (1992) observed and it does not explain the paradox.

The paradox only arises if degree of differentiation is related to time since divergence. If it is not, a given vicariant event will lead to different degrees of divergence in different groups, depending on their genetic potential. This would allow general, community-wide vicariance and imply "relaxed" clocks in individual clades. Just how relaxed can a clock be? A clock that is relaxed enough is no longer a clock. Fitting tectonic events to nodes suggests that clocks can be very relaxed indeed, with extreme rate variation, and this would account for Avise's (1992) paradox.

Many studies claim to use a relaxed molecular clock, but accept only minor rate variation. Bouetard et al. (2010) used relaxed molecular clock methods to test vicariance versus dispersal in the orchid genus *Vanilla*. They found that vicariance in *Vanilla* following the breakup of Gondwana at 95 Ma would be incompatible with the age of the orchids, as it would imply that the orchid family was 310 m.y. old (Carboniferous). But this would not be true using a more relaxed clock. For example, with extreme changes in rates, most of the evolution leading to the modern subfamilies, tribes, and genera of orchids could have occurred over just a few million years in, say, the Cretaceous.

The Possibility of Extreme Variability in Evolutionary Rate

If evolutionary rates can show such drastic changes, a distinction must be made between the *time* involved in the evolution of a group and its *age*. A group may have undergone most of its evolution in 10 million years, 100 million years ago. Rapid phases of evolution are often reported in plants and animals introduced to islands, for example, with morphological differences observed in a century or less (e.g., Mathys and Lockwood, 2011). Rapid, ancient radiations are often inferred (e.g., Fishbein et al., 2001), and rapid phases of evolution mediated by tectonics and climate change could have modernized whole communities over continental or global scales. These would mean that evolutionary or molecular clock hypotheses could be in error by orders of magnitude, as suggested below in many cases on the basis of biogeography.

Most evolution may have occurred during general phases of modernization, following which there have been long periods of evolutionary stasis or only parallel evolution in which morphological and genetic distance between groups remains similar. This is already well documented in some groups. For example, there has been marked stasigenesis in the main form of some arachnids for 200 million years (Turk, 1964). A fossil bee from the Late Cretaceous is "astonishingly similar" to living species of *Trigona* (Engel, 2000). The moss *Pyrrhobryum mnioides* occurs in Australia, New Zealand, and South America. It has been attributed to Gondwanan vicariance followed by morphological stasis for over 80 million years (McDaniel and Shaw, 2003). This pattern may be much more common than it appears to be, as most current methods of dating tend to give ages that are too young.

The idea of a strict, universal evolutionary clock is rejected by many authors (e.g., Rodríguez-Trelles et al., 2003) but using a clock that is only slightly relaxed does not solve the fundamental problems with the whole idea. Anderson et al. (2005: 1744) concluded: "[O]ur methods rely on the assumption that there is an autocorrelation of evolution rates in adjacent lineages and that some kind of smoothing is reasonable, but could it be that rates change abruptly rather than gradually?" As Near and Sanderson (2004: 1482) observed, "With respect to rate heterogeneity, once the model of molecular evolution departs from a simple one-rate molecular clock, the divergence time problem enters a realm of model selection in which the number of models is effectively infinite." If evolution is not clock-like, which mode or modes does it conform to?

Accepting non–clock-like evolution means that extreme rate variability among and within groups and genes is possible or even likely. The evolutionary "clock" may be so "relaxed," with great changes in rate within and between groups, that it can no longer be described as even vaguely clock-like. Thus, the differential evolution of different groups seen at a biogeographic boundary may reflect a single, ancient tectonic event. This in turn means that the spatial relationship between phylogenetic/biogeographic nodes and tectonic structure can be used to date nodes on a phylogeny and avoid relying on fossils or fossil-calibrated clocks to give maximum ages of clades.

In a simple clock model of evolution, the degree of divergence of a group is related to its *age*. In a slightly different model, a "punctuated-equilibrium" clock, degree of differentiation is proportional to the *time* involved in an ancient evolutionary differentiation; since then there has been no or little evolution. In a strictly non-clock model of evolution, degree of differentiation is proportional to neither the *age* of a clade, nor the *time* involved in its evolutionary origin. Instead, it is related to prior genetic propensity for differentiation. Overall, evolutionary divergence may be related to time—a phylum is older than a species—but at smaller scales, evolution may often be non–clock-like.

CALIBRATING THE EVOLUTIONARY CLOCK

Molecular clock theory adopts assumptions about the *mode of evolution*— for example, whether or not it is clock-like—and uses another set of assumptions in the process of *calibration*. This establishes the age of at least one node on a phylogeny as a "known," which is then used as a basis for calculating the ages of the other nodes by secondary cross-calibration. Whether or not a strict clock is assumed, at least one node on a tree must be calibrated to give evolution a time dimension. It is a crucial step with many potential problems.

Apart from cross-calibrating from another node with a known age, there are only three ways to date evolutionary development:

· Use the oldest fossil of a group. But this only gives a minimum age for the group.
· Use the age of the island or the strata that a group is endemic to. But young islands and strata often have old taxa.
· Correlate the geographic distribution of a group with associated tectonic events. But tectonic features can be reactivated at different times.

All three methods are flawed, but the first two have serious, inherent limitations, while the third suggests possible lines of research.

Fossil Calibrations

Fossils only give minimum ages for their clade. Statistical analysis of the clade's fossil record as a whole has been used to estimate the actual age of a group, although this is not possible for the many groups in which the record is sparse. Even in groups with better fossil records, estimations of the accuracy of the fossil record using the record itself are not convincing.

Although fossils only give minimum ages, these are often mysteriously transmogrified into maximum ages. For example, the monocot families Bromeliaceae and Rapateaceae occur in the New World and in addition each has a single African species. Givnish et al. (2004) calibrated a molecular clock for the families using "Cretaceous fossil information to place minimum ages on eight monocot nodes." Based on this, Givnish and Renner (2004a) found that the African species are the result of divergence "*no earlier than* the past 7–12 million years" (italics added). Although this estimate was based on a calibration which gives a minimum age, it was presented as a maximum age. This was then used to rule out earlier events (vicariance by Atlantic opening) and thus support trans-Atlantic long-distance dispersal.

For Malvaceae on Madagascar, Koopman and Baum (2008) used fossil calibrations to give minimum dates. By transmogrifying these, they were able to conclude that "our molecular dating analyses support a radiation of Hibisceae in the early to mid-Miocene (ca. 19.5–15.5 Ma) which is *too late* to be explained by vicariance" (p. 372; italics added).

The problem of transmogrification has been pointed out by Olmstead and Tank (2008), who described it as "an error in logic." As they observed, "Providing an error estimate for a node based on the variance in substitution rate, or based on multiple calibration points on the tree, does not provide an upper and lower bound on the *time of origin* of the node, but rather upper and lower bounds on the *minimum age* estimate of that node" (italics added).

Nevertheless, many molecular studies are based on transmogrified dates and so attribute all the biogeographic patterns found in a large group to dispersal, none to vicariance. For example, Hedges (2010: 353) wrote that "Most research suggests that the entire living biota of the Caribbean islands arrived by dispersal." For the Rubiaceae of Madagascar (95 genera, 650 species) Wikström et al. (2010: 1095) wrote: "More recently, an increasing emphasis has been put onto oceanic dispersal as a mechanism of origin for elements of the Malagasy flora. . . . This partly results from a growing understanding that many plant groups with large Malagasy elements are *too young* to be affected by the break-up of Gondwana . . . , but also from a resurrection of dispersal as an important mechanism in historical biogeography" (italics added). This suggests that there are two separate sources of evidence, one concerning the age of clades and the other concerning the importance of dispersal. But dispersal has only been resurrected because of the young ages inferred, and the young ages are all based on transmogrified dates.

Wikström et al. (2010: 1095) proposed that "Rubiaceae separated from remaining families in the Gentianales during the Late Cretaceous, c. 78 Ma (Bremer et al., 2004), and this places any origin of Rubiaceae in Madagascar well after the final break-up of Gondwana." But although the Bremer et al. (2004) paper is titled "Molecular Phylogenetic Dating," their dating is calibrated solely with fossils and so again provides minimum ages only.

For the Malagasy Rubiaceae, Wikström et al. (2010: 1108) concluded: "Notwithstanding topological and mapping uncertainties in our analyses, it is clear that the four tribes arrived in Madagascar at least 11 times, and that they arrived via dispersal from both Africa and Asia." But dispersal is only supported if fossil-calibrated minimum dates are transmogrified into maximum dates. The authors suggested that "These results are highly consistent with the conclusions drawn by Yoder and Nowak (2006)," but the two studies are "highly consistent" only because they both rely on the same method of transmogrifying minimum dates into maximum dates.

While critics sometimes make the mischievous claim that panbiogeography ignores the fossil record (Cox and Moore, 2010: 18), this is not correct. Biogeographical analyses of a group should incorporate all known records, living and fossil, and aim at an integration of the two. Matthew's (1915) treatment provided excellent mapped examples that graphically integrated fossil and living distribution in many groups. Fossils are a tremendous source of information on minimum ages, on the phylogeny of many groups, and on general changes in evolutionary level through time. The broad change in form seen between the biotas of the Mesozoic and Cenozoic is concrete evidence of evolution. But deducing the chronological details of individual clades by a literal reading of the fossils' stratigraphic distribution can be very misleading; in A.B. Smith's (2007: 231) careful wording, "the fossil record provides direct evidence but it cannot be taken at face value." The age of a fossil is not the same as the age of its group, and relying on the oldest fossil has produced the house of cards that modern phylogeography has become. The consensus and "corroboration" seen among phylogeographic studies exist only because authors use the same assumptions and methods, and then cite each other's conclusions as supporting evidence.

The Assignment of Fossils to Points in Phylogenies: Another Problem with Molecular Dating

The time course of evolution as shown in fossil-calibrated molecular clocks assumes that fossils have been correctly assigned to points in phylogenies, using morphology. Many molecular biologists seem to accept the identifications and phylogenies of the fossils without question, and this is surprising, as molecular studies have shown that the morphological classification of so many living groups is untenable. Despite this, molecular workers rely on morphological assignment of fossil clades to lineages in order to calibrate molecular clocks. The morphology of fossil groups is never as well understood as that of living groups because of the inherent limitations of the material. Molecular systematists would not necessarily have great confidence in the assignment of living clades to a group using morphology, so basing their entire biogeographic analysis of a group on the assignment of a fossil clade to a group using morphology does not make sense.

Calibrations Using the Age of Islands and Mountains

Some authors have assumed that clades endemic on islands cannot be older than the islands, and so island-endemic clades have been used to calibrate phylogenies. Yet many island-endemic clades have been dated as older than their islands, even in clock studies using conservative fossil calibrations (Heads, 2011). For example, Parent et al. (2008) reviewed dating studies of Galápagos Islands taxa and found that many of the endemics have been dated as older than the islands themselves (4–5 Ma). *Galapaganus* weevils have been dated at 7.2 Ma, the giant tortoise *Geochelone nigra* at 6–12 Ma. Differentiation between the endemic marine iguana (*Conolophus*) and its sister the endemic land iguana (*Amblyrhynchus*) is estimated to have occurred at 10 Ma; other Galápagos iguanids (the lava lizards *Microlophus*, formerly *Tropidurus*) and also geckos (*Phyllodactylus*) have been dated at 9 Ma. Wherever their differentiation took place, these currently endemic taxa are older than the current islands.

Volcanism in the Galápagos has not progressed in narrow, time-progressive lines of seamounts and ridges as predicted by the hotspot/mantle plume hypothesis. There are notable exceptions to the age progression, and the causes of these remain enigmatic. The complex tectonic history has been attributed to the location of the islands near both a mid-ocean ridge—the Galápagos spreading center (the margin of the

Cocos and Nazca plates)—and a plume (O'Connor et al., 2007; Geist and Harpp, 2009). Werner and Hoernle (2003) studied the volcanoes between the Galápagos and central and South America, on Cocos Island, and the Carnegie, Malpelo, and Coiba ridges. Flat-topped seamounts (guyots), paleobeach structures, and intertidal wave-cut platform deposits imply that islands have existed continuously above the Galápagos hotspot for at least the past 17 million years. Werner and Hoernle (2003: 904) concluded that "These new data significantly extend the time period over which the unique endemic Galapagos fauna could have evolved." As they concluded, the data provide "a complete solution to the long-standing enigma of the evolution of Galapagos land and marine iguanas" and, they might have added, the Galápagos biota as a whole. To summarize, the current Galápagos Islands are 4–5 m.y. old, there is direct evidence for prior islands back to 17 Ma, but the history of island production over this hotspot could extend back 80–90 m.y., the estimated age of the Galápagos hotspot (Parent et al., 2008).

Biologists sometimes describe isolated mountains as ecological islands, and the principles just discussed for real islands apply here in the same way. Endemics on young volcanoes are often assumed to be no older than the volcano they currently survive on. Nevertheless, counter-examples are known from the East African rift volcanoes. The volcanic edifice of Mount Kilimanjaro in Tanzania has built up since ~2.5 Ma (Nonnotte et al., 2008), while the cricket *Monticolaria kilimandjarica* (Tettigoniidae: Orthoptera), endemic to the mountain, was dated as 7–8 m.y. old (Voje et al., 2009). A population on a single volcanic edifice can survive more or less *in situ* by constantly colonizing younger lava flows from older ones. As the volcano increases in elevation, so does the population.

Calibrations Based on Tectonics

Smith et al. (2010: 5901) expressed concern about changes in evolutionary rate and wrote: "It is increasingly clear that there may be extreme differences in molecular rate . . . and current methods may be unable to cope." The method of dating used in this book does not assume an evolutionary clock, even a relaxed one. Instead, it fits multiple tectonic events (rather than multiple fossils) to a phylogeny. This indicates a chronology in which rates can show extreme changes within and among lineages and genes at different times and places.

R. T. Pennington et al. (2004a) discussed the option of using geological events or fossils to calibrate nodes on a phylogeny and concluded in support of fossils. They argued that the high frequency of long-distance dispersal "highlights the danger" of using distributions and geological events, especially "old" ones, to date clades because patterns will have been obscured. But these authors only assumed that long-distance dispersal is frequent because many nodes in many papers are dated as recent (e.g., papers in R. T. Pennington et al., 2004b and Givnish and Renner 2004b), and they are only dated as recent because they were calibrated with fossils. The authors have continued to depend on fossil dating and fossil-based calibrations and have not been able to suspend judgment long enough to give tectonic-based calibration a fair test, or indeed any test. The reliability of fossils and the "danger" of using tectonic calibrations may both be more imaginary than real, and this can be assessed by trying out the alternative methods of calibration in different cases and comparing the results.

For example, the opening of the Atlantic is a well-studied phase of Earth's history and can be used to date clades. *Thurnia* is a herb of sandy lowlands of the Guiana Shield and adjacent Amazon lowlands and is sister to *Prionium* of intermittent watercourses on sandstone in South Africa. The pair form the family Thurniaceae, sister to Cyperaceae plus Juncaceae. Givnish et al. (1999) regarded this trans-Atlantic affinity as surprising, although three other families in the same order, Poales, are also trans-Atlantic disjuncts: Mayacaceae, Bromeliaceae, and Rapateaceae. Givnish et al. indicated that the distribution of *Thurnia-Prionium* and their habitat on granite-derived sand and sandstone "might argue for the origin of their lineage in western Gondwana before the rifting of the Atlantic" (1999: 371), and this is accepted here. For the three other trans-Atlantic families, Givnish et al. (1999) suggested that several facts "argue persuasively" for long-distance dispersal, but these facts were all oldest fossils or fossil-calibrated clocks. The trans-Atlantic disjunctions could instead indicate that all four clades in Poales evolved before being rifted apart by the opening of the Atlantic.

Tectonic and Fossil Calibrations of Cichlid Evolution

The cichlid fishes have high diversity in African lakes, and this is often cited as the classic example of recent, "explosive" radiation. But the cichlid phylogeny has been calibrated using the fossil record and the age of the lakes, both of which may be very misleading. New work has instead calibrated the evolution of the cichlids with tectonic events (Sparks, 2004; Sparks and Smith, 2004, 2005; Genner et al., 2007;

Azuma et al., 2008). This approach has given interesting results, including bold, new hypotheses that contradict the young ages usually accepted for the family.

There are two main clades of cichlids, one in Madagascar, Africa, and America, and one in Madagascar, India, and Sri Lanka. Sparks (2004) concluded that these relationships "are congruent with prevailing hypotheses regarding the sequence of Gondwanan fragmentation and a vicariance scenario to explain the current distribution of cichlid fishes." He observed that

> A vicariance scenario for cichlids is disputed by some paleontologists, given that acanthomorphs ["higher teleosts"] do not appear in the fossil record until the Upper Cretaceous . . . , that there are no unquestionable Cretaceous perciforms . . . and that representatives of Madagascar's extant freshwater fish groups are absent from Cretaceous deposits examined to date on the island. . . . However, there is a global paucity of Early Cretaceous through mid-Paleocene freshwater teleost fossils which is presumably attributable to a scarcity of fossil-bearing freshwater rocks of Cretaceous age. . . . Given that there is a *notable gap* in the acanthomorph fossil record extending from the Late Cretaceous to the Late Paleocene, by which time an incredibly diverse fauna has evolved . . . there is no reason to dismiss the possibility that the fossil record is likewise *misleading to date* with respect to the origin of percomorph fishes. (italics added)

Recent discoveries of well-preserved fossil cichlids from the Eocene of Tanzania "illustrate just how quickly and substantially our notions have changed regarding a time of origin for cichlid fishes" (Sparks, 2004). These fossils are dated at 46 Ma, and prior to their discovery cichlids were only known back to 36 Ma. The fossil taxa "appear to be very derived and similar to modern African lineages." Sparks (2004) continued:

> Rapid diversification of cichlid lineages (especially East African forms) has been reported time and again in the literature; however, here we have a well-documented example of morphologically conserved forms persisting in eastern Africa for nearly 50 Myr. What do these recent fossil discoveries imply about the age of cichlid fishes? If 46 Myr fossils appear similar to modern forms and are nested well within the African assemblage, certainly cichlids are much older even than these fossils (i.e., Cretaceous in age). Moreover, topologies recovered for Cichlidae and aplocheiloid killifish, the only two groups of freshwater fishes in Madagascar with broad Gondwanan distributions, are not only congruent with each other, but also with prevailing hypotheses regarding the sequence of Gondwanan breakup in the Mesozoic. These repeated patterns are intriguing and may well point to a common cause.

Neither Sparks (2004) nor Sparks and Smith (2004, 2005) presented a calibrated molecular clock. Instead, their emphasis was on the topology of the tree, the distribution of the clades, the regional tectonics, and a critical reading of the available fossil record.

Potential Problems with Tectonic Calibration: The Isthmus of Panama and Geminate Atlantic/Pacific Clades

The tectonic method of calibration is not without its own problems. Many tectonic features, such as fault zones and belts of uplift, have been active over long periods or have been reactivated at different times, and some applications of the tectonic method have been over-simplistic. For example, authors often assume that continental rifting is the only process that could have led to vicariance in, say, the Cretaceous. But Earth and its life may also evolve together during phases of uplift, subsidence, erosion, volcanism, metamorphism, juxtaposition by terrane accretion, marine transgression, and many other processes.

Another example where correlations between distribution and paleogeographic events have been simplistic concerns the final rise of the Isthmus of Panama at about 3.5 Ma. This is a key date in dispersal theory, as it is supposed to mark the beginning of the "Great American Interchange" of terrestrial faunas (Wallace, 1876; Stehli and Webb, 1985). The rise of the isthmus is also assumed to have brought about vicariance between marine taxa, dividing them into Atlantic and Pacific clades, and so the 3.5 Ma age of the isthmus is often used in clock calibrations. Nevertheless, many Atlantic/Pacific pairs ("geminate" taxa) probably diverged well before the final rise of the isthmus. In biogeographic work, Croizat (1975) questioned the significance of the modern Isthmus of Panama and instead emphasized the complex geological history of the Colombia–Central America region since the Mesozoic. Other geologists and biologists have also suggested that Pacific/Atlantic differentiation of marine taxa occurred in the late Mesozoic and early Cenozoic. Differentiation of the main centers of endemism for reef corals, the Atlantic and Indo-Pacific regions, was often attributed to the Pliocene emergence of the isthmus, but Rosen (1988) pointed out that this event long postdates faunal differentiation of reef corals as shown in the fossil record. The actual date, based on the age of fossils, was at least early Cenozoic and possibly much older.

Likewise Knowlton and Weigt (1998) suggested that at least some of the geminate species pairs in the snapping shrimp *Alpheus* may have diverged before the final closure of the Panama seaway. Knowlton and Weigt estimated times of separation of trans-Panama pairs of *Alpheus* ranging from 3 to 18 Ma for 15 species pairs. They assumed that the divergence of the pair with the least difference was due to isthmus closure, but there is no real evidence for this.

In bivalve molluscs (Arcidae), Marko (2002) dated differentiation in geminate pairs at up to 30 Ma. In gastropod molluscs, Lessios (2008) studied six geminate species pairs in *Echinolittorina* and wrote: "Despite a preference of the genus for upper intertidal habitats [suggesting that populations would only be separated with final closure], an assumption of a geminate split contemporaneous with the closure of the isthmus would suggest that the genus is only 9 My old, whereas fossil evidence suggests it goes back to 40 My" (p. 83).

In the opisthobranch *Bulla*, fossil-calibrated estimates indicated a (minimum) Miocene date for geminate speciation across the Panamanian isthmus, long before the uplift of the isthmus (Malaquias and Reid, 2009).

These conclusions have been ignored by the many authors, who continue to rely on the closure of the Panama seaway to calibrate clocks. As Knowlton and Weigt (1998: 2257) concluded, "Many past studies may have overestimated rates of molecular evolution [and underestimated ages] because they sampled [trans-Panama] pairs that were separated well before final closure of the Isthmus." What are the alternatives? Muss et al. (2001) suggested that because the Atlantic and Pacific species of *Ophioblennius* probably diverged prior to the closure of the Isthmus of Panama, "a geologically calibrated clock is unavailable." But there is more to geology than the Isthmus of Panama, and contemporary studies may be taking too narrow an approach to calibration. It is probably best to take a broad, tectonic approach to the paleogeographic development of Panama rather than a stratigraphic one; as Wörner et al. (2009: 183) wrote, "The geological development of Panama's isthmus resulted from intermittent magmatism and oceanic plate interactions over approximately the past 100 m.y.," that is, since the Cretaceous.

Differentiation in and Around the Andes

The Andes are similar case to the Panama isthmus in that they often separate sister taxa to the east and the west of the range. Biologists have used the uplift of the range to date the taxa, and the age of the uplift is critical. Many biologists have assumed that the Andes began to rise only in the Miocene (Pirie et al., 2006; Linder, 2008), but in fact uplift began as the Atlantic started to open, in the Cretaceous. The first ("Peruvian") phase of uplift took place at 90–75 Ma (Late Cretaceous), although there is some evidence for deformation in the Early Cretaceous (Cobbold et al., 2007).

It is now accepted that the plant and animal populations of a mountain range may have existed in the region since before uplift of the mountains and then been passively uplifted with them, rather than colonizing them from surrounding lowlands after uplift has taken place (Chapter 1; Heads, 2006b; Ribas et al., 2007; Thomas et al., 2008; Losos and Mahler 2010: 394). This means that montane taxa can be older or much older than the mountain range and the elevational belt that they now occupy, and that a calibration based on their present elevation will give dates that are underestimates.

Critics have commented on the "strong and persistent desire" (Near and Sanderson, 2004) and the "great thirst" (Graur and Martin, 2004) to know the divergence dates of clades. This is natural, but in many molecular studies impatience has clouded judgment and led to rushed conclusions. Much of the molecular-clock work is based on assumptions that are dubious or even irrational, and normal scientific logic, caution, and scepticism have been rejected. Instead of critically examining the basic assumptions in the methods used, authors have applied these unthinkingly and any problems have been swept under the carpet. The first question when assessing a study on chronology should be: What is the calibration based on? If it is based on fossils, the second question is: Have the dates been treated as minimums or have they been transmogrified into absolute dates?

The case studies discussed next illustrate some possible approaches to dating evolution.

CASE STUDY: INTERTIDAL SNAILS (*NERITA FUNICULATA* AND ALLIES; GASTROPODA)

Trans-Atlantic connections occur in many plants and animals, including the widespread mollusc *Nerita* (Gastropoda: Neritidae). *Nerita* is a marine group but it occupies intertidal, benthic habitat (on rock, sand, mud, and mangrove trees) and so its history may have more in common with terrestrial animals than with those of the deep ocean. The biogeography and phylogeny of the trans-Atlantic *Nerita funiculata* clade are indicated in Figure 2-2 (based on Frey, 2010) The "basal" species in the clade, *N. funiculata*, occurs in the Pacific, and so this region

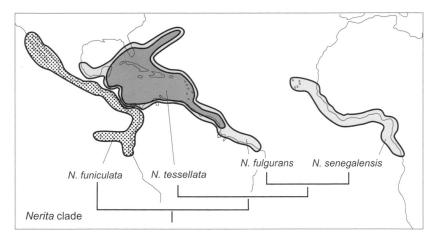

FIGURE 2-2. The *Nerita funiculata* clade (Gastropoda) (Frey, 2010).

could be taken as a center of origin. Alternatively, the phylogeny could be interpreted as a sequence of differentiation events—not dispersal events—that has proceeded eastward. The first break, between eastern and western Mexico and Central America, may be related to tectonic events along this boundary in the Early Cretaceous. This was a period of overall crustal extension in the region, and western Mexico was a multiple arc system or a single arc with intra-arc/back-arc rifting (Centeno-García et al., 2008) (see Glossary for geological terms). The second break in the *N. funiculata* clade occurred somewhere around the Caribbean (or its precursor), between a Caribbean group and a Caribbean–African group. Assuming original allopatric speciation between the Caribbean species, *N. tessellata* and *N. fulgurans*, their overlap has developed subsequently. Dissecting out the history of the pair would require more information than the simple distributions. The final break in the clade is the trans-Atlantic split between *N. fulgurans* and *N. senegalensis*, and if this occurred with the geological split, it can be dated to the Early Cretaceous, as with the first node. The genus has a fossil record back to the Late Cretaceous, giving a useful minimum date. The possible effects of Early Cretaceous rifting on other groups in Mexico, Central America, and around the Alantic are discussed below. Distributions similar to that of the *Nerita* clade are well-known. For example, two clades in the rocky shore gastropod *Echinolittorina* each have the pattern: (East Pacific (West Atlantic (East Atlantic))) (Williams and Reid, 2004, Figs. 6B and 6F).

CASE STUDY: THE FRESHWATER PLANKTON GENUS *DAPHNIA* (CRUSTACEA: CLADOCERA)

Most molecular clock calculations are not accepted here, but the molecular work itself is revealing a spectacular level of biogeographic structure in many groups that is of great inherent interest and is not predicted in dispersal theory. In groups such as marine taxa with planktonic larvae, this level of structure was completely unsuspected and has led to a paradigm shift in marine bioigeography (Heads, 2005b). The same thing has happened in aquatic biology. In freshwater zooplankton, the morphological similarity of species inhabiting different continents, combined with observations of strong dispersal mechanisms, also led to the early conclusion of cosmopolitan distributions. High dispersal rates were inferred, along with limited opportunities for allopatric speciation. Yet extensive molecular evidence and detailed morphological revision now indicate that a reversal of this view is needed (Adamowicz et al., 2009), as freshwater zooplankton show biogeographic patterns similar to those of terrestrial groups such as birds and angiosperms.

Most aquatic invertebrate species are confined to single continents, and studies now stress the provincialism of clades. Detailed distributional information indicates that allopatric divergence is likely to be an important mechanism of diversification. A good example of this is *Daphnia* (Crustacea: Cladocera), a genus of freshwater zooplankton with a global distribution. Adamowicz et al. (2009) wrote that the dominance of the subgenus *Ctenodaphnia* in the southern hemisphere and of subgenus *Daphnia* in the north, as well as the confinement of *Australodaphnia*, the basal subgenus, to southwestern Australia, "suggests ancient splits that most likely correspond to the break-up of Pangaea into Gondwanaland and Laurasia. . . . The biogeographic patterns are complex, with several ancient clades detected among samples from each of South America, Australia, and the Mediterranean region, indicating that *Daphnia* had already diversified to a certain extent by the time of the break-up of Gondwanaland" (p. 432). Fossil remains (ephippia) of daphniids and other anomopods are known from the Mesozoic, supporting this chronology.

Different Degrees of Evolution in Different Crustacean Orders

Patterns of continental endemism similar to those seen in *Daphnia*, a member of order Cladocera, are also observed in the fairy shrimps, order Anostraca, but at much higher taxonomic levels, even including

families. Adamowicz et al. (2009) concluded that "These contrasting patterns in depth of endemism may be due to differing ages of
families in the two groups. However, the similarity of branch lengths
among families in the two orders . . . suggests that the difference in
the geographic scale of diversification is probably the result of differing dispersal ability *or some other biological feature*" (p. 434; italics
added). The different branch lengths for groups showing the same
geography do not necessarily indicate different ages; instead of representing time, branch lengths could indeed reflect a "biological feature." This could be the genetic potential or propensity for divergence
referred to above. During the last phase of evolution, Anostraca and
Cladocera may have responded to many of the same tectonic events
in similar ways, but Anostraca may have had a greater potential for
evolution than Cladocera. In Anostraca, regional endemics in the different continents became families, whereas in Cladocera, the same
events only provoked differentiation to the level of subgenera, with
shorter branch lengths. Since then there may have been only minor or
parallel evolution, and as Adamowicz et al. (2009) concluded, "The
general long-term morphological stasis observed in this genus, and
in many other freshwater organisms, is fascinating considering the
prolific cladogenesis" (p. 434). This is probably true for modern life in
general, although in many groups without a Mesozoic fossil record it
is obscured in accounts that transmogrify minimum dates into absolute clade ages.

*The Paradox of Endemism in Weedy Taxa; the Persistence of Ancient
Patterns; Phases of Mobilism*

Clades that are weedy in their ecology but also show regional or even
local endemism, as in several groups of *Daphnia*, present an enigma
unless the concepts of normal ecological dispersal and theoretical "chance
dispersal" are distinguished. For *Daphnia*, Adamowicz et al. (2009: 432)
wrote, "The facts that new habitats are rapidly invaded . . . and that some
genotypes have vast geographic distributions . . . suggest that dispersal
potential is high, which is seemingly paradoxical given the high degree of
genetic structuring among populations even at local scales." With respect
to the main clades, "The persistence of the ancient, subgeneric north–
south split presents an apparent paradox. Although this pattern appears
to have arisen during the breakup of Pangaea, it remains strong despite
the evidence for high vagility in the genus *Daphnia*." Recent dispersal

"should have acted to erode the continental affinities established by the fragmentation of Gondwanaland and Laurasia"—but it has not. Thus, while any group can in theory expand its range, as all groups have the power of movement, in practice many groups do not, at least for long periods of geological history. "Means of dispersal" seem to turn on and off, rather than being a constant factor, and what activates them is not chance, but geological or climatic change.

Adamowicz et al. (2009) acknowledged that the deeper phylogenetic patterns in *Daphnia* are consistent with vicariance scenarios linked to continental fragmentation. They discussed the possible reasons for the "intriguing" and "puzzling" persistence of these ancient patterns in light of the eroding force of dispersal. How are they maintained? Perhaps the most convincing of the explanations suggested by Adamowicz et al. (2009) is "home advantage," or priority effect: Propagules entering the territory of another group would usually be entering habitats occupied by similar but well-adapted local forms already utilizing available resources.

For *Daphnia*, Adamowicz et al. (2009) concluded that the extent of intercontinental range expansion (dispersal) has shown marked changes over time. Geological history is compatible with this episodicity in phases of dispersal: "There were several brief intervals enabling 'free for all' events as new habitats were created, in which incoming migrants could gain a foothold with greater success than had previously been possible" (p. 433). Examples could include the period following the great marine transgressions of the Cretaceous.

Molecular Clock Dates in Daphnia

While some clades in *Daphnia* may represent ancient intercontinental vicariance, in other intercontinental affinities, "The shallow divergences often observed are strongly suggestive of dispersal rather than ancient vicariance, if mitochondrial molecular clocks . . . are correct even to within an order of magnitude: (Adamowicz et al., 2009: 424). But due to calibration error and dramatic change in divergence rates, molecular clock dates may often be incorrect by one or more orders of magnitude. Adamowicz et al. (2009: 432) cited a rate calculated for arthropods (Brower, 1994) and argued that within-species distances (<3% in 12S) and the shallower of the between-species intercontinental splits "can only be explained by dispersal." What is the basis for Brower's rate?

CASE STUDY: THE CLOCK RATE FOR ARTHROPODS CALCULATED BY BROWER (1994)

Brower (1994) derived a rate for arthropods that is often used to calibrate phylogenies in clock studies. The rate (2.3% sequence divergence per million years) was an average derived from rates given in other studies. These included two rates that were personally communicated to Brower and presented without the mode of calibration (these are not discussed further here), and four published rates derived for the following taxa. None of the calibrations used fossils.

Magicicada septendecim (Martin and Simon, 1990). In the eastern United States, a boundary between northern and southern clades occurs at 33° latitude. This was assumed to be due to Pleistocene events, in line with traditional theory, although there is no particular evidence for this.

Drosophila silvestris (DeSalle and Templeton, 1992). This study assumed that distinctive populations on Kilauea volcano in southeastern Hawaii Island dispersed there from other parts of the island and diverged only after the formation of the volcanoes Kilauea and Mauna Loa at ~100,000 BP. *Drosophila silvestris* has two clades, eastern and western, that correspond to the two volcanic arcs that run through Hawaii and the other the Hawaiian Islands (see Chapter 7). The arcs and possibly the taxa on them predate the individual volcanoes.

Drosophila "*melanogaster* subgroup" (8 species) (Caccone et al., 1988). These authors used the divergence times suggested by Lemeunier et al. (1986). The split between *D. mauritiana* of Mauritius and its sister *D. sechellia* on the Seychelles on one hand and the cosmopolitan *D. simulans* on the other was proposed to have occurred at 0.4–1 Ma; this triad split from the cosmopolitan *D. melanogaster* at 0.8–3 Ma. But endemic taxa on Mauritius, including the dodo, the plant *Monimia*, and the snake family Bolyeridae, indicate that taxa on the islands can be much older than the islands themselves (Heads, 2011).

Alpheus species (Decapoda) (Knowlton et al., 1993). In geminate pairs of species on Pacific and Caribbean sides of the Panama isthmus, degree of divergence varied considerably, and the authors inferred evolution at different times. In a key step, the minimum level of divergence observed was attributed to formation of the isthmus at 3.5 Ma. The authors ruled out isthmus formation as the cause of at least some geminate differentiation, but they did not show why it must have caused *any* of it, they only assumed it. As discussed already, most geminate distributions may have existed prior to the final isthmus formation.

To summarize, none of the divergence rates used to find the average rate are well substantiated.

CASE STUDY: LEPIDOPTERA

In Lepidoptera, 95% of the species feed (as larvae) only on angiosperms, and many are host-specific on particular angiosperm clades. In their ecology, evolution, and biogeography the two groups are intimately related. The fossil record of Lepidoptera is poor (there are only 19 specimens from the Mesozoic; de Jong, 2007) and entomologists have been tempted to rely on the angiosperm fossil record for timing evolutionary events in Lepidoptera, including the origin of the group. De Jong (2007) wrote that the angiosperm fossil record is "incomparably better" than in Lepidoptera and that divergence times of the families are "known with some reliability," but while the first statement is true, the second is questionable. For example, de Jong (2007) cited the Old World butterfly *Libythea*. A species on the Marquesas Islands, southeast Polynesia, has been regarded as sister to the rest of the genus and the pattern attributed to Jurassic tectonics. De Jong rejected this idea as the food plant, *Celtis*, has been dated at 25 Ma and this is "much too young for whatever vicariance event" (p. 327). He accepted the age without question as an absolute (not a minimum) date and built a complex dispersal scenario for *Libythea* around this, but the age may be incorrect and the dispersal unnecessary.

CASE STUDY: PERISSODACTYLS

In mammals, there are many disparities between the age of a clade as indicated by the oldest fossil and the age estimated in fossil-calibrated molecular phylogenies. Norman and Ashley (2000) utilized the extensive fossil record of Perissodactyla to calibrate molecular clocks. They used a recent fossil calibration point (divergence of two equid species at 3 Ma) and an ancient fossil calibration point (divergence of Hippomorpha, including equids, and Ceratomorpha, including rhinocerotids and tapirids, at 50 Ma). Application of these produced very different estimates of evolutionary rates and divergence times for the two genes they studied. Neither calibration point produced estimates of divergence times consistent with a literal reading of the fossil record. The older calibration point placed the separation of the two equid species at greater than 13 Ma, which the authors suggested is "incompatible" with the fossil evidence. But an early age is not truly "incompatible" with the fossil record, as this only gives minimum ages for clades. Using the recent calibration point gives a date that really is incompatible with the fossil record: a separation of Ceratomorpha and Hippomorpha at 8–11 Ma, and Rhinocerotidae and Tapiridae at 6–8 Ma, although fossils of these groups are known before these dates. Norman and Ashley (2000) criticized the use of molecular clocks, but there may also be large gaps in the perissodactyl fossil record that they did not suspect, for example, in the early record of the equids.

CASE STUDY: PHYLOGENETIC BREAKS AROUND MADAGASCAR AND NEIGHBORING ISLANDS

This region is the site of important phylogenetic and biogeographic breaks in many groups, and the evolution of several of these is described here.

Cyprinodontiformes Suborder Aplocheiloidei

This group, the aplocheiloid killifishes, was cited above because of the biogeographic similarities it shares with the cichlids. The group includes ~350 species that occur mainly in freshwater, especially small, ephemeral ponds and streams. They range widely through the tropics, including Madagascar (Fig. 2-3). Murphy and Collier (1997) wrote that the phylogeny "strongly indicates the role of vicariance in the diversification of these fishes in spite of their definition as secondary freshwater fish," with a few found in more saline water around estuaries. Murphy and Collier emphasized that the phylogeny is congruent with the breakup sequence of Gondwana. First, India/Madagascar/Seychelles separated from Africa (at 160 Ma, in the Jurassic). Later, America separated from Africa, and Madagascar–Seychelles split from India. The Seychelles are old, granitic islands northeast of Madagascar. The absence of aplocheiloids from areas such as Borneo is another aspect of the distribution that is inexplicable in a center of origin/dispersal history but is compatible with vicariance.

No fossils of the Aplocheiloidei are known (Costa, 2010), and the oldest fossil of the entire order Cyprinodontiformes is from the Oligocene (30 Ma), long after the Mesozoic rifting events. But Murphy and Collier

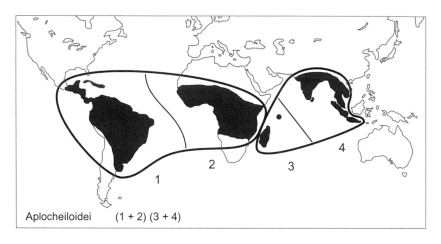

FIGURE 2-3. Cyprinodontiformes suborder Aplocheiloidei (Murphy and Collier, 1997). The phylogeny is also shown.

(1997) and Costa (2010) stressed that the oldest fossil only gives a minimum age for the group, and so the Oligocene *fossil age* is compatible with the Jurassic *clade age* that Murphy and Collier proposed for Aplocheiloidei.

Despite the absence of fossils, Parenti (1981) also interpreted the distribution pattern of Aplocheiloidei as the result of Gondwana breakup. This hypothesis was rejected by Lundberg (1993) based on the absence of early fossils and on the incompatibility between the sequence of continental drift and earlier phylogenies (in which African taxa were more closely related to Asian taxa than to American taxa). Lundberg concluded that the ecology and physiology of cyprinodontiform fishes do not contradict a model of transoceanic dispersal. In fact, though, American and African aplocheiloid fishes inhabit shallow, freshwater swamps, with the exception of a rivulid clade containing three species of estuarine swamps, and none of these exhibit any ability to survive in undiluted sea water (Costa, 2010). In addition, phylogenetic analysis of aplocheiloid fishes based on molecular data (Murphy and Collier, 1997) provided a well-corroborated phylogeny in which aplocheiloid lineage relationships fit the Gondwanan continental breakup. Sparks and Smith (2005) also attributed the aplocheiloid distribution to a Gondwanan origin.

Among the mainly freshwater aplocheiloids, the *Rivulus marmoratus* species group is unusual in that it inhabits estuarine mangroves throughout the Gulf of Mexico and along the coast south to Rio de

Janeiro. *R. caudomarginatus* is found in similar habitat around Rio de Janeiro. Murphy and Collier (1997) pointed out that *Rivulus* occupies a relatively terminal (nested) position in the phylogeny, and so they suggested that the saltwater habitat is secondarily derived (cf. Fig. 1-6). But this is not necessary. If the ancestor was widespread globally, as the phylogeny and biogeography indicate, it may have had a diverse ecology as well. Most of the suborder may have attained its range through being stranded inland following recession of inland seas in the Cretaceous, while the small mangrove group has retained the early ecology.

Madagascar and Trans–Indian Ocean Disjunction

The aplocheiloid killifishes indicate that Madagascar is part of a broader biogeographic region made up of the Indian Ocean basin and its margins. The region was formerly part of Gondwana and is one of the main global centers of endemism. Distributions of groups there show many disjunctions, both across the basin and with other areas.

Groups based around the Indian Ocean include the passerine family Campephagidae, the cuckoo-shrikes. Jønsson et al. (2010) proposed that the family represents a "convincing case of colonization over a significant water gap of thousands of kilometers from Australo-Papua to Africa." But this conclusion depended on molecular clock dates, in which calibration points were based on island ages (for Mauritius and Réunion), passerine mitochondrial substitution rates, and secondary calibration points for passerine birds. These are all dubious, and so the dispersal scenario is not convincing. In any case, if members of the group were able to disperse such long distances, they would be expected to occur in America or on the islands east of Fiji, where they are completely absent.

Many trans–Indian Ocean disjunctions involve Madagascar. In the palms, for example, *Tahina* of Madagascar is sister to *Kerriodoxa* of southern Thailand (Dransfield et al., 2008). Connections between Madagascar and Malesia are well documented in traditional taxonomy. Groups endemic to Madagascar and Borneo include the plant *Allantospermum*; groups endemic to Madagascar and Papua New Guinea include the amphipod *Pseudocyphocaris* (Myers and Lowry, 2009). Similar connections can be predicted in molecular studies.

Many Indian Ocean clades are present on Madagascar but absent from Africa. In some cases this pattern may be due to extinction on Africa, but often the Indian Ocean group has an allopatric sister in Africa, as in aplocheiloids (Fig. 2-3). In these groups, the absence from Africa is relative only and may be due to phylogeny rather than extinction.

The genus *Nepenthes* (Nepenthaceae) has a classic Indian Ocean disjunct distribution (Meimberg et al., 2001). Its basal clades are in Sri Lanka and the Seychelles, and there are diverse other species in Madagascar, northern India (Khasi Hills, Meghalaya), Southeast Asia, Malesia, northeastern Australia, and New Caledonia. Possible fossil pollen is recorded in Europe. To discuss the origin of *Nepenthes* means to compare it with its sister group (Drosophyllaceae, Ancistrocladaceae, Dioncophyllaceae), which is present in Africa (two families), Europe (one family), and Asia (one family) but notably absent from Madagascar. The break between the two clades lies along the Mozambique Channel and can be attributed to the rifting that took place here.

Rifting around East Africa may also be related to the main break in *Exacum* (Gentianaceae); Yuan et al. (2005) drew the interesting comparison between this genus and *Nepenthes*. *Exacum* is recorded around the margins of the Indian Ocean, with one clade in Africa and Madagascar, and the other distributed between Socotra and northern Australia/New Caledonia (including the Arabian Peninsula, Sri Lanka, India, the Himalayas, southern China, and mainland Southeast Asia; Yuan et al., 2005). The authors wrote that "The distribution of this genus was suggested to be a typical example of vicariance resulting from the breakup of the Gondwanan supercontinent. The molecular phylogeny of *Exacum* is in principle congruent with morphological conclusions and shows a pattern that resembles a vicariance scenario with rapid divergence among lineages." (Following this promising beginning, Yuan et al., 2005, relied on a center of origin program and the use of fossil ages to give maximum clade dates. They deduced a Madagascar center of origin and long-distance dispersal. "Yet," they concluded, "the process and mechanism of these long-distance dispersals are not well understood.") Trans–Indian Ocean connections now appear regularly in molecular work. For example, the passerine known as *Rhipidura hypoxantha*, recorded from the Himalayas and southern China, is sister to the South African *Stenostira* (Stenostiridae) (Fuchs et al., 2009).

The bird family Caprimulgidae (Caprimulgiformes) has a wide distribution in Africa, Asia, and the Americas. The basal branches are *Eurostopodus*: India to New Caledonia, and then *Caprimulgus enarratus* of Madagascar (Han et al., 2010), indicating basal differentiation (a paraphyletic group) at nodes around the eastern and then western margins of the Indian Ocean. As usual, the basal clades do not necessarily represent a center of origin and could instead reflect a phase of phylogenetic fracturing associated with rifting. This is also indicated in chameleons.

Basal Breaks in and Around Madagascar: Chameleons

Chameleons are a distinctive family of lizards known from Africa to Spain and India, with about half of the 150 species in Madagascar. They also occur on islands around Madagascar, including the Comoros, the Seychelles, and the Mascarene Islands. In traditional biogeography the diversity on Madagascar would suggest a center of origin there, but at least some studies have supported a vicariance model for the group (Klaver and Böhme, 1986, 1997; Hofman et al., 1991). A molecular analysis of 52 species (Raxworthy et al., 2002) found the phylogeny: (Madagascar (Madagascar (Africa (Africa (Africa, Madagascar, and India))))).

Raxworthy et al. (2002) invoked a center of origin, "out-of-Madagascar" model and, given the basal grade in Madagascar, this would be a standard interpretation. But the basal grade in Madagascar is also compatible with initial differentiation of a widespread Africa–India ancestor somewhere in or around Madagascar, with subsequent overlap there (see Fig. 1-6).

Further work has shown that the basal grade in Madagascar is probably a monophyletic clade, the genus *Brookesia* (Townsend et al., 2011). The next basal clade may comprise the only chameleon on the Seychelles (formerly treated as *Calumna tigris*, now as the genus *Archaius*) plus the small genus *Rieppeleon* (formerly in *Rhampholeon*). *Rieppeleon* occurs between northern Mozambique and Kenya, along the coast and at varying distances inland.

Thus, following basal differentiation in chameleons around Madagascar, differentiation took place around central East Africa/Seychelles, and after this shifted east to central Africa with divergence between the two African genera, *Rhampholeon* and *Bradypodion*.

The last and most widespread clade in the phylogeny (Raxworthy et al., 2002) comprises three subclades distributed in Madagascar (*Calumna* in part), Africa–India (*Chamaeleo*), and Madagascar–Comoros–Réunion (*Calumna* in part + *Furcifer*). The partial vicariance among the three clades is distinctive and probably significant, although the phylogeny of the group needs much more work (see also Townsend et al., 2011). Africa and Madagascar have several chameleon clades each, so overlap has probably occurred in these areas before the opening of the Mozambique Channel in the Jurassic.

Raxworthy et al. (2002) proposed that the branching sequence in the chameleon phylogeny cannot be reconciled with vicariance (which they assumed would be due to Gondwana breakup), and so they concluded that the distributions must be due to chance dispersal. But vicariance with Gondwana breakup and chance dispersal are not the only possibilities. The basal nodes in the chameleon phylogeny indicate that tectonic vicariance and overlap had *already* occurred around Madagascar, Seychelles, and central East Africa, forming the five main clades, before differentiation in the widespread group took place, and only this last phase of differentiation is due to final Gondwana breakup. There was extensive and profound tectonic activity in the Africa–Madagascar–India region leading up to final breakup, and this may have been involved with the basal differentiation (see Chapter 5).

Raxworthy et al. (2002) found what they regarded as "corroborating evidence" for oceanic dispersal in the fact that the volcanic Comoro Islands, dated at 0.13–5.4 Ma, have endemic chameleons whose direct ancestor "could only" have reached the archipelago by means of oceanic dispersal. This assumes that there has been no prior land in the Comoros area, but the Comoros have formed at a hotspot which can be linked to volcanism around the Seychelles Islands dated at 63 Ma (Ito and van Keken, 2007). McCall (1997) suggested emergent land occurred in the Mozambique Channel through Eocene–Miocene time along the Davie Fracture Zone. Rieppel (2002) agreed that the study by Raxworthy et al. (2002) "confirms that dispersal is important." Yet he also wrote that the occurrence of groups such as chameleons, freshwater fishes, and terrestrial mammals in Madagascar "has been seen as paradoxical." This is because they are supposed to be younger than the separation of Madagascar and so they must have dispersed there, but these groups are considered to be poor dispersers. Rieppel concluded that "how chameleons managed to disperse across the ocean must remain a matter of speculation."

Thus, modern phylogeography, insofar as it is based on Matthewian principles, does not solve biogeographical problems but only leads into a maze of unresolved mysteries, paradoxes, and speculations. Instead, oceanic dispersal in chameleons may never have occurred and vicariance, not dispersal, may be the mode of speciation in the family. This is compatible with differentiation in the African genus *Rhampholeon*, where Matthee et al. (2004) reported "a close correlation between geographical distribution and phylogenetic relatedness . . . indicating that vicariance and climate change were possibly the most influential factors driving speciation in the group."

Basal Breaks in and Around Madagascar: Tenrecs (Tenrecidae)

The great phylogenetic and biogeographic break that occurs between Madagascar and mainland Africa in aplocheiloid killifishes, chameleons, and other groups discussed above is one of the best-known patterns in biogeography. In some cases it may have been caused by one of the best-known of all tectonic events, the rifting between Madagascar and Africa in the Jurassic.

The profound break between African and Malagasy groups is also seen in mammals, including tenrecs, a suborder in the order Afrosoricida. The single family comprises four subfamilies, with three endemic to Madagascar and their sister, Potamogalinae, in Africa. Douady et al. (2002) used a molecular clock and estimated that the split between the African and Madagascar clades occurred at 53–51 Ma, much older than the oldest fossil tenrecid (20 Ma). No information on the calibration of the tree was given, but Douady et al. argued that "dispersal events" are required. They concluded that "the mechanism by which tenrecids arrived in Madagascar remains unclear," and these burrowing animals seem unlikely candidates for long-distance transoceanic dispersal. The phylogenetic placement of the tenrecs in the correct order (rather than Insectivora, where they had been placed) was only achieved with molecular study. Thus it is likely that fossil mammals from the Mesozoic, known from much poorer material than living groups, have also been placed in the wrong order.

Madagascar–Seychelles: Hyperoliidae

The Madagascar–Seychelles connection is seen in one of the four main clades of aplocheiloid killifishes (Fig. 2-3). It also occurs in the frog family Hyperoliidae, where a clade from Madagascar (*Heterixalus*) and the Seychelles (*Tachycnemis*) is the only representative outside Africa. No fossils of the family are known. The overall distribution of hyperoliids has been explained by continental drift and vicariance (Richards and Moore, 1996). But although Vences et al. (2003: 213) admitted that "vicariance often offers more appealing explanations," for hyperoliids they proposed island hopping and rafting from Africa to Madagascar and from there to the Seychelles. They argued that the "presence of the Hyperoliinae on the Madagascar-India continent implies a very early age of their evolution [and vicariance]. Such an assumption, however, meets with several contradicting facts. The first problem is the absence of these frogs in South America. . . . Madagascar was apparently connected with South America via the Kerguelen plateau and Antarctica in the late Cretaceous. . . . It is difficult to understand why such a vagile group would not have been able to colonize [South America]" (p. 212).

The absence of these Madagascar–Seychelles frogs from South America only contradicts vicariance if it is assumed that the Gondwana biota was uniform throughout the supercontinent. There is no reason to assume this or to assume that groups endemic to one part of Gondwana would have invaded other parts of Gondwana. Just because a "vagile" group can in theory expand its range does not mean that it will. For example, the birds of paradise on New Guinea should be able to colonize the large islands of the nearby Bismarck Archipelago easily, but for some reason they never have. The argument that Vences et al. (2003) presented follows a line of reasoning often used in dispersal theory. In an early example, Wallace (1881) argued that Madagascar cannot have been connected with southern Asia at the time that squirrels, deer, and antelope existed in Asia; otherwise they would have migrated into Madagascar.

Vences et al. (2003) also emphasized that the Madagascar–Seychelles hyperoliids were not in India, but again there is no need to assume that groups on the Madagascar–Seychelles–India continent ranged throughout the region. None of the current continents have anything like a homogeneous biota.

Finally, Vences et al. (2003) concluded that the "low" genetic distance observed between the Malagasy and Seychelles frogs means they "may not be well explained by vicariance," as this would require "unprecedentedly low substitution rates" (p. 213). On the other hand, the authors admitted that "molecular clock estimates are often of limited value" (p. 213). In a different approach, Bossuyt and Milinkovitch (2001) calibrated a Madagascar versus India–Seychelles split in Ranidae s.lat. (i.e., between Mantellinae and Rhacophorinae) using the date of the geological split between these areas (88 Ma). This assumes rather than tests vicariance for this event, but the predictions made using the calibration, including the ages and locations of other nodes, can then be investigated.

Seychelles–Asia Connections: Sooglossidae

The hyperoliid frogs that Vences et al. (2003) investigated are on Madagascar and the Seychelles, while the ranid group Rhacophorinae is on the Seychelles and India (Bossuyt and Milinkovitch, 2001). As in Madagascar, the Seychelles biota is a biogeographic composite and represents an early boundary. It has inherited at least two different regional biotas that already existed before the breakup of Gondwana; one around the Seychelles and regions to the southwest, one around the Seychelles and areas to the northeast. In another example of the latter connection, the frog families Sooglossidae of the Seychelles and Nasikabatrachidae of the Western Ghats in India form a clade that is sister to the large global clade Hyloidea plus Ranoidea (Biju and Bossuyt, 2003). The group's ancestors may have occupied the Seychelles–Western Ghats sector long before the Seychelles existed as islands or the Western Ghats existed as mountains.

Giant Pill-Millipedes (Diplopoda Order Sphaerotheriida)

The giant pill-millipedes occur around the margins of the Indian Ocean basin, in Africa (southeastern coast), Madagascar, India, Southeast Asia, Australia, and New Zealand. In morphological studies, Wesener and VandenSpiegel (2009) found that all genera from southern India and Madagascar form a monophyletic group, the new family Arthrosphaeridae. The authors concluded that the phylogeny of the families "mirrors perfectly the suggested break-up of Gondwana fragments 160–90 Ma. No evidence for a dispersal event could be found." The allopatric clades (Wesener and VandenSpiegel, 2009; Wesener et al., 2010) are:

1. Procyliosomatidae: eastern Australia and New Zealand.
2. Zephronidae: Seychelles (*Sechelliosoma*) and the Himalayas–Philippines/Sulawesi.
3. Arthrosphaeridae: Madagascar–southern India/Sri Lanka.
4. Spaerotheriidae: southern Africa, from Cape to Malawi.

The molecular phylogeny is not yet resolved but the morphological tree is: (1 (2 (3 + 4))). In this sequence, differentiation moves westward. The Seychelles are again connected with Southeast Asia rather than with Madagascar. They are allopatric with a group in Madagascar–India and its sister in southern Africa, the two having the same break at the Mozambique Channel as in aplocheiloid killifishes.

The Mascarene Islands–Seychelles

The Mascarene Islands (Mauritius and Réunion) are a classic case of young islands with endemics that are older than the islands themselves. These endemics include plants (*Monimia* and *Hyophorbe*), birds (the dodo and its sister *Pezophaps*), and snakes (Bolyeridae) (Heads, 2011). The Bolyeridae have their closest relatives in Borneo. Other distinctive plants on the Mascarenes include a clade in the *Ixora* group (*Myonima* and *Doricera*: Rubiaceae) that is sister to a diverse, pantropical complex (Mouly et al., 2009). Many widespread Indo-Pacific groups reach their western limit at the Mascarenes and are absent from Madagascar or Africa (e.g., the monocot *Astelia*; Craw et al., 1999: Fig. 5-2). The plants and animals on the Mascarenes also show interesting links with the Seychelles, 2000 km to the north. *Euploea* butterflies and *Aerodramus* swiftlets are distributed from India to Australia and also in the Seychelles and the Mascarenes, but are absent from Madagascar and Africa. At least in *Aerodramus*, the species on the Mascarenes (*A. francicus*) and the Seychelles (*A. elaphrus*) are sister species (Price et al., 2005). Warren et al. (2010: 9) wrote: "Based on present-day geography, it seems incongruous that winds should have brought these genera to both the granitic Seychelles and Mascarenes, but not to Madagascar." If the distribution of the butterflies and the birds had been determined by physical movement, the pattern would indeed be surprising. Nevertheless, the link between the Seychelles and the Mascarenes that excludes Madagascar is seen in other groups and is probably the result of tectonics and evolution around the Mascarene Plateau, not the physical movement of populations.

In the passerine *Zosterops* (Zosteropidae), Warren et al. (2006) found a clade of "ancient white-eyes" endemic on islands that form an arc around Madagascar: the Mascarene Islands, the Seychelles, and the Comoros (Grand Comore). The authors wrote that "The relatively close relationship between *Z. semiflavus*, *Z. mouroniensis* and the Mascarene white-eyes is not only unexpected based on existing taxonomy, but is also biogeographically surprising. The islands occupied by these ancient Indian Ocean white-eyes—the granitic Seychelles, Grande Comore and Mascarenes—are disparate island groupings, while the intervening islands of the eastern Comoros (Mohéli, Anjouan and Mayotte) as well as the much larger landmass of Madagascar are devoid of ancient

Indian Ocean white-eyes, and are solely occupied by white-eyes of the *maderaspatanus* clade" (Warren et al., 2006: 3779).

Thus dispersal theory finds it anomalous that these birds of "disparate" offshore islands should be more closely related to each other than to the mainland (Madagascar) that they skirt. The connection is seen as surprising and incongruous, but instead it may be an important clue. In other groups, the giant scorpion *Chiromachus ochropus* is only known from the Seychelles and the Mascarene Islands (an early record from Zanzibar is believed to be an error; Gerlach, 2005). In the passerine *Nectarinia* (Nectariniidae), Warren et al. (2003) found the phylogeny: (northwestern Comoros (Seychelles (southeastern Comoros (central Comoros (Madagascar and Aladabra Islands)))))). All the clades are allopatric and there is no need for any dispersal. The sequence of differentiation began around the Comoros and Seychelles before affecting the Madagascar populations.

In the mangrove crab *Neosarmatium* (Sesarmidae) a clade in Africa-Madagascar is sister to one in the Seychelles and the Mascarenes (Ragionieri et al., 2009).

In the *Drosophila* "*melanogaster* subgroup," *D. mauritiana* of Mauritius is sister to *D. sechellia* of the Seychelles, according to Caccone et al. (1988), while Lachaise et al. (2000) gave the phylogeny as (*D. mauritiana*: Mauritius (*D. sechelliana*: Seychelles (*D. simulans*: cosmopolitan))). The new phylogeny does not imply a great change in the biogeography but suggests a sequence of differentiation first around Mauritius and then around the Seychelles.

Although the biogeographic arc in *Zosterops* surrounding Madagascar may be unexpected and even surprising, it is not meaningless and it could reflect aspects of former geography not preserved in current bathymetry or exposed stratigraphy. This could be tested by seeing whether other groups show similar patterns.

Madagascar–Socotra: Colubrid Snakes

Groups such as *Exacum* show major breaks between Africa/Madagascar and Socotra–India, and so on, while in other groups, Madagascar is linked with Socotra and the Horn of Africa. A good example is seen in the snakes, where 14 colubrid genera in Madagascar form a clade that is sister to *Ditypophis* of Socotra (Nagy et al., 2003). This confirms a connection suggested in earlier accounts. For example, the shrub *Coelocarpum (Verbenaceae)* is native to Madagascar, Socotra, and Somalia (Marx et al., 2010). At least in its morphology, the coralliform tree *Euphorbia arbiuscula* of Socotra has a "homologue" in *E. fihirensis*, found on the calcareous plateaux of southwestern Madagascar (Thomasson and Thomasson, 1991). Nagy et al. (2003) commented that Madagascar, Socotra, and India may have all separated from Gondwana about the same time. Their molecular-clock estimates suggested that Malagasy and Socotran colubrids diverged from their noninsular sister groups later, between the Eocene and Miocene. But the clock was calibrated by assuming African and Asian sister species of colubrids could be no more than 16–18 m.y. old, and this is questionable.

Nagy et al. (2003) provided an "admittedly speculative" scenario for the Madagascar–Socotra colubrids in which both islands were colonized independently from the African plate and there was subsequent extinction of the relatives in Africa. Invoking extinction in Africa is ad hoc and unnecessary. It would also need to be invoked for many other groups on Madagascar and other Indian Ocean islands but not in Africa, even though many of these have an allopatric sister group in Africa. Instead, the Madagascar–Socotra connection is probably related in tectonics and biogeography to the Socotra–India and Madagascar–Seychelles tracks discussed above.

CASE STUDY: MARINE BIODIVERSITY IN THE MALAY ARCHIPELAGO

The Malay archipelago, comprising Indonesia, Malaysia, the Philippines, and New Guinea, is well known for its high levels of biodiversity; it is the major hotspot for marine biodiversity. This has often been attributed to Pleistocene evolution. Renema et al. (2008) reviewed fossil and molecular data for different marine groups over the past 50 m.y. and concluded instead that "The strong correlation between the presence of [biodiversity] hotspots and major tectonic events suggests that the primary drivers may operate over time scales beyond those traditionally used to examine diversity (p. 656) (cf. Heads, 2005b). Renema et al. (2008) found evidence that main diversity centers have developed along the Tethyan sector between New Caledonia and Spain, a region characterized by the encroachment of Africa, India, and Australasia on Asia and related subduction. Renema et al. wrote that the "accumulation of diversity as a result of the juxtaposition of communities and accretion of tectonic terranes is most likely to occur in a compressive tectonic setting" (p. 656). They also concluded that the modern taxa are "much older than previously thought" and show "surprising antiquity" (p. 654). Finally, "the critical role of tectonic events emphasizes the importance of abiotic factors in shaping the world's biotic realm. They drive and underpin the birth, life and senescence of biodiversity hotspots" (p. 657). This view contrasts with the idea that all differentiation is due to chance, long-distance dispersal, as inferred in clock studies, but agrees with a biogeographic analysis of the plant family Ericaceae in the Malay archipelago (Heads, 2003).

CASE STUDY: THE PHILIPPINES

Several "old endemic" Philippines clades that have widespread sister groups were discussed in Chapter 1. The endemic clades are distinctive in a global context and indicate that the Philippines biota is much older than has been assumed by most authors. Studies focused at a smaller scale, within the Philippines, have tended to stress young events such as Pleistocene sea-level change (islands in the archipelago were joined and then separated again) and have neglected earlier tectonic history. But interesting molecular-clock analyses of several Philippines bird taxa (Jones and Kennedy, 2008) give at least minimum dates and have refuted the hypothesis of a major role for Pleistocene events. The authors concluded that "new hypotheses" are needed, in particular, models that incorporate ideas on pre-Pleistocene geography.

In a later study, Siler et al. (2010) examined the gecko *Cyrtodactylus* in the Philippines to see if patterns of inter- and intra-specific diversity could be explained by the "Pleistocene aggregate island" model of diversification. They concluded in a way similar to Jones and Kennedy (2008):

> Contrary to many classic studies of Philippine vertebrates, we find complex patterns that are only partially explained by past island connectivity. In particular, we determine that some populations inhabiting previously united island groups show substantial genetic divergence and are inferred to be polyphyletic. Additionally, greater genetic diversity is found within islands, than between them. Among the topological patterns inconsistent with the Pleistocene model, we note some similarities with other lineages, but no obviously shared causal mechanisms are apparent. Finally, we infer well-supported discordance between the gene trees inferred from mitochondrial and nuclear DNA sequences of two species, which we suspect is the result of incomplete lineage sorting. This study contributes to a nascent body of literature suggesting that the current paradigm for Philippine biogeography is an oversimplification requiring revision. (Siler et al., 2010: 699).

The authors concluded: "we are left with many unanswered questions. If the PAIC [Pleistocene Aggregate Island Complexes] model does not suffice, roughly what proportion of Philippine biodiversity has been generated by PAIC-relevant processes? If the last century's paradigm does not suffice, how 'wrong' is it?"

If an accreted terrane model of Philippines biogeography is correct, rather than a Pleistocene model, the latter could err in its chronology by more than an order of magnitude. In a clock study, Blackburn et al. (2010) found that the frog *Barbourula* of the southern Philippines (Palawan) and Borneo diverged from its sister taxon *Bombina*, of Europe and eastern Asia, in the Eocene (47 Ma). The authors described the date as "unexpected" and "surprisingly ancient," with the dates indicating "great antiquity" for the endemism. Blackburn et al. suggested that some portion of Palawan has been above water since it rifted from southern China and that components of the terrestrial fauna have survived in this way.

VICARIANCE VS. DISPERSAL: OLD QUESTIONS AND NEW MOLECULAR DATA

Modern systematics became recognizable in the 16th century when Renaissance scholars such as Cesalpino left off writing accounts of the useful plants of a region (herbals) and began to write accounts of all the plants, whether they were useful or not ("floras") (Heads, 2005c). In the same century, the cartographer Ortelius mapped the world and suggested that America and Africa/Europe had been joined but were separated. By this time Europe was aware of the strange flora and fauna of the New World, and it is not surprising that the vicariance versus dispersal debate is an old one. Gilbert White cited it in his 1789 classic *The Natural History of Selborne* when discussing trans-Atlantic affinities, and already he regarded it as a difficult question (White, 1977: 65). Over two centuries later the question remains controversial.

Many workers have attempted to decide between vicariance and founder dispersal in particular cases by dating the clades concerned. But the molecular clocks have been calibrated using methods that can give drastic underestimates of clade age. For example, assuming that *Lactoris* was no older than its island, Juan Fernández, would give a date that is too young by over 100 m.y. (Heads, 2011). Tectonic dating is better, although uncritical correlations have been made in applying this method, for example, around the Isthmus of Panama. This rose at ~3.5 Ma, but assuming that all taxa currently separated by the isthmus differentiated at this time will, again, give drastic underestimates of clade age. The use of fossil age or island age to date clades can be rejected. Instead, the geographic distribution of the molecular clades and the precise geometry of the vicariance provides a growing body of more or less uncontroversial evidence and a solid foundation for comparative biogeography.

The molecular phylogenies so far published have already resolved countless problems in biogeography and evolution that had held up progress for decades. The vast amount of complex allopatry that the molecular studies have revealed is of particular interest and has led to complete paradigm shifts in marine biogeography and microbial biogeography. All these advances involve the *branching sequence* or topology of the phylogeny, not the degrees of difference—the *branch lengths*—of the clades. As discussed in this chapter, the dates of nodes inferred in many molecular studies are based on branch length, and because of the flawed evolutionary models and calibrations used, they are probably unreliable.

The molecular phylogenies and the beautiful distributions of the clades they reveal are the raw material for this book. Sampling is improving all the time, and the only real problems with the molecular work are the interpretations of evolution in space (reading the sequence of nodes as a dispersal sequence) and time (calibrating the nodes with fossil or island age and assuming that branch length reflects age).

PHYLOGENY, BIOGEOGRAPHY, AND BRANCH LENGTH

Many molecular studies now attribute all patterns of distribution and speciation to dispersal and none to vicariance (e.g., Wikström et al., 2010, on the Malagasy Rubiaceae, cited above). This idea is a new development, as the dispersal theorists of the last century usually accepted that at least a minor component of allopatric differentiation (the "paleoendemic element") was caused by vicariance rather than dispersal. Likewise, most textbooks on evolution refer to allopatric speciation caused by the uplift of a mountan range or formation of a sea, a process affecting the whole community. But the neodispersalism of modern phylogeography suggests that vicariance may be very rare or may not even occur. This is based on interpretations of two main sets of evidence.

The first set involves the *differences in the degree of differentiation (branch lengths)* of groups in any pattern. All biogeographic phenomena, such as centers of endemism or disjunctions, involve taxa with widely differing degrees of taxonomic rank or genetic differentiation.

If an evolutionary clock is accepted, all the clades are a different age; the different branch lengths seen in any pattern will imply a separate evolutionary event for all taxa and a more or less continuous sequence of differentiation (or stream of immigrants). In this way, all distribution is interpreted as the result of chance dispersal events and none due to community-wide vicariance. All patterns would be interpreted as pseudopatterns, with all of the component taxa being the result of the same unlikely event happening, by chance, many times.

The second set of evidence comprises the *young clade ages* based on branch length, as given in clock studies of different groups. Yet the two methods used to produce most of the clade ages, based on fossil age or island age, both give systematic underestimates. By treating the young clade ages as maximum (not minimum) dates, these can be used to show that biogeographic distributions have been caused by chance, one-off dispersal events at different times through the Cenozoic.

This approach is simple and popular but always concludes with a mystery—how does the process actually work? Why exactly does dispersal start and then stop? In panbiogeography, the differentiation is due to geological change: Earth and life evolve together. This approach stresses that global distribution is orderly, not chaotic, and can be analyzed with reference to a few main phylogenetic and geographic breaks (nodes) that occur repeatedly in different groups at different levels of differentiation.

Although the Earth's biosphere is much thinner than the crust or mantle, it shows a much more complex geographic structure. One of the clearest examples of this is the clumping of basal nodes in certain geographic regions (Heads, 2009a). This and other patterns are not explained by simply appealing to "chance." Although chance dispersal has been favored in many molecular studies, earlier work questioned the efficacy of means of dispersal in many cases, and this problem has not been resolved. Molecular-clock workers often admit that the evident means of dispersal in the studied organism do not appear sufficient to explain the geographic pattern, but then support chance dispersal anyway. Sometimes authors suggest that more study is required to elucidate the means of dispersal, but the dispersal events suggested are very rare, one-off events that are not correlated with any others. and it is hard to imagine how they could ever be studied empirically, whether by examining means of dispersal or anything else. The means of dispersal in a clade must be effective, carrying an organism 10,000 km across the Pacific, for example. On the other hand, they cannot be too efficient as they operate just once in the immensity of geological time. Most authors who infer founder dispersal are discreet enough to avoid discussing the details of the process.

With millions of species and now DNA sequencing, biology has access to an immense amount of information on differentiation in space, infinitely more than geology, with its few hundred major minerals. Despite this, the use of chance dispersal means that most modern biogeography has nothing to offer its sister science, and it is symptomatic that biology lags far behind geology in most aspects of mapping. Ideally, biology should be able to make fundamental contributions to debates in tectonics, as in the work of Wegener (1912, 1915).

Instead of trying to prove or disprove the vicariance/dispersal argument in theory, it may be more productive to see what the results are if different sets of assumptions are tried out in practice. Can dispersal or vicariance provide a coherent scenario for world distribution? Of the two possible options, a dispersal scenario is easy to describe. Any pattern at all can be explained by chance dispersal, and global biogeography would be seen as the result of endless, one-off dispersal events, each one unrelated to any other aspect of biology or geology. In the other option, there is an orderly development by vicariance of entire communities, with the main phylogenetic and biogeographic breaks of different groups recurring in the same locations. Alternating with phases of vicariance are phases of mobilism and overlap, with both being determined by geological and climatic events, as discussed through the rest of this book.

3

Evolution and Biogeography of Primates

A New Model Based on Molecular Phylogenetics, Vicariance, and Plate Tectonics

The last two chapters considered general aspects of evolution in space and time. The rest of the book is a study of biogeography and tectonics in the tropical regions. The tropics are well known for their stupendous biodiversity, and a survey of all groups is obviously impossible. On the other hand, discussions that are not based on concrete examples often become theoretical and unrealistic. As a compromise, the members of a single clade, the primates, were selected as the focus of a case study of tropical America, Africa, and Asia. The primates are suitable for this because they are diverse, widespread, and one of the best-known tropical groups. Birds are also well studied in terms of field observations, but primates have a more extensive fossil record and their molecular phylogeny has been researched in considerable detail.

Primates are widespread in most tropical areas of the New World and the Old World east to Sulawesi and the Philippines, but are absent from Australasia and the Pacific (apart from humans and human introductions). There are about 400 extant primate species, and these are most diverse in lowland tropical rainforest, including mangrove and freshwater swamp forest. Many taxa have local distributions and are threatened by habitat loss and commercial hunting, while others are abundant. With their intelligence and agility, primates can dominate

This chapter is reprinted from *Zoologica Scripta* (Heads, 2010a) with corrections and additions, and with permission from John Wiley and Sons.

lists of pests that damage crops in rural areas (Pienkowski et al., 1998). A tendency to a "weedy" ecology and ability to survive in disturbed sites occurs in many primates, such as macaques (Richard et al., 1989), and its significance in the evolution of the group is discussed below.

A great deal is now known about the phylogeny and the distribution of the primates (Goodman et al., 2005; IUCN, 2009). Nevertheless, there are fundamental disagreements about the group's evolution, beginning with where and when it originated. The lack of any consensus on this has held up progress in understanding many other aspects of primate evolution.

The basal primate in a phylogenetic sense—the sister of all the others—may be the fossil *Altanius*, from the Eocene of Mongolia, and the oldest primate fossil may be *Altiatlasius*, from the Late Paleocene of Morocco (Fleagle and Gilbert, 2006). Still, the interpretation of these fragmentary fossils, in particular their phylogenetic status, is controversial. (Different interpretations and phylogenies are available for many primate fossils.) Center-of-origin analyses of the living primates are also inconclusive, and the study by Heesy et al. (2006: 420) begins and ends with the unanswered question: "On which continent did primates originate?" Biologists have been debating whether the true center of origin of primates was in Africa, Asia, or America for more than a century and yet the argument has never been resolved.

The problem of primates' geographic origin is related to the vexed question of their origin in time. While the oldest fossils indicate an origin for primates in the Paleocene, at ∼56 Ma, fossil-calibrated molecular clocks calibrated with fossils from other groups give Cretaceous dates, at ∼90 Ma (e.g., Janečka et al., 2007). Despite the fact that both these estimates are minimum dates, in many accounts they are "transmogrified" into maximum or absolute dates. Although the molecular clock dates are still only minimum, not absolute dates, they are important as they show that fossil-based dates could underestimate ages by tens of millions of years.

Several authors have recognized the problem of transmogrification in fossil-calibrated molecular clock analyses. In primate studies, Wilkinson et al. (2011: 28) wrote that Chatterjee et al. (2009) "implicitly assume that the true age is close to the minima [the calculated dates based on fossil calibrations] and unlikely to be much older than those minima. This assumption, we feel, is unlikely to be warranted, as it does not take account of the sizable gaps that exist in the primate fossil record." Wilkinson et al. (2011) attempted to estimate the error in the primate record using the details of the record itself, but this approach is fraught with problems.

TABLE 3-1   THE MAIN CLADES OF EXTANT PRIMATES

Clade	Geographical Range
Strepsirrhines	
Lemurs	Madagascar
Lorises and galagos	Mainland Africa and Asia
Haplorhines	
Tarsiers	Southeast Asia
Anthropoids:	
Old World monkeys (catarrhines)	Mainland Africa and Asia
New World monkeys (platyrrhines)	Tropical America

SOURCE: Goodman et al., 2005.

Masters (2006: 112) concluded: "Where, then, did primates originate, and how did they come to occupy their current distribution? This remains the single most puzzling aspect of primate evolution." Most reconstructions require one to several over-water dispersal events on vegetation rafts, and many authors have raised serious objections to this. Simons (1976: 50), for example, was unequivocal: "[N]o explanation involving transport across wide reaches of ocean is tenable in accounting for the distribution of any primate."

Most writers on biogeography have found the center of origin concept easy to justify in theory, but in practice they have had difficulty achieving consistent results. For primates, Fleagle and Gilbert (2006) supported Asia as center of origin, although this was tentative and they considered the issue "far from resolved" (p. 385); Rasmussen (2002) favored Africa or India; Silcox (2008) suggested Europe; Arnason et al. (2008) proposed South America; Bloch et al. (2007) supported North America (for Primates s.lat.). One multiauthor study (Miller et al., 2005) epitomized the problem: Of the three authors, "two are more strongly inclined to identify the African origin hypothesis (E.R.M. and G.F.G.) as the clear front runner, the other (R.D.M.) favors the Indo-Madagascar hypothesis" (p. 87). The primates are already quite well known, and it seems that the more data are accumulated, the more confusing the search for the center of origin becomes. The origins of the main clades within primates (Table 3-1) are also obscure. Wright (1997: 129) concluded that there is "no convincing explanation" for the origin of the New World monkeys and that the origins of the primates on Madagascar, the lemurs, are equally "enigmatic."

It is suggested here that the endemism of major primate clades in Madagascar and in America is a valuable clue for deducing the evolution

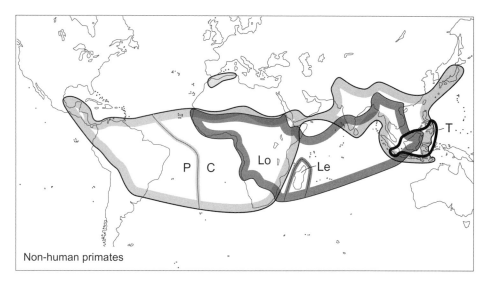

Non-human primates

FIGURE 3-1. Distribution of extant primates (humans and human introductions are not shown). Haplorhines: P = platyrrhines, C = catarrhines, T = tarsiers. Strepsirrhines: Le = lemurs, Lo = lorises and galagos (Africa and Asia).

of the group as a whole. The only primates, living or fossil, in Madagascar are lemurs, while the only ones in South and Central America are platyrrhines (Fig. 3-1). Apart from platyrrhine fossils in Patagonia and the Caribbean, neither of the two groups has fossils outside its extant distribution. What is the reason for this classic allopatry? In a dispersal framework, the two groups are considered as separate issues: How did the lemurs and the platyrrhines each migrate to their respective areas and from where? Yet Lehman and Fleagle (2006) emphasized that a group can occupy an area either because it moved there or because it evolved there. In the vicariance model proposed here, lemurs, platyrrhines and the other groups did not migrate to their respective localities, but instead they evolved there during a single process of differentiation. Each is the local, vicariant representative of a widespread common ancestor.

THE MAIN CLADES OF PRIMATES

Strepsirrhines and Haplorhines

The two main primate clades differ in the morphology of the oro-nasal region: In strepsirrhines, such as lemurs, the nose has a moist, glandular rhinarium similar to that of a dog, a cow, and many other mammals; in haplorhines, such as monkeys and humans, the nose is dry and without a rhinarium. The strepsirrhine–haplorhine split is supported mainly by soft-tissue characters (e.g., the rhinarium and other nasal features) that are seldom preserved in fossils (Gunnell and Rose, 2002: 46).

Haplorhines occur in South America but not Madagascar, while strepsirrhines are in Madagascar but not South America (Fig. 3-1). The two main clades of aplocheiloid fishes show the same features (Fig. 2-3). Haplorhines are widespread in the New and Old World tropics, but are absent, living and fossil, from both Madagascar and Australia/New Guinea. (Other large clades distributed in this way include the woodpeckers, Picidae, and the squirrels and their allies, Sciuromorpha.) Conversely, no strepsirrhine, living or fossil, has ever been found in South or Central America.

The New World monkeys and the lemurs each have their sister group on mainland Africa and Asia, and haplorhines and strepsirrhines overlap there. Yet the respective distributions of the two main groups within Africa are quite different. Haplorhines are especially diverse in the west and center, and occur throughout South Africa, whereas strepsirrhines are diverse in East Africa, and in South Africa do not range west of KwaZulu-Natal. The proportion of strepsirrhine species in local primate faunas declines markedly from East Africa to West Africa—for example, from 60% in the Rufiji delta, Tanzania (six strepsirrhine species, four haplorhines; Doody and Hamerlynck, 2003), to 27% in the Niger delta/Sanaga River area of Nigeria and Cameroon (six strepsirrhines, 16 haplorhines; Oates et al., 2004). This difference is examined a little more closely in Chapter 5.

The distributions of strepsirrhines and of haplorhines also differ in southern Asia. Haplorhines have seven subspecies in Sri Lanka and, as might be expected, many more (26) in the much larger India. In contrast, strepsirrhines have three subspecies in Sri Lanka and the same number in all of India, and so have a higher proportion of their diversity in Sri Lanka than haplorhines.

In Southeast Asia, strepsirrhines are known east to the Sulu Archipelago, off northeastern Borneo, but do not occur on the main Philippine islands such as Mindanao. In contrast, haplorhines (macaques and tarsiers) are present on the main Philippine islands and also in Sulawesi. Again, as in Africa and India, strepsirrhines are located on the Indian Ocean side of the region.

Following Matthew (1915), some authors have interpreted the strepsirrhines as older than haplorhines and their distribution in Madagascar and Sri Lanka as "marginal" and relictual. In this evolutionary model, strepsirrhines originated in the north and were later pushed out to the margins by the new, more advanced and more competitive haplorhines. While this idea is not accepted here, the differences between the distributions of haplorhines and strepsirrhines are real and important, and could be the result of initial vicariance between the two groups. The centers of strepsirrhine diversity, such as Madagascar and Sri Lanka, may appear to be "marginal" from a north temperate (Holarctic) viewpoint, but in an Indian Ocean or Gondwanan perspective they are central.

Tarsiers

Tarsiers (Tarsiiformes s.str.) comprise living *Tarsius* species in Sumatra, Borneo, Sulawesi, and the Philippines (Fig. 3-1), recently divided into three genera (Groves and Shekelle, 2010), and there are also fossils on mainland Southeast Asia. There is an Eocene *Tarsius* from Jiansu (near Shanghai) and another tarsiid, *Xanthorhysis*, from Shangxi (near Beijing). The identity of *Afrotarsius* from the Paleogene of Egypt and Libya is disputed; it could be a tarsier or an anthropoid (Fleagle and Gilbert, 2006); Jaeger et al. (2010) concluded it was an anthropoid.

For many years the small-brained tarsiers were linked with the small-brained lemurs and lorises as "prosimians," precursors of the large-brained anthropoids or "simians." In contrast, most nuclear DNA studies place tarsiers with anthropoids, supporting the strepsirrhines and haplorhines as the two main groups of primates, rather than prosimians and anthropoids (Goodman et al., 2005; Schmitz et al., 2005). Nevertheless, the position of *Tarsius* remains controversial. Some studies of mitochondrial DNA sequences place it with strepsirrhines to give the traditional prosimian clade (Eizirik et al., 2001, 2004; Horner et al., 2007), although Matsui et al. (2009) sequenced complete mitochondrial genomes and found support for the tarsier–haplorhine clade. One tarsier individual had a strepsirrhine nasal notch on one side and a typical haplorhine opening on the other side (Simons, 2003). The two alternative positions of *Tarsius* are integrative, not exclusive, and may result from retention and incongruent recombination of ancestral characters. Placing tarsiers with either anthropoids or strepsirrhines should not obscure the fundamental differences that tarsiers show from both. Even apart from the huge eyes, each one the size of the brain (Rosenberger, 2010), and the ability to turn its head 180° either way—both features unique in mammals—*Tarsius* is distinguished from other primates by its brain, which shares features with carnivorans, its almost totally carnivorous diet, and other characters which make it an unlikely ancestor for anthropoids (Wright et al., 2006). Its biogeography is also distinctive, as it has populations (and endemic species) further east than all other primates.

Fossil Clades of Primates: Adapiforms and Omomyiforms

The extant clades of primates have fossils dating back to the Early Eocene. Two additional groups, adapiforms and omomyiforms, are only known as Cenozoic fossils from the U.S. and Eurasia. Adapiforms were traditionally regarded as Eocene lemurs and omomyiforms as Eocene tarsiers (Matthew, 1915), but disagreements with this interpretation have emerged and are discussed below.

PRIMATES AND THEIR RELATIVES—THE ARCHONTA

The term "primates" as used here refers to the order in the narrow sense, excluding the extinct plesiadapiforms. (Primates in this narrow sense are sometimes referred to as "euprimates," but a new name is not necessary just because the delimitation of a group changes). The orders Primates and Plesiadapiformes, along with two small Southeast Asian orders, Dermoptera and Scandentia, make up the superorder Archonta (Fig. 3-2A). (In the pre-molecular era, Archonta included bats; bats are now excluded but, again, there is no need for the new name, euarchonta.)

Plesiadapiforms have most of their known diversity in North America, where they occur in the U.S. and Canada as far north as Ellesmere Island, by northern Greenland; no primate fossils are known this far north. Plesiadapiforms are also known from Europe and China.

Scandentia or tree shrews comprise four Asian genera ranging between Borneo and India/Nepal (not Sri Lanka). Fossils (Asia only) are known back to the Eocene.

Dermoptera include the two Asian genera of "flying lemurs," *Galeopterus* in Southeast Asia (Thailand, Vietnam, Sumatra, Java, and Borneo) and *Cynocephalus* in the Philippines. Extralimital fossils are known from Myanmar and Pakistan (Marivaux, Bocat et al., 2006). Fossil groups such as Mixodectidae and Plagiomenidae are sometimes attributed to Dermoptera, but the relationships are uncertain. Horner et al. (2007) retrieved Dermoptera in an interesting, alternative position, as sister to anthropoids (and tarsiers were placed with strepsirrhines). In contrast with the Janečka et al. (2007) arrangement shown in Fig. 3-2, Horner et al. (2007) found Scandentia to fall outside primates

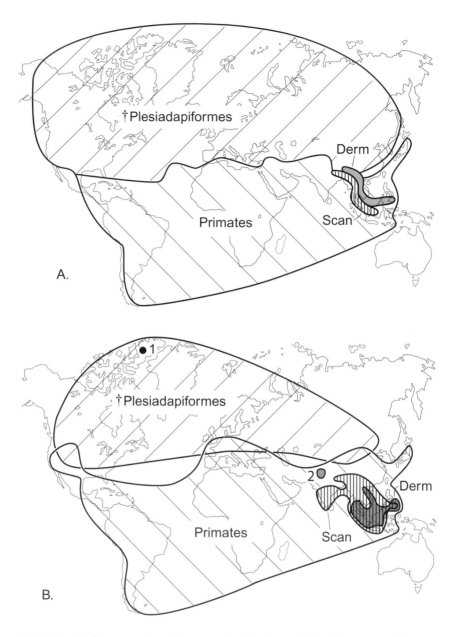

FIGURE 3-2. A. Reconstruction of Archonta distributions at 185 Ma (Early Jurassic),
showing the initial allopatric differentiation of an already widespread Archonta
ancestor into the four orders.
Derm = Dermoptera, Scan = Scandentia.
B. Total known range (living and fossil) of the four orders in
Archonta. 1 = Ellesmere Island, the northernmost record of
Archonta. 2 = Sivalik Hills, northern India and Pakistan, the locality of extralimital
Dermoptera fossils.
C *(facing page)*. Phylogeny of Archonta (Janečka et al., 2007).

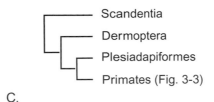

C.

FIGURE 3-2. *(continued)*

TABLE 3-2   CLASSIFICATION OF ARCHONTA

1. Order Scandentia: Tree shrews
2. Order Dermoptera: Flying lemurs or colugos
3. Order [†]Plesiadapiformes
4. Order Primates

	S.O. Strepsirrhini	S.O. Haplorhini
"Prosimians" (paraphyletic)	I.O. [†]Adapiformes	I.O. [†]Omomyiformes
	I.O. Lemuriformes (lemurs)	I.O. Tarsiiformes
	I.O. Lorisiformes (lorises and galagos)	
Simians = anthropoids	I.O. Simiiformes Catarrhini Platyrrhini	

NOTE: S.O. = suborder. I.O. = infraorder. [†] = extinct. The primate classification follows Groves (2001), except Chiromyiformes (*Daubentonia*) are treated under Lemuriformes (Goodman et al., 2005). There is no formal name for Lemuriformes + Lorisiformes; the clade is often referred to as the "tooth-comb" primates.

and even Glires. Although Dermoptera and Scandentia are diverse in and around Borneo, neither group occurs in Sulawesi.

Archonta are widespread globally, yet they have never been recorded east of Sulawesi or in parts of western America (e.g., western Mexico), whereas their sister group Glires (rodents and lagomorphs) has major diversity and endemism in both areas. The groups of Archonta and primates are listed in Table 3-2.

The topology often given for Archonta is: (Scandentia (Dermoptera (plesiadapiforms, primates))), although the statistical support for this is generally weak (Janečka et al., 2007). The four orders are treated here as more or less equivalent clades that were originally allopatric, and this is discussed next.

THE CENTER OF ORIGIN THEORY OF PRIMATE EVOLUTION

Matthew's (1915) account of mammalian evolution remains widely accepted as a basic framework. This proposed that:

1. The northern, fossil groups of primates and other orders represent ancestors that evolved in the Holarctic (north temperate) center of origin.

2. Tropical and southern members of the orders dispersed from the north.

3. The orders are Cenozoic, as shown by the fossils.

4. The orders originated on, and dispersed over, a more or less modern arrangement of land and sea.

Many contemporary workers still accept Holarctic centers of origin. For mammals, Gheerbrant and Rage (2006: 224) concluded that "all successful and typical African radiations" have resulted from origins in Laurasia followed by dispersal into Africa. While modern interpretations of primate evolution differ widely in where they locate the center of origin, all agree with Matthew (1915) that centers of origin do exist. Beard (2006: 439) defended this assumption, arguing that "The continuity of phylogenetic descent requires that sister taxa originate in the same place and at the same time." For example, if we accept that tarsiers and anthropoids are each other's closest relatives, "we must also assume that both lineages originated in the same place (since speciation, like politics, is local)" (Beard, 2004: 13–14). Likewise, Godinot and Lapparent de Broin (2003) suggested that because primates have sister groups (Dermoptera and Scandentia) in Southeast Asia, this was their center of origin. In fact, there is no need to assume this. The process may have only involved normal, allopatric (dichopatric) evolution in a global group that divided into the two small clades in Southeast Asia, and proto-primates/plesiadapiforms everywhere else. Here there is no center of origin of primates and no dispersal (i.e., founder speciation), just differentiation in a widespread ancestor around breaks in Southeast Asia followed by range expansion and secondary overlap (cf. Chapter 1).

Beard (2006: 439) (cf. Fleagle and Gilbert, 2006) suggested that "The order Primates is hierarchically nested within an exclusively Asian branch of the mammalian family tree, suggesting that primates originated in Asia." This is not quite correct. Dermoptera and Scandentia do not form a branch (a clade) and primates are not nested in an Asian group; they are nested in an *otherwise* Asian group. In the simplest model, the northern plesiadapiforms, the southern primates, and the two Southeast Asian orders all originated from an almost global ancestor which divided into the four geographical groups (one possible solution is given in Fig. 3-2A). While the phylogeny and relative timing indicate that the breaks could have occurred at rifting around the central North Atlantic and Asia/Tethys, this does not mean that primates had a localized "center of origin" in any of these localities.

Neontologists often assume that a "basal group," such as Scandentia or Dermoptera in Archonta, is "primitive" and occupies the center of origin of the whole clade. As stressed in Chapter 1, a "basal group" is just a conventional term for a small sister group; the other sister is just as basal (it is the node, rather than any of the clades, which is basal). With respect to fossils, many paleontologists also assume that a basal group—whether defined by phylogeny, stratigraphy, or both—occupies a center of origin. They read the phylogenetic tree or the stratigraphic sequence as a sequence of dispersal events from a center of origin, rather than a sequence of differentiation in a widespread ancestor. For example, the omomyid *Teilhardina* is the oldest known primate fossil in Europe and in North America, and also occurs in China. Smith et al. (2006) depicted cartoon figures of omomyids leaping from a center of origin in China to Europe, and from there to America. Beard's (2008) figure instead has them leaping from Asia across Beringia to America, and from there to Europe. The two "routes" are interesting alternatives, but as sequences of differentiation events in a widespread northern ancestor, not as series of dispersal events.

Fleagle (1999) acknowledged that many of the key areas for primates have a "remarkably meager" geological record and lack strata that are even potentially fossil bearing. Nevertheless, he followed Matthew (1915) in accepting a center of origin/dispersal model for spatial evolution. He also agreed with Matthew's fossil-based chronology, writing that "The evolution of primates, like that of most other groups of modern mammals, has occurred almost totally within the Cenozoic era, the Age of Mammals" (Fleagle, 1999: 317). Based on this literal reading of the fossil record, Fleagle and Gilbert (2006: 375) concluded that "the evolutionary history of primates seems to have involved a wide range of traditional dispersal mechanisms, including land bridges, chance dispersal over open ocean, and intermediate island hopping." In Europe, Asia, and America, the oldest primate fossils all "appear to be immigrants with no clear ancestors in underlying deposits" (Fleagle and Gilbert, 2006: 382). The authors interpreted these absences in the fossil record as due to real absence and so accepted that they were evidence for dispersal. Instead, the absences could be due to normal sampling error in the sparse fossil record of early Cenozoic primates.

Plesiadapiforms as Ancestors

The main primate clades were thought to represent ancestors (prosimians) and descendants (simians) of each other, but now it is sometimes suggested that the clades are instead derivatives of common ancestors. The northern fossil group, Plesiadapiformes, is often regarded as the sister group of the primates (e.g., Janečka et al., 2007) and, traditionally, as the ancestor (Gingerich, 1976). On the other hand, Godinot (2006a: 85), noting the "numerous and contradictory" hypotheses about relationships in these groups, contended that plesiadapiforms lie outside the clade Scandentia + primates and are not ancestral to primates. The plesiadapiforms' sister group may be Dermoptera, rather than primates, or these three may form a trichotomy (Kay et al., 2004). In any case, plesiadapiforms are probably "too derived to tell us much about primate evolution" (Ross, 2003; cf. Fleagle, 1999, Martin et al., 2007). In other words, there is no need to interpret the plesiadapiforms as ancestral to primates, even though their oldest fossils are somewhat older (latest Cretaceous/earliest Paleocene vs. later Paleocene in primates). A literal reading of the fossil record would interpret the primates as evolving from the plesiadapiforms and so emerging in the first place from a center of origin at Garbani Channel, Montana (the location of the oldest plesiadapiform fossil, *Purgatorius*; Clemens, 2004); this is not supported here. Instead, the considerable allopatry between the distributions of plesiadapiforms and primates (which is seldom mentioned in the literature) is interpreted a resulting from an original vicariance event.

Adapiforms and Omomyiforms as Ancestors

These two groups are only known as fossils from northern hemisphere countries. Adapiforms are recorded from the Eocene to Miocene of the United States, the Mediterranean, and Asia. Omomyiforms are known from the Eocene to Oligocene and have a similar geographic range. For many years adapiforms were regarded as Eocene lemurs, omomyiforms as Eocene tarsiers. These links are now being questioned, and the precise relationships of the fossil groups with each other and with the extant groups are the subjects of endless debate. The fossils show mosaic and kaleidoscopic recombinations of features, and there are almost as many different views on the groups' affinities as there are logical possibilities (Kay et al., 2004). In traditional analyses, omomyiforms and adapiforms both appear as paraphyletic complexes with many early offshoots, and modern strepsirrhines arose from within "adapiforms," haplorhines from "omomyiforms" (Matthew, 1915: Fig. 8; Kay et al., 2004: Fig. 7). In other studies (Bajpai et al., 2008) adapiforms are sister to lemuriforms, and omomyiforms are sister to anthropoids. Other authors have supported an adapiform–anthropoid connection; "several interesting similarities" in upper dentition of the two groups are admitted, although cranial and postcranial structure does not show this affinity (Gebo, 2002: 23). Franzen et al. (2009) hypothesized that a well-preserved adapiform from Germany (*Darwinius*) and, by implication, the other adapiforms, were haplorhines rather than strepsirrhines. On the other hand, Godinot (2006b) suggested a relationship between early anthropoids and stem lemuriforms, excluding adapiforms and omomyiforms. He noted that while tarsal characters contradict this, convergences in locomotor adaptations have occurred in other groups.

Ross (2003) pointed out that some omomyiforms and adapiforms are "almost indistinguishable." He suggested they may have nothing directly to do with the origins of haplorhines or strepsirrhines, but instead could be an independent radiation and an evolutionary dead end. Anthropoids show marked divergence from basal adapiforms and omomyiforms, and anthropoid fossils may be just as old as these groups (Bajpai et al., 2008), so it is "extremely unlikely" (Ross, 2003: 199) that anthropoids arose from within them. Martin et al. (2007) interpreted the adapiforms and omomyiforms themselves as early northern "offshoots." They are probably vicariants, not ancestors, of extant clades. Gaps in the southern hemisphere fossil record mean that adapiforms and omomyiforms may once have been more widespread further south, for example, in the mountains of Gondwana. Eocene adapiforms from Gujarat, India (Rose et al., 2009) may predate the India–Asia collision.

Other groups have been suggested as "the ancestor" of the anthropoids, such as tarsiiforms and fossil groups such as eosimiids, but these scenarios are also not convincing. Beard (2004: 27) pointed out that studies have become "fixated on the issue of direct ancestry" and the "ladder paradigm" of evolution. As Rasmussen (1994) observed, "To many researchers a tarsier-anthropoid clade suggested that anthropoids arose directly from a tarsier-like prosimian" and the possibility that the two groups may be sisters (or more distant cousins) is often overlooked.

Breaking with the "search for the ancestor" approach, Miller et al. (2005: 60) presented evidence for a "deep time origin of anthropoids" and argued for "an ancient, Gondwanan, nonadapiform, nonomomyiform, nonstrepsirrhine, nontarsiiform origin of anthropoids." While this proposal has been described as "little more than a claim for ignorance regarding the origin of the group" (Fleagle, 1999: 421), the idea that the main primate clades are derived from a common ancestor older than them all, rather than one member being the ancestor of the others, is a critical step in understanding anthropoid origins. (Fleagle himself gave an "agnostic" diagram of phylogeny for primates, showing five separate groups rather than ancestor–descendant lineages; Fleagle, 1999: 345.)

Ross and Kay (2004: 712) also suggested "the possibility that extant anthropoids might not be derived from any of the currently known groups of primates"; the tarsier–anthropoid clade "might not have anything to do with omomyiforms. . . . The possibility still remains that the major clades of extant primates (Strepsirrhini, Tarsiiformes, and Anthropoidea) are not derived from either adapiforms or omomyiforms and that the latter taxa are completely independent, dead-end radiations of primates." Miller et al. (2005: 67) probably revealed the mundane truth in suggesting that "the search for an anthropoid ancestor, at least among the adapiforms and omomyiforms, has more to do with the lack of fossils from Paleogene localities on southern continents than with any special features of adapiforms or omomyiforms."

Center of Origin Explanations for the Primates of South America

As Fleagle (1999: 444) pointed out, "The most unsettled question surrounding platyrrhine origins is the geographical one: how did platyrrhines get to South America?" Most of the debate has focused on whether North America or Africa was the source and on the possible means of dispersal. Fleagle (1999: 446) inferred rafting: "[R]egardless of how unlikely rafting may seem, it is presently the only suggested mechanism for transporting terrestrial animals between continents separated by open ocean. If South America was indeed an island continent during the period in question, we must assume that primates rafted from some other continental area. *Only a revision of the paleocontinental maps could eliminate the need for rafting in the origin of platyrrhines*" (italics added). But there would be no need for either rafting or changing the maps if primates were older than their oldest fossils, and the molecular clock studies discussed below indicate that primates are indeed much older than their fossil record.

Fleagle (1999: 447) concluded: "At present there is no convincing explanation for the origin of South American monkeys, but dispersal across the South Atlantic from Africa seems to be the least unlikely method." The question then is: If monkeys were able to raft to America, why were strepsirrhines not able to do the same? Fleagle and Gilbert (2006: 395) wrote that rafting across the Atlantic is "clearly a chance event, an example of 'sweepstakes' dispersal. One can only speculate that by a stroke of good luck anthropoids were able to 'win' the sweepstakes while lorises and galagos did not." The model described below provides an alternative explanation for platyrrhine divergence that does not rely on chance events but instead on the opening of the Atlantic Ocean.

Center of Origin Explanations for the Primates of Madagascar

The unsettled question of the American platyrrhines and their origin is matched by a similar lack of clarity concerning the primates further east, around the Indian Ocean region. Yoder (1997: 13) wrote that "One of the most perplexing problems in strepsirrhine evolutionary history is the derivation of a realistic biogeographic model to explain the presence of lemuriforms on the remote island of Madagascar. Because they are the only primates other than humans to have lived there, it is unlikely that the primate clade originated on Madagascar. Thus, the ancestral lemuriform must have come from somewhere else." Conversely, the fact that lemurs are the only primate clade on Madagascar and are not known, living or fossil, from anywhere else could be taken as good evidence that they are autochthonous and formed *in situ*, as an allopatric vicariant, not by arriving from elsewhere. If lemuriform ancestors (strepsirrhines) did disperse over the sea to Madagascar, it seems strange that they never colonized America, Australia, or New Guinea. Likewise, Masters, Lovegrove, and de Wit (2007) observed that monkeys have a far greater chance of dispersing to Madagascar than strepsirrhines, yet there is no evidence that they ever did. Sweepstake dispersal does not account for these particular aspects of the problem or for the overall problem: the great difference between the distributions of strepsirrhines (in Madagascar but not South America) and haplorhines (in South America but not Madagascar).

For Madagascar, Tattersall (2008: 405) argued that "The strongly filtered nature of the island's mammal fauna clearly implies some degree of overwater crossing by the founding stocks." But a fauna that includes, for example, most of the tenrecs and the endemic aye-ayes, sloth lemurs, sucker-footed bats (Myzopodidae), and pseudo-aardvarks is not simply a "filtered" or "imbalanced" version of a "normal" one; it is a distinctive fauna in its own right (cf. Heads, 2009a). The pseudo-aardvarks (the order Bibymalagasia, extant until about 1,000 years ago) are known only from Madagascar and their affinities are unclear (MacPhee, 1994). The absence of any fossil mammals from the Tertiary rocks of Madagascar means that there could also have been basal taxa, now extinct, in other groups. Tattersall (2008: 398) stressed that the Mesozoic fossil record in Madagascar "lacks a plausible ancestor" for the mammals now found there, implying dispersal, but this absence may not be significant given the great gaps in the fossil record of the region and the poor understanding of morphology (and fossil phylogeny) in mammals. Tattersall also argued that "Madagascar has been widely separate from Africa and India since well before the beginning of the Age of Mammals." But the old idea that the Cenozoic was the only Age of Mammals is not quite accurate; dramatic discoveries over the last decade (for example, Hu et al., 2005; Luo and Wible, 2005; Ji et al., 2006) show that Mesozoic mammals and mammaliaforms are much more diverse than was thought. Tattersall (2008: 398) concluded: "It seems certain that dispersal must have been involved . . ." for mammals in Madagascar. Nevertheless, given the doubt surrounding the age of lemurs, for example, this confidence may not be warranted.

The idea that the Malagasy mammals arrived by over-water dispersal from Africa has been accepted since Victorian times and is often cited as a well-established precedent for chance, "sweepstake" dispersal (Wallace, 1876; Simpson, 1940; Rabinowitz and Woods, 2006; Tattersall, 2008). In this model, improbable dispersal events become possible or even probable given enough time. Despite its popularity, the idea of sweepstake dispersal in lemurs had not been examined by scientists familiar with the wide range of relevant factors—geological, oceanographic and meteorological, phylogenetic, ecological, and behavioral—until Stankiewicz et al. (2006) carried out a detailed analysis. The study considered a wide range of possible variants, including rafts of vegetation with or without "sail effects" produced by trees on the raft, dispersal by transport of animals in cyclones, and others. The authors concluded that while the rafting theory "currently enjoys wide support [it] is not valid at either the theoretical or applied level" (p. 221). For example, a raft coming from Africa would drift back to the mainland. Animals would not survive transport by cyclones even if it did happen. The chances of successful dispersal are "ludicrously small" (p. 231) and, in particular, they are "so small that even vast tracts of time cannot compensate" (p. 232).

Torpor or hibernation by heterothermy is observed in two lemur genera and another strepsirrhine, *Loris*, and it has been suggested that in the Africa–Madagascar crossing, "entire groups of animals survived the weeks or, even months . . . without food or water sleeping in a hollow tree while rafting across the sea" (Kappeler, 2000: 423). Roos et al. (2004: 10653) took this idea seriously and thought it the "most plausible" hypothesis. In contrast, Masters, Lovegrove, and de Wit (2007: 21) rejected it outright (their paper was titled "eyes wide shut") and concluded that "alternative explanations should be sought" for primates on Madagascar.

Many aspects of the primates are repeated in the other mammal orders there. For example, apart from a few species of shrews, subfossil Bibymalagasya and subfossil hippopotamus, there are five orders of terrestrial mammals on Madagascar: primates, tenrecs, bats, carnivorans, and rodents. Four of these (not bats) show a distinctive pattern, with just a single clade on Madagascar (this was established for the last two only with molecular work). This pattern is not because the Madagascar clades are relictual, as in all four cases the Madagascar clade is diverse. The four Malagasy clades are recognized at a high rank (above family level in primates, at family or subfamily level in the others). In addition, the four Malagasy clades have sister groups that are widespread in Africa or the Old World, not at a localized area of, say, coastal Mozambique or Tanzania, as dispersal theory would predict. The repeated pattern of a single, high-level, diverse clade in Madagascar with a widespread sister group is compatible with an early origin by vicariance. In bats, there are seven families represented, including the endemic Myzopodidae. This family is basal in a worldwide complex comprising Vespertilionidae, Molossidae, and others, not a minor, low-level derivative of African clades (Eick et al., 2005).

Ali and Huber (2010) provided a dispersal model for the biogeography of Madagascar mammals that is more interesting than many dispersal interpretations, as it provides a geological explanation (changing paleocurrents) for the cessation of dispersal. Nevertheless, it does not explain, or even fit with, several key facts. If Paleogene currents were effective in dispersing terrestrial mammals to Madagascar, why are the four orders each represented there only by a single clade? If primates could migrate across the Mozambique Channel and also the Atlantic, why have they never been able to cross the 25-km strait between Sulawesi and the western Moluccan Islands (and thus reach New Guinea and Australia)? Why did strepsirrhines (lemurs and relatives) migrate to Madagascar, when haplorhines (monkeys) did not? Monkeys would be just as likely, if not more so, to make the crossing. Why did strepsirrhines colonize Madagascar but not America, whereas haplorhines colonized America but not Madagascar? This symmetry is spatially correlated with higher values of haplorhine diversity/strepsirrhine diversity in West Africa than in East Africa.

All these patterns are explained easily in a simple vicariance model.

A VICARIANCE MODEL FOR PRIMATE EVOLUTION

The model given here for the origin of primates (Fig. 3-2) and their main clades (Fig. 3-3) begins with vicariance of a widespread ancestor that was already in Africa, Madagascar, Asia, and America before the extant clades existed. Long-distance founder dispersal of the modern groups across open ocean is not required. The model is simple and has not been accepted only because the clades are thought to be too young (Cenozoic) and so transoceanic dispersal is assumed to have taken place.

The model aims to minimize change in distributions while assuming that groups originated as allopatric vicariants. The chronological calibration is based on the simple biogeographic divisions and associated tectonics at nodes IV (Mozambique Channel) and V (Atlantic Ocean). The

---

FIGURE 3-3. A. Molecular phylogeny of primates (topology from Goodman et al., 2005); age of nodes (I to V) calibrated with associated geographic-tectonic events.
B. Reconstruction of distributions at 180–170 Ma, around the time that primates were differentiating into haplorhines and strepsirrhines (at node II, 180 Ma) and haplorhines were differentiating into anthropoids and tarsiers (at node III, 170 Ma). All three clades may have split around the Borneo precursor terranes (the three clades overlap today in southern Sumatra and Borneo, a tectonic composite). The phylogenetic/biogeographic nodes (II–V) are those shown in Figure 3-3A. The total known distribution of the extinct adapiforms and omomyiforms is also shown (stippled); both groups have a similar range in Europe, East Asia, and North America. Their phylogenetic relationships with the other clades are unresolved.
C. Total distribution of the primates, living and fossil (except adapiforms and omomyiforms; see Fig. 3-3B). Breaks within strepsirrhines (at node IV, 160 Ma) and within anthropoids (V, at 130 Ma) are indicated. Range expansion of haplorhines, strepsirrhines, and tarsiers following their initial allopatry (Fig. 3-3B) has led to secondary geographic overlap in Africa and Asia, and distribution north of the current range (with fossil records). Subsequent extinction in peripheral areas (x symbols) has produced the current distribution (Fig. 3-1).

A.

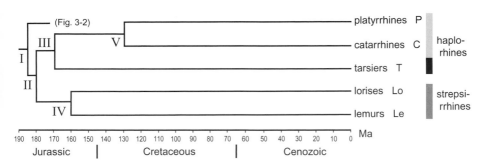

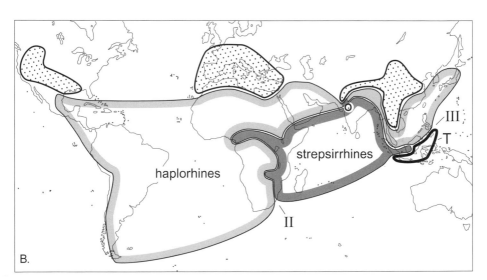

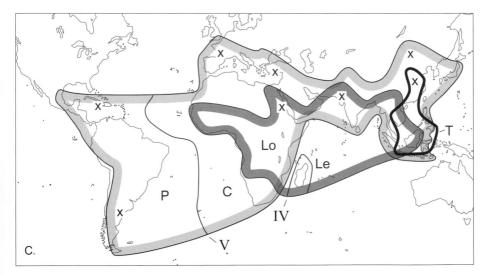

dates for nodes I and III, involving earlier opening of the central Atlantic and complex Tethys events, were accepted following the first two calibrations. Node II involves a split between haplorhines, mainly in West and Central Africa, and strepsirrhines, mainly in East Africa. Given this geography and the relative timing (earlier than node IV at the Mozambique Channel), a likely explanation for this node is tectonic activity along the Lebombo monocline. This volcanic rifted margin is a great warp running north–south in eastern South Africa, Zimbabwe, and Mozambique and is one of the major structures in the region. Rifting here remained incomplete and was a precursor to successful rifting in the Mozambique Channel. The Lebombo monocline is already known as an important biogeographic boundary for taxa in the region (Craw et al., 1999: 77–79).

Node I. Early Jurassic (~185 Ma)

1. A global Archonta ancestor **divided** into plesiadapiforms in the north, primates in the south and east, and Scandentia and Dermoptera on terranes that became Southeast Asia. The division is correlated with the initial breakup of Pangaea, in which the central Atlantic opened and Africa/South America separated from North America. The date (from Veevers, 2004) is based on rifting in the Atlas Mountains and mid-ocean ridge basalts on the Canary Islands.

2. Marginal **overlap** between plesiadapiforms and primates developed in the north and complete overlap of Dermoptera, Scandentia, and primates took place in Southeast Asia. This may be related to the rotation of Africa–Arabia and its approach to Asia at the Persian Gulf.

Node II. Early Jurassic (~180 Ma)

3. Global primates **divided** into strepsirrhines, mainly east of the Lebombo Monocline (in Tanzania, Madagascar, etc.) and haplorhines west of this line, in South America, Africa (not Madagascar), and Asia (north of the strepsirrhines). This is correlated with flexure and volcanism on the Lebombo monocline and final Karoo faulting in East Africa.

4. Partial **overlap** developed between haplorhines and strepsirrhines with range expansion of both in central Africa and Asia. Fossil

"tooth-comb" primates from Egypt and Oman indicate that the modern distribution of strepsirrhines around the Arabian Peninsula has been reduced through extinction.

At some stage (possibly between 3 and 4) the northern adapiforms may (the phylogeny is debated) have split from the other, southern strepsirrhines, and the northern omomyiforms may have split from the other, southern haplorhines (possibly around the same time). This could be correlated with Late Jurassic rifting in the Mediterranean. While adapiforms and tooth-comb strepsirrhines have remained vicariant, anthropoids have subsequently expanded their range northward into Europe and Asia (but not north of Mexico), more or less completely overlapping the range of omomyiforms and adapiforms there.

Node III. Early Jurassic (~180 Ma)

5. Haplorhines **divided** into tarsiers and anthropoids at the same time as, or soon after, the strepsirrhine/haplorhine split. Tarsiers are diverse on Sulawesi (where the only other primate group is the haplorhine *Macaca*) and also occur east of other primates, with endemic species on eastern Sulawesi islands. This distinct distribution suggests that the tarsiers may have evolved on one or more of the Sulawesi precursor terranes originally located some distance from anthropoids on other Sulawesi terranes. Tarsiers and anthropoids currently overlap in Sulawesi, Borneo, and Sumatra. These three islands are geological composites, and tarsiers, anthropoids, and their ancestors could have occupied different sets of "Sulawesi," "Borneo," and "Sumatra" terranes long before the modern islands formed as geographic entities.

   Until recently, most reconstructions portrayed eastern Tethys as a wide sea, devoid of any land, between India/Australia (Gondwana) and Asia. Instead, Aitchison et al. (2007) suggested that two intraoceanic island arc systems existed between pre-collision India and Asia, and concluded that the Tethys sea "was obviously more complex than originally envisaged" (p.6). Pre- and proto-primates probably occurred widely throughout the archipelagos formed by these arc terranes, and the populations would have been incorporated into Asia during arc–continent accretion.

Node IV. Middle Jurassic (~160 Ma)

6. Strepsirrhines **divided** into lorises and galagos on mainland Africa and Asia, and lemurs on Madagascar. This is correlated with the opening of the Mozambique Channel and the split of East Gondwana (Madagascar, India, Antarctica, Australia) from West Gondwana (Africa–South America). There was crustal extension between Madagascar and India in the Early Cretaceous, 140–118 Ma, and the two regions separated in the Late Cretaceous (96–84 Ma) (de Wit, 2003 Fig. 4).

Node V. Early Cretaceous (~130-120 Ma)

7. Anthropoids **divided** into catarrhines (Old World) and platyrrhines (New World). This is correlated with the opening of the South Atlantic. The South Atlantic rift propagated northward, the Central Atlantic rift (node I) propagated southward, and a transform domain formed between them, separating West Africa and Brazil.

In this model, none of the clades has a localized center of origin. While there has been some range expansion and secondary overlap, the only phylogenetic process has been subdivision (vicariance). There is no requirement for any center of origin, founder speciation, or transoceanic rafting, and the topology of the phylogeny represents a sequence of differentiation in a widespread ancestor, not a radiation by founder events.

Miller et al. (2005) suggested that plesiadapiforms may have been Laurasian ecological vicariants of primates, and this is compatible with the origin of the two clades as phylogenetic and biogeographic vicariants. Marginal overlap of plesiadapiforms and primates in Laurasia, and within primates, overlap of adapiforms/omomyiforms and anthropoids in Laurasia, does not mean there was a center of origin in the north. Instead, it suggests that differentiation and then overlap of Laurasian and Gondwanan groups has taken place in two more or less distinct phases.

Seiffert, Simons, Clyde et al. (2005) accepted *Altiatlasius* of the Moroccan Paleocene as the oldest known crown primate and a stem anthropoid. They argued that this implies either immigration into Africa of anthropoid ancestors or "the presence of ancient parallel radiations of anthropoids in Asia and Afro-Arabia." The second alternative (along with a third center in South America) is compatible with the model presented here.

The pioneer Argentine paleontologist Ameghino (1906) recognized the great diversity of South American fossil mammals, including primates, and saw the region as a center of origin—a "garden of Eden"—for mammals. This interpretation has been rejected by most North American writers (e.g., Romer, 1966), yet McKenna (1980: 44) noted that ever since Matthew (1915), Simpson and many other authors have advocated a *North* American garden of Eden, "with nearly Ameghinian fervour." In a vicariance model, the North Americans and the South Americans are both right. Northern and southern groups of primates both evolved *in situ*, with anthropoids being autochthonous south of the Mexico/United States border, adapiforms and omomyiforms autochthonous north of there.

INTEGRATING THE PRIMATE FOSSIL RECORD AND THE MOLECULAR CLOCK STUDIES

A vicariance model explains several aspects of primate evolution better than a dispersal model and also explains the disagreement among the dispersal theories about the location of the true center of origin. The only real objection is that it implies older ages than are generally accepted, and so this is discussed further here. Traditional paleontological accounts and also fossil-calibrated molecular clock studies are dependent on the fossil record. Nevertheless, rainforest taxa seldom fossilize, and Soligo et al. (2007) stressed the great geographic biases of the mammal record.

Many workers now accept that two sister groups are about the same age. If one of the two groups has much earlier fossils than the other, this implies the second group existed as a "ghost lineage" that has left no record. Analyses based on this concept of cross-calibration have already indicated many large gaps in the primate fossil record (Yoder and Yang, 2004; Soligo et al., 2007; Steiper and Young, 2008). Nevertheless, all molecular studies of primate evolution in time have been based on fossils (these provide the initial calibration). The cult of the fossil holds that "Fossil evidence is the only direct source of information about long-extinct species and their evolution" (Wilkinson et al., 2011). In this view, other sources of evidence about the past, such as the inherited information of the genome and the biogeography, only provide "indirect" evidence that requires interpretation. But in practice there is often profound disagreement about the identification and phylogeny of early fossils. They are fragments only (soft parts are never preserved), they belong to animals that are otherwise unknown, sample sizes are often very small, and the trends in the evolution of the fossil groups are not clear. Thus, setting up a neat dichotomy between "direct" fossil evidence and all other, "indirect" evidence is unjustified. Fossils can provide valuable evidence, for example, in providing minimum clade ages and information on morphology, but they do not constitute a fundamentally distinct category of scientific evidence. Fossil evidence, as with geographic distribution or any other data, needs to be interpreted and, as noted in the last chapter, cannot be taken at face value.

Not only do advocates of the fossil record argue that it is the *best* evidence (the only direct source of information) about ancient groups, they also accept it as the *only* possible basis for establishing the time course of evolution. For primates, Wilkinson et al. (2011: 17) proposed that "Fossil calibrations are essential when dating evolutionary events," and that "Many methods and models have been proposed [to establish divergence times], but *all* rely in some way on fossil evidence, as an external source of information must be used to calibrate the substitution rate" (Wilkinson et al., 2011: 16; italics added). These claims are misleading, as the time course of evolution can be calibrated using geological data other than the fossil record (Fig. 3-3).

Molecular Clock Studies Show That Species Can Be Mid-Cenozoic

Fossil-calibrated molecular clock studies give useful minimum ages indicating that mammal species are older than was thought. In particular, the Pleistocene refugium hypothesis has been rejected in many studies on primates such as chimpanzees (*Pan*), guenons (*Cercopithecus*), and spider monkeys (*Ateles*). These indicate that species and even subspecies diverged well before the Pleistocene (Lehman and Fleagle, 2006). Divergence between Sumatran and Bornean populations of orangutan (*Pongo*), regarded by different authors as species or subspecies, has been dated with a fossil-calibrated clock at ~10 Ma (Arnason et al., 1996), giving a useful minimum age. Amazon diversity has often been explained as the result of evolution in Plestocene refugia, but ideas on this are changing too. Boubli and Ditchfield (2000) concluded that the Amazonian platyrrhines *Cacajao calvus* and *C. melanocephalus* s.lat. had already diverged by the Pliocene.

Cenozoic Fossils Are Already "Modern"

The oldest fossils suggest Cenozoic, Holarctic centers of origin for mammal orders (Matthew, 1915) and many authors have assumed that primates originated somewhere in the northern continents. Most of the oldest primate fossils are from the Early Eocene of North America, Europe, and China. Recent discoveries further south, such as a possible omomyiform from the Paleocene of Morocco (*Altiatlasius*) and a possible anthropoid from the Early Eocene of India (*Anthrasimias*; Bajpai et al., 2008) change the picture somewhat. But in any case, a literal or "direct" reading of the fossil record always involves "the questionable inference that [a group's] origin is located in or close to the geographical region that has yielded the earliest known fossil representative" (Martin et al., 2007: 281).

In other mammal orders, clearly recognizable bats, artiodactyls, and perissodactyls also appear first in the northern continental fossil record, and again at the base of the Eocene. This might be interpreted as a burst of evolution. But in most cases, forms transitional to the early fossils are unknown, and this implies the former existence of lineages that have left no record. Thus, the oldest fossil indicates neither the time nor the place of origin. In primates, Covert (2002: 19) cited the "striking similarity" between Middle Eocene tarsier fossils from China and modern tarsiers. Rossie et al. (2006) described a Middle Eocene Chinese *Tarsius* species from a facial material "virtually identical" (p. 4381) to the corresponding anatomy in living species. They concluded that "[v]irtually modern tarsier-like facial morphology" (p. 4384), with greatly enlarged orbits and a haplorhine oronasal region, was already present at 45 Ma. Other extant mammal genera with Eocene fossils include the bats *Hipposideros* and *Rhinolophus* (M. McKenna, pers. comm. in Simons, 2003). Fleagle (1999: 557) cited *Tarsius*, *Aotus*, and *Macaca* as primate genera "that seem to have persisted for tens of millions of years with very little change." In Dermoptera, the close relative of primates, Marivaux, Bocat et al. (2006) named a new Oligocene species *Dermotherium chimaera*, as it exhibits a mosaic recombination of characters seen in the two living species.

At La Venta, central Colombia, Miocene beds (12–14 Ma) have yielded fossil monkeys "remarkably similar to modern platyrrhines" and "clearly related" to living *Aotus*, *Saimiri*, *Alouatta*, and others (Fragaszy et al., 2004: 29; cf. Setoguchi and Rosenberger, 1987; Rosenberger et al., 2008: 99). Fleagle and Kay (1997) concluded that "many of the fossil monkeys from La Venta are strikingly similar to modern platyrrhines" (p. 9), and they attributed the morphological "modernness" of the fauna (p. 11) to its "relatively late age."

Fossil Evidence for Modern Orders in the Mesozoic

The earliest recognized fossils of most extant mammal orders are from the Eocene or Paleocene, and so a literal reading of the fossil record indicates that they originated after the Cretaceous. Extant bird orders were also assumed to be the result of a Cenozoic radiation. This in turn led to the idea that the rise of the mammals and birds was related chronologically and ecologically to the extinction of the dinosaurs. In this interpretation, Mesozoic history has no direct relevance to the evolution and biogeography of modern birds and mammals; the Cenozoic is the Age of Mammals (Simpson, 1937) and also the Age of Birds.

Thus the birds are usually interpreted in the same way as the mammals; although the oldest bird fossil, *Archaeopteryx*, is Late Jurassic, the traditional view is that evolution of all the extant orders (except perhaps the ratites) is confined to the Cenozoic, with an "explosive radiation" after the Cretaceous/Cenozoic boundary. Yet new fossil discoveries, such as a Cretaceous member of the extant order Anseriformes (ducks and geese; Clarke et al., 2005) challenge the idea that the Cenozoic is the only Age of Birds.

In the mammal order Monotremata, an Early Cretaceous fossil of Ornithorhynchidae (the platypus family) implies that the platypuses and their sister group, the echidnas, were distinct by then (Rowe et al., 2008). Other Early Cretaceous mammal fossils from Australia include a possible erinaceid (Rich et al., 2001). As indicated above, the known diversity of interesting Mesozoic mammals continues to grow, with the discovery of larger forms that fed on young dinosaurs (Hu et al., 2005), a swimming and/or burrowing species with a beaver-like tail (Ji et al., 2006), and a group with digging forelimbs (Luo and Wible, 2005).

Molecular Clock Evidence That Modern Vertebrate Orders Are Mesozoic

In addition to the new fossil evidence, another challenge to the idea that the extant orders are Cenozoic comes from molecular clock studies. These have shown that many modern mammal orders, including primates, could have existed tens of millions of years before their first appearance in the fossil record. In contrast with the earliest fossil dates in the Paleogene, molecular clock studies indicated that primates split from the other orders of mammals at ~95 Ma, in the mid-Cretaceous (Kumar and Hedges, 1998). Clock studies have even dated the lemur/loris split as Cretaceous (Fabre et al., 2009). One fossil-calibrated clock study that found all the mammal orders to be pre-Cenozoic cited the "increasing difficulty" of reconciling the clock dates with the Cenozoic explosion model of radiation (Springer et al., 2003).

Primatologists Confront a New Chronology

Godinot (2006b: 458) described the (fossil-calibrated) molecular clock dates for primates as "absurd." This response is understandable from a paleontologist who interprets the record as a literal representation of evolution. Yet this involves transmogrifying minimum (fossil) dates into maximum dates and to a non-paleontologist this is, if not absurd, not completely logical.

While some workers have dismissed the new ideas on the time course of primate evolution, others have used them as a stimulus to reexamine the primate record and also the methods used to interpret it (Miller et al., 2005; Soligo et al. 2007; Martin et al., 2007). Soligo et al. (2007: 30) criticized the "common procedure of dating the origin of a group by the earliest known fossil representative, perhaps adding a safety margin of a few million years." Instead, as Martin et al. (2007: 280) recognized, "It is vital to recognize that [using oldest known fossil to infer age] can yield only a minimum estimate." The idea of a long time interval between the origin of a group and its initial appearance in the record is not accepted by all researchers. Still, the molecular work and also studies of preservation show that in groups such as primates, "palaeontologists are likely to have substantially underestimated the true time of divergence" (Miller et al., 2005: 67).

Sometimes it seems as if the molecular clock workers do not quite believe their own results and they make genuine efforts to square their ancient dates with the fossil-based chronology. Both molecular workers and paleontologists refer to a molecular clade date of, say, 100 Ma as "incongruent" or "incompatible" with an oldest-fossil date of, say, 50 Ma. But a difference of this magnitude is standard, and some difference would always be expected due to the limited rock record available, aspects of preservation, and geographic sampling. There is no logical incongruence between these two dates and no special explanation or inquiry is needed. It is quite normal for groups that are known to be ancient through their phylogeny to have no fossil record at all, or for a group with no fossil record to then turn up in, say, the Early Cretaceous (see Chapter 2).

No other biology-related discipline has escaped the seismic upheavals of the molecular revolution, and paleontology will be affected too. While the subject may appear to be immune, as DNA is not preserved in older fossils, a new synthesis of molecular phylogeny, developmental genetics, and morphology should lead to new ideas on homology and different interpretations of fossil phylogeny. As Szalay and Delson (1979: 5) wrote: "In a study of primate evolutionary history, it is clear we must concentrate on fossil taxa, but their interpretation is not possible without a firm grasp of extant relatives and other relevant living species."

For most times and places, there is no fossil record and, in agreement with studies on mammals in general (Bininda-Emonds et al., 2007), Soligo et al. (2007: 33) concluded that "early evolution of the primates has simply remained undocumented [by fossils]." They indicated undeniable gaps in the record. For example, the extant lemurs make up five of the 13 families of extant primates (Goodman et al., 2005) but have no fossils older than 2800 BP. (There are no Tertiary mammal fossils in Madagascar; Godinot, 2006b). If the other strange lemur families only known from recent subfossils on Madagascar are included as "extant," 50% of the extant primate families have no fossil record. The *Daubentonia* branch may have diverged from the other lemurs long before the platyrrhine/catarrhine split (Arnason et al., 2008; Steiper and Young, 2008; Fabre et al., 2009), and the other lemur families could be older than any of the anthropoid families (Steiper and Young, 2008). Thus, the lemur families are all ancient clades with no fossil record.

Many paleontologists have overestimated preservation rates and have even argued that the more a scenario differs from a direct reading of the fossil record, the less likely it is to be real (references in Soligo et al., 2007; Martin et al., 2007). One devout literalist calculated the probability of primates existing at 80 Ma (Late Cretaceous) as one in 200 million! (cited in Soligo et al., 2007). These fundamentalist paleontologists have carefully assessed the completeness of the fossil record but only through analysis of the fossil record itself, and Soligo et al. (2007) criticized the narrowness and circularity of this approach. In this important paper the authors suggested that underlying any direct reading of the primate fossil record is the unstated assumption that most of primate evolution has by now been unearthed and described.

Instead, Soligo et al. inferred poor sampling of primates and other mammal orders in the record and concluded that Cretaceous (rather than Cenozoic) divergence in these orders "should now be considered the more likely scenario, in which case the influence of continental drift has probably been considerable" (Soligo et al., 2007: 46). Eizirik et al. (2001) and Arnason et al. (2008) also discussed possible Mesozoic ages for primate clades. They cited Gondwana breakup and referred in general terms to the possibility of vicariance models, although their models still used centers of origin and dispersal.

CRITIQUE OF THE MOLECULAR CLOCK: FOSSIL-CALIBRATED CLOCKS GIVE MINIMUM AGES ONLY

The idea that fossils only provide a minimum date is now accepted as an important principle in theory, yet in practice it is still often ignored in fossil-calibrated clock analyses. Cross-calibration takes advantage of the fact that one group has a better fossil record than another; for example, "Contrary to primates, glires [rodents and lagomorphs] have a very good record" (Godinot, 2006b: 459). But what exactly does "very good" mean? The glires record is only very good compared with other fossil groups, and this does not mean that the glires fossils give absolute ages. The molecular clock dates were accepted above as more realistic than raw fossil dates. Nevertheless, they are still calibrated with fossils and so are probably too young.

According to Matthew (1915), fossils prove that primates are not old enough to have been affected by major tectonic change. Most modern biologists would agree and observe that fossil-calibrated molecular clocks prove the same thing. In fact, both methods produce fossil-calibrated minimum ages only and cannot rule out earlier ages, only later ones. While Springer et al. (2003) described several shortcomings of previous molecular clock work on mammals, they did not mention the most important one—the use of fossil calibrations. Springer et al. themselves found Late Cretaceous ages (85 Ma) for primates, yet this relied on proposing explicit *maximum* dates based on the fossils.

All the molecular clock studies of primates have been calibrated using fossils, and so all the calculated dates are minimum ages. Despite this, in many studies the dates are presented transmogrified into maximum ages, with suggestions that primates evolved *at*, for example, ~90 Ma. Cox and Moore (2010: 18) suggested that panbiogeography neglects the fossil record, but this is not correct, as it incorporates fossil data where available (cf. Figs. 3-2, 3-3) and uses fossil ages to provide minimum dates. It is not the fossil record that panbiogeography rejects, only the validity of transmogrification.

Apart from the significance of the fossil dates, as noted in Chapter 2, it is natural to be skeptical of morphological phylogenies of fossil clades—on which the cross-calibration depends—when morphologists did not retrieve the clades of extant taxa such as Archonta s.str. or Afrotheria, despite having much better (living) material. The tendency to regard fossil groups as direct ancestors (stem taxa) will also lead to underestimates of age (see Chapter 2). In any case, many fossil primate clades are known only from fragmentary material, so reconstructing the phylogeny is a difficult if not impossible task; the results remain controversial.

Some authors have acknowledged that molecular clocks calibrated with fossils "are not wholly independent tests of fossil evidence" (Miller et al., 2005: 73; cf. Raaum et al., 2005). Thus, the great value of the clock is not that it gives the right dates (they are still based on fossils and so are minimum ages) but that it shows the fossil record, as an indication of evolutionary dates, could be wrong by tens of millions of years. Molecular workers try to fit the phylogeny to the fossil dates. Instead, the phylogeny of primates is integrated here with the chronological sequence of rifting events which broke up Pangaea and Gondwana at rifts in the central-North Atlantic, Tethys, Lebombo monocline, Mozambique Channel, and Atlantic Ocean.

In their perceptive critique, Steiper and Young (2008: 180) emphasized that: ". . . fossils 'set' the molecular clock. Therefore, calibrations can only establish a lower bound. . . . For this reason molecular clock dates are best considered minimum bounds for divergence dates, whether stated explicitly or implicitly. Because fossil-based calibrations are biased towards younger dates, [fossil-calibrated] molecular clock dates estimated elsewhere in a phylogeny should also tend toward underestimation."

Primatologists sometimes suggest that "Molecular divergence date estimates are dependent on calibration points gleaned from the paleontological record" (Raaum et al., 2005: 237) or that the fossil record "is our only source of information for calibrating phylogenies" (Godinot, 2006b: 455). This is not correct, and tectonics offers an alternative which has been used for calibration, more or less successfully, by workers in many other groups (see the work on cichlids discussed in Chapter 2, e.g., Genner et al., 2007; Azuma et al., 2008). Molecular phylogenies define groups in space, and the spatial breaks between the groups are often obvious and clear-cut, as in the primates. These breaks can be correlated with geological history, and the radiometric techniques used to date tectonic events are much more reliable than fossil-calibrated molecular clocks.

To summarize, the inferred dates for primates using the different methods are:

- fossils: Cenozoic, 56 Ma (a minimum date),
- fossil-calibrated molecular clock: Cretaceous, 80–120 Ma (a minimum date),
- tectonics-calibrated molecular biogeography: Jurassic, 185 Ma (an absolute date).

There is no real conflict between these dates, as the first two are estimates of minimum age, while the third is an estimate of absolute or maximum age.

Molecular clock studies, with minimum ages, have shown that most species in groups such as primates are older than the Pleistocene and are often mid-Cenozoic, while the mammal orders are older than the Cenozoic: They are Mesozoic. The total absence of many large groups in the fossil record of the Late Cretaceous–Paleocene is probably an artifact of preservation, even in groups that were thought to have a good record, or misidentification. With the new clock dates for primates, it is now much easier to accept that the fossil record could be misleading about the existence of groups in the Jurassic, identified in this paper as the main period of primate evolution. A vicariance model implies that the fossil-calibrated ages (both molecular and non-molecular) are incorrect if interpreted as absolute ages. The fact that fossil dates only give minimum ages for clades need not constitute a real problem in theory or practice, but it does undermine dispersal models that are based solely on the treatment of fossil dates and their derivatives as maximum dates. This approach has not led to any coherent synthesis of primate evolution, only into a morass of fundamental, unresolved problems about the origins of the group and all its main clades.

Shifting the age of the primates from the Cretaceous to the Jurassic may seem a radical step, but workers in other groups have had to come to terms with much greater changes. Angiosperms have an excellent fossil record, better than that of primates, and on the basis of the fossils were assumed to be Cretaceous. They are now thought to be much older than this, as molecular phylogeny indicates they are sister to one or more groups known from Permian/Triassic fossils (see Chapter 2). Botanists are now in the process of reassessing the identification of fossil pollen from the Triassic and Jurassic (Doyle, 2005).

Fossil Age Versus Clade Age

In their critique of a the model presented here, Goswami and Upchurch (2010) wrote:

> While certainly some people may mistakenly treat fossil or fossil-calibrated molecular dates as maximum divergence estimates, the vast majority of workers realize that they are in fact minimum divergence dates.... However, we can constrain a realistic range of divergence times for a clade provided we have a suite of phylogenetic nodes where the minimum divergence time has been estimated. For example, if the earliest known euprimate fossil is the approximately 56 million years old (Mya) *Altiatlasius koulchii* from Morocco . . . , and the earliest known eutherian mammal is the 125 Mya *Eomaia scansoria* from China, *it seems probable that the first true primate originated somewhere between 56 and 125 Mya.* (italics added)

This is not strictly logical. If both the fossil dates are minimum dates, as Goswami and Upchurch acknowledged, there is no reason to assume that the first primate evolved between 56 and 125 Ma. The authors pointed out that the line of reasoning they cited "has led to the development of sophisticated Bayesian methods. . . ," but this does not address the basic flaw in the argument.

Goswami and Upchurch (2010) also suggested that the ages of tectonic events proposed by geologists were less accurate than the ages of clades based on fossils. They wrote: ". . . substantial margins around the dates of tectonic events contrast with many fossil-based calibration points that can be dated accurately to the nearest million or even a few hundred thousand years" (p. 3). But the fossil-based clade ages (and nodes in phylogenies calibrated with them) are minimum dates only, whereas the tectonic dates (and nodes calibrated with these) are estimates of absolute dates established using radiometric methods, not fossils. Fossil-calibrated clade dates can only be compared with tectonic dates if the former are transmogrified into estimates of absolute dates, and this is what Goswami and Upchurch have done here, implicitly. Earlier in their note they described my suggestion that fossil-calibrated divergence times are generally treated as absolute, rather than minimum, dates as "something of a 'straw man'", but here they are reasoning in the same way as the straw man.

Using Molecular/Tectonic Biogeography to Date Clades and Assess Gaps in the Fossil Record

Fossil-calibrated molecular clocks do not provide an independent test of the fossil record, and so claims that dates from the molecular studies "corroborate" parts of the fossil record can be misleading. In contrast, tectonics-calibrated dates for phylogenetic and biogeographic nodes do provide an independent test of the fossil record. Glazko et al. (2005) dated the Archonta/Glires split at 84–121 Ma. The authors emphasized that the clock was fossil-calibrated, and so these are minimum dates. Divergence between Archonta and Glires may have occurred less than 5 m.y. prior to the origin of primates (Janečka et al., 2007), and so the 121 Ma date for Archonta suggests a (minimum) date of ~116 Ma (Early Cretaceous) for primate divergence (Janecka et al., 2007: Fig. 2, indicated ~86 Ma). It is suggested here that the primates originated at 185 Ma, in the Early Jurassic. The 69 m.y. difference between the fossil-based minimum date (116 Ma) and the biogeographic estimate (185 Ma) probably reflects the gap in the record of the fossil lineage that Glazko et al. (2005) used to calibrate their clock. As indicated, demonstrable gaps of 100 m.y. or more are already known in the fossil record.

Many modern advocates of fossil-based molecular chronology aim to test vicariance versus dispersal by using a relaxed clock and transmogrifying minimum, fossil-calibrated dates into maximum, absolute dates (Yoder and Nowak, 2006; Poux et al., 2006). Instead, the clock dates and the fossil record used for the calibration can themselves be tested by assuming vicariance. This seems reasonable; recent evidence from population genetics indicates that while vicariance is likely to be a mode of speciation, there is little or no evidence for founder effect speciation (chance dispersal) (see Chapter 1).

EVOLUTION BY VICARIANCE OF POLYMORPHIC ANCESTORS WITH INCOMPLETE LINEAGE SORTING

Many studies of primate taxa focus on reconstructing the single "ancestral morphotype" in a group and assume a uniform ancestor. Opposed to this idea is the fact that incomplete lineage sorting is now well documented in primates, for example, in lemurs (Heckman et al., 2007), colobines (Ting et al., 2008), and hominids (Salem et al., 2003; Caswell et al., 2008). This indicates that the ancestor was *already* polymorphic before differentiation of the taxa under study began and that this polymorphism was passed on in the descendants. If the differentiation of the mammal orders and suborders was rapid and if it occurred in ancestral complexes that were widespread and polymorphic, as suggested here for primates, it probably involved retention and recombination of ancestral polymorphism. Relevant examples of incongruent character combinations in morphology include striking similarities in the visual system of primates and fruitbats (Martin, 1990; Pettigrew et al., 2008), the ears of lemurs and bats, the teeth of some lemurs and those of fruitbats (Kingdon, 1974), and the continuously growing incisors of the basal lemur *Daubentonia* and those of rodents, some marsupials, and some Afrotheria. The mammals of Madagascar can be interpreted as a remnant of a former biota based in and around the Indian Ocean, and their morphologies represent recombinations from an old pool of early mammalian and pre-mammalian characters.

Retention of ancestral polymorphism is often seen as a problem in the search for the true phylogeny. On the other hand, each incongruent gene tree usually shows a distinct biogeography. Hopefully, former genetic connections may be revealed by searching genetic palimpsests for incomplete lineage sorting. For example, connections between primates in particular regions of northeastern Brazil and in particular areas of West Africa might be traceable, just as pre-Atlantic shear zones in Brazil can be followed into Nigeria and Cameroon. The unrelated melanic forms *Chiropotes satanas* of northeastern Brazil (a platyrrhine) and *Colobus satanas* of Gabon (a catarrhine) would be interpreted in traditional theory as convergent adaptations; instead, there may be a biogeographic and genetic basis to the pattern which predates the platyrrhine/catarrhine split.

Primates show many parallel tendencies shared with various other mammals and birds, including binocular vision and a larger brain, and there may be only one morphological synapomorphy for the group, an auditory bulla formed by the petrosal bone (Rasmussen, 2002). This could also be a parallelism, one that has developed in all the main groups, and a direct ancestor, by definition without the character, would be difficult to identify. Although the extant primates (and extant mammals) are monophyletic with respect to other extant groups, it is not certain that this is the case with respect to the many fossil groups. In addition to having only a single morphological synapomorphy, primates are unique among mammal orders in that both the extreme types of placentation (noninvasive epitheliochorial and highly invasive haemochorial) are present in a single order (Martin, 2008). Strepsirrhines have the former, haplorhines the latter.

If parallelism is as universal as molecular phylogenies indicate, the strange extant mammals of southern Africa and Madagascar—lemurs, aardvarks, pseudo-aardvarks, and many others—may be related to evolution in Gondwana and the Karoo basin of southern and East Africa. This interpretation predicts a wide range of morphology in Mesozoic mammaliaforms, for example. The Karoo fossils provide a remarkable sample of the transition from "reptile-like" early mammals to true mammals. Unfortunately, sedimentation in the Karoo beds ended with the great flood-basalt eruptions of the mid-Jurassic. After this there is an almost continuous gap of 120 m.y. in the mammalian fossil record of Africa that extends through much of the Jurassic and all the Cretaceous. This "surely represents an enormous omission in our knowledge of mammalian evolution rather than real absence of mammals from Africa" (Miller et al., 2005). Even in the Karoo there will be many groups still unsampled (or misidentified), and these may include pre- and proto-primates.

Molecular Clock Studies and the Biogeography of Madagascar

Masters et al. (2006: 400) described the origin of the Madagascar vertebrates as "one of the most tantalising enigmas facing biogeographers. . . . Simpson's (1940) prediction that sweepstake dispersal events should occur at random intervals is not borne out by the mammal data. According to the molecular divergence dates estimated by Poux et al. (2005), sweepstakes dispersal events seem to have occurred very early in the history of a clade, and never again." Masters et al. (2006: 414) suggested that a reconsideration of the geophysical and molecular data is needed to address the "apparent paradoxes," and this data is the basis of the model presented here.

Yoder and Nowak (2006) gave a thorough review of the molecular clock literature on Malagasy taxa. In every study of plants, the fossil-calibrated clocks dated the Madagascar clades as younger than 80 Ma, and so they were all attributed to post-Gondwana dispersal, none to vicariance. With a single exception, studies of animal taxa showed the same result. All molecular dating studies of Malagasy invertebrates, reptiles, and mammals have concluded in favor of dispersal, as the inferred (fossil-calibrated) divergence times were post-Mesozoic. The only sequenced group in Yoder and Nowak's review (2006) whose presence on Madagascar has been attributed to vicariance was the cichlid fishes. Molecular dating studies of this group avoided the use of fossil calibrations completely (Sparks, 2004; Sparks and Smith, 2004; see Chapter 2). Instead, the vicariant distributions of the two main molecular clades, Madagascar–Africa–South America and Madagascar–India–Sri Lanka, was correlated with tectonics—the opening of the Mozambique Channel—and this was used as a calibration. (The same method is used here for primates.) So although Yoder and Nowak (2006: 416) concluded that the importance of dispersal "cannot be denied," really, the only thing the cited studies show is the importance of the calibration method. Recent molecular studies on cichlids have continued to use the rifting of Gondwana rather than the fossil record to calibrate molecular phylogenies (the oldest fossil is Eocene), and this has produced the intriguing results discussed in Chapter 2.

The freshwater fishes called snakeheads, Channidae, comprise one clade in Africa and one in Asia (India to Java). Li et al. (2006) calibrated a clock for the family by using the opening of the Atlantic to date neotropical and African cichlids. They supported an early Cretaceous age for the Africa/Asia split in Channidae and a vicariance origin for the break. The oldest fossils of the family are Eocene, and so the gap between fossil age and clade age that they accepted in both Channidae and in Cichlidae is not dissimilar to that proposed here for primates and by other authors for aplocheiloid fishes (see Chapter 2).

As discussed, fossil-based "molecular" clocks can only give maximum, absolute ages by transmogrification of minimum ages, and this is seen in a clock study of the four main Malagasy mammal orders (Poux et al., 2005). The authors found that lemurs and their African sister group diverged first (at 60 Ma), followed by equivalent Africa/Madagascar splits in tenrecs, carnivorans, and finally rodents. The clock calculations were calibrated using six "well-established" (p. 721) fossil dates which were transmogrified into maximum dates; Poux et al. (2005) assigned minimum and *maximum* dates for five fossil clades and a minimum date for a sixth. The primate radiation was assigned a minimum age of 63 Ma and a *maximum* of 90 Ma, but this is not logical. There are no maximum clade dates in the mammalian fossil record. In a similar study, Poux et al. (2006) again used transmogrified data and inferred young dates for trans-Atlantic disjunctions in primates and in hystricognath rodents. Based on this, they eliminated Mesozoic vicariance and accepted trans-Atlantic rafting.

In hystricognath rodents, Rowe et al. (2010) found a sister-group association between the South American Caviomorpha and the Old World Hystricidae (Asian and African porcupines). In timing the divergence, they wrote on page 306 that "fossils themselves generally serve only as indicators of the minimum ages of lineages," but on page 310 wrote: "An upper bound (i.e. maximum age) for MULTIDIVTIME analyses was most reliably designated at the node associated with the diversification of caviomorph lineages (i.e. Caviomorpha), given the more widespread geographical distribution and comparative abundance of fossil information (i.e. both presence and absence of fossils) available from South America." Only by transmogrifying the fossil-calibrated clock dates into absolute dates in this way were Rowe et al. (2010) able to reject Gondwana vicariance as a possible cause of the Caviomorpha/Hystricomorpha divergence. Their paper is titled "Molecular Clocks Keep Dispersal Hypotheses Afloat," but by now the clocks—the weakest aspect of molecular biology—are the *only* things keeping dispersal afloat.

Matthew (1915), Yoder et al. (2003), and Poux et al. (2005) all concluded that the different degrees of differentiation shown between the Malagasy mammals and their mainland sister groups indicate different times of divergence and hence colonization. Instead, in a vicariance model different degrees of difference reflect prior aspects of genome architecture. Molecular clock analyses that depend on degree of difference suggest that every endemic clade in Madagascar is the result of a separate event, unrelated to any other, and that community-wide vicariance has not occurred. But all the terrestrial mammal orders in Madagascar are represented there with just a single, diverse clade that has a widespread, allopatric sister group, the typical signature of early vicariance. This suggests that their spatial differentiation has resulted from exposure to the same Earth history and that the different degree of differentiation in each of the four is not related to time.

The primates are closely related to the rodents, and the biogeography of this larger group, with about 40% of all mammal species, is relevant. The trans-Atlantic disjunctions have already been mentioned. Rodents have three primary clades, and each occurs in America, Africa, and Asia (Horner et al., 2007). Despite this overlap, there are significant differences among the three clades, especially with respect to Madagascar:

- Hystricomorpha (porcupines, etc., the basal group): east to Borneo (cf. strepsirrhines); not in Madagascar.
- Sciuromorpha (squirrels): east to Sulawesi (cf. haplorhines); not in Madagascar.
- Myomorpha (rats and mice): east to Australasia, with a diverse clade in Madagascar.

Many authors have commented on the surprising absence of the first two groups, especially the otherwise widespread squirrels, from Madagascar and Australasia. The same pattern occurs in haplorhines. Although the distribution of the rodents is different from that of the primates, the main phylogenetic/biogeographic breaks (nodes) are the same: Atlantic Ocean, Mozambique Channel, Makassar Strait (Wallace's line) east of Borneo, and the Banda/Molucca Sea east of Sulawesi.

Horner et al. (2007) suggested that rodents diverged from lagomorphs and that the three main rodent clades diverged from each other at around 60 Ma (a minimum date), with all this evolution taking place within just 3.1 m.y. A similar "early, rapid" differentiation is accepted here for the main primate clades; the proposed age of the groups is much older than has been thought, although the time involved in their differentiation may be much less.

THE EVOLUTION OF PRIMATES

Traditional, fossil-based models and the new fossil-calibrated molecular clocks both support an origin of primates after the Early Cretaceous, in which case the group would be too young to have been affected by Pangaea–Gondwana breakup. Both models infer dispersal over a more or less modern geography. The problem is that dispersal models all leave many serious problems unresolved. For example, how did primates cross the major barriers of the Atlantic Ocean and Mozambique Channel? And if they could disperse across, why did this only happen once, early on, in each case? Why are New World monkeys sister to the Old World monkeys, not deeply nested within them, as a dispersal model would predict? How could primates disperse across the Atlantic and Mozambique Channel when they have never crossed the stretch of sea (Salue Timpaus Strait, 20 km wide) that separates Sulawesi from the Sula Islands and Australasia? What is the significance of the narrow strip where the only nonhuman primates are tarsiers? Studies have focused on means of dispersal and island hopping, but it seems these may not be relevant. Most primates have excellent powers of movement, running, climbing, and leaping on cliffs and trees. Some Old World primates, such as *Nasalis* and *Macaca*, are well known to be strong swimmers; in America, *Alouatta* has been observed swimming in the wild (Eisenberg et al., 2000), and most neotropical forms can probably swim to some extent (Eisenberg et al., 2000: 231). Many primates, including several Amazon species, undertake extensive annual migrations through different forest types (Barnett and Brandon-Jones, 1997; Boubli et al., 2008). On the other hand, primate species also exhibit strong philopatry and maintain strict geographic limits. The problem lies in dovetailing the excellent powers of movement with the observed strong phylogenetic differences among the primate faunas of the different regions. The platyrrhines in South and Central America and the lemurs in Madagascar both show a complete lack of geographic overlap with any other group. This, together with the widespread overlap of nearly all genera *within* South America, Africa, and Asia, indicates that while there has been no transoceanic dispersal (as expected from their observed ecology), there have been phases of mobilism on the continents, leading to overlap of the genera by normal range extension. Within most genera, nearly all the species are allopatric, indicating a later phase of immobilism and differentiation.

In center of origin/dispersal theory, the distribution of a group develops *after* its origin in phylogeny, through a separate process—range expansion from the center of origin. In vicariance, distribution is produced *by* the phylogeny, and so distribution and phylogeny develop at the same time. This means the distribution can be informative about the phylogeny and vice versa. When considering the biogeography of any taxon, traditional biogeography looks for a center of origin; a vicariance analysis looks for the break between the group and its sister. The main problem is not explaining how each individual group managed to migrate to its own area. It is explaining how and why migration *ceased* and the rifts between the groups formed, that is, how an originally widespread group could break up and evolve different members in different areas. The mechanism suggested here is Earth history and plate tectonics.

Rose (1995: 170) noticed that "Conjecture about the place of origin of both primates and anthropoids seems to change with each new discovery—a sure indication that existing evidence is simply insufficient." Instead, the situation (unchanged since 1995) could indicate that there is no center of origin and that primates originated as a southern vicariant of plesiadapiforms, while anthropoids developed as a western vicariant of tarsiers.

Critical analyses of sweepstake dispersal (Stankiewicz et al., 2006; Masters, Lovegrove, and de Wit, 2007), new interpretations of the fossil record and phylogeny (Miller et al., 2005; Martin et al., 2007; Soligo et al., 2007), and molecular clock studies indicating a much older, Mesozoic chronology (Glazko et al., 2005) together represent a major break with Matthew's (1915) center of origin/dispersal model of primate evolution. The new work is instead compatible with a vicariance model in which the spatial distribution of molecular clades is correlated with radiometrically dated tectonics. In the new chronology, fossils are used to give minimum ages only and dates are not transmogrified. While the method gives dates that are older than those accepted in fossil-based clock scenarios, many clock studies (and fossils themselves) have already demonstrated massive gaps in the fossil record.

Yoder and Yang (2004: 768) concluded their fossil-calibrated molecular clock study by writing that

> The initial radiation of lemuriform primates . . . is estimated to have occurred approximately 62 Ma, near the onset of the Tertiary. This is a surprisingly ancient date, as it precedes the appearance of euprimates in the global fossil record. Indeed, if we were to base our judgements of primate antiquity on a strict interpretation of the known fossil record, this estimate of lemuriform antiquity would be considered incredible. Instead, increasing numbers of primatologists and paleontologists concur that *the fossil record is far too scant and "frighteningly incomplete" (Fleagle, 2002) to impose strict limits on our interpretation of the temporal context for primate evolution.* (italics added)

This reasoning is correct and important, and the argument should be extended to a critique of fossil-calibrated molecular clock dates. If workers have already dismissed the fossil record as a basis for chronology, as Yoder and Yang (2004: 768) have in accepting the "surprisingly ancient" dates for lemurs, tens of millions years older than the oldest fossil, it would be logical to take one more step and not base clock dates on fossils at all. It makes sense to avoid staking the entire chronological analysis on a "frighteningly" scant record, when the distinctive biogeography of molecular clades and accurate radiometric dates can be used instead (cf. Li et al., 2006; Genner et al., 2007; Azuma et al., 2008).

In primates, as in many other groups, biologists have assumed that the geographic distributions were fluid and ephemeral compared with the morphological distinctions, yet distribution now appears to have a special phylogenetic significance. The main phylogenetic/geographic breaks reflect a simple tectonic sequence. If Cenozoic fossil-based clock dates are replaced with Mesozoic dates based on tectonics and molecular clade distribution, it should be possible to move beyond the legacy of Matthewian zoogeography: paradoxes and enigmas, endless, fruitless debate about center of origin, and reliance on theoretical processes of dispersal that in practice have proved to be unfeasible.

APPENDIX

Supplementary Notes on Plesiadapiforms and Primate Fossil Groups

PLESIADAPIFORMS

35 genera and 75 species (Fleagle, 1999) are known from western North America (latest Cretaceous/earliest Paleocene to Late Eocene; Clemens, 2004), western Europe (Paleocene to Early Eocene), and China, north to Eureka Sound, Ellesmere Island (*Ignacius*) (Fleagle and Gilbert, 2006).

Fleagle (1999: 333) wrote that "the evidence linking plesiadapiforms with primates is no stronger than that linking primates with either Scandentia or Dermoptera." Other studies have proposed Dermoptera as sister to plagiomenids (a fossil group of North America), to Scandentia (the two orders comprising Sundatheria), or to anthropoids.

Altanius, the "basal primate," is sometimes regarded as a plesiadapiform or an omomyiform.

Altiatlasius, the "oldest primate fossil," comprises isolated teeth which have also been identified as belonging to a plesiadapiform, a basal primate, or an anthropoid (Fleagle and Gilbert, 2006). While the teeth of *Altiatlasius* "do indeed look quite a bit like the teeth of true primates," so do the molars of plesiadapiforms and several other extinct non-primate groups (Rasmussen, 2002: 9). Bloch et al. (2007) suggested that *Altiatlasius*, not *Altanius*, may be the phylogenetically basal primate.

ADAPIFORMS AND OMOMYIFORMS

There is some evidence to suggest that both adapiforms and omomyiforms shared a strepsirrhine nasal configuration, although it is also possible that neither group did (Rasmussen and Nekaris, 1998).

Fleagle (1999) noted that adapiforms lack the tooth-comb of the living strepsirrhines and that the similarities of the two groups are shared with other taxa. Miller et al. (2005: 68) wrote that an adapiform–strepsirrhine link "has always been tenuous." Adapiforms "may" share "a few" derived characters with strepsirrhines (Fleagle, 1999: 387), and "a few" characters link omomyiforms and tarsiers, but other interpretations are often proposed and there is no sign of a consensus. Covert (2002: 19) also stressed that "while it is widely assumed that the extant tooth-combed primates are derived from the adapoid radiation, none of the adapoids clearly shows a series of traits linking them exclusively to the more recent strepsirrhines" (cf. Gebo, 2002).

In haplorhines, the major preservable features linking tarsiers and anthropoids, namely the postorbital closure and accessory cavity in the ear, have not been documented in any omomyiforms. "It is conceivable that the strepsirrhine-tarsier-anthropoid divergence was, for all practical purposes, a trichotomy" (Miller et al., 2005: 71). Several authors compare adapids with anthropoids. Others view omomyids as the sister group of adapiforms (the earliest known representatives differ very little from one another).

Adapiforms and omomyiforms are a challenging problem for biogeography, as their distributions show so much mutual overlap, at least in Europe and America. There is no obvious difference comparable to that seen between living haplorhines and strepsirrhines (except possibly in Southeast Asia, where the enigmatic Sivaladapidae may partially replace omomyids). Unless there was sympatric evolution, there must have been an original allopatry somewhere, possibly related to the haplorhine/strepsirrhine split in the south, although it has been largely obscured by subsequent migration. Any slight differences between the distributions of adapiforms and omomyiforms may be significant, and a detailed, mapped study comparing the distributions and phylogeny of adapiforms and omomyiforms would be of great interest.

Rooneyia of Texas has usually been treated in omomyiforms, although Rosenberger (2006: 139) suggested that it was the sister group of anthropoids. In a center of origin analysis he wrote: "If the sister group of anthropoids occupied North America as part of a Laurasian distribution during the Paleogene, as some primate genera did [cf. omomyiforms], ancestral anthropoids may likewise have occurred [only] across Laurasia, prestaging them to enter Africa and Central/South America in two independent episodes of dispersal." This is very similar to the ideas of Matthew (1915) and Simpson (1940). The interpretation of *Rooneyia* does not have major implications for biogeography, as anthropoids occur in southern Mexico and (fossil) in the West Indies. A sister relationship between *Rooneyia* and anthropoids would not necessarily imply any dispersal and could instead indicate a break between Texas and Cuba/southern Mexico.

OTHER FOSSIL STREPSIRRHINES

The fossil Amphipithecidae (Pakistan, Burma, Thailand) belong with adapiforms (e.g., Gunnell et al., 2002; Gunnell and Ciochon, 2008) or anthropoids (e.g., Beard et al., 2009). The Pakistan record is north of the extant distribution limit of strepsirrhines (central west India–northeastern India) (Kay et al., 2004; Marivaux et al., 2005). Fossils of possible stem lorisids (*Karanisia*) and stem galagids (*Saharagalago, Wadilemur*) from the Egyptian Late Eocene indicate these two groups had diverged by this time (Seiffert, Simons, Clyde et al., 2005; Seiffert, Simons, Ryan, and Attia, 2005; Seiffert, 2006), and *Omanodon* from Oman is "probably" a related lemuriform/lorisiform (Godinot, 2006b: 451). These new oldest fossils are interesting in their own right, but do not prove an African or Arabian (rather than Asian) center of origin for strepsirrhines or even Lorisidae. There is no confirmed record of lemurs outside Madagascar. *Bugtilemur* from the Oligocene of Pakistan has been identified as closest to cheirogaleid lemurs (Marivaux et al., 2001), yet it is only known from isolated teeth and there is no direct evidence of a tooth-comb, one of the main uniting and identifying features of lemuriforms and lorisiforms. This makes its identification as a lemur "very suspect" (Fleagle and Gilbert, 2006: 390) and it "will need better substantiation before it is wholly accepted" (Miller et al., 2005: 74). Godinot (2006b) did not accept it as a lemuriform. The discovery of a similar form, *Muangthanhinius*, in Thailand led Marivaux, Chaimanee et al. (2006: 432) to conclude that "the possibility exists that *Bugtilemur* is not closely related to the lemuriforms, because *Muangthanhinius* clearly is not. In this case, *Bugtilemur* might alternatively be interpreted as a very specialised adapiform" (cf. Godinot, 2006b).

Azibiidae (*Azibius* and *Dralestes*) of westernmost Algeria (the Hammada du Dra by the Anti-Atlas Mts.) were identified as plesiadapiforms by Tabuce et al. (2004), although Beard (2006) did not accept this and Godinot (2006a) thought they were probably euprimates. Tabuce et al. (2009) concluded that *Azibius* and *Algeripithecus* (incl. *Dralestes*) were strepsirrhines. *Algeripithecus* had been interpreted as an anthropoid, and Tabuce et al. suggested their new identification challenges the idea of an ancestral homeland of anthropoids in Africa, but this argument assumes that the oldest fossil of a group indicates its center of origin.

OTHER FOSSIL HAPLORHINES

Catarrhine fossils are known from Africa and Eurasia to Southeast Asia, with extralimital records north to southern England (*Macaca*, Pliocene) and south of Lake Baikal (colobines, Pliocene). Platyrrhines have extralimital fossils on Cuba, Jamaica, and Hispaniola, and in Patagonia including the Chilean Andes. The fossil groups Eosimiidae (Pakistan, China, Burma), Parapithecidae (Egypt) and Oligopithecidae (Egypt) may belong here (Kay et al., 2004; Marivaux et al., 2005), and mainly lie within the extant range.

Parapithecus, Apidium, and allies are early anthropoids from the Eocene–Oligocene of Egypt (Fayum). They show resemblances with platyrrhines that Hoffstetter (1974: 346) regarded as "astonishing," and this author derived platyrrhines directly from parapithecids. While Szalay and Delson (1979: 308) described Hoffstetter's idea as "intriguing, especially given the simplistic equality in premolar number [in the two groups]," they found that details of morphology showed no "special relationship" between the two groups.

Subsequently, a new anthropoid family from Fayum, Proteopithecidae, was accepted for *Proteopithecus*, and again "many features resemble those of platyrrhine monkeys" (Simons, 1997: 14970). Still, Simons also described the group as "the most generalized well-known anthropoidean" (p. 14974); Proteopithecidae "could be part of the basal radiation that produced the New World platyrrhine primates, or it could be unrelated to any subsequent lineages" (p. 14970). *Proteopithecus* "shows a greater resemblance to living and extinct platyrrhines [than do *Parapithecus* and the other Fayum anthropoids], but it is unclear whether these features are of particular phylogenetic significance" (Simons and Seiffert, 1999: 921).

Kay et al. (2004: Fig. 7) favored a phylogeny: (proteopithecids + parapithecids) (oligopithecids (catarrhines + platyrrhines)) and this is more or less compatible with the ideas of Szalay and Delson (1979) and Simons (1997, 2001). Kay et al. (2004: 123) concluded that their analysis "does not support the hypothesis that African Parapithecidae are the stem group out of which platyrrhines arose."

Parapithecids and proteopithecids seem to be relics of a sister group of crown anthropoids: catarrhines, oligopithecids (another Fayum group), and platyrrhines. (For one *Parapithecus* species, Simons, 2001: 7897, described the "mosaic nature of the anatomy of the skull, combining as it does a mixture of primitive, platyrrhine and perhaps even catarrhine characteristics"). These Paleogene anthropoid families may or may not have once been widespread; currently they are all known only from Fayum and central Libya (Jaeger et al., 2010).

4

Biogeography of New World Monkeys

The more we study the distribution of organized life on
the globe, the more we tend to abandon the hypothesis of
migration.

—Humboldt, 1814–1825 (1995: 139).

Detailed information on the molecular phylogeny of New World
monkeys (Opazo et al., 2006; Osterholz et al., 2009) and on their
distribution and ecology (IUCN, 2009) is now available, and so the
group is an excellent subject for biogeographic and evolutionary study.
Researchers now agree on many descriptive aspects of the phylogeny
and distributions, although the interpretation of the data remains controversial. As yet, there is little agreement on how, when, or where the
different primate clades evolved in America or even how primates came
to be there in the first place. Most previous work has suggested that
primates colonized America from somewhere else, for example, by rafting. Instead, it is suggested here that New World primates "arrived" in
America by evolving there, *in situ*, from earlier generalized ancestors
that were widespread in South and Central America before the breakup
of Gondwana.

In Chapter 3 it was suggested that the American primates, the platyrrhines, were isolated by the opening of the Atlantic. In a similar way, the
lemurs may have been isolated in Madagascar with the opening of the
Mozambique Channel. Both these well-dated tectonic events were used
to calibrate the time course of primate evolution, rather than relying on
the scanty fossil record of early primates.

The geology of the South American plate is outlined in Figure 4-1.
The plate is delimited by a divergent plate margin to the east (the

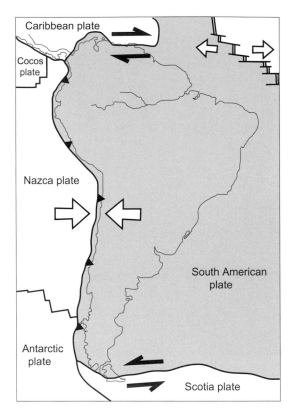

FIGURE 4-1. The South American plate (gray) and its current boundaries: the divergent margin at the Mid-Atlantic ridge, strike-slip transform faults (black arrows) at the northern and southern boundaries, and the convergent margin (subduction zone) to the west, along the Peru–Chile trench. The rise of the Andes is linked to the opening of the Atlantic.

mid-ocean ridge in the Atlantic), a convergent margin to the west (the Chile–Peru trench), and transform faults to the north and south. As the South American plate moved west, beginning in the Cretaceous, it collided with the plate in the Pacific and the continent underwent great deformation. The most obvious result was the uplift of the Andes along the convergent plate margin. South America is composed of older cratons that are exposed as the Guiana and Brazilian shields, the Andean orogenic belt, and an intermediate zone (Fig. 4-2). The latter comprises a fold and thrust belt (the Subandean hills) and a foreland basin (in front of the craton) and has special significance for biogeography.

Apart from the marginal deformation—the Andes—the Cretaceous events led to significant intraplate rifting and magmatism throughout the continent. Figure 4-3 (from Costa et al., 1996) shows the effects in the Amazon region: normal (extensional) faults in the northeast, reverse (thrust) faults indicating convergence in the southwest, and strike-slip

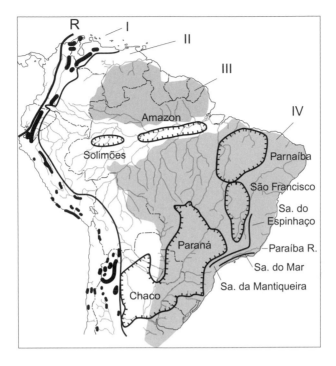

FIGURE 4-2. Geology of South America. R = Romeral fault zone, the boundary between accreted terranes in the west and allochthonous terranes in the east; I = Andean orogen; II = foreland basins; III = Guiana shield; IV = Brazilian shield. Paleozoic–Mesozoic basins: Amazon, Solimões, Parnaíba, São Francisco, Chaco-Paraná.

faults striking close to 90° found throughout. The course of the rivers is determined by the faults, which also give rise to lines of hills with an elevation of about 200 m. The other figures in Costa et al. (1996) should be consulted for many cases of fault control at local scales. It is well known that many platyrrhine distributions reach limits at rivers, but biogeographers often neglect the factors that have caused the rivers to flow where they do. The geometry of the stress regime responsible for the Amazon faults was initiated with the opening of the Atlantic in the Cretaceous and has remained more or less the same until now. For example, in 1983 there was an earthquake of magnitude 5.5 at Codajás, 200 km upstream of Manaus (Costa et al., 1996). Because the landscape in the interior of the cratons is so flat, the slightest inclines and the smallest faults can cause significant changes in vegetation and

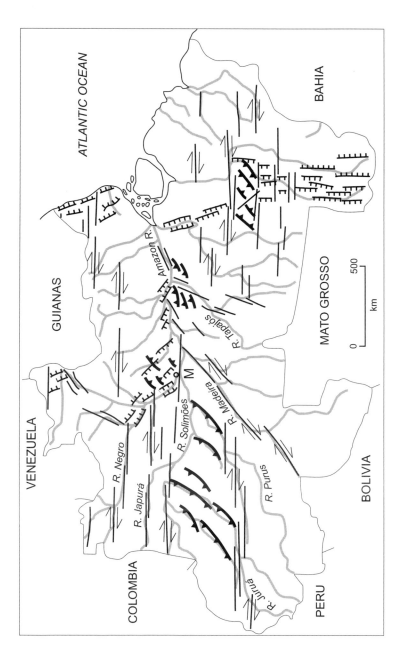

FIGURE 4-3. The Amazonian region of Brazil, showing some of the main faults (adapted from Costa et al., 1996). Normal faults as hatched lines, reverse faults as barbed lines, strike-slip faults as plain lines with arrows. All the strike-slip faults are dextral, but for clarity this is only indicated on some.

biogeography over large areas. In the alpine zone of the Andes a fault with a throw of 1 m has little significance for local ecology, but in the intracratonic lowlands it can determine the vegetation over hundreds of square kilometers. In savanna, lower-lying land in fault angles and synclines has standing water in the wet season and is occupied by an open vegetation very different from that on higher ground. In rainforest, the lower-lying areas are river courses.

Fossil-calibrated molecular clock studies in neotropical groups have calculated Miocene ages for many clades, and so the later stages of Andean orogeny are often cited as important. Nevertheless, these fossil-based molecular dates are minimum ages only, based in many cases (as in the primates) on a fragmentary record restricted to the Cenozoic. For lowland groups such as primates that have centers of diversity in western Amazonia, the most important phase of the Andean orogeny may have been the initial uplift. This changed a landscape of shallow deltas and lagoons into a terrestrial system of lowlands, which in turn developed into the modern Andes.

As discussed in Chapter 3, molecular clock studies have inferred Cenozoic dates for divergence between New and Old World monkeys, and also between New and Old World clades of hystricomorph rodents. Based on this, authors have accepted trans-Atlantic rafting, but the dates themselves were calibrated using fossils and so are estimates of minimum age. The treatment of these as maximum dates, used to rule out earlier (not later) events, is rejected here. In addition, the idea of primate rafting has been dismissed in a detailed study (Masters Lovegrove, and de Wit, 2007). At their eastern boundary in Sulawesi, primates reach their limit at Salue Timpaus Strait, which is just 20 km across. Instead of relying on hypothetical rafting, the rift between American and African primates is attributed here to plate tectonics.

More or less detailed molecular phylogenies are available for most platyrrhine genera, although molecular study of groups such as *Callicebus* and *Cebus* is just beginning.

DISTRIBUTION OF PRIMATES IN AMERICA

The molecular phylogeny of the platyrrhines (Fig. 4-4, from Opazo et al., 2006; Osterholz et al., 2009; Wildman et al., 2009) shows the 15 genera (more, if *Callithrix* is split) arranged in three main clades, the families. Primate distribution in South and Central America is similar to that in Africa and Asia: Most of the genera are widespread and have

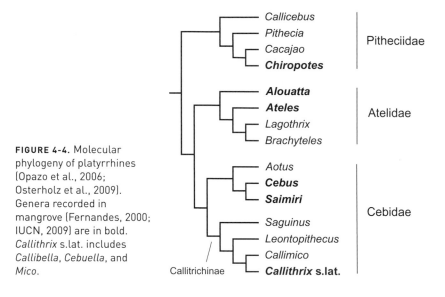

FIGURE 4-4. Molecular phylogeny of platyrrhines (Opazo et al., 2006; Osterholz et al., 2009). Genera recorded in mangrove (Fernandes, 2000; IUCN, 2009) are in bold. *Callithrix* s.lat. includes *Callibella*, *Cebuella*, and *Mico*.

quite distinct ecology, while within the genera most of the species are allopatric. These are broad generalizations and in fact the only primate genera in the Amazon basin that appear to be ubiquitous are *Alouatta* and *Cebus* (Voss and Emmons, 1996). (*Ateles* is absent from large parts of the central, flooded area. *Saimiri* is often absent from extensive areas of upland, non-flooded forest far from rivers and lakes.) Other exceptions to the pattern include significant overlap in several congeneric species (e.g., in three *Cebus*, two *Saguinus*, and two *Callicebus* species.)

To summarize, there is vicariance at the level of parvorder (platyrrhines and catarrhines) and at species level, but at genus level there is widespread overlap. This overlap is not complete, and some interesting differences among the genera and tribes may represent the traces of former allopatry.

For trees, central and western Amazonia are much richer in species than eastern Amazonia and the Guiana shield area (Ter Steege et al., 2000). Primate species diversity is also higher in western Amazonia than in the east, with maximum levels around the Peru/Brazil border between the Ucayali and the Purus Rivers (Voss and Emmons, 1996), around the Sierra del Divisor and the Fitzcarrald arch. Hoorn et al. (2010: Fig. 3) showed mammal richness to be greatest in western Amazonia (Ecuador–eastern Peru–northern Bolivia) and tree diversity highest in northeastern Peru.

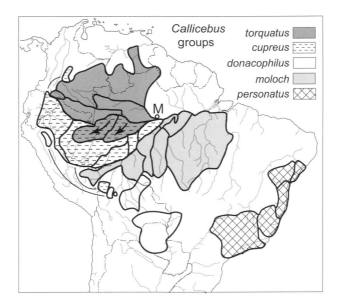

FIGURE 4-5. Distribution of *Callicebus*. Arrows indicate inferred range expansion of current species. m = Manaus.

PITHECIIDAE

Callicebus

This is sister to the three other genera in the family and is one of eight widespread clades of primates in tropical South America. The groupings within *Callicebus*, as mapped here (Fig. 4-5), are still based on morphology (van Roosmalen et al., 2002). The five groups are almost entirely allopatric, with some overlap in southwestern Amazonia. The putative connections among outliers of the *cupreus* group (e.g., *C. ornatus* at the Serranía de La Macarena in Colombia) and the *donacophilus* group in Peru (*C. oenanthe* in northern Peru) form outer arcs. These may be relictual, formed by the stranding of populations during retreat of inland seas.

The basal split in the genus seems to be a break between the *torquatus* group, mainly north of the Amazon, and the others, mainly south of the Amazon (Groves, 2005; Fabre et al., 2009; Chatterjee et al., 2009). The second node may be a break between the southeast Brazilian *personatus* group and the rest (Fabre et al., 2009; Chatterjee et al., 2009).

The only significant overlap among any of the five species groups in *Callicebus* is between the *torquatus* and *cupreus* groups in southwestern Amazonia. This overlap is interpreted as secondary range expansion of the *torquatus* group from north of the Amazon southward (see arrows in Fig. 4-5). The boundary between the *cupreus* group and the *moloch* group follows the course of the Madeira River (meeting the Amazon just east of Manaus), a pattern that also occurs in other groups and has a tectonic basis in a series of large strike-slip faults (Fig. 4-3).

The primate distribution maps produced by IUCN (2009) are the best available and are followed here with one exception: The map of *Callicebus discolor* (*cupreus* group) shows the species only in Ecuador and Peru north of the Marañón, and this does not include the type locality, whether this is taken to be Tabatinga, that is, Leticia (as in Groves, 2005), or Sarayacu on the left bank of the Ucayali (as in van Roosmalen et al., 2002). A wider range for the species south of the Marañón (cf. van Roosmalen et al., 2002) is shown here. The main habitat of *Callicebus* species is in unflooded forest ("terra firme" forest is the Portuguese term), where they favor gaps, edges of streams and lakes, and liane forest. Despite being unable to swim, they visit flooded forest for fruit in the high-water season. *Callicebus* is mainly in lowland forest but occurs at higher elevation in the Huallaga area, where *C. oenanthe* is known from 750–1,000 m. *C. donacophilus* is in drier forest, and the *moloch* and *cupreus* species groups, distributed along the south bank of the Amazon, are tolerant of, or sometimes prefer, habitat that is disturbed by either annual flooding or human activity (van Roosmalen et al., 2002; IUCN, 2009).

The Pithecines: Cacajao, Chiropotes, and Pithecia

Cacajao (Fig. 4-6) is landlocked in the western Amazon, west of Manaus. It is most characteristic of various types of freshwater flooded forest and riparian forest. In Peru, though, most sightings are from terra firme forest (including palm swamp) or mixed habitats with both terra firme and flooded forest (Heymann and Aquino, 2010). *Cacajao* is a parallel case to *Allenopithecus*, landlocked in the Congo basin. Both genera are interpreted here as stranded mangrove associates. The sister group of *Cacajao* is *Chiropotes* (Fig. 4-6), which replaces *Cacajao* east of Manaus, the Madeira River, and the Negro River, and occurs in terra firme forest and mangrove. Overall in the *Chiropotes–Cacajao* group there is high species and subspecies diversity along the main Amazon valley. The importance of a biogeographic boundary at Manaus is often emphasized (see de Oliveira and Daly, 1999, for trees), and both the Madeira and Negro Rivers flow along faults. The western *Cacajao* and the eastern

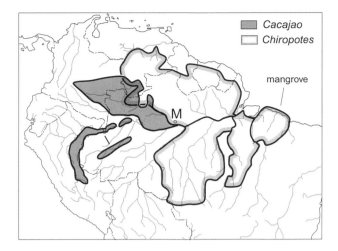

FIGURE 4-6. Distribution of *Cacajao* and *Chiropotes*. M = Manaus.

Chiropotes also meet at Mount Neblina on the Venezuela/Brazil border (Boubli et al., 2008), another standard boundary for eastern and western groups.

In a vicariance model, a common *Chiropotes–Cacajao* ancestor was widespread through the Mesozoic swamp forests and mangroves of the Guiana and Brazilian cratons before the rise of the Andes. With Cretaceous uplift and retreat of inland seas, the ancestor has been subdivided into the modern genera. Each has inherited its own geographic sector of the region, along with its particular substrate and forest type. Any populations of *Chiropotes–Cacajao* caught in the Cretaceous–Cenozoic uplift of the Andes range itself have died out, although one has survived uplift to 1,500 m by Mount Neblina (Boubli et al., 2008).

As indicated, at the western limit of its range, in Peru (Ucayali River, etc.), *Cacajao* is not restricted to flooded forest. In this area the highest encounter rates for *Cacajao* were instead recorded from the Sierra de Contamana, a site that is not only occupied by terra firme forest, but also has a much higher elevation (600–700 m) than any of the other *Cacajao* sites (Heymann and Aquino, 2010). Populations in the Sierra de Contamana (part of the Contaya arch) may have been uplifted along with these hills during Andean orogeny.

The clade *Cacajao* + *Chiropotes* is sister to *Pithecia*, and together the three genera form another widespread clade (Fig. 4-7). If Amazonia is divided into four sectors, around Manaus, the two clades are mainly allopatric in the northwest, southwest, and southeast, but they overlap widely

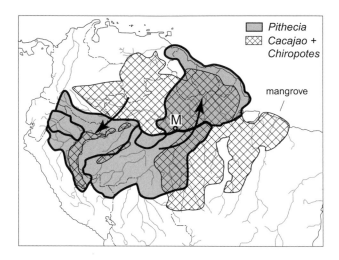

FIGURE 4-7. Distribution of *Pithecia* and *Cacajao/Chiropotes*. Arrows indicate inferred range expansion of current species. M = Manaus.

in the northeast. Although the area of generic overlap is extensive, only one species (*Cacajao calvus*) is restricted to it, and the considerable allopatry between the two clades may be a relic of their original distribution. The arrows on Fig. 4-7 indicate possible range expansion of *Pithecia* from southwestern Amazonia northeast into the Guianas, with *Cacajao* expanding southward into southwestern Amazonia. Part of the old boundary is preserved in the allopatry around Manaus, the key phylogenetic/biogeographic break point among the three genera and the only locality where all three meet. From here the original boundaries may have extended west along the Solimões and Japurá Rivers and southwest along the Madeira.

The pitheciids are absent from Panama and the rest of Central America, unlike the main clades of cebids and atelids, and may never have existed much west of their present limits, in particular west of the Romeral fault zone (Fig. 4-2). This is a linear feature that runs along the western flank of the Colombian Central Cordillera, passing near the cities of Cali and Medellín. It is also an important biogeographic boundary in *Callicebus* and other groups, as discussed below. Pitheciids are also absent in northern Colombia and Venezuela, unlike cebids and atelids. *Callicebus* is on the coast in southeastern Brazil, whereas the other pitheciid clade—*Cacajao*, *Chiropotes*, and *Pithecia*—is instead along the coast in northeastern Brazil and the Guianas. This allopatry may be relictual from the original phylogenetic break in the family, the one that took place between these two widespread clades.

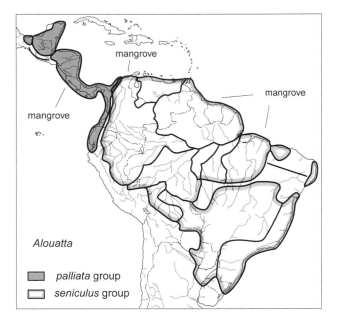

FIGURE 4-8. Distribution of *Alouatta*.

ATELIDAE

Alouatta

The howler monkeys, genus *Alouatta*, form another widespread group. Molecular studies confirm that there are two clades (Fig. 4-8; Cortés-Ortiz et al., 2003; Chatterjee et al., 2009). The Central American/Chocó clade (with a higher-pitched howl) is replaced by the South American clade (with a deeper howl) at the Romeral fault zone, indicating that the differentiation could be Late Cretaceous, when the fault was initiated. Cortés-Ortiz et al. (2003) instead dated the phylogenetic split at only 6.8 Ma, but the molecular clock was calibrated using a human–chimp split at 7 Ma and a New World/Old World monkey split at 40 Ma, both of which are fossil-based minimum dates. Dates derived using these calibrations will also be minimum dates, and the actual dates for *Alouatta* differentiation could be much older.

The Atelines

In this subfamily, *Ateles* (Fig. 4-9) is in Central America but not southeastern Brazil, while its sister, *Lagothrix* + *Brachyteles* (Fig. 4-10), is absent from Central America but present in southeastern Brazil. Initial allopatry

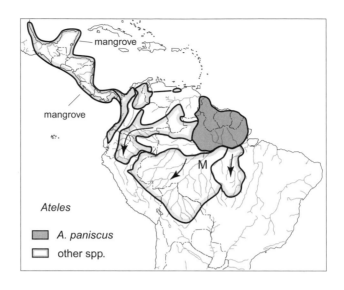

FIGURE 4-9. Distribution of *Ateles*. Arrows indicate inferred range expansion of current species. M = Manaus.

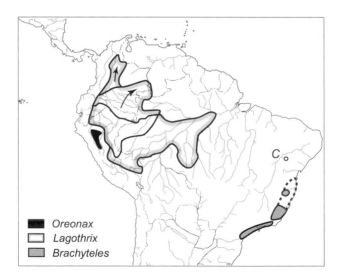

FIGURE 4-10. Distribution of *Lagothrix* (incl. *Oreonax*) and *Brachyteles* (extinct range as broken line; from Mendes et al., 2005), *C* = *Caipora* (fossil). Arrows indicate inferred range expansion of current species.

between the two clades is inferred, followed by range extension of *Ateles* (southward) and *Lagothrix* (northward) as indicated in the figures. (The maps in Collins, 2008, show *Ateles* much less widespread south of the Amazon than the IUCN maps reproduced here in Fig. 4-9.) The two clades together make up the atelines, another widespread clade.

Within *Ateles* (Fig. 4-9), molecular data indicate a basal clade (*A. paniscus*) in the Guianas region (Collins and Dubach, 2000; Collins, 2008; Fabre et al., 2009), and so Collins and Dubach (2000) inferred a center of origin for the genus there. But this does not account for the geographic relationship between *Ateles* and its sister group, *Lagothrix* + *Brachyteles*, and there is no need to assume such a restricted center of origin. Instead, the basal node is interpreted here as a break between the Guianas in the northeast and the other three quadrants of Amazonia. At the second basal node in the phylogeny, the two species south of the Amazon are separated from those to the north (Fabre et al., 2009). Chatterjee et al. (2009) found a different arrangement for *Ateles*, with the main break between the Central American/western Colombian species on one hand, and the South American species on the other. This break, whether it is at species or subgenus level, follows the Romeral fault zone, as in *Alouatta* (Fig. 4-8).

Brachyteles occurs in montane southeastern Brazil (at 600–1,800 m) and is the largest platyrrhine. Its range, already narrow, has diminished in historical times (Fig. 4-10). The Pleistocene fossil *Caipora* of eastern Brazil (Bahia) is related to the atelines (Cartelle and Hartwig, 1996). *Lagothrix* s.lat., the woolly monkeys, are distinct in their dense, woolly coat and stocky build. A species of the Peruvian Andes has been treated under *Lagothrix* (Rosenberger and Matthews, 2008) or as a monotypic genus *Oreonax* (Groves, 2005). Cranium and dental characters in *Oreonax* are "suggestively different" from *Lagothrix* (Hartwig et al., 1996) and "in many respects, these characters resemble *Brachyteles*." The distribution of *Oreonax*, in the northern Huallaga–Marañón area, is shared with taxa such as the monkey *Callicebus oenanthe* and one of the most distinctive hummingbirds, *Loddigesia*. The region corresponds to the gap between the pre-Andean highs, a region of old lowlands, coasts, and mangroves.

CEBIDAE

Aotus

Aotus includes the nocturnal owl monkeys, and the genus is usually included with the cebids (cf. Chatterjee et al., 2009: in their genus tree Fig. 1; their species tree has it instead as sister to Atelidae). In *Aotus*,

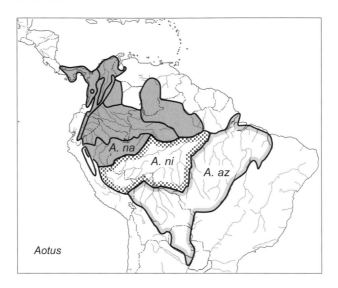

FIGURE 4-11. Distribution of *Aotus. A. na = Aotus nancymaae;
A. ni = A. nigriceps, A. az = A. azarae.*

variations in morphology and karyology indicated two "well founded" groups (Hershkovitz, 1984; cf. Groves, 2005; Fig. 4-11). Their mutual boundary is a transcontinental split running along the Amazon and Marañón valleys, with the "gray-necked" group in the north and the "red-necked" group in the south. This arrangement has not been confirmed in molecular studies. Instead, Plautz et al. (2009) found *A. nigriceps* (south-central Amazon) to be basal to all the others and *A. nancymaae*, south-western Amazonia, to be part of the northern group (see Fig. 4-11). Fabre et al. (2009) and Chatterjee et al. (2009) found *A. azarae* then *A. nigriceps* to be the basal branches. Further study, with sampling of species such as *A. miconax*, is needed. Unlike the pitheciids, *Aotus* is present in Panama, with endemic species there and in western Colombia (Chocó) west of the Romeral fault zone (Fig. 4-2). Other species have typical Andean ranges in Colombia and in Peru, up to 3,200 m.

Saimiri

Saimiri and one clade of *Cebus* are all present west of the Romeral fault zone and absent in southeastern Brazil (as in *Aotus* and *Ateles*). The other clade of *Cebus* is absent west of the Romeral fault zone and is present in southeastern Brazil. Rylands et al. (1996: 43) referred to

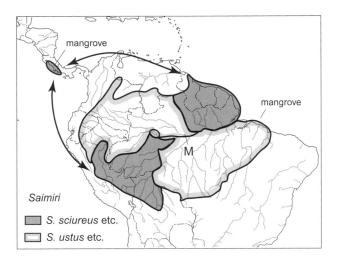

FIGURE 4-12. Distribution of *Saimiri*. M = Manaus.

the lack of *Aotus* in the Atlantic forest as "surprising." The absences of this genus, *Saimiri*, and the *Cebus capucinus* group from southeastern Brazil are probably related to phylogeny and ancestral distribution further west; this would explain the ecological anomalies.

Saimiri (Fig. 4-12), like *Cacajao*, is a specialist occupying floodplain or inundated forests (Rylands et al., 1996: 43). In *Saimiri*, the clades again divide up Amazonia into four quadrants around Manaus, with the local endemic *S. vanzolinii* on the Amazon/Solimões near Tefé, and the final species disjunct in Costa Rica/Panama. D-loop sequences show *S. boliviensis* of southwestern Amazonia basal to the rest, while cytochrome *b* sequences show either the Costa Rica/Panama *S. oerstedii* or *boliviensis* + *oerstedii* as basal (Cropp and Boinski, 2000). Lavergne et al. (2010) studied full-length cytochrome *b* sequences in *Saimiri* and found the slightly different arrangement depicted in Figure 4-12. The northeastern *S. sciureus* is sister to the Costa Rica/Panama species, and their sister is in southwestern Amazonia. This group surrounds the other main clade, comprising the northwestern and southeastern Amazonian species. Chiou et al. (2011) found that *S. boliviensis* is basal to the rest and also confirmed the Panama – Guianas link. (none of the three studies cited here sampled *S. vanzolinii*.) Lavergne et al. (2010) discussed the great disjunction between the northeastern (Guianas) species and the Costa Rica/Panama species and compared this with other disjunctions among northeastern Brazil/Guianas, the Caribbean (and the islands), and Central America (cf. *Ateles* disjunct in north-central Venezuela; *Cebus* on the islands off Venezuela). The gap in *Saimiri*, across northern Venezuela, northern Colombia, and most of Panama, is filled by its sister genus, *Cebus*. The Costa Rica/Panama/Guianas clade may have a disjunct connection with the southwestern Amazonian species across the Gulf of Panama, instead of directly along the Amazon. These two outer connections (lines with arrowheads in Fig. 4-12) both follow belts of instability in the Pacific and Caribbean around the margins of the cratonic core. The cause of this double biogeographic connection of Costa Rica/Panama—with the Pacific and also the eastern Caribbean—is one of the key problems in Central American biogeography. Apart from these breaks, the main breaks in *Saimiri* are along the western Amazon (west of Manaus) and along the eastern Amazon (east of Manaus).

Lavergne et al. (2010) made the important observation that while the breaks among the species and subspecies of *Saimiri* often coincide with rivers, the species are preferentially distributed along watercourses, and so the water *per se* is unlikely to have provided the barrier. Instead, tectonic *change* may have been the trigger. Lavergne et al. (2010) cited floodplain dynamics and ecological heterogeneity as likely factors, and if this is so, the pattern of faulting, established in the Cretaceous, has an underlying significance.

Cebus

Cebus is the sister group of *Saimiri*. The range of *Cebus* completely overlaps that of *Saimiri*, presumably due to ancient range expansion in one or both genera if these evolved in allopatry. By now it is difficult to see where the original break between the two was located. Molecular studies of *Cebus* are still preliminary and as yet the trees are incongruent (Fabre et al., 2009; Chatterjee et al., 2009). The genus comprises two morphological clades, the "capuchin" and "tufted" groups (Groves, 2005; Figs. 4-13 and 4-14), which are interpreted here as the result of original vicariance around the Amazon. Range expansions (arrows in Figs. 4-13 and 4-14) may have led to secondary overlap following initial allopatry around the Amazon. *C. albifrons* of the capuchin group (Ruiz-García et al., 2010) may have expanded its range south, while *C. macrocephalus* of the tufted group in western Amazonia migrated north into Colombia. The last species inhabits lowland to submontane forest, favoring areas dominated by palms, and is a serious agricultural pest in Colombia (IUCN, 2009). This weedy ecology is consistent

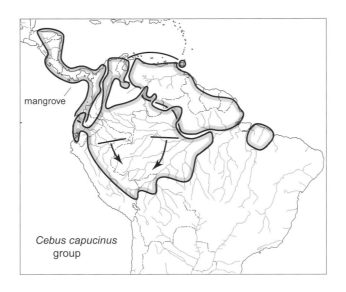

FIGURE 4-13. Distribution of *Cebus capucinus* group. Arrows indicate inferred range expansion of current species.

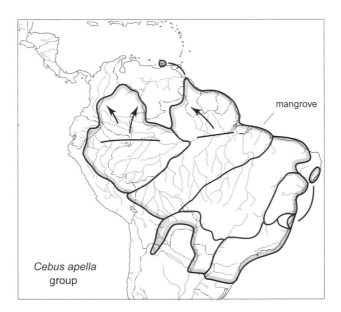

FIGURE 4-14. Distribution of *Cebus apella* group. Arrows indicate inferred range expansion of current species.

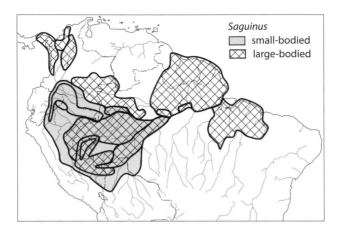

FIGURE 4-15. Distribution of *Saguinus*. There is only one small-bodied species in southwestern Amazonia.

with a range expansion during the incursion and recession of the last (Miocene) inland seas in the area.

The Callitrichines

This clade, the marmosets and tamarins, includes the world's smallest anthropoids and is also notable for its diversity (Figs. 4-15, 4-16, and 4-17). It is the eighth and last widespread New World clade. The phylogeny and distributions have been reviewed by Cortés-Ortiz (2009) and Rylands et al. (2009) (see also Ferrari et al., 2010). Van Roosmalen et al. (2003) showed that the primary molecular distinction in the group is between *Saguinus* of Central America + Amazonia (Fig. 4-15) and the other genera in Amazonia + southeastern Brazil (Fig. 4-16), and so the split resembles that in *Cebus* and the atelines. *Saguinus* occurs on the accreted terranes of Central America west of the Romeral fault zone and is absent in southeastern Brazil, while its sister group, the other callitrichines, is absent west of the Romeral fault zone and present in southeastern Brazil. For callitrichines, rather than proposing a center of origin in the north or south, Rylands et al. (1996: 42) proposed that a callitrichine ancestor was widespread throughout a broad belt from the Amazon basin to the Atlantic. Evolution led to the allopatric differentiation of *Saguinus* in the north, *Callithrix* s.lat. in the south, and at some later period the two overlapped. This general process is accepted

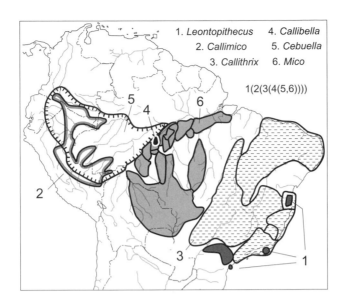

FIGURE 4-16. Distribution of callitrichines (except *Saguinus*).

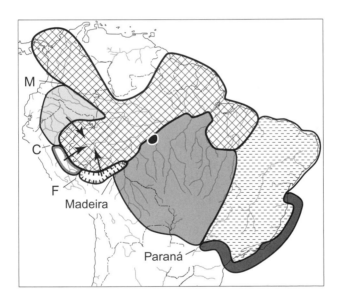

FIGURE 4-17. Diagrammatic reconstruction of original vicariance among callitrichine genera and subgenera (clade symbols as shown in Figs. 4-15 and 4-16). Main phylogenetic/biogeographic breaks occur at M = Macarena (Vaupes) uplift; C = Contaya arch; F = Fitzcarrald arch; Madeira River/lower Madre de Dios River; and Paraná basin. Arrows indicate subsequent range expansion of *Saguinus fuscicollis*, *Callimico goeldii*, and *Cebuella pygmaea*.

here. One of the key pieces of evidence is the geographic relationship between the two main clades.

Within *Saguinus*, the two molecular clades are the "small-bodied" group in western Amazonia and the "large-bodied" group widespread in the Amazon and Panama (Jacobs Cropp et al., 1999; Araripe et al., 2008). The two clades overlap, but the overlap only occurs in southwestern Amazonia and is due to a single species of the small-bodied clade, *S. fuscicollis*; apart from this the two groups and all the species and subspecies within the groups show neat, dovetailing vicariance, and so *S. fuscicollis* is assumed to have expanded its range. The small-bodied *Saguinus* group is centered around the oil-rich basins of eastern Ecuador and adjacent Colombia and Peru (Putumayo and Napo Rivers). In the large-bodied group, the usual break at the Romeral fault zone is expressed only at species level.

In the *Callithrix* group, the phylogenetic sequence of nodes (Fig. 4-16) makes a series of "jumps," with differentiation in the far east (*Leontopithecus* vs. the rest), then far west (*Callimico* vs. the rest), then east again (*Callithrix* s.str. vs. the rest), ending with breaks along the Madeira River (among *Callibella*, *Cebuella*, and *Mico*). There are long distances involved in the "jumps," which are interpreted here as reflecting a sequence of differentiation events in a widespread ancestor rather than a sequence of founder-dispersal events.

In addition to being more or less allopatric, the callitrichine genera occupy most of the vegetation types in their range: *Callimico* occupies primary and secondary forest, preferring sites by streams and in mixed bamboo forest; *Saguinus* and *Mico* favor forest edge, gaps, and secondary forest; *Cebuella* is in inundated and riverine forest. *Callithrix* occurs in mangrove, xerophytic shrubland (caatinga), savanna (cerrado), closed-canopy forest, and disturbed sites such as urban parks and gardens (IUCN, 2009). *Leontopithecus* species occur in mature primary forest, in lower, littoral forest on sandy soil (*restinga*), and in inundated *Tabebuia* (Bignoniaceae) forest on the coast (*caxetal*). They survive well in degraded and secondary forest. The main habitat of the southernmost *Leontopithecus* species is a small island (IUCN, 2009). This diversity in ecological tolerance helps explain how the callitrichine lineage has managed to survive more or less *in situ* through geological time and despite major changes in tectonics, climate, and vegetation.

The interesting problem in callitrichine biogeography is explaining the sympatry of *Cebuella*, *Callimico* (both monospecific; Fig. 4-16), and different *Saguinus* species, in particular the only widespread small-bodied species, *Saguinus fuscicollis* (Fig. 4-15). (*Callimico* often travels and forages with *Saguinus* species; Porter et al., 2007, IUCN, 2009). As with *Callicebus* (cf. pitheciids and *Cebus*), the overlap is restricted to southwestern Amazonia. The distributions of the overlapping clades are not identical, though. For example, *Cebuella* has a western limit at the Shira Mountains in central Peru (cf. *Cacajao*), whereas *Callimico* ranges west of the Shira Mountains (cf. *Aotus*), and these differences can be accommodated in a model of the original allopatry (Fig. 4-17). This assumes that the callitrichine genera and subgenera were all allopatric in origin. In the reconstruction shown here, *Cebuella pygmaea* has expanded from an original home in southern Peru (Madre de Dios basin), *Callimico goeldii* from one north of here west of the Shira Mountains (western Ucayali and Huallaga basins), and *S. fuscicollis* from north of here (Oriente–Marañón basin). Of the 45 callitrichine species, these are the only three in which the ranges show significant overlap. In this reconstruction *Callimico*, *Cebuella*, *Callithrix*, and *Callibella* formed an arc in southwestern Amazonia (along the Ucayali and Madeira Rivers) surrounding *Saguinus* in central southwestern Amazonia and separated from it by the Contaya arch and the Sierra del Divisor. In this way the original differentiation of the groups could have been by simple vicariance.

Other arrangements are possible, but one alternative, the usual break along the Amazon–Solimões, would not account for the diversity of *Saguinus* in southwestern Amazonia (where *Cebuella* and *Callimico* are monospecific). As another alternative, there may have been sympatric evolution. Finally, there may be microallopatry, and *Callimico* in particular has very low densities and discontinuous populations (Porter and Garber, 2004). A detailed map of what are assumed here to be widely sympatric taxa would be of special interest. The IUCN maps are excellent but only give outlines; dot-maps of actual records (including seasonal migrations) are needed for detailed analysis. *Callimico* is small and black, and tends to forage in the low understory (Porter and Garber, 2004). This and its cryptic nature make it the least understood of all platyrrhines. Its morphology represents a classic case of the recombination of characters, as *Callimico* has the mandible and feet of *Callithrix* or *Mico*, the tail and teeth of *Cebus*, and the skull of *Callicebus* (Pitheciidae) (Miranda Ribeiro, 1940; Simpson, 1945: 185). Some aspects are unusual; for example, about one-third of its diet is fungi (jelly fungi), and it is more of a mycophage than any other primate, although the high-elevation *Rhinopithecus* of China often feeds on lichen.

FOSSIL PLATYRRHINES

Platyrrhine fossils are only known from a few scattered locations (Rosenberger et al., 2008). The oldest is from the Late Oligocene of Bolivia (*Branisella*). In platyrrhines, paleontologists have cited the "impressive number of genera and lineages that have deep phylogenetic origins and little changed adaptive histories" (Rosenberger et al., 2008: 73), and they have emphasized the evidence for "morphological stasis," at least over the past 15–20 m.y. This argument is developed here and the absence of earlier fossils accepted as an artifact of sampling.

Fossils from the Antilles

Primates are represented on the Caribbean islands by three Neogene fossil genera on the Greater Antilles. As is so often the case with fossil mammals, the affinities of the three genera are controversial. Rosenberger et al. (2008) identified *Xenothrix* (Jamaica) as a pitheciid, *Paralouatta* (Cuba) as an atelid, and *Antillothrix* (Hispaniola) as a cebid (the latter is "an intriguing mosaic whose primitive characters are consistent with an early origin"; Rosenberger et al., 2011). In contrast, MacPhee and Horovitz (2002) and MacPhee (2005) accepted the three genera as a monophyletic group, closest to *Callicebus* (Pitheciidae). The problem is a classic one in biogeography: A group in a well-defined region is regarded as monophyletic by some authors but as a polyphyletic assemblage by others, even if the convergent characters are striking and confined to the region. In many cases—for example, the Madagascar rats—molecular phylogeny indicates that enigmatic regional groups are monophyletic. But whether the West Indian primates are monophyletic or not, they represent a recombination of characters that are found in all three American families, and so both mainland and West Indian groups may have been derived from a common platyrrhine ancestor. This would have been widespread on the Caribbean plate, on the associated terranes that accreted with western Colombia–Costa Rica, and on the South American mainland.

In the usual interpretation, West Indian groups are assumed to have invaded the region, and both MacPhee (2005) and Rosenberger et al. (2008) accepted this. On the other hand, many endemics on the Greater and Lesser Antilles are basal to pan-American groups (e.g., the butterfly *Antillea* cited in Chapter 1) or even global groups (Heads, 2009a); this suggests that the region has been occupied by a diverse biota since the Mesozoic and well before the current islands or the Caribbean plate itself existed (cf. Doadrio et al., 2009). Taxa endemic to young volcanic islands, such as lizards on the Lesser Antilles, have been dated as older or much older than the volcanic rock itself (Heads, 2011), indicating the persistence of taxa there through tectonic and geographic revolutions.

MacPhee et al. (1995) noted that "In platyrrhine studies there has been a longstanding prejudice against accepting substantial antiquity for [these] island clades." MacPhee argued that although almost all the Antillean fossil evidence of primates is Quaternary, unusual specializations of the taxa imply a lengthy period of occupation. The oldest Antillean fossil is from the Cuban Miocene, earlier than or coeval with the oldest continental atelid (MacPhee, 2005), but, as he stressed,

> it seems unlikely that these first appearances have much bearing on real origins. Rather, it suggests that the fossil basis for interpreting the origin and earliest diversification of "South American" clades during the latest Cretaceous/early Cenozoic is even scantier than generally realized. In particular, the Antillean record strengthens arguments that some crown-group continental lineages are considerably older than fossil evidence currently allows—a point increasingly (if unevenly) supported by molecular studies of many of the same clades. (MacPhee, 2005: 551)

Fleagle (1999) agreed that the Antillean genera, in particular *Xenothrix*, are "quite different from any extant platyrrhines on the mainland," and this is "compatible with a view that the Caribbean primates may have been separated from other platyrrhines for many millions of years" (Fleagle, 1999: 443). The main question concerns how they have been separated. The Caribbean plate and the Atlantic Ocean did not exist before Early Jurassic time, and North America, South America, and Africa were united. In some reconstructions, Yucatán, Florida, Venezuela/Guianas, and Guinea were juxtaposed (Meschede and Frisch, 2002; Pindell et al., 2006; Pindell and Kennan, 2009). Rifting developed from the mid-Jurassic to Cretaceous along a single spreading center between Mexico/North America and South America/Africa, and this opened the North Atlantic and a seaway between North and South America. At the same time, more localized rifting occurred along a separate spreading axis in the Gulf of Mexico. Rifting in the region is associated with major transcurrent movement on transform faults and also large-scale volcanism. Eruption of mid-Cretaceous flood basalts produced the Caribbean plate and the modern tectonic arrangement, with Caribbean coasts following plate boundaries. The history is complex. Before the opening of the Atlantic, Cuban basement terranes were adjacent to Guyana/Trinidad (160 Ma) (Pindell et al., 2006; Pindell and Kennan, 2009). In contrast,

Cuban arc terranes developed in the Pacific, west of the Galápagos in the Cretaceous (Mann et al., 2007; Pindell and Kennan, 2009), and these were only accreted, along with ophiolites, to the Cuban basement later. Thus the Cuban fold belt and the Greater Antilles region in general have a hybrid origin, as with the larger fold-belt islands of the west Pacific (New Zealand, New Caledonia, New Guinea, Taiwan) and Indonesia. The same ambiguity seen in the tectonics of the Antilles appears in the biogeography, as the flora and the fauna have strong relationships with both the Pacific in the west and with the Guianas, northeastern Brazil, and Africa in the east.

Fossils from Colombia and Patagonia

Apart from the Antillean groups, most of the remaining fossil platyrrhines come from Miocene beds at La Venta, Colombia, dated about 13 Ma. These include one or two species in each subfamily of Cebidae (Callitrichinae and Cebinae), one relative of *Aotus*, one of *Alouatta*, and four early members of the Pitheciidae (Delson and Tattersall, 2002). All the main clades in the modern fauna are represented in the Miocene by more or less clear precursors. Rosenberger (2002: 152) described the La Venta *Neosaimiri* and *Cebupithecia* as "remarkably similar to living forms." Rosenberger et al. (2008) listed these, together with *Mohanamico*, *Miocallicebus*, *Stirtonia*, and *Aotus* from La Venta, and *Proteropithecus* from the Miocene of Neuquén, Argentina, as "living fossils," that is, "forms closely related to modern genera" (p. 77).

In the Miocene of Patagonia, *Dolichocebus* may be close to *Saimiri*, *Tremacebus* close to *Aotus* (Rosenberger, 1984; Rosenberger et al., 2008). Kay et al. (2008) disputed this and interpreted *Dolichocebus*, *Tremacebus*, and others as "stem platyrrhines." In any case, the biogeographic interest lies in the fact that while platyrrhines have been found (as fossils) south to 52°, near the Straits of Magellan (an excellent map of fossil localities is given in Rosenberger et al., 2008: Fig. 4.4a), there is no record, living or fossil, north of Mexico. Here platyrrhines are replaced by two other groups, both extinct, the adapiforms and omomyiforms. These are abundant in U.S. beds but have never been found in the Latin American range of platyrrhines, and so the break can be interpreted as simple phylogenetic allopatry (cf. Chapter 3).

The following discussion reviews tectonics and biogeography in different regions of tropical America.

ROMERAL FAULT ZONE (WESTERN COLOMBIA–ECUADOR–PERU)

South America comprises the stable Brazilian and Guiana cratons, the very unstable Andean orogen, and the basins between them, which have intermediate levels of stability (Fig. 4-2). The Andean orogen is complex, though (as with Cuba), and the western Andes of Colombia and Ecuador had a separate history from the rest of South America before they docked with the mainland. At the suture where this took place there is a major biogeographic boundary for primates, seen between Central America/western Colombia (Chocó) on one hand, and eastern Colombia, Venezuela, and Amazonia on the other (Goldani et al., 2006). In *Alouatta*, *Ateles*, *Saimiri*, and *Saguinus*, Chatterjee et al. (2009: 10) noted that "One consistent feature is the separation of Central American from South American groups." The formation of the Panama isthmus is often thought to have permitted dispersal and range expansion between Central and South America (the Great American Biotic Interchange model), but it cannot explain a *break* in the region. The break in both geology and biology is aligned with the Romeral fault zone, running along the western flank of the Colombian Central Cordillera. The fault zone marks the tectonic boundary separating allochthonous, accreted terranes in the west (the Western Cordillera, Chocó, Panama, Central America, etc.) from continental basement to the east (eastern slopes of the Central Cordillera, Eastern Cordillera, etc.). The allochthonous terranes formed in the Pacific and included island arcs and their products. Together with the Caribbean plateau the terranes collided with the South American margin and were incorporated into the continent, beginning in the Late Cretaceous (Campanian–Maastrichtian, 75–65 Ma; Kerr et al., 1997; Chicangana, 2005; Luzieux et al., 2006; Vallejo et al., 2006). The zones of terrane accretion and ophiolites are reflected in the primate biogeography here, and the same phenomenon will also be discussed for primates in Sumatra, Borneo, and Sulawesi (Chapter 5). These islands, as with Colombia and Ecuador, are all tectonic and biogeographic composites.

THE ANDES FROM CENTRAL BOLIVIA TO VENEZUELA

This northern sector of the Andes is the main region of endemism and diversity for many montane groups, while the eastern foothills and adjacent lowlands are important for primates. The rise of the Andes has been proposed as the cause of differentiation between eastern and western primate faunas (Lehman and Fleagle, 2006). (For early work on vicariance caused by Andean uplift, see Eigenmann, 1920, discussed by Lundberg et al., 1998; and Eigenmann, 1921, quoted in Heads, 2005c).

While some groups have been separated by uplift, other groups in the main Andean cordilleras such as *Oreonax* (*Lagothrix* s.lat.), at 1500–2,700 m in northeastern Peru, have instead been elevated during the orogeny. The woolly pelage of *Oreonax* is also found in the low-land species of *Lagothrix* and may have been a preadaptation enabling populations in the rising Peruvian Andes to survive the uplift.

Biogeographers sometimes regard the Andean orogeny as a Miocene event, but the Neogene (Quechua) phase of uplift is just the latest of three main orogenic phases. These also involved deformation in other parts of the continent. The first ("Peruvian") phase of uplift took place at 90–75 Ma (Late Cretaceous), although evidence is accumulating for an onset of compressional deformation in the Aptian (Early Cretaceous) (Cobbold et al., 2007). The Peruvian phase events in particular would have been significant for lowland/back mangrove taxa as shallow seas and embayments in the pre-Andean landscape were transformed into deltas, alluvial systems, and, eventually, dry land. Hoorn et al. (2010: Fig. 1) showed "mountains/hills" of the proto-Andes already present in Peru, Ecuador, and Colombia at the Cretaceous/Cenozoic boundary. The second (Incaic) phase of Andean orogeny took place at 50–40 Ma (Middle Eocene), and in addition to reverse faulting, strike-slip faulting was common.

The pre-Andean basins and old cores of land would have already had a well-differentiated flora and fauna before the rise of the Andes. With Cretaceous orogeny, there would have been concomitant extinction and evolution as whatever survived of the original lowland and coastal biota was transformed into the modern Andean biota by passive uplift (Croizat, 1975; Lynch, 1986; Lundberg et al., 1998; Ribas et al., 2007; Thomas et al., 2008). Orogenic uplift of populations may be accompanied by speciation, as in *Oreonax*, but this is not always the case, as the following widespread primates indicate.

> *Cebus capucinus*: mangrove, deciduous forest, secondary forest, palm forest, and humid forest, up to 2,000 m (IUCN, 2009),
>
> *Alouatta seniculus*: mangrove (northern Colombia), drier forest, várzea flooded forest, rainforest, up to 3,200 m (Ecuadorean Andes). Related vicariants in the Amazon lowlands include *A. puruensis* of the Purus and Madeira regions, more common in flooded than in unflooded forest (IUCN, 2009), and

Alouatta pigra: from mangrove (Yucatán–Belize) up to 3,350 m (Guatemala) (Baumgarten and Williamson, 2007).

The great ecological range of these species indicates a flexibility and a weediness that enable them to survive great disturbance, whether this is caused by orogeny, human activity, or periodic flooding, which may be semidiurnal (in mangrove) or annual (in inland flooded forest).

SOUTHEASTERN BRAZIL

The coastal ranges of eastern Brazil reach 2,800 m elevation (in the Serra da Mantiqueira; Fig. 4-2), and the many endemics in the forests there include monkeys such as *Leontopithecus* and *Brachyteles*. The ranges lie between the coast and the Paraná/São Francisco valleys. The area, together with the coastal shelf that is currently submerged by the sea, is a rifted continental margin, where Jurassic rifting culminated in the Early Cretaceous opening of the Atlantic (Cobbold et al., 2007). This process was not a simple one, and there was a long phase of deformation before complete rifting of the continents finally occurred. The details of the magmatism, uplift, scarp retreat, repeated marine incursions, and rifting would have provided many opportunities for biological evolution.

Cretaceous Basins and Lakes

The Cretaceous history of the continental margin of eastern Brazil is preserved in the Santos, Campos, and Espírito Santo basins (Mohriak et al., 2008). These began to form on land with rifting and extension in the Early Cretaceous (Barremian; Fig. 4-18) and filled with alluvial and lacustrine material. They now lie out to sea off the eastern Brazilian coast in a belt along the continental margin that has been extended and thinned. The basins are most famous for their giant oil fields. The alluvial and lacustrine sediments, the source of the petroleum, became flooded more and more with episodic marine incursions; this was the first appearance of the east Brazil–West Africa basin. Periodic influxes of saltwater covered the Barremian freshwater sediments under layers of Aptian salt (evaporite). The rifting that began at 135 Ma was complete by 120 Ma, when it was converted into drifting and the opening of the Atlantic. The rifting was a complex process and, as discussed next, there was substantial uplift along the margins of the rift.

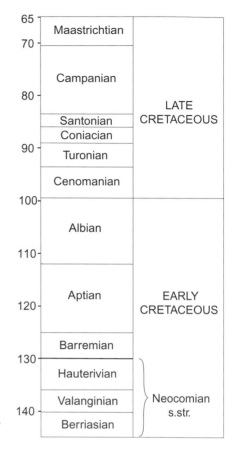

FIGURE 4-18. The ages of the Cretaceous period.

Rift Shoulder Uplift in the Early Cretaceous

It may not be intuitive for uplift to occur at the sides of a rift—a zone of extension, not compression—but this occurs. Subsidence in a rift can cause elastic rebound along its margins, or "shoulders," leading to the uplift of major mountain ranges there. Cretaceous rift shoulder uplift associated with the formation of the Atlantic produced the topography and drainage patterns of modern eastern Brazil. Ribeiro (2006) gave a detailed analysis showing that the rift shoulder uplift also produced many of the biogeographic patterns seen in the modern freshwater fish fauna. These include distribution breaks as well as elevational anomalies.

In addition to the direct significance of Brazil–Africa rifting for vicariance between the two continents, Ribeiro (2006) observed that there was associated vicariance within Brazil; coastal clades of the rifted margin are often basal to clades that are widespread in inland South America. This pattern is seen in many groups. In primates, *Brachyteles* is endemic to southeastern Brazil and is sister to *Lagothrix/Oreonax*, widespread through the central and western Amazon (Figs. 4-9 and 4-10). *Leontopithecus* of coastal southeastern Brazil is sister to a widespread Amazonian/eastern Brazil clade (*Callimico/Callithrix*) (Fig. 4-16).

In this pattern the southeastern Brazilian region is basal and so is often taken to represent a center of origin, with taxa dispersing from there (e.g., Velazco and Patterson, 2008, on phyllostomid bats). Instead, as Ribeiro (2006) emphasized, uplift along the rift shoulder would have caused vicariance between coastal and inland clades. There is no need for any dispersal, as the phylogeny, distribution, and elevation of communities can all be accounted for by the direct effect of tectonics.

The margin of southeastern Brazil was reactivated in three main phases (Ribeiro, 2006), beginning with the events of the Barremian–Aptian. The three phases seem to have been separated by quieter intervals that coincide with the Peruvian (90–75 Ma), Incaic (50–40 Ma), and Quechuan (25–0 Ma) phases of Andean orogeny. During these times, there was more rapid convergence at the Andean margin. This alternation between tectonic activity in the Andes and at the Atlantic coast may be related to the geographic "jumps" referred to above in the phylogeny of callitrichines, from east to west and then back again. The relationship is complex, though. Mohriak et al. (2008) noted that the tectonic and magmatic episodes in the offshore Santos and Campos basins at 100–90 Ma are coeval with tectonic and magmatic episodes of the Mirano diastrophism of southwestern South America, the first phase of mid-Cretaceous deformation. They also compared the Eocene orogeny in the Andes with coeval magmatism in eastern Brazil. In any case, there have been phases of deformation in both the Andes and eastern Brazil, including through the critical time of the mid-Cretaceous.

Scarp Retreat

The uplift in the rift shoulder originated 100 km off the current coast along the landward margin of the offshore basins. Since then it has retreated with erosion some 200 km inland from the original hinge to the line of the current Serra da Mantiqueira (Mohriak et al., 2008).

During parallel scarp retreat, erosion of a landscape takes place from the side, not from the surface. A scarp, often with distinctive topography, soil, climate, and vegetation, retreats horizontally, along with its biota. The evidence for this process can be seen in the savanna regions of Africa and Brazil. In this way, the seaward edge of the continent was eventually drowned, but its biota, having survived phases of marine inundation, may have evaded drowning by surviving on the new mountains and migrating inland on the retreating scarp.

The Barremian–Aptian, with both rifting and uplift, is a phase of history 18 m.y. long that may be crucial for primate evolution. What was happening to the primates of these alluvial, lacustrine environments as their habitats were undergoing repeated phases of marine inundation in some areas, orogeny in others? The platyrrhines as a group and also some of the main subgroups could have diverged at this time. There may also have been phases of range expansion in any weedy forms of the rifting margin, and also hybridism, as is typical of disturbed sites. Finally, the paleogeographic history of the continental margin may be related to disjunctions along the Atlantic coast in groups such as *Leontopithecus* (Fig. 4-16). These suggest the loss of former populations in regions that became completely submerged.

In post-Albian time, in the Late Cretaceous (from 100 Ma), the marginal belt of the continent across which the scarp has retreated has been inundated and affected by seaward gravity gliding of thick layers of Early Cretaceous (Aptian) salt and associated Barremian and Tertiary sediments, 60 to 160 km seaward (from the margin to the center of the basins). This slide (with detachment at the salt layers) was caused by the uplift. The onshore erosional unloading resulted in further flexural uplift of the continental margin and could have led to a further phase of biological evolution.

Late Cretaceous/Paleogene Events

The mountains of southeastern Brazil include the different maritime ranges grouped as the Serra do Mar (2,200 m) and, further inland, the belt formed by the Serra da Mantiqueira (2,800 m) and Serra do Espinhaço (2,000 m). The two mountain belts are separated by the Paraíba do Sul valley and other depressions, which mark a large feature, the "continental rift" (= Serra do Mar rift system) of southeastern Brazil (Ribeiro, 2006).

In the Late Cretaceous (85 Ma), alkaline volcanism began in the region, and in the Paleocene the Serra do Mar was uplifted west of the Santos fault; this runs parallel with the coast, just off it, and passes through São Sabastião Island. Subsequently, the scarp retreated westward, in the same way as the earlier uplift. Also in the Paleogene, extension reactivated the old continental rift along the Paraíba do Sul valley, separating the seaward mountain ranges from the inland belt (Ribeiro, 2006; Tello Saenz et al., 2003).

Cobbold et al. (2001) attributed Late Cretaceous and Cenozoic reactivation of older structures on the southeastern Brazil rifted margin to local activity and also far-field stresses caused by Andean deformation. These rift- and orogeny-related events of the Late Cretaceous may have caused further divergence between southeastern Brazil clades and their inland sister groups, as occurred earlier in the Early Cretaceous.

Mohriak et al. (2008) emphasized that following continental breakup, the continental margin did not behave as a typical passive margin. Instead, both tensional and compressional stress events operated after breakup, and this tectonic complexity is typical of belts that are reflected in phylogeny and biogeography. (In geology, "strain" is the deformation caused by "stress"—force per unit area. Shortening or compaction is the strain counterpart of compressional stress; extension is the counterpart of tensional stress.) As noted, extension led to the reactivation of the continental rift and the development of several small rift basins along it, between the Serra da Mantiqueira and Serra do Mar ranges. The largest of these is the northeast/southwest trending Taubaté Basin, drained by the Paraíba do Sul River (Fig. 4-2), and others are occupied by the cities of São Paulo and Curitiba.

Padilha et al. (1991) suggested a transtensional model for the Taubaté Basin, while Mohriak et al. (2008) cited transpression, and both authors accepted strike-slip rifting along the basin. Strike-slip movement in this mobile zone may have been important in the evolution of *Leontopithecus* and *Brachyteles* (cf. the break between the two species of the latter; Fig. 4-10).

The aplocheiloid fishes were discussed in Chapter 2 and their main break attributed to the opening of the Mozambique Channel (Fig. 2-3). One of the diverse American clades is the tribe Cynolebiasini of family Rivulidae. Three lineages (*Nematolebias*, *Xenurolebias*, and *Ophthalmolebias*) are endemic to the eastern Brazilian coastal plains and form a basal grade in the tribe. The rest of the genera occur through the Brazilian shield area. Costa (2010) wrote: "The basal cynolebiasine node is hypothesized to be derived from an old vicariance event occurring just after the separation of South America from Africa, when the terrains at the passive margin of the South American plate were isolated from the remaining interior areas . . . the diversification of the tribe Cynolebiasini in north-eastern South America was first caused by vicariance events in the Paraná–Urucuia–São Francisco area." This is a good example of evolution during rift shoulder uplift.

The repeated bouts of differentiation that occurred in eastern Brazil at the base of the cynolebiasine phylogeny reflect the complex tectonics of the mid-Cretaceous rifting and orogeny.

THE BREAK BETWEEN SOUTH-ASTERN BRAZIL AND AMAZON: PARANÁ, SÃO FRANCISCO, AND PARNAÍBA BASINS

The Atlantic forest of Brazil is "surprisingly rich" in primates (Peres and Janson, 1999), although several main clades are "conspicuously absent." Endemics in southeastern Brazil include *Leontopithecus*, *Brachyteles*, and the *Callicebus personatus* species group; absentees include *Ateles*, *Aotus*, *Saimiri*, *Saguinus*, and the *Cebus capucinus* group. Rift shoulder uplift has been important in some groups, such as *Leontopithecus*, but in others the break between southeastern Brazil and the Amazon basin groups seems to have occurred further inland. The breaks involve the Paraná, São Francisco, and Parnaíba basins (Fig. 4-2) and volcanism around these.

The main part of the Paraná basin is filled with Paleozoic–Mesozoic sediments and capped with Early Cretaceous (134–129 Ma) tholeiitic flood basalts up to 1,700 m thick. These are equivalent to the Etendeka basalts in Namibia and the two are connected by the Rio Grande Rise–Tristan–Walvis Ridge high. Together they make up one of the largest continental igneous provinces, and the magmatism was linked to the northward opening of the South Atlantic Ocean.

In Brazil and Paraguay, alkaline magmatism was widespread around the margins of the Paraná basin, both during (135 Ma) and after (85–55 Ma) the flood basalts. Late-phase eruptions, for example, along the Alto Paranaiba arch between the Paraná and San Franciscan basins, were contemporaneous with Late Cretaceous, kilometer-scale uplift across southern Brazil and Paraguay. Thompson et al. (1998) and Gibson et al. (1999) supported a mantle plume hotspot model for the magmatism, whereas Comin-Chiaramonti et al. (2007) found the hotspot model "not compelling" and supported rifting as an explanation. In any case, the timing and tectonic style of Late Cretaceous events in Brazil are similar to those in the Late Cretaceous central African rift system.

In the northeastern Paraná basin, the Bauru basin has filled with Late Cretaceous strata (Aptian–Maastrichtian) following Early Cretaceous Serra Geral magmatism at 137 Ma. Sediments indicate alluvial plains and braided rivers; semi-arid to arid conditions alternated with periods of heavy rain. The vertebrate fossils include a remarkable diversity of tetrapods, with both "austral Gondwanan" and "boreal Gondwanan" groups of mammals (Candeiro et al., 2006).

THE AMAZON BASIN

The southern and northern Atlantic opened with mid-Cretaceous sea-floor spreading on the mid-ocean ridges, and at the same time in the central Atlantic region West Africa was pulling away in a right-lateral sense from South America along transform faults. (Movement is right-lateral or dextral if the block on the other side of the fault from an observer moves to the right.) At the same time, strong right-lateral wrenching acted along the Amazon basin. The basin occupies an old Palaeozoic rift between the Guiana and Brazil cratons. The wrenching in the Amazon caused strike-slip faulting, reverse faulting, and large folds (Cobbold et al., 2007). Right-lateral rifting occurred in the Amazon basin east of Manaus (and the Purus arch) and along a separate fault system in the Solimões basin, west of Manaus. The plate configurations and general directions of motion around continental South America have not changed greatly since these mid-Cretaceous events (Cobbold et al., 2007), and underlying patterns in primate distribution may also have been established during this time.

After the rifting along the Amazon, the first non-deformed strata are those of the mid-Cretaceous (Albian–Cenomanian) Alter do Chão Formation. West of Manaus, this occurs throughout the Solimões basin in subsurface strata. East of Manaus, surface exposure is widespread in an elongate belt along the Amazon extending to the coast (Rossetti and Netto, 2006; Rossetti and Toledo, 2007). At a site 50 km east of Manaus, most of the sandstones have been deposited in a high-energy braided to anastomosed fluvial systems, but some formed in muddy, lacustrine–deltaic environments. The authors propose a deltaic system that prograded into a marine-influenced basin. They compared this with other marine transgressions at this time in the Subandean basins, central Africa, and the Western Interior basin of the United States. Sea levels reached a peak in the mid-Cretaceous (120–80 Ma), about 150 m above present levels, and this led to widespread continental flooding. Since that time, global (eustatic) sea levels have fallen and inland seas have regressed

(Müller et al., 2008). The Cretaceous flooding would have been a critical opportunity for animals able to survive in mangrove with high levels of disturbance, as it would have permitted great range extension.

In the Alter do Chão Formation, marine facies have also been described in subsurface strata, west of the Purus arch (by the Purus River, 200 km west of Manaus) (Mapes et al., 2006). These suggest a Cretaceous cross-continent connection in which the Amazon flowed west. As result of Andean orogeny, the Amazon basin was split in two (before mid-Miocene), with westward drainage (west of Purus) and eastward drainage (east of Purus). When the western basin was filled with sufficient sediment, perhaps in the late Miocene, modern, east-directed drainage began from the Andes to the Atlantic.

The history of the Amazon and Congo Rivers may reflect each other, if they both flowed away from the America–Africa margin (the Atlantic region) in the Early Cretaceous and drained into the Pacific and Indian Oceans. This was caused or reinforced by uplift of the rift shoulders in Cameroon–Angola and Brazil. With uplift in the Andes and in East Africa, both rivers changed direction in the Cenozoic and began to drain into the Atlantic.

Paleoarches in Amazonia

Several old structural highs or arches in the basement traverse the Amazon basin and delimit subbasins. Their involvement with foreland basin dynamics—caused by Andean uplift affecting the adjacent foreland basins (Fig. 4-2)—has imparted a tectonic imprint on the floodplain geomorphology (Mertes et al., 1996). Many clades reflect this basin evolution, but it is ichthyologists who have integrated molecular phylogenies most closely with evolution of the Amazon paleoarches (Lundberg et al., 1998). Activity on these may have led to differentiation of sections of the river (e.g., Hubert and Renno, 2006). From west to east, the arches include the Vaupés (by the Macarena and Chiribiquete Mountains), the Iquitos (near the city of Iquitos), the Carauari (at the Jutaí River), the Purus, and the Gurupá (delimiting the Marajó basin). In piranhas, Hubert et al. (2007) found that tectonic boundaries at the paleoarches caused by Andean foreland basin dynamics have led to five main vicariant events at the Vaupés, Gurupá, Purus, Carauari, and Iquitos arches, in turn. Sympatry has been caused by range expansion and/or lineage duplication by sympatric evolution. Most of the extant species predate the Pleistocene.

The Iquitos arch region, and southwestern Amazonia in general, is rich in primates. Some endemics here, such as *Saimiri boliviensis* (Fig. 4-12), are sister to widespread Amazonian taxa; others, including *Callimico* (Fig. 4-16), are sister to widespread Amazonian/southeastern Brazil taxa. Similar patterns occur in birds. For example, the *Mionectes oleagineus* group of tyrant flycatchers comprises one clade in southwestern Amazonia (*M. macconnelli*) basal in a widespread clade of central America, Amazonia, and southeastern Brazil (Miller et al., 2008). The primary breaks in these widespread groups are interpreted here as Cretaceous events. Southwestern Amazonia is also notable as an area of overlap, but at least in the case of certain species there this may be the result of later, Miocene events.

Miocene Marine Incursions in Western Amazonia: Lake Pebas

A long-lived Miocene wetland system or "marine megalake," Lake Pebas, occurred around the Iquitos area repeating the mid-Cretaceous (Albian–Cenomanian) marine incursions, and may have connected eastern Peru with the Caribbean via the Llanos basin. There may also have been a connection with the South Atlantic (Gingras et al., 2002); Miocene deposits in Madre de Dios catchment, south of the Fitcarrald arch in southern Peru, indicate tide-dominated, estuarine conditions similar to those in the Pebas lake, and so seaway connections with the inland Paraná Sea to the southeast are possible (Hovikoski et al., 2005; Antoine et al., 2007).

The Pebas system was lacustrine but with occasional salinity increases to mesohaline, and more brackish-water faunas reflect marine incursion (Gingras et al., 2002; Vonhof et al., 2003). One Miocene palynological assemblage from the Colombian Amazon (Apaporis) indicated vegetation dominated by *Rhizophora* mangroves and *Mauritia* palm (Hoorn, 2006). Occasional decreases of *Rhizophora* and increases in *Mauritia* indicate a fluctuating coastline; marine incursions were related to global sea-level rises (much lower than that of the Cretaceous) and also subsidence. The Miocene emergence of the Iquitos arch in the foreland basin induced the retreat of the Pebas "marine megalake" near the Iquitos area (Roddaz et al., 2005).

Neotectonics in Amazonia

Costa et al. (1996) showed the Madeira fault as a long zone of dextral strike-slip. The Madeira River is a major biogeographic boundary, and the history of the fault is of special interest. Saadi et al. (2002) described the conspicuous structural control of Amazonian rivers (Amazon, Madeira, Juruá, Negro, etc.) by faults. Both the cited studies were concerned with the Neogene history of faults, but the tectonic regime that the faults reflect (Fig. 4-3) was established in the Cretaceous. Many of the individual faults are older structures that have been reactivated at different times. In the mid-Cretaceous, sea-floor spreading in the Atlantic and Andean deformation produced a new tectonic regime through Amazonia. This new regime caused the faulting, the faults caused the rivers, and the rivers separated the primate taxa. The trigger for differentiation was not the rivers that keep the taxa separate now, but a *new* river or physiographic change at a critical phase in primate history, and the geographic break has persisted as an ecological and phylogenetic memory. The effects of this new fault regime in western Amazonia were complex, as they were imposed on a low, pre-Andean landscape at the same time as the high mid-Cretacous sea levels.

WESTERN AMAZONIA AND THE ANDEAN FOOTHILLS: FORELAND BASIN DYNAMICS, GIANT OIL-FIELDS, AND MAXIMUM PRIMATE DIVERSITY

As shown in Fig. 4-2, South America comprises stable cratons in the east, the unstable Andean orogen in the west, and an intermediate region, a fold and thrust belt with adjacent foreland basins, where there have been intermediate levels of disturbance. The main region of foreland basins is in western Amazonia and includes southeastern Colombia, eastern Ecuador, eastern Peru, and western Brazil (Fig. 4-19). This is where the primate communities with the highest species diversity in America are found (Voss and Emmons, 1996).

The land in Peruvian Amazonia is at low elevation (<500 m) throughout, but the western part has rougher topography related to Andean tectonics, while the northeastern part is flat due to the stability of the craton. The lowlands of eastern Peru/Ecuador include endemism in primates (e.g., species of *Pithecia*, *Callicebus*, and *Aotus*), other placentals, and also marsupials such as *Marmosa andersoni*, *M. quichua*, and *M. rubra* (Didelphidae; IUCN, 2009). This diversity is associated with an area of Cretaceous coastlines that disappeared as such when the old topography of eastern Peru was incorporated into the Andes and their foothills. That revolution in the landscape could have been accompanied by the last major evolution in the primates of the region.

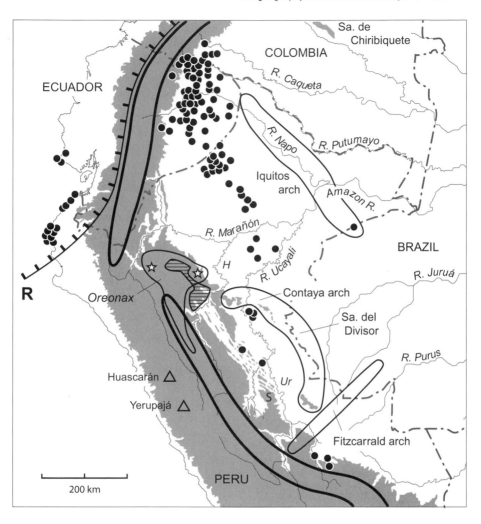

FIGURE 4-19. Western Amazonia, showing aspects of tectonics and biogeography. Andean orogen (500 m altitude) in gray; oil fields as dots (from Schenk et al., 1997; Higley, 2001); pre-Andean Mesozoic axial highs (Marañón geanticline) in bold lines; R = Romeral fault zone; paleoarches (Iquitos, Contaya, Sierra del Divisor, Fitzcarrald) are also indicated. Huascarán (6,768 m) and Yerupajá are the highest mountains in Peru and are both in the Western Cordillera. H = Huallaga River, S = Shira mountains, Ur = Urubamba River. Horizontal hatching = the monkey *Callicebus oenanthe*; stars = the hummingbird *Loddigesia*.

In Peru and Ecuador, the Andes comprise two parallel cordilleras. The Western Cordillera is the higher; in Peru its summit ridge is at ca. 5,000 m and reaches 6,768 m at Huascarán, formed of Cenozoic granite. This Western Cordillera is dry, and there are no primates. In Peru the Eastern Cordillera has substantial areas above 3,000 m, but only a few summits are above 4,000 m. East of the Andean cordilleras is the "intermediate" region comprising the Subandean foothills and the adjacent lowlands, with high biodiversity in many groups including primates. East of here, on the craton, the tectonics are much more stable.

Lundberg et al. (1998) related foreland basin dynamics to possible vicariance events in freshwater fish evolution. They criticized the usual reliance on late Cenozoic or even Pleistocene events in biogeographic models and wrote: "The long and complex history of South America's landscape and river systems must be considered" (p. 13). They continued: "It is now known that the Andes were built by compressional tectonics during the last ~90 Myr or even longer. It is therefore simplistic to view Andean vicariance as a singular event occurring with the Miocene uplift" (p. 16). Lundberg et al. (1998) concluded that ". . . Pliocene and Pleiostocene Earth history events had nothing to do with creating the higher taxonomic diversity of Neotropical fishes" (p. 43) and cited "Some higher endemic clades extending back into the Cretaceous" (p. 13). Likewise, Antonelli et al. (2010) reviewed molecular studies of 50 Amazonian groups of tetrapods and concluded that "The origin of most clades clearly predates the Pleistocene by a considerable margin."

The Pre-Andean Phase of Back-arc Basin Formation

The Andean orogeny in the broadest sense can be dated to the onset of subduction beneath western South America, in the Middle Jurassic. Uplift did not occur until the Cretaceous; instead, there was extension, normal faulting, and back-arc basin formation (behind the magmatic arc, with respect to the subducting plate, and toward the mainland). The generation of a volcanic arc at a subduction zone is understandable, but the development of extension in a zone of overall plate convergence seems counterintuitive. One proposed mechanism involves the sinking backward or "roll-back" of the subducting slab and the trench. This induces back-arc extension in the crust above and may have led to the separation of Japan from the mainland, for example. Slab roll-back is an intriguing idea, but whatever the tectonic mechanism is, the development of extension at a convergence zone is of great biogeographic interest. Areas of extension, convergence, and, especially, both together, are usually regions of great allopatric differentiation. The system of pre-Andean back-arc basins remained active until the mid-Cretaceous, when regional tectonic processes began inverting the basins, forming new foreland basins, and, in the west, producing the main range of the orogen. Populations in the pre-Andean region would have been pulled apart by back-arc spreading (causing vicariance), then consolidated and exposed to further vicariance (during basin inversion).

Along the central axis of the orogen, the uplift could have led to overlap of populations and also further vicariance. Although the overall regime is one of shortening and thickening, different processes can also cause local areas of extension, normal faulting, and subsidence within the growing mountain range (Molnar, 1990: Fig. 9.8). This will have some biological significance, but for lowland groups may be less important than the earlier phase of large-scale extension in which the back-arc basins formed.

Andean Orogeny and Development of Foreland Basin Tectonics

Andean uplift is the result of shortening and thickening of the crust. This was not caused simply by subduction—the uplift occurs far inland of the trench and subduction began well before orogeny—but by the westward movement of South America during the opening of the Atlantic (and also terrane accretion from the west). If an overriding plate, here the South American plate, is stationary, rifting and a back-arc basin can develop behind the volcanic arc. But if the overriding plate begins to move toward the subducting plate (Fig. 4-19), contraction develops in the back-arc zone (van der Pluijm and Marshak, 2004). The Albian onset of convergent tectonism in the central Andes was caused by the Aptian onset of North Africa/South America separation in the equatorial Atlantic (Pindell and Tabbutt, 1995). Since then the Atlantic has continued to open and the Brazilian craton has been thrust westward above the subducting Nazca plate of the East Pacific. With the Cretaceous revolution, normal (extensional) faulting in what would become the Andean region gave way to reverse and thrust faulting. Uplift occurred, and the earlier basins were inverted (biologists might understand the process of basin inversion as *eversion*). External horizontal, rather than isostatic vertical forces, are required for inversion—isostatic rebound by itself will not cause inversion. In most cases basin inversion is caused by transpression, a combination of compression and strike-slip (Lowell, 1995).

As the South American plate was thrust westward, the Nazca (East Pacific) plate was also being thrust eastward under it with seafloor spreading in the Pacific. The cordilleras of the Andes developed on top of the convergent margin and along it, as a sedimentary pile intruded by a magmatic arc. The pile is thrust both west (by the South American plate) and east (by the Nazca plate) as the two plates move beneath it. Along the eastern margin of the Andes, the thrusts have produced the Subandean foothills, a belt of folds and thrust faults. In addition, the increasing load of the Andes has caused downflexures and elastic rebound further inland, across the foreland basin. The foothills and the adjacent basins comprise an extensive foreland basin system that has been active since the first mid-Cretaceous inversion of earlier rift faults.

Other foreland basin systems relevant to primate evolution include the Western Interior basin of the United States, the Karoo, Ganges, and Sichuan basins, and perhaps Makassar Strait. A foreland basin forms on continental crust in front of a mountain belt by loading, downflexing, and elastic rebound, not by extension as in a rifted basin. The Subandean foreland basin occurs behind the arc with respect to the subducting plate, but the tectonics are dominated by compression and lithospheric flexure (it is a retro-arc basin), rather than extension as in a back-arc basin. Foreland basins typically form a sequence of four depocenters; reading away from the orogen/fold-thrust belt, these are termed wedgetop, foredeep, forebulge, and back-bulge (DeCelles and Giles, 1996). The sequence has been recognized in the Subandean basins; for example, the Iquitos arch has been identified as a forebulge.

The Transition from Pre-Andean to Modern Geography

Before the rise of the Andes, the presence of an old axial swell is indicated by a belt in the Eastern Cordillera that does not contain Cretaceous sediments. This structural high, the Marañón geanticline, may represent an area of land extending north to the bend in the Marañón River (Fig. 4-19). North of the break at the river a similar pre-Andean core occurs in Ecuador (Jaillard, 1994). Pindell and Tabbutt (1995) mapped the Marañón geanticline through mid-Jurassic and Cretaceous, along with the basins that developed on its eastern and western flanks. With Cretaceous orogeny, this Andean precursor, along with its flora and fauna, was incorporated into the Eastern Cordillera of the modern Andes.

In the Cretaceous, shallow basins developed in eastern Peru between the Marañón massif in the west and the Brazilian shield in the east. At the beginning of the Late Cretaceous, these were transformed from shallow sea into a mixed marine–terrestrial environment and finally into dry land. This initial phase of Andean orogeny involved a tectonic change from extension to transpression that led to the inversion of prior topography and of individual structures such as rift faults. During the process, formation of anticline structures produced seals which trapped petroleum as it migrated upward. This anticline formation also led to physiographic changes and possibly evolution in the local primates.

Six maximum flooding events are recognized in the Late Cretaceous of Ecuador and northern Peru (Jaillard et al., 2005). Most can be correlated with global, eustatic rises in sea level, but inconsistencies show that the Andean margin was already being deformed by the "Peruvian phase" of the Andean orogeny. The paleogeographic upheaval is well documented in the strata of the eastern basins, where the onset of fine-grained clastic sedimentation in the Turonian (Fig. 4-18) is taken as the first event of the Peruvian phase. A major transgression in the Campanian conceals the continued effects of the deformation.

In Ecuador, mid- to late Cretaceous Amazon basin sediments derived from the primordial eastern Andes have eroded from non- or very low grade metamorphic source rock, but the increasing metamorphic grade of the grains since the Maastrichtian indicates general, rapid exhumation of deeper crust in the Eastern Cordillera (Ruiz et al., 2004; Winkler et al., 2005). A phase of tectonic activity began in the Turonian (~90 Ma), and rapid exhumation is indicated from Santonian to Paleocene (85–60 Ma). At 70 Ma an eastern sediment source is turned off, indicating a change in paleocurrent direction at these Ecuador sites from westward to eastward, due to the rise of the Andes. The main period of orogenic growth was in the Eocene, correlated with the docking of the Macuchi terrane from the west, and additional, moderate uplift occurred in the Miocene. Sobolev and Babeyko (2005) found that the most important factor controlling the intensity of the tectonic shortening was the rate of westward drift of the South American plate.

Putumayo Basin (Colombia), Oriente Basin (Ecuador), and Marañón Basin (Peru)

These three basins together form an asymmetric foreland basin bounded in the north by the Macarena uplift/Vaupés arch (separating it from the Llanos basin) and in the south by the Contaya arch and the Ucayali basin (Higley, 2001) (Fig. 4-19). In contrast with the open marine environment that the basins west of the Marañón geanticline experienced, the eastern basins are marked by stable tectonic conditions, lower subsidence rates, shallow, restricted environments, and by short-lived, but widespread, marine transgressions.

There are some 172 oil fields in the basin (Fig. 4-19; Schenk et al., 1997; Higley, 2001). Most of the oil here is derived from Late Cretaceous source rocks deposited in shallow marine and mixed marine–terrestrial environments. Since the change at the end of the Cretaceous from a passive to convergent margin environment, tectonic events through to Pliocene have inverted normal fault structures and created thrust faults, subtle folds, and low-relief anticlines. This has caused structural enhancement of stratigraphic reservoirs and seals for petroleum. In the Oriente basin of eastern Ecuador, the east/southeast oriented Agaurico and Cononaco paleoarches are associated with giant oil fields and also with some of the highest levels of biodiversity on Earth. The paleogeography of the Oriente Basin has been portrayed as a north–south oriented paleoshore zone, but it may have been more complex than this. The first phase of convergence in the Oriente basin occurred with terrane accretion from the west (Late Jurassic–Early Cretaceous) (Lee et al., 2004). A second phase of convergence in the region that became the Eastern Cordillera began in late Cretaceous and lasted until the early Tertiary. Transpression, with both strike-slip faulting and convergence, "created a network of low relief basement reverse faults and structural highs in the Oriente Basin . . . , providing lines of basement weakness for later reactivation. All of the major oil and gas fields in the Oriente basin are related to this basement deformation" (Lee et al., 2004).

Late Cretaceous events in western Ecuador involved the accretion of converging oceanic plateaus to the Andes pile by tectonic underplating beneath the continental margin. Jaillard et al. (2008) confirmed that each accretion event correlates with a phase of uplift and erosion in the Eastern Cordillera and with a sedimentary hiatus (uplift) in the Oriente basin. Thus, the eastern areas did not behave as simple, flexural foreland basins (caused by orogenic loading) during late Cretaceous–Paleogene times, as their evolution was also influenced by the impact of terrane accretion. Winkler et al. (2005) is another study attributing phases of exhumation in the nascent Ecuador Andes to terrane accretion. The accreted blocks include the Carnegie Ridge, a large terrane that extends from Ecuador to the Galápagos.

In the Oriente–Marañón basin, the main oil fields correspond to old grabens and half-grabens inverted along three NNE–SSW right-lateral/convergent wrench-fault zones (Baby et al., 1999). In the western part of the basin, the Subandean foothills are formed by positive flower structures, the Napo and Cutucú uplifts. ("Flower structures" are characteristic medium-scale features of zones under transpression. Sheets of rock form radiating, nested anticlines that resemble the petals in a flower and often create petroleum traps.) In the center of the basin, the Sacha–Shushufindi wrench-fault zone is the most productive petroleum play of Ecuador and corresponds to another flower structure. There have been three stages of inversion and trap formation. The first and longest was Late Cretaceous (Turonian–Maastrichtian, 90–65 Ma), and this is well illustrated in the Shushufindi giant field. Here the Upper Napo Formation (shallow marine, 90–78 Ma) and Lower Tena Formation (continental–shallow marine, 78–65 Ma) sealed the structures and were themselves covered with the continental sediments of the Paleocene Upper Tena Formation (62–55 Ma). The sequence represents the initial, transpressive deformation of Andean orogeny. In the third petroleum play of the system, east of Shushufindi, Eocene (Incaic phase) transpressive inversion sealed the structures. Further trap formation in the system occurred in the Miocene–Recent Quechua phase of orogeny, but Baby et al. (1999) concluded "It is obvious that the more productive fields correspond to the Upper Cretaceous traps."

Huallaga Basin: A Thin-skin Fold-and-Thrust Belt

This basin lies well into the Subandean belt in the strict sense (the wedge-top depozone), west of the main foreland basin (the foredeep) but east of the main basement thrusts of the central Andean cordilleras. The area corresponds to a thin-skin fold-and-thrust belt superimposed on inverted and halokinetic structures (Hermoza et al., 2005). Layers of salt are very weak, and this leads to characteristic salt tectonics, an important aspect of many petroleum systems. The Jura Mountains in France are another thin-skin fold-and-thrust belt, with detachment (décollement) of the "skin" along strata of Triassic salt. The thin skin slithers and folds on the layer of salt like a carpet on a polished floor (Fortey, 2004). This deep detachment creates (a) geological structure, (b) surface topography, and (c) a specific type of habitat for plants and animals. In western Amazonia it is also associated with very high levels of biodiversity and may have been a causative factor of this. In the Andes, thin-skinned thrust belts such as the Huallaga basin and the Subandean Santiago basin north of the Marañón show crustal shortening of 40–70%; further east, the Contaya arch and the Shira Mountains are belts of foreland basement thrusts, but this involves shortening of less than 10% (Kley et al., 1999). In the Huallaga basin, thrust propagation occurred in a deltaic environment which evolved into an alluvial system linked with the Amazon. The western part of the Huallaga basin is structured by thrust-related anticlines spaced ~25 km apart. These are evident in the modern topography, both here and in the Ucayali valley, as parallel belts of low hills about 500 m high (Fig. 4-19). The less deformed Marañón basin further east corresponds to the foredeep depozone.

Ucayali Basin

In its early development, the Ucayali basin was part of a much longer basin extending along eastern Peru that was divided into subbasins during Andean orogeny (Sanchez Alvarez, 2007). The Ucayali basin is bound by the Contaya arch to the north and the Fitzcarrald arch to the south (the topography here can be seen in very clear images on Google Earth). The current morphology of these arches results from the new working of older faults (Dumont et al., 1990). The Contaya arch is the greatest structure in the Marañón basin; it is the result of inversion tectonics and started to develop in the Late Cretaceous (Hermoza et al., 2005). The Fitzcarrald arch (equated with the Pisco-Juruá fault by Ching, 1999) has been attributed to the subduction of the Nazca ridge, although this is controversial and implies a greater coupling between subducting and overriding plates than has been thought possible. Clift and Ruiz (2007) instead suggested the arch originated with reactivation of Paleozoic structures.

Following a long Paleozoic/Mesozoic phase of sedimentation in the Ucayali basin, there was a phase of no sedimentation in the Early Cretaceous, and a major regional unconformity developed. Following this, there was continental fluvial deposition over the basin. Marine and restricted-marine sedimentation was reinitiated in the Aptian/Albian and lasted 50 m.y., until the Campanian; this sea may have been confined to the western part of the basin. The Agua Caliente Formation (late Albian) in the central part of the basin has sediments derived from a shallow marine environment at its base, with a transition at the top to coastal deltaic and fluvial deposits.

In the Late Cretaceous, there was marine deposition (Chonta Formation) and a long period of tectonic quiescence, followed by renewed uplift with a sequence of marine, deltaic (Vivian Formation) and continental fluvial sediments. These mark the Peruvian phase of orogeny. The Cachiyacu Formation (Campanian) defined the end of a long period of oceanic sedimentation that lasted through most of the Cretaceous. It was deposited on the flanks of the Shira high as wedges that pinch out toward the mountains, indicating that early uplift of these mountains coincides with this compressive episode. Basin inversion began during the Cretaceous, but the Paleogene Incaic phase of orogeny was the main compressive event affecting the basin. During this phase, the Shira Mountains fault and others became active as thrust surfaces and the Shira Mountains/ Vilcabamba uplift region was brought to its present elevation. (A key western limit at the Shira Mountains in *Cebuella* is used in the reconstruction for callitrichines; Fig. 4-17.)

PLATYRRHINE ECOLOGY

Mangrove, Flooded Forest, Terra Firme Forest

Although by now several platyrrhines have been observed in mangrove (Fig. 4-4), primate (and mammal) use of mangrove and back-mangrove remains poorly known. In one of the very few studies in America, Fernandes (1991) found that intertidal resources exploited by *Cebus* include oysters. Apart from causing the anaerobic, saline soil, tidal flooding also means mangrove areas suffer a high level of physical disturbance. Mangrove and back-mangrove forest often grade into secondary formations, either natural or anthropogenic, and many mangrove and mangrove-associated taxa are weedy forms. Feeding behavior of most primates inhabiting mangrove in Maranhão, northeastern Brazil, is similar to that in other marginal habitats (Fernandes, 1991) and the back-mangrove is interpreted here as a low-elevation, "common denominator" vegetation type.

Lehman and Fleagle (2006) provided a useful summary of the types of vegetation inhabited by primates, but did not distinguish mangrove, which was included in the general category of "swamp forest." A worldwide review cited 111 mammal species recorded in mangrove (Fernandes, 2000). In northeastern Brazil, Andrade et al. (2008) added 18 more from Pará alone and the world list, already substantial, is far from complete.

Of the different types of tropical rainforest, mangrove and terra firme forest (non-flooded forest) are well known, but another category, freshwater flooded forest on alluvium, is also important in Amazonia. Lehman et al. (2006) stressed the importance of floodplain forests in primate ecology and distribution. Alluvial floodplain soils are often very fertile, and in Brazil primate density and total biomass/area in floodplain forests can be twice that of terra firme forest (Peres, 1999). During the high annual flooding, when the trees are in fruit, many species migrate there from terra firme forest. Large areas in the tropics are covered in "periodically inundated forest," a broad, synthetic concept that is useful in understanding primate distribution, ecology, and evolution. Based on tree distributions, Prance (1979) considered all the flooded forests of Amazonia, both freshwater and mangrove, together as one entity. In primates, the *Saimiri–Alouatta* association, for example, is a "signature" of gallery, riparian, tidal, and flooded forest (Peres and Janson, 1999), and both genera are also well known in mangrove.

The primates are most diverse by far in "rainforest" of hot, wet tropical lowlands, but in practice it is difficult to define this vegetation type, either through its structure or its taxa, and its origins are debated. It is suggested here that modern terra firme rainforest and flooded forest have both been derived from mangrove-associated forests that were uplifted or left stranded inland following retreat of epicontinental seas in Cretaceous Africa, Asia, and America (cf. Craw et al., 1999). The outer, seaward edge of mangrove is often species-poor, but the inner, landward margin of the mangrove can be a complex, diverse mosaic of rainforest types: mangrove, beach forest, palm forest, freshwater swamp forest, peat-swamp forest, riparian forest, "normal" terra firme forest, lower-canopy sclerophyll or heath forest on white sands, and drier and secondary variants of all of these (Heads, 2006b). Even species that are more typical of montane forest can be found in or near the back-mangrove as "anomalies" in many areas, such as the Pacific coast of Colombia and parts of Borneo. Characteristic trees of each of these forest types may occur in just a few square meters of back-mangrove on different microtopography. One large population of orangutan, *Pongo*, occurs on a floodplain in Borneo that is covered with a patchwork of mangrove, riverine forest, swamp forest, seasonally flooded forest, *Nypa* palm forest, and dry dipterocarp forest on unflooded land (Goosens et al., 2005).

Although it is seldom discussed by ecologists, this back-mangrove mosaic or ecotone maintains a high diversity of primates and other groups throughout the tropics and could, on its own, provide the main taxa for most of the extensive inland forest types (Fig. 4-20). Inland flooded forest in the Amazon is interpreted here as an example of a mangrove-associated community stranded following marine regression in the Cretaceous.

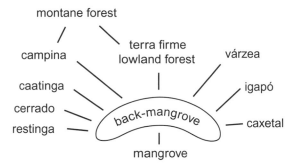

FIGURE 4-20. Geographical/ecological relationships among some forest types. Periodically inundated forests are on the right.

The flooded habitats are important because of their areal extent, but their diversity is limited by the severity of the environment. Mechanical disturbance and anoxic conditions during the flooding season prevent the establishment of many species in periodically flooded forest, and forests on swampy ground are always poorer in species than those on terra firma. Nevertheless, for groups of primates and pre-primates that are tolerant of mangrove or back-mangrove conditions, a marine incursion provides much new habitat. Even for weedy forms, though, the arms of the sea constitute barriers, and the coasts of shallow seas may show rapid change due to tectonic movement, eustatic sea-level change, or both.

Nypa palm forest was mentioned above as orangutan habitat. *Nypa* is a large palm with an unusual horizontal trunk. It forms dense stands (the leaves are about 8 m tall) in many Southeast Asian mangrove swamps, and in earlier times was widespread from Australia to Africa, Europe, and the U.S. In palm phylogeny, it is basal to all other clades except the rattans (Calamoideae) (Asmussen et al., 2006). Near Laredo, Texas, fossils indicate that a tropical mangrove/estuarine community dominated by *Nypa* existed in the Eocene. Westgate (2008) recorded 32 mammal species there, including four new primates (omomyiforms), one a new genus. Westgate and Salazar (1999) gave a detailed account of this diverse community (although the reconstruction of the Eocene *Nypa* swamp depicted on the cover shows a sparse, temperate-looking vegetation). The oldest known fossil primate in North America, the Eocene omomyid *Teilhardina magnoliana* from the coastal plain of Mississippi, was recovered from sands deposited in estuarine tidal channels of a former coastline (Beard, 2008).

Inland Flooded Forest in the Amazon Basin

Apart from mangrove, primates are diverse in swamp and riparian forest that undergoes either periodic or permanent inundation. As noted already, these are an important feature of the Amazon basin; between dry and wet seasons, river levels vary up to 11 m, and vast floodplains are under water for months each year with the "annual tide." Areas of flooded forest along the larger rivers are often 10 km wide or more. Terra firme forest covers 83% of the central Amazon basin; 12% is in flooded forest (Haugaasen and Peres, 2005).

Great variations in the flooded forests of the Amazon basin are associated with the nutrient status of the flood water, which may be very high or very low. Arboreal groups such as epiphytes, birds, and primates occupy the mangrove and other flooded forest without being affected in a direct way by the water quality, but the primates are dependent on the fruit trees present, and these differ in the various types of forest. Rivers that drain the Andes, such as the Solimões and Madeira, are high-nutrient, "white-water" rivers, with a café-au-lait color caused by the high levels of mud and silt, and a pH of 7–8. The forest on the fertile floodplains of these rivers is termed "várzea." Rivers draining the sterile white sands of the Guiana shield, such as the Rio Negro, and many smaller rivers in Amazonia, are low nutrient, "black-water" rivers, in which the water is dark and acidic due to the presence of tannin and humic acids. A type of forest called "igapó" occurs along these rivers; it has distinct structure and floristics (similar to vegetation on oligotrophic white sands; Kubitzki, 1989) and often forms narrower, more homogeneous bands than várzea. Rivers with clear water and neutral pH occur in eastern Amazonia (Xingu and Tapajós). Differences among forest types associated with mangrove vary in a similar way, with marked differences in substrate pH between outer mangrove (basic) and inner mangrove/*Nypa* swamp (acidic), grading into peat-swamp forest (acidic). In this way, várzea can be compared with mangrove, igapó with peat-swamp forest.

In flooded forest, Prance (1979) recognized seasonal and permanent várzea, seasonal and permanent igapó, mangrove, and tidal várzea. "Mixed-water" ecotone forests, in which species typical of black- and white-water rivers occur together, have been described near Manaus (do Amaral et al., 1997).

In an area of terra firme–igapó–várzea forest on the lower Purus River, the forest types replaced each other on a local scale (Haugaasen and Peres, 2005, 2006, 2008). In terra firme forest, Lecythidaceae were the most important trees, while in both várzea and igapó Euphorbiaceae were most important (Haugaasen and Peres, 2006). For trees, terra firme forest was the most species-rich, with várzea next and igapó least rich. Of the large birds, toucans were the most abundant in terra firme and igapó; macaws were the most abundant in várzea (Haugaasen and Peres, 2008).

For primates at the Purus site, terra firme forest was more species-rich (12 species) than igapó (11) and várzea (8) (Haugaasen and Peres, 2005). On the other hand, density and total biomass was much greater in várzea, where the soil is much more fertile than in terra firme forest or igapó. *Cebus apella* was the most common species in terra firme forest and the truest generalist, with high numbers in all forest types. *Saimiri*, *Alouatta*, and *Cebus apella* had their highest levels of biomass/km^2 in várzea, *Pithecia* in igapó, and the others in terra firme. Some populations of *Cebus*, *Saimiri*, and *Aotus* are restricted to várzea. *Callicebus cupreus* was abundant only in secondary forest throughout the study area and was often recorded in areas that were disturbed by human activity. *Cebuella* occurred only in fringe forest with high levels of disturbance.

For the primates in these flooded forests, even a 50-m-wide channel provides an effective barrier to dispersal. Some primates can swim, others cannot, but in any case predation is a real threat (Haugaasen and Peres, 2005). The authors also emphasized that primates make extensive use of flooded forests at fruiting time, and the forests "should not be seen as marginal habitats that occasionally subsidize terra firme forest specialists" (p. 252). In the model suggested here, modern primates are derived from ancestors that already occupied flooded forest and its environs but in some cases were left stranded on dry land in inland areas.

Palm swamp forest covers large areas and has been mentioned as an important habitat for *Cacajao*. In Amazonian Peru, T. D. Pennington et al. (2004) recognized six tree formations:

- lowland rainforest on non-flooded land;
- varillal or pole forest on white sands;
- swamp forest on permanently flooded land;
- várzea on periodically flooded land;
- igapó on periodically flooded land; and
- savanna woodland in areas of flat topography, poor drainage, and periodic fires.

Here swamp forest in areas of permanent swamp is distinguished from seasonally flooded várzea and igapó. Permanent swamp forests are known as aguajales, as they are dominated by the palm *Mauritia flexuosa* (aguaje). This is a tree up to 25 m tall, with a trunk 60 cm in diameter and usually bearing pneumatophores ("breathing roots") (T. D. Pennington et al., 2004). It is widespread through lowland Amazonia up to 1,000 m elevation and often forms pure stands in permanently flooded swamps, in wet savanna, and along rivers.

The pneumatophores in *Mauritia* and another, purely coastal, palm, *Euterpe*, were described by de Granville (1974). Pneumatophores and "aerial roots" are typical features of mangrove trees and are usually interpreted as recent, secondary adaptations of normal roots for swamp habitat. But following anatomical studies of the aerial "roots" of *Rhizophora*, Menezes (2006: 213) concluded instead that "These peculiar branches are rhizophores or special root-bearing branches, analogous to those found in Lepidodendrales and other Carboniferous tree ferns that grew in swampy soils."

Variation in Terra Firme Forest: Campina Forest on White Sand

Amazon rainforest includes flooded forest on the alluvium of the floodplains and terra firme forest on the higher ground that is seldom if ever flooded. Terra firme forest itself is far from homogeneous. In the terra firme forest of Peru, Fine et al. (2005) distinguished three soil types: white sands, terrace soils (brown, sandy soils with some silt and clay, deposited by Pliocene and Pleistocene rivers), and clays derived from Cretaceous metamorphic rocks that were exposed with Andean uplift. The authors studied 35 species of trees in the Burseraceae tribe Protieae and found that 26 of these were associated with only one of the soil types.

White sand forest is of special interest for biogeography and ecology. It is more open than normal terra firme forest and has a much lower canopy (as low as 5–10 m). It is referred to as campina in Brazil, varillal and irapayal in Peru, and wallabah in Guyana. It occurs on podzolized white sand soils that are sterile and drought-prone because of the porosity. The forest is often found in areas of uplift where late Tertiary sands, derived from the cratonic rock, have been exposed by erosion. These are surrounded by more fertile Quaternary sediments (Alvarez Alonso, 2002). The campina includes pole forest (the varillal) and also a short, rather open forest in which the trunks of the trees are often gnarled, twisted, and much branched, and the foliage is scleromorphic

(Smithsonian Institution, 2009). The trees carry a high epiphyte load, and the forest can resemble the stunted woodland of "tepui" (mesa) summits in southern Venezuela (Whitney et al., 2004). "Campinarana" is transitional between campina and normal terra firme forest. Campina is related to the open *Eperua*-dominated forest of northeastern Amazonia (caatinga) and the Guianas (wallabah), and also to the kerangas or heath forest of Asia. The latter is a sclerophyll forest that occurs on white sands between the coastal peat-swamp forest and the terra firme dipterocarp forest.

Amazon campina forests are well known for their endemic bird species and are also important primate habitat. For example, *Callicebus lucifer* (along the Putumayo River on the Peru/Colombia border) prefers white sand forest (IUCN, 2009). The campina forest biota is phylogenetically distinctive, probably ancient, and its taxa have far-flung affinities. The Amazonian plant *Potalia* (Gentianaceae) has several species endemic to white sands, and these share morphological characteristics with *Anthocleista*, the African–Malagasy sister of *Potalia* (Frasier et al., 2008).

Disjunct affinities between white sand forest taxa of western Amazonia and relatives in the Rio Negro/Guiana shield area are well known in birds. For example, the antshrike *Thamnophilus divisorius* is endemic to white sand forest on the Serra do Divisor (Fig. 4-19) and is closest to *T. insignis*, endemic to the tepui area of southern Venezuela (Whitney et al., 2004), supporting the idea of an ancient, disjunct distribution surviving as metapopulation. Alvarez Alonso (2002) suggested a plausible metapopulation model, in which the endemic bird taxa (and so also, for example, primate populations), survive and evolve over long periods on "islands" of white sands that may each be short-lived in geological time. A similar model may also explain survival of bird endemics and primate populations on river islands in the Marañón area.

The antbird *Percnostola arenarum* (Thamnophilidae) is restricted to the white sand forests of northeastern Peru (Marañón–Napo interfluve) (Isler et al., 2001). Its sister is *P. rufifrons*, which occurs elsewhere in central Amazonia in habitats with dense understory, including forest edge, secondary growth, low stands around rock outcrops, and tree falls and brush piles within forest. *P. rufifrons* is also in savanna forests on the sandy coastal plain of Suriname and in mature mangrove stands in French Guiana. The entire range of habitat—mangrove/tree fall/white sand forest, as seen in the *Percnostola* species—may have already been occupied by their common ancestor before the differentiation of the modern species.

In primates, the pithecines (Figs. 4-6 and 4-7) are a good example of a mangrove and swamp forest clade that also occurs in vegetation on white sands. *Chiropotes* is in terra firme forest, mangrove, inundated forest, and forest–savanna margins. *Pithecia* is in terra firme forest, várzea, igapó, and "swamp forests of the [Amazon] mid-delta" (IUCN, 2009). In *Cacajao*, the *C. melanocephalus* complex is in the northwestern sector of the Amazon and so is associated with black-water rivers (Boubli et al., 2008). The species are:

> *C. melanocephalus* s.str.: black-water flooded forest (igapó) and campinarana (low-canopy forest transitional to open campina) on white sandy soil. Occurs from the interfluve of the Japura, Negro, and Guaviare Rivers, northwest almost to the Serranía de La Macarena, Colombia.
>
> *C. ayresi*: igapó in large groves of *Mauritia* palm and terra firme forest interspersed by an open, herbaceous vegetation. Occurs along the Rio Araçá, a tributary of the Rio Negro.
>
> *C. hosomi*: terra firme forest, campinarana, also montane forest to 1,500 m in the Mount Neblina area. Occurs from the Rio Negro to southern Venezuela.

The only other clade of *Cacajao* is *C. calvus* of southwestern Amazonia, south of the other species. It occurs in várzea around white-water rivers (Solimões, Içá) and in terra firme forest (Ucayali) (Barnett and Brandon-Jones, 1997).

The landlocked *Cacajao* thus occupies a wide range of forest types, but all of them can be derived from close equivalents present at the coast. This is a standard pattern in many primates. Seasonal migrations occur in *C. calvus*, with populations residing in várzea at high flood when many trees produce fruit, but moving to *Mauritia* palm swamp for the dry season. Several other species of New World monkeys are known to migrate between várzea and terra firme forest, and this may reflect an old pre- or proto-platyrrhine affinity with both forest types. Similar seasonal migrations between habitats also occur in Old World primates. These migrations are cyclical and remain within the area of endemism, and so they differ from both chance dispersal and range extension dispersal. Foraging movements are yet another form of dispersal, and "riverine refuging" in Old World primates is one interesting form of this; the animals move inland to forage through the day but in the afternoon return to the riverbank trees, where they spend the night.

EVOLUTION IN PLATYRRHINES

Polymorphic Ancestors

Ancestors of primate clades are often assumed to have been single species that were uniform in their morphology and ecology. For any given character, analyses have focused on establishing the (single) character state of the ancestor. For example, "the ancestral primate" is often assumed to have been small, even shrew-sized. Soligo and Martin (2006) emphasized that "Interpretation of the adaptive profile of the ancestral primates has been constrained for decades by general acceptance of the premise that the first primates were very small." But this may not be correct; the first primates may have already had a diverse range of sizes, inherited from a diverse ancestor. The idea of a polymorphic ancestor is now accepted in a growing number of studies, both morphological and molecular (Heads, 2009c).

In the Southeast Asian gibbons, Mootnick and Groves (2005: 974) concluded that in morphology, "generic characters in the Hylobatidae are mosaic in nature, presumably reflecting differential loss of aspects of the original hylobatid morphotype." This "polymorphic ancestor" idea accounts for the geographic vicariance seen among three of the four gibbon genera better than ancient hybridism.

Many clades in primates and related groups have undergone ancient, rapid evolution. With large, widespread, polymorphic populations as ancestors, and rapid, regional-scale differentiation, as suggested here, evolution is likely to involve retention of ancestral polymorphism and fixing of different morphs in different clades. This process, termed "incomplete lineage sorting" by geneticists, results in incongruence among different gene and character trees, and parallelism by recombination of ancestral characters. It is recorded in many primate groups, such as lemurs (Heckman et al., 2007) and hominids (Salem et al., 2003), and is probably a general phenomenon. Thus vicariance with incomplete lineage sorting may be a standard mode of evolution and can account for both allopatry and regionally restricted parallelism.

Calibration of the Time Course of Evolution

As far as timing is concerned, for primates it seems safer to use spatial breaks in molecular clades and associated tectonics to calibrate molecular phylogenies rather than relying on the sparse early fossil record. Tectonic–phylogenetic dating has not been used to calibrate primate evolution before, but it has been used in studies of many other taxa (Chapter 2). Use of the method is straightforward in primates, as the phylogeny and distribution are well documented, although molecular phylogenies are still needed for several genera. More detailed investigation of *Cebus* and *Callicebus*, for example, and the Amazon break predicted in both on morphological grounds, would be of special interest.

The Eight Widespread Clades in Platyrrhines

The two main groups of primates, strepsirrhines and haplorhines, are vicariant in South and Central America (with only haplorhines) and in Madagascar (with only strepsirrhines) but overlap in mainland Africa and Asia. The two main clades of haplorhines show complete vicariance, with platyrrhines in America and catarrhines in Africa and Asia. In platyrrhines, the eight "widespread clades" all show extensive overlap, but within the widespread clades most species and subspecies are vicariant. Overall, vicariance and continued immobilism are evident in most taxa, and the "widespread clades," at about the level of genus, are the exception. How did they evolve?

The platyrrhines differentiated at 130 Ma and the widespread American clades may have evolved at about this time or soon after. Depending on how much overlap is permitted, there are about eight "widespread clades" in America (counting pithecines as one), ten in Africa, and nine in Asia. The overlap of these within each of the three continents—but never among them—is assumed here to be due to a more or less simultaneous, more or less global phase and is attributed to the mid-Cretaceous rifting and marine transgressions. Because of the extensive overlap among the widespread clades, it is difficult to reconstruct the original allopatry among them. But differentiation *within* them shows interesting patterns, and the breaks can be correlated with tectonics. This suggests that the main breaks in the widespread clades are mid- to late Cretaceous.

A Model of Platyrrhine Evolution

The following model is suggested for platyrrhine evolution:

1. Early Cretaceous (135 Ma). With the continuing breakup of Gondwana, complete rifts began to develop between South America and Africa. Volcanism was among the first signs of this.

Haplorhine primates differentiated into platyrrhines in America and catarrhines in the Old World.

2. About this time, the eight platyrrhine groups that are now the widespread clades originated with rifting and uplift associated with the opening of the Atlantic.

3. Early Cretaceous (120 Ma). The eight widespread clades developed almost complete overlap in a period of widespread mobilism. This is attributed to one or more of the great marine transgressions that occurred in the mid-Cretaceous from 120 Ma (Müller et al., 2008). At this time the sea may have reached its highest eustatic level since the Carboniferous. Since the Cretaceous highs, global sea level has continued to drop. With the retreat of the epicontinental seas, mangrove and swamp-associated primate taxa were left stranded far inland in the Amazon and Congo basins.

4. Mid- to Late Cretaceous. The primary differentiation within the widespread clades took place, producing the modern genera and subgenera. This involved terrane accretion and transcurrent faulting at the Romeral fault zone (western Colombia and Ecuador), the initial "Peruvian" phase of Andean orogeny, rifting on the Amazon, and volcanism, rifting, and rift shoulder uplift in southeastern Brazil. The vicariance resulting from this phase of differentiation is still more or less intact, although a number of individual species have expanded their ranges.

5. Eocene. The Incaic phase of Andean orogeny and its far-field effects led to continued differentiation of subgenera and species.

6. Miocene. Range expansion of several species with transgression and retreat of Lake Pebas led to overlaps in distributions in southwestern Amazonia.

7. Miocene to Recent. The Quechua phase of Andean orogeny led to continued differentiation within species by fault control of river courses and low-lying belts of hills.

To summarize, the tectonic revolution that brought the modern, Cenozoic geography into existence also brought about the last main reorganization in the primates. In particular, the main phylogenetic/biogeographic nodes of primates correlate with major events of the Jurassic–Cretaceous: massive inter- and intra-continental rifting, sea-floor spreading, transcurrent strike-slip faulting, basin formation, uplift, down-warping, and large-scale magmatism. The widespread clades in each of America, Africa, and

Asia all became widespread during the same global marine transgressions in the mid-Cretaceous. The American widespread clades have been split by mid-Cretaceous rifting on the Romeral fault, Amazon, and southeastern Brazilian faults, while the African widespread clades have been split by mid-Cretaceous rifting on the Benue Trough and Central African rift system. Other tectonic rifts differentiating primate clades—those between Africa and America and between Africa and Madagascar—are used here to calibrate the primate phylogeny.

Tectonic processes that may produce topographic change and separate populations include strike-slip faulting, crustal extension with normal faulting, anticline formation with reverse faulting, and combinations of these. Strike-slip movement can lead to extension and subsidence (under transtension) or to compression and uplift (under transpression) as at the Romeral fault zone and the Peruvian Andes, and the effects of strike-slip faulting on biogeography have been discussed elsewhere (Heads, 2003, 2008a, 2008b; Heads and Craw, 2004). Rifted margins that have been converted into foreland basins are often the site of petroleum systems that have developed, along with centers of diversity in primates, at turning points in paleogeographic history.

Tectonic convergence involving terrane accretion and associated transcurrent faulting is associated with phylogenetic breaks and high levels of diversity in western Colombia, as well as in the Himalayas/northern Burma, western Sumatra/Mentawai Islands, Borneo, and Sulawesi. With subduction, terrane accretion, crustal shortening, and tectonic erosion, the distributions of organic species may also have been "piled up" and shortened, as with the diverse forms of *Aotus*, *Cebus*, and *Saguinus* in western Colombia. Final Andean uplift above 3,000 m in the Miocene has little direct relevance to lowland groups such as primates. But in pre-Andean western Amazonia, the Cretaceous transition from shallow seas to deltas, mangrove, and low hills would have been critical during the phase of evolution that produced the genera and subgenera.

An important ecological effect of tectonics is the uplift of entire communities. Mangrove and montane primate faunas are seen here as end-members of an ecological series, with montane groups derived from lowland or mangrove ancestors by uplift. The whole series is seen in primate clades of Africa (e.g., *Pan*) and Asia (e.g., the odd-nosed clade of colobines). In America, the mangrove–montane series occurs even in single species of groups such as *Alouatta* and *Cebus*. Instead of invading the montane zone by dispersal from a center of origin in the lowlands, or vice versa, the direct ancestors of *Alouatta*, *Cebus*, and others already inhabited the areas that became the Andes before the Andean orogeny and were lifted up with it.

Most primates occur in different types of lowland tropical rainforest. They inhabit unflooded terra firme forest, palm-swamp forest, and periodically inundated forest, including mangrove and floodplain swamp forest, as well as their drier and secondary equivalents. All these forest types occur behind the mangrove, often in a mosaic. It is suggested that many early primates had a back-mangrove or mangrove-associated ecology, as many still do. This, together with a widespread tendency to weedy ecology, meant that the early ancestral primates could take advantage of the new coastlines that developed with the Jurassic breakup of Pangaea and Gondwana and the Cretaceous incursion of epicontinental seas. When these seas retreated, primate taxa were left stranded inland throughout the Amazon in a new phase of immobilism that has lasted, with notable exceptions in western Amazonia, until the present.

5

Primates in Africa and Asia

Several decades of work on the molecular biology of primates, together with extensive field studies by biologists, paleontologists, and conservation ecologists, mean that a great deal is now known about the phylogeny and distribution of the order. Nevertheless, the interpretation of this data and proposals about how, where, and when the different groups developed remain controversial. This chapter develops a new model of primate evolution in the Old World by integrating the distributions of the molecular clades with plate tectonics.

Primates are widespread in the tropics, but apart from humans and one introduced species in New Guinea, they are absent in Australia, New Guinea, and the Pacific islands, despite extensive suitable habitat there. This absence is a valuable clue indicating that historical factors, not just current ecology, have been involved in shaping the group's distribution.

Primates can be abundant and aggressive, and are often pests for farmers in rural areas. Forest edge and secondary forest are important habitats for primates and many "weedy' species live in disturbed sites associated with humans. Primates also occur, but with much less diversity, in drier sites with open vegetation and in cooler habitat on mountains in the tropics and in China and Japan.

The molecular phylogeny of primates used here (Fig. 5-1) is based on that of Goodman et al. (2005), Fabre et al. (2009), Chatterjee et al. (2009), Perelman et al. (2011), and other studies cited below. The topology of the trees is not questioned in this chapter, but the calibration of the

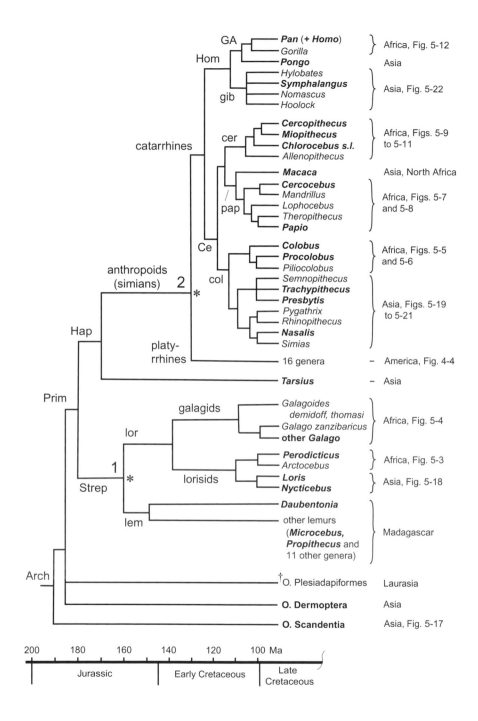

GA — **Pan** (+ **Homo**)
    — *Gorilla*
Hom
    — **Pongo**   } Africa, Fig. 5-12
        Asia
   — *Hylobates*
   — **Symphalangus**
gib
   — *Nomascus*
   — *Hoolock*
} Asia, Fig. 5-22

catarrhines

cer   — **Cercopithecus**
     — **Miopithecus**
     — **Chlorocebus s.l.**
     — *Allenopithecus*
} Africa, Figs. 5-9 to 5-11

— **Macaca**   Asia, North Africa
pap
   — **Cercocebus**
   — *Mandrillus*
   — *Lophocebus*
   — *Theropithecus*
   — **Papio**
} Africa, Figs. 5-7 and 5-8

Ce

col   — **Colobus**
     — **Procolobus**
     — *Piliocolobus*
} Africa, Figs. 5-5 and 5-6

anthropoids
(simians)   2
     *

— *Semnopithecus*
— **Trachypithecus**
— **Presbytis**
— *Pygathrix*
— *Rhinopithecus*
— **Nasalis**
— *Simias*
} Asia, Figs. 5-19 to 5-21

Hap

platy-
rrhines   — 16 genera   − America, Fig. 4-4

— **Tarsius**   − Asia

Prim

galagids   — *Galagoides*
       demidoff, thomasi
     — *Galago zanzibaricus*
     — **other Galago**
} Africa, Fig. 5-4

lor

1   *
Strep

lorisids   — **Perodicticus**
     — *Arctocebus*
} Africa, Fig. 5-3

     — **Loris**
     — **Nycticebus**
} Asia, Fig. 5-18

lem   — **Daubentonia**
     — other lemurs
       (**Microcebus**,
       Propithecus and
       11 other genera)
} Madagascar

Arch

†O. Plesiadapiformes   Laurasia

O. Dermoptera   Asia

O. Scandentia   Asia, Fig. 5-17

200   180   160   140   120   100 Ma

Jurassic   Early Cretaceous   Late Cretaceous

different nodes with time is problematic and is a main focus of inquiry. Fossil-calibrated molecular clock dates only estimate minimum ages for clades. In the model of early primate evolution adopted here (Chapter 3), fossils are not used to calibrate the tree. Instead, the endemic clades of South America (platyrrhines) and Madagascar (lemurs) are used to estimate absolute (not minimum) dates for two of the basal nodes of the molecular phylogeny (Fig. 5-1). The related tectonic events, namely the opening of the Mozambique Channel (160 Ma) and the Atlantic Ocean (130 Ma), are both well dated and correlate spatially with the phylogenetic/biogeographic breaks (nodes). Based on these primary calibrations, geographic comparisons of the other molecular clade distributions with tectonics are made here for the other main nodes and evolutionary correlations suggested.

Together with the focus on geographic breaks and the chronology of evolution, an ecological theme in this chapter is the widespread association of primates and related orders with mangrove and ecotone "back-mangrove" forests at the inner edge of mangrove. This is not well documented, but many clades of primates in the New World (Chapter 4) and the Old World (Fig. 5-1) occur in true mangrove.

PRIMATES IN AFRICA

The different primate clades in Africa are considered together for convenience, although they do not form a monophyletic group (Fig. 5-1; see also Fig. 3-1). Chapter 3 presented a model for the evolution of primates as the allopatric vicariant of its relatives—the plesiadapiforms in the north and the Dermoptera and Scandentia in Southeast Asia. Following the initial differentiation, there has been a degree of overlap between primates and plesiadapiforms and complete overlap among primates, Dermoptera and Scandentia. Within primates, the main groups show considerable overlap but also an intriguing component of allopatry.

FIGURE 5-1. Phylogeny of extant primates (topology from Goodman et al., 2005, with modifications from Xing et al., 2007; Masters et al., 2007; Ting, 2008; Arnason et al.; 2008; Matsui et al., 2009). Node 1 = Mozambique Channel; node 2 = Atlantic Ocean. Taxa recorded from mangrove are in **bold** (Fernandes, 2000; Butynski, 2003; Oates et al., 2004; IUCN, 2009; Gonedelé Bi et al., 2008; R. Binns, T. Butynski, J. Masters, and K. Nowak, pers. comm.). Arch = Archonta, Prim= primates, Hap = haplorhines, Strep = strepsirrhines, lor = lorisiforms, lem = lemuriforms, Hom = Hominidae, Ce = cercopithecidae, cer = cercopithecines, pap = papionines, col = colobines, GA = great apes, gib = gibbons (lesser apes).

STREPSIRRHINE–HAPLORHINE DIVERGENCE AND THEIR DIFFERENT DISTRIBUTIONS IN AFRICA

The main phylogenetic division in primates is between strepsirrhines, which have a well-developed snout and a rhinarium (a patch of moist, hairless skin around the nostrils, as seen in a dog or a cow), and haplorhines, in which the snout is more or less reduced and there is no rhinarium. Strepsirrhines are completely absent from America, haplorhines are absent from Madagascar, but both occur together in Africa and Asia (Fig. 3-1). Nevertheless, in Africa there are interesting differences in the distributions of the two groups. Strepsirrhines are concentrated in the east and haplorhines in the west (Chapter 3), and this may indicate that part of the original division separating the two groups was in central Africa. Similar differences between the distributions of the two groups occur in Asia. The relative timing and phylogeny, together with the distribution of strepsirrhines (concentrated in the east of Africa and absent in western South Africa), suggest that in Africa the break occurred around the Lebombo monocline (Fig. 5-2). This volcanic rifted margin was the major tectonic feature in eastern South Africa–Zimbabwe–Mozambique active during the relevant time. The magmatism and folding along this margin was an important precursor of the complete rifting that later formed the Mozambique Channel.

On the east coast of Africa, strepsirrhines are diverse. For example, the Rufiji River delta and flood plain in Tanzania (Fig. 5-2) has six strepsirrhine species and four haplorhines (Doody and Hamerlynck, 2003). The area includes the largest area of estuarine mangrove in East Africa as well as other types of forest, woodland, and wetland. In contrast, on the west coast of Africa, a somewhat larger area of equatorial forest between the Niger delta in Nigeria and the Sanaga River in Cameroon has six species of strepsirrhines and 16 haplorhines (Oates et al., 2004). At the East Africa site, strepsirrhines make up 60% of the fauna, at the West African site only 27%.

PRIMATES IN THE RUFIJI DELTA REGION, TANZANIA (FROM DOODY AND HAMERLYNCK, 2003)

Strepsirrhines: *Otolemur garnetti, Otolemur crassicaudatus, Galago moholi, Galago senegalensis, Galagoides granti, Galagoides zanzibaricus.*

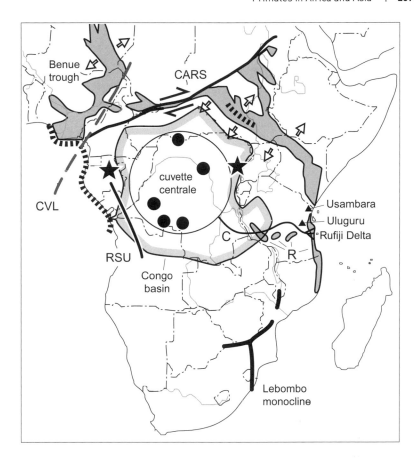

FIGURE 5-2. Cretaceous rifts and other features in Africa. Line with gray border = present Congo basin watershed. Dots = Late Jurassic–Cretaceous fossiliferous marine beds in the Congo basin (from Giresse, 2005). C = former course of Congo River (heavy line) with former outlet at the Rufiji delta. RSU = rift shoulder uplift (Cretaceous) (Stankiewicz et al., 2006). R = Jurassic–Cretaceous rift basins (from west to east: Rukwa, Usangu, Kilosa, and Ruvuma) (Nyblade, 2002). CARS = Central African rift system. Stars = sites with maximum diversity of primate species. Broken line = belts of giant oil fields.

Haplorhines: *Colobus angolensis, Papio cynocephalus, Cercopithecus mitis, Cercopithecus aethiops.*

PRIMATES IN THE NIGER DELTA–SANAGA RIVER REGION (FROM OATES ET AL., 2004)

Strepsirrhines: *Arctocebus calabarensis, Perodicticus potto, Euoticus pallidus, Galago alleni, Galago demidoff, Galago thomasi.*

Haplorhines: *Cercocebus torquatus, Mandrillus leucophaeus, Lophocebus albigena, Cercopithecus preussi, Cercopithecus erythrogaster, Cercopithecus sclateri, Cercopithecus erythrotis, Cercopithecus mona, Cercopithecus pogonias, Cercopithecus nictitans, Procolobus verus, Procolobus pennantii, Colobus satanas, Colobus guereza, Gorilla gorilla, Pan troglodytes.*

On a larger scale, in West Africa west of the Niger, 18 of the 74 African haplorhine species (27%) are present and 8 (11%) are endemic there. In contrast, strepsirrhines have only four of their 25 African species (16%) in West Africa and none are endemic there (figures from Groves, 2005). The greater representation of haplorhines in West Africa correlates with their presence in Brazil and the absence of strepsirrhines there. In the south of the continent, haplorhines include three extant clades (plus fossil colobines and hominids in western South Africa), while strepsirrhines occur only in eastern South Africa. In one of the widespread African clades of haplorhines (*Procolobus + Piliocolobus*), a clade endemic west of the Niger is basal. There is no strepsirrhine clade with this pattern.

In East Africa, from South Africa to Kenya there are only five endemic haplorhine species out of 74 total in Africa (7%); another five are endemic to the Ethiopia/Red Sea area. In contrast, in the South Africa–Kenya area there are seven endemic strepsirrhines out of the 25 African species (28%); no strepsirrhines are endemic to the Ethiopia/Red Sea area. In addition, a coastal East African strepsirrhine (*Galago zanzibaricus*, perhaps with the coastal *G. granti*; see below) is basal to a widespread African clade; no haplorhine has this pattern.

All these statistics and examples point to an underlying allopatry in the distributions of haplorhines and strepsirrhines. The split seen here between western and eastern sectors of Africa is one of the most important patterns in the biogeography of the continent and does not coincide with an ecological break. In other taxa it is seen in the pantropical plant family Combretaceae, for example. This comprises three groups (Maurin et al., 2010):

- the basal group in west tropical Africa;
- a group comprising a mangrove in America and *West* Africa (*Laguncularia*), a mangrove in *East* Africa to the Pacific (*Lumnitzera*), and two Australian genera;
- a large pantropical clade.

THE TEN WIDESPREAD CLADES OF AFRICAN PRIMATES

The five main clades of primates all have different distributions, despite some overlap (Fig. 3-1). In Africa, as just discussed, haplorhines are concentrated in the center and the west, strepsirrhines in the east and in Madagascar,

What are the patterns *within* African strepsirrhines and haplorhines? The African primates comprise ten widespread and widely overlapping clades (nine are shown in Figs. 5-3 to 5-12, and one of these, *Cercopithecus*, itself comprises two widespread clades, the *mona* group and the *cephus–mitis* groups, not shown here). The ten widespread groups all show extensive sympatry and dissecting out their possible original allopatry is not attempted here. The ten African clades can be compared with the eight widespread clades in tropical America.

The overlap among the ten African widespread clades could be due either to sympatric evolution or to initial allopatry followed by range expansion ("dispersal" in the sense of normal ecological dispersal, not founder dispersal). Given the high degree of allopatry among the five main primate clades and also *within* the widespread African clades, it seems likely that the widespread African clades themselves were also originally allopatric, and so the second model is accepted here. This implies that there was a phase of mobilism widespread throughout sub-Saharan Africa that led to the overlap of the widespread clades.

The extensive sympatry among the widespread clades means that their origin is obscure, but in contrast, *within* the widespread clades the biogeographic patterns are striking. The phylogenetic and biogeographic breaks in nine of the ten widespread clades are indicated in Figs. 5-3 to 5-12, where the two main groups in each clade are shown (in gray and in bold outline, respectively; the three main clades are shown for galagos, Fig. 5-4). As it turns out, most of the widespread clades comprise pairs of sister genera. The patterns of differentiation within each of the widespread clades are very similar, suggesting that a single set of events has affected the ten ancestral complexes of the clades when they were already widespread. Thus, the phase of mobilism that led to the overlap of the widespread clades occurred after ~130 Ma, the age of Atlantic rifting and the inferred age of catarrhines, but before the differentiation within the widespread clades. The geography of the breaks in the widespread clades is studied here in order to date the breaks.

In three of the widespread African clades (Figs. 5-3, 5-11, 5-12) the two main groups in each clade show complete overlap. In six of the widespread clades (Figs. 5-4, 5-5, 5-6, 5-7, 5-8, 5-10) there is at least some vicariance between the two main groups, and this is compatible with original allopatry in all the widespread clades.

STREPSIRRHINES IN AFRICA

Strepsirrhines comprise the lorisiforms of mainland Africa and Asia and the lemurs (lemuriforms) of Madagascar. Lorisiforms are among the most poorly known primate groups and are sometimes regarded as a minor offshoot of lemurs. Authors such as Godinot (2006b) still include lorisiforms under lemuriforms, but this downplays some important distinctions. Molecular clock (branch length) studies concluded that the two groups diverged long before the Old World catarrhines split from the New World platyrrhines (Janečka et al., 2007; Steiper and Young, 2008; Arnason et al., 2008; Matsui et al., 2009).

Lorisiforms comprise the lorisids of Africa and Asia and the galagos, known only from Africa. In Africa the lorisids form the first of the widespread clades (Fig. 5-3), with *Arctocebus* distributed between the Niger and Congo Rivers and its sister, *Perodicticus*, widespread from West Africa to Kenya.

In galagos (Fig. 5-4), *Galagoides demidoff* and *G. thomasi* are widespread from West Africa to Uganda/Zambia. In the remaining clade, *Galago* (or *Galagoides*) *zanzibaricus* of eastern Tanzania is sister to a large, widespread clade of 15 species (treated variously by different authors in *Galago*, *Galagoides*, *Otolemur*, *Sciurocheirus*, and *Euoticus*; Roos et al., 2004; Masters et al., 2005; Masters, Boniotto et al., 2007). This group of 15 species is mainly eastern, and in Somalia it is recorded from the driest, thorniest vegetation inhabited by African primates (Masters et al., 2005; Masters, Boniotto et al., 2007).

Galago zanzibaricus (Zanzibar and Mafia Islands, Rufiji delta, and the Usambara, Uluguru, and Udzungwa Mountains) defines both an area of endemism and an important phylogenetic/biogeographic break, the Usambara–Uluguru node. Many taxa endemic here (for example, groups of mosses, land snails, and plants) have sister groups in Madagascar (Burgess et al., 2007). The significance of the Usambara–Uluguru region lies not only in the number of its endemic taxa, but also in their far-flung affinities and the phylogenetic status of many of the endemics as sister to large, widespread clades. Apart from the Madagascar connection, endemics at the node, such as *Galago zanzibaricus* (perhaps with the coastal *G. granti*), have sister groups that are

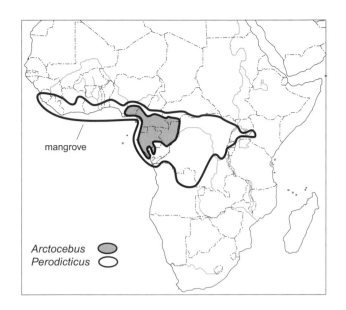

FIGURE 5-3. Distribution of *Arctocebus* and *Perodicticus* (Lorisidae).

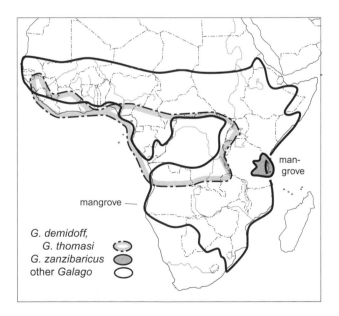

FIGURE 5-4. Distribution of *Galagoides demidoff* and *G. thomasi* (broken line). *Galago zanzibaricus* (gray) and its sister group (a clade including *Galago*, *Euoticus*, *Otolemur*, and *Sciurocheirus*).

widespread through much of sub-Saharan Africa. This indicates that a phylogenetic break near the base of galago phylogeny in or around the Usambara–Uluguru region fractured a widespread clade. Prior to rifting in the Mozambique Channel, the Tanzania center was adjacent to Madagascar, the other main center of strepsirrhine diversity.

Phylogenetic study of the galagos is progressing but is still far from resolved. In the study by Chatterjee et al. (2009), for example, it is not *G. zanzibaricus* alone, but the pair *G. zanzibaricus* plus *G. granti* of coastal Mozambique to coastal Tanzania that is sister to a widespread African complex. This would not change the overall pattern of a basal eastern split in the widespread galago clade.

HAPLORHINES IN AFRICA

Within the widespread clades of haplorrhines, the boundaries of the main subclades occur around the Nigeria/Cameroon border–Gabon area (as in the lorisids) and the eastern Congo.

Colobines

In colobines, there are two widespread clades. In the first, *Procolobus* s.str. of West Africa (west of the Niger) is basal to *Piliocolobus*, widespread across Africa (Fig. 5-5; phylogeny from Ting, 2008). Both genera occur in mangrove in West Africa. *Piliocolobus* is also in "marsh forest" of the Niger Delta, montane forest to 1,400 m in northwestern Cameroon, flooded forest in the central Congo basin (the "cuvette centrale"), miombo woodland in the Udzungwa Mountains of Tanzania, and mangrove on Zanzibar. At the last locality, *Piliocolobus* and another monkey, *Cercopithecus*, obtain freshwater from treeholes and coral-rock crevices (Nowak, 2008). Miombo woodland covers a large area of southern and eastern Africa that was uplifted through the Cenozoic. Many miombo taxa have their closest relatives in west and central African rainforest, especially swamp forest and mangrove (Craw et al., 1999), and *Piliocolobus* in Tanzania is a typical example.

In the second widespread clade of colobines, *Colobus* (Fig. 5-6; cf. *Piliocolobus*, Fig. 5-5) is widespread across Africa. *C. satanas* of Cameroon, Bioko, and Gabon is basal (Ting, 2008); this species ranges from coastal swamp forest to montane heathland on Bioko Island, and the genus is recorded up to 2,400 m in Kenya.

In *Colobus angolensis*, Groves (2006: 17) noted an "odd hiatus" in eastern Congo between the Lindi and Ulindi rivers (lowland Kivu).

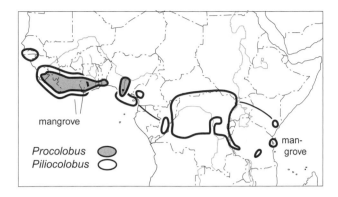

FIGURE 5-5. Distribution of *Procolobus* and its sister *Piliocolobus* (Ting, 2008).

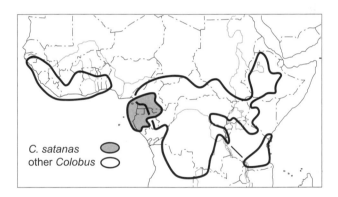

FIGURE 5-6. Distribution of *Colobus satanas* and its sister group, the remaining *Colobus* species (Ting, 2008).

He also cited a similar but smaller gap between Maiko and Ulindi in *Piliocolobus* (Fig. 5-5). Groves (2006) observed that the gap is "precisely" the area where *Gorilla bereingei graueri* extends into the lowlands toward the Lualaba and regarded this as "especially interesting." Groves (2006: 20) asked, "What determines which species shall exist where? Is it chance, or is it competition?" It may be neither: Competition may maintain a pattern, but the pattern itself has probably been caused in the first place by historical factors of paleogeography. In such a dynamic region as Kivu, there are many possible scenarios that could explain local allopatry.

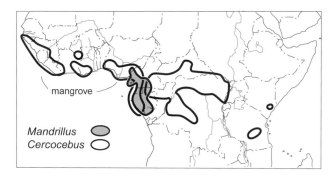

FIGURE 5-7. Distribution of *Mandrillus* and its sister *Cercocebus*.

Papionines

African papionines (the baboons and their allies) comprise two clades. In the first (Fig. 5-7), *Mandrillus* of the Gulf of Guinea is sister to the widespread *Cercocebus* (in mangrove in West Africa, seasonally inundated forest in central Africa, and submontane forest in Tanzania).

In the other papionines (Fig. 5-8), *Lophocebus* of Nigeria to Uganda is sister to *Theropithecus* + *Papio* (including *Rungwecebus*), found throughout most remaining parts of sub-Saharan Africa. *Lophocebus* is in swamp forest; *Papio* is present in mangrove, miombo, and similar communities; *Theropithecus* survives in montane grassland and moor in Ethiopia at 1,800–4,400 m. The mountainous plateau there, along with much of eastern and southern Africa, is the result of epeirogenic uplift in the Tertiary. *Theropithecus* is also known from fossils in east and South Africa, and in India (Uttar Pradesh).

In African strepsirrhines, differentiation has occurred in the west (between *Arctocebus* and *Perodicticus*), but in addition, strepsirrhines have a basal node at Usambara–Uluguru. In contrast, there are no obvious basal haplorhines in the South Africa–Kenya region. Olson et al. (2008) proposed *Rungwecebus* (southwestern Tanzania) as sister to *Papio*, although the only molecular phylogeny available at that time including more than one *Papio* species (Davenport et al., 2006: Fig. 2) showed *P. papio*, other *Papio* species, and *Rungwecebus* in a tritomy. Mitochondrial data indicate that *Rungwecebus* is closest to *Papio* species in the same area (*P. cynocepehalus* and *P. ursinus*); nuclear data are inconclusive (Zinner, Arnold, and Roos, 2009; Burrell et al., 2009). Despite this, Zinner, Arnold, and Roos (2009) concluded that "due to its morphological and ecological uniqueness *Rungwecebus* more likely

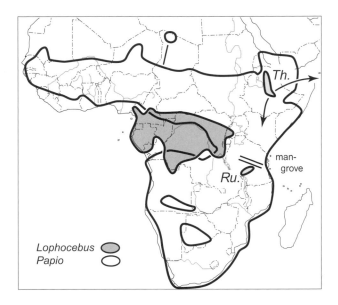

FIGURE 5-8. Distribution of *Lophocebus* and its sister *Papio* (incl. "*Rungwecebus*"; *Ru.*) + *Theropithecus* (*Th.*). *Theropithecus* is also known from fossils in East Africa, South Africa, and India. The main break in *Papio* is indicated by a double line (Zinner Groeneveld et al., 2009).

represents a sister lineage to *Papio* and experienced later introgressive hybridization" (p. 1) (cf. Roberts et al., 2010). Burrell et al. (2009) concluded instead that *Rungwecebus* is not a basal, eastern group, but an old hybrid of *Papio cynocephalus* and a *Lophocebus* species.

There is little molecular support for the traditional species delimitations in *Papio*; instead there are two main clades, one distributed north of a boundary at the Usambara–Uluguru node and one found to the south (Fig. 5-8; Zinner, Arnold, and Roos, 2009; Zinner, Groeneveld, et al., 2009; Keller et al., 2009). Zinner Groeneveld et al. (2009) proposed a center of origin for *Papio* in southern Africa, although there is no real evidence for this. The main problem in the group as a whole is explaining the large degree of allopatry between *Lophocebus* and *Papio*.

Cercopithecines

The cercopithecines (guenons and relatives) are a diverse African group of about 36 species with a phylogeny as shown in Figure 5-1 (Xing et al., 2007; see also Chatterjee et al., 2009). Some studies show other

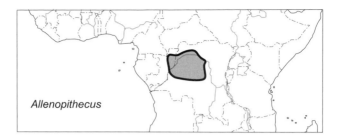

FIGURE 5-9. Distribution of *Allenopithecus.*

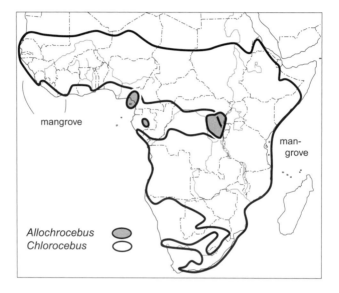

FIGURE 5-10. Distribution of *Allochrocebus* (*Chlorocebus lhoesti* group) and its sister *Chlorocebus* s.str. plus *Erythrocebus.*

arrangements (for example, with *Miopithecus* basal to *Chlorocebus* + *Cercopithecus*; Tosi et al., 2004; McGoogan et al., 2007), but most agree that Allen's swamp monkey, *Allenopithecus*, is basal. It is landlocked in the cuvette centrale of the Congo and occurs in swamp forest (Fig. 5-9). Barnett and Brandon-Jones (1997) regarded it as an Old World equivalent of *Cacajao*, the swamp forest genus landlocked in Amazonia, and both genera can be interpreted as former mangrove-associates that have been stranded inland with the retreat of former inland seas.

In the other cercopithecines, *Chlorocebus* plus *Allochrocebus* (Fig. 5-10) is widespread but allopatric with *Allenopithecus* (Fig. 5-9) (these clades

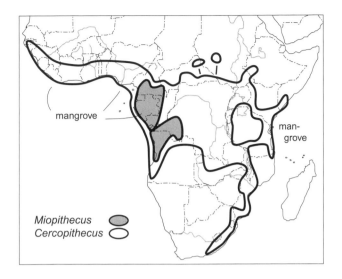

FIGURE 5-11. Distribution of *Miopithecus* and its sister *Cercopithecus* s.str.

are all terrestrial). *Allochrocebus* (formerly the *C. lhoesti* group; Groves, 2006) is disjunct around the Congo basin and is sister to the widespread *Chlorocebus* (incl. *Erythrocebus*; Fig. 5-10). Tosi (2008) explained the disjunction in *Allochrocebus* as the result of dispersal from one side of the Congo basin to the other, perhaps around the northern rim. But he did not mention the critical fact that *Allochrocebus* is vicariant with its sister, and so the two probably evolved by allopatry. Dispersal is required only to produce the minor overlap of *C. lhoesti* and the other *Chlorocebus* near the Congo/Uganda border. *Chlorocebus* is recorded in mangrove in both West and East Africa.

In the last group of cercopithecines, *Miopithecus* of Cameroon–Angola (often recorded in mangrove) is sister to *Cercopithecus* (Fig. 5-11). The latter is a diverse, arboreal genus comprising the widespread *mona* group (Senegal to Uganda) and a second widespread clade made up of the *cephus* group (mainly in West Africa) and the *mitis* group (mainly in East Africa).

The formation of the widespread clades in Africa, their overlap, and finally the differentiation of their main groups may all have occurred with the repeated Cretaceous marine incursions and regressions discussed below. The more local swamp-forest clades (*Allenopithecus* and *Miopithecus*) and the widespread ones (in both swamp forest and

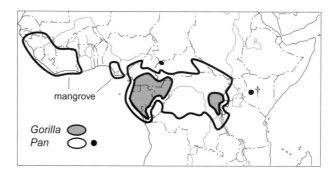

FIGURE 5-12. Distribution of *Gorilla* (gray) and one member of its sister group, *Pan*. (The other member of this clade, *Homo*, is not shown.)

non-flooded forest) may all be relictual, having been stranded following minor Cretaceous uplift and recession of the inland seas. The only reason *Allenopithecus* does not occur in true mangrove is that it is landlocked, probably for historical reasons.

In the last group of African primates, the hominids, *Gorilla* of central Africa (Fig. 5-12) is disjunct across the Congo basin in a pattern similar to that of *Allochrocebus* (the *Chlorocebus lhoesti* group) (Fig. 5-10). *Gorilla* is sister to the more widespread *Pan* (in mangrove in West Africa) and *Homo*. For a mapped treatment of fossil hominids, see Grehan and Schwartz (2009).

TECTONICS AND PRIMATE EVOLUTION IN AFRICA

The regions of Africa where primate clades have basal biogeographic breaks and centers of diversity are of special interest, and the structural history of these areas may shed light on the divergence events. The relevant areas include the Gulf of Guinea region (southeastern Nigeria, Cameroon, and Gabon), the eastern Congo, and, in strepsirrhines, eastern Tanzania. With respect to tectonics, the Cenozoic rift system of East Africa is well known, but other large areas of Africa have also been dominated by regional extension at different times. Rifting in the Early Jurassic (Lebombo monocline) and Middle Jurassic (Mozambique Channel) has already been cited as a possible cause of differentiation between major clades of primates. In the Late Jurassic, widespread rifting occurred in western, central, and northern Africa, with crustal extension culminating in the Cretaceous. By this time, a transcontinental rift system had developed, and this was connected directly with early rifting in the Atlantic Ocean (Fig. 5-2). Reactivation of the African basins continued in the Paleocene and Eocene, and new rifts formed in the Red Sea and western Kenya. In the Oligocene and Early Miocene, rifts in Kenya, Ethiopia, and the Red Sea linked and expanded to form the current East African rift system (Bosworth, 1994).

EARLIEST JURASSIC: TETHYAN RIFTING IN NORTH AFRICA

The breakup history of Pangaea and Gondwana was long and complex. The initial phase involved intense magmatic activity around what would become the margins of the Central Atlantic, in the southeastern United States, northeastern South America, and from Liberia to Spain. This was one of the greatest volcanic eruptive events known and produced the Central Atlantic "large igneous province." It may also have led to the divergence of primates from their sister group, the plesiadapiforms. In geological terms, the main period of magmatism was more or less instantaneous, taking place within just 1.5 m.y. at 200 Ma (Triassic–Jurassic boundary) (Hames et al., 2000; Nomade et al., 2006). Continental flood basalts ("traps") and oceanic plateaus are grouped together as large igneous provinces. The volcanism in these usually lasts for ~10 m.y., but most of the volume is erupted in ~1 m.y. (Courtillot and Renne, 2003). Over short periods, fluxes may exceed global production at mid-ocean ridges, and the emplacement of a large igneous province is a major geodynamic event. If not the largest in volume, the Central Atlantic magmatic province covers the greatest area of any continental large igneous province (McHone, 2003). The magmatism may have been caused by mantle plumes or by convection effects interacting with prior lithospheric structure (McHone, 2000).

The Triassic–Jurassic boundary represents one of the three (White and Saunders, 2004) or five (Tanner et al., 2004) greatest mass extinctions (along with end-Cretaceous and earlier ones), and it has been linked to the Central Atlantic magmatism (Hesselbo et al., 2002). The link is controversial, and other great igneous events, such as the southern Atlantic (Paraná–Etendeka) flood basalt eruptions in the Early Cretaceous, were not associated with extinction (Wignall, 2005; Nomade et al., 2006). In any case, there was a significant disturbance of the global carbon cycle at the Triassic–Jurassic boundary, and there is a close association between large igneous province formation and sudden environmental change (Tanner et al., 2004; Kelley, 2007). Whatever the cause, there was great extinction at the Triassic–Jurassic boundary, and this would have provided new ecological opportunities for any weedy taxa that survived. Weedy groups tolerant of great disturbance may have spread (post-extinction fossil faunas were cosmopolitan; Pálfy, 2003), and it is suggested here that these groups included ancestral Archonta, the precursors of primates and related orders.

Following the Central Atlantic magmatism, extension and eventually rifting took place between northwestern Africa/South America and North America at ~185 Ma, in the Early Jurassic (Veevers, 2004). Mid-oceanic ridge basalts on the Canary Islands that were erupted at this time represent the earliest sea-floor spreading in the area (Steiner et al., 1998) and the initial breakup of Pangaea into the southern Gondwana and the northern Laurasia. Either the Central Atlantic magmatism, the rifting, or both may have led to one of the primary differentiation events in Archonta, an allopatric split between the northern plesiadapiforms in Laurasia and their southern sister group, the primates, in Gondwana. Later, at some time in the Jurassic–Cretaceous, primates expanded north into the U.S., Europe, and northern China, as shown by the fossil primates there (omomyiforms and adapiforms, from Eocene to Miocene, and also anthropoids), but they have never expanded into the far north, which was, as far as fossils indicate, occupied by plesiadapiforms alone.

Major intracontinental extension and rifting in north Africa during the Triassic–Jurassic produced basins in which the sediments that form the present Atlas Mountains accumulated (Teixell et al., 2003; Arboleya et al., 2004). The complex fault pattern seen there developed under a NW–SE tensional field, either as a strike-slip (transcurrent) system, an extensional system, or a combination of the two—a transtensional (oblique) rift system. In the west, the basin was open to the proto-Atlantic; to the east there were several epicontinental troughs connected to the Tethys. Africa/South America rotated away from North America and toward Europe/southwestern Asia. With the eventual collision of Africa and Eurasia in the Cenozoic, the Mesozoic extensional troughs evolved into compressional belts with the inversion of the Mesozoic faults and the rise of the Atlas Mountains (Ellouz et al., 2003).

Thus, tectonic change in the same region of north Africa may have been involved in primate evolution during different phases. In the first, a zone of extension, subsidence, and rifting divided plesiadapiforms from primates. In the second, following range expansion of primates northward, northern primates (omomyiforms and adapiforms) differentiated from their southern relatives (extant clades) along a Tethyan boundary. In a third, Cenozoic, phase, a thick-skinned thrust and fold belt produced high mountains in the Atlas fold belt and may have been part of the margin that separated *Macaca* of north Africa and Asia from its relatives, the other papionines in sub-Saharan Africa and, fossil, in India.

EARLY JURASSIC: THE LEBOMBO MONOCLINE, SOUTHERN AFRICA

Extant haplorhines are widespread throughout South Africa, but strepsirrhines (living and fossil) occur in South Africa only east of Lesotho and north in Transvaal (Figs. 5-3, 5-4). Similar biogeographic boundaries around the Lebombo monocline are common (Craw et al., 1999: Fig. 3-10, mapped examples in ferns, angiosperms such as *Dianella*, and birds). Many taxa have a distribution break in the region resembling that of strepsirrhines; these groups occur in Madagascar, East Africa, and elsewhere, but have western limits near the Lebombo monocline and, in particular, are absent from Western Cape province. Plant examples mapped in Stevens (2010) include Cycadaceae (replaced in South Africa by Zamiaceae, which are not on Madagascar), Gesneriaceae, Kirkiaceae, Buxaceae + Didymelaceae, Aphloiaceae (replaced in Western Cape by Geissolomataceae), Basellaceae, Lecythidaceae, Primulaceae subfam. Maesoideae (replaced in Western Cape by subfam. Theophrastoideae), Podostemaceae subfam. Tristichoideae, and Ericaceae subfam. Vaccinioideae. The last group is in South Africa only in the east (in the Drakensberg Range), unlike subfam. Ericoideae, which has a center of high diversity in Western Cape province. The boundary in all these families is attributed here to differentiation at the Lebombo monocline.

The dragonfly *Hemicordulia* ranges east from the Lebombo monocline (St. Lucia, Maputo, etc.) to Madagascar, the Mascarenes and Seychelles, India, Southeast Asia, and Australasia to French Polynesia. Dijkstra (2007: 28) invoked wind dispersal for *Hemicordulia*: "As is clearly shown by its distribution, the genus has most decided powers of dispersal." But this does not explain the precise limit in Africa. Why would the genus blow halfway around the world and then stop at the Lebombo monocline? The African mainland species is endemic; if the powers of dispersal are so effective, what caused the break with congeners in Madagascar?

Mesozoic sediments in the extensive Karoo basin of southern Africa include many vertebrate fossils, and these are of special interest as they show a sample of the transition from "pre-mammalian" to "mammalian" structure. Deposition in the basin ended in the Early Jurassic with a phase of extensive volcanism, in which flood basalts erupted over much of southern Africa. This Karoo volcanism, along with coeval flood volcanism in Patagonia–West Antarctica (Chon Aike province) and Antarctica–Tasmania–New Zealand (Ferrar province), marks the first stage in the breakup of Gondwana.

Features carved out of the Karoo basalts by later erosion, such as the Drakensberg and the Victoria Falls, are spectacular highlights in the current landscape, but the most impressive tectonic feature in the system is the Lebombo monocline (Watkeys, 2002; Jourdan et al., 2007). This is a linear, volcanic rifted margin which runs roughly north–south in eastern South Africa, Mozambique, and Zimbabwe. It connects with the Sabi monocline and the Southern Malawi dike swarm, giving a belt of deformation at least 1,500 km long (Fig. 5-2). The Lebombo monocline represents a zone of intense volcanism and is also a great fold. Monoclinal flexing and faulting took place here, together with rhyolite eruption, at ~180 Ma (late Early Jurassic), and all the beds dip eastward, toward the Mozambique Channel. To the west of the monocline, the crust is of normal thickness; to the east the crust has been thinned by extension (Miller and Harris, 2007). The monocline was formed before the Mozambique Channel but runs parallel with it and is a precursor to it. Extension has also occurred along what is possibly a branch of the monocline, the rift arm that forms the Limpopo valley; this and the Zambesi valley are the only gaps in the rim formed by the Lebombo monocline. The monocline may also be contiguous with the Explora Escarpment in the Antarctic Weddell Sea (Marks and Tikku, 2001).

Volcanic rifted margins and ocean basins, such as the Lebombo Monocline and the Mozambique Channel, develop when continental crust is thinned or torn. The process involves magmatism and extension, and also uplift along the rift shoulders (Menzies et al., 2002). This may be just as important for biological evolution as the rifting itself, as discussed for southeastern Brazil in Chapter 4. Another characteristic feature of continental rifted margins is flood volcanism. This can occur at the same time as breakup extension or predate it, sometimes by tens of millions of years. The breakup of Gondwana was associated with the Karoo flood basalts and later those of Paraná–Etendeka volcanism in the Atlantic.

MIDDLE JURASSIC—THE MOZAMBIQUE CHANNEL AND ASSOCIATED RIFTING

Rifting between South America/Africa (West Gondwana) and Madagascar/India/Australia/Antarctica (East Gondwana) began in the Middle Jurassic (160 Ma) along the Mozambique Channel. The process that ended with Madagascar separating from Africa was not just a single split along one great fault; large-scale faulting occurred *within* both land masses as well as along the channel between them. At the end of the long Karoo phase of African history (Carboniferous–Jurassic) and at the beginning of Gondwana breakup, rift faulting produced Jurassic–Cretaceous horsts and grabens over much of East Africa and eventually the coast itself. No later faulting in the region has approached this magnitude (Griffiths, 1993). Lake Malawi, in central East Africa, has more fish species (over 400 cichlids alone) than any other lake. The primate *Rungwecebus* is only known from the Rungwe Mountains at the head of the lake and in the Udzungwa Mountains of central Tanzania (Fig. 5-8; Olson et al., 2008). It may be an old hybrid (*Papio × Lophocebus*, see above), and anomalous hybrids, along with boundaries, endemism, and absences, are a typical feature of biogeographic nodes. The basin that Lake Malawi occupies originated as a rift in the Karoo basin and developed through the Mesozoic and Cenozoic. The Rungwe and Udzungwa Mountains have biogeographic connections around the margin of the Tanzania craton with the Usambara–Uluguru node, and the Uluguru Mountains themselves had started to form as a distinct fault-bound unit by the end of the Karoo phase (Griffiths, 1993).

The monocot tribe Kupeeae is basal in the pantropical Triuridaceae (Heads, 2009) and occurs only at Cretaceous rifts – in the Bakossi Mountains at the junction of the central African rift system and the Cameroon volcanic line, and in the Udzungwa Mountains by the Kilosa basin (Fig. 5-2; the nearby Rukwa rift basin includes Cretaceous mammal and dinosaur fossils; O'Connor et al., 2006).

MADAGASCAR

The classification of extant primates by Goodman et al. (2005) attributes highest family diversity to Madagascar, with five families. Other forms here include the extinct sloth lemurs Paleopropithecidae and monkey lemurs Archaeolemuridae, both much larger than living lemurs and possibly extant until 100 years ago. In contrast with the seven families of primates in Madagascar, there are four extant families in Southeast Asia, four in mainland Africa, and three in America. These figures incorporate new molecular data and show subtle but important differences from earlier treatments (Groves, 2001). In particular, they highlight the great concentration of primate diversity in Madagascar, and this may reflect biogeographic heterogeneity already present there before the rifting.

The western third of Madagascar is composed of Carboniferous–Cretaceous continental sequences equivalent to the Karoo system of Africa and the Gondwana system of India (de Wit, 2003). The sequences filled large, distinct basins that were precursors to the complete rifts that finally developed at Africa/Madagascar breakup. The largest, Morondava basin, is 1,000 km long. The basins are half-grabens formed by normal (tensional) faults and have also been affected by major dextral movement along strike-slip faults. The record indicates a slowly subsiding environment of meandering rivers, swamps, and deltas that only ended in the mid-Jurassic when there was a sudden change to regional marine conditions. This marks the end of a period of rifting and terrestrial sedimentation that lasted for ~120 m.y. The long phase of survival and evolution in the swamp forests and back-mangroves of a rifting margin, culminating in complete separation, would have been critical for the mammal populations in the region, including the pre- and proto-primates. Arnason et al. (2008) and Matsui et al. (2009) concluded that the basal node in lemurs, that is, the split between the aye-aye *Daubentonia* and the others, occurred before the catarrhine/platyrrhine division (calibrated here with the Early Cretaceous opening of the Atlantic). Distribution of the lemurs within Madagascar is not examined here, but it is relevant that *Daubentonia* (the basal clade) and two other lemur genera, *Microcebus* and *Propithecus*, are recorded in mangrove (J. Masters, pers. comm.).

De Wit (2003) described the biogeographical origin of the extant Madagascar vertebrates as one of the great unresolved questions of natural history. He also pointed out a related problem with the Late Cretaceous fossil vertebrates there. Some of these indicate Gondwanan affinities with South America and Africa long after Madagascar/Antarctica/India had separated from Africa/South America in the mid-Jurassic. De Wit (2003) suggested that "The paradox is important to solve, for it probably holds clues to similar enigmas concerning the modern endemics in Madagascar." The "paradox" of the Late Cretaceous fossil forms and the "enigma" of the extant forms are different aspects of the same problem. Some of the groups that were fractured by the opening of the Mozambique Channel, such as the dinosaurs, died out later, while others, including the primates, have survived until now. The Gondwana biota was not homogeneous, and clades that had differentiated before final breakup may well show affinities, such as Madagascar–South America, that are incongruent with the breakup sequence. In primates, though, the phylogeny conforms with the breakup sequence.

EARLY CRETACEOUS: RIFTING IN WEST AFRICA, CENTRAL AFRICA, AND THE ATLANTIC

The area between the Niger delta in Nigeria and the Sanaga River in Cameroon, including Bioko Island, is one of the main centers of primate diversity in Africa. Twenty-two primate species are present in the region, and individual forests support up to 14 sympatric species (Oates et al., 2004). At least six of the species are endemic. Major phylogenetic breaks in one strepsirrhine clade (Fig. 5-3) and in seven African haplorrhine clades (Figs. 5-5, 5-6, 5-7, 5-8, 5-10, 5-11, 5-12) occur around the Nigeria/Cameroon border. In addition, a species-level boundary occurs in at least seven primate clades at the Sanaga Valley (Gonder et al., 2006). The important biogeographic region around the Gulf of Guinea has long been known as a boundary separating taxa in the "Upper Guinea" forest of West Africa from their relatives in the "Lower Guinea" forest of central Africa. For a molecular study of another group in which an Upper Guinea clade is separated from a Lower Guinea clade at the Sanaga River, see McBride et al. (2009), on butterflies in the genus *Cymothoe* (Nymphalidae).

Despite its importance, the boundary at the Nigeria/Cameroon border does not correspond to any obvious discontinuity in rainforest environment, and this has puzzled ecologists. The only obvious ecological break in the African rainforest occurs much further west, in Togo and Benin, where the savanna reaches the coast, but this limit has no biogeographic significance above species level for primates or most other groups. On the other hand, the biogeographic break at the Nigeria/Cameroon border also has special tectonic significance, as it is the site of large-scale Cretaceous rifting (Fig. 5-2).

In the Early Cretaceous, flood volcanism around the opening Atlantic (the Paraná–Etendeka large igneous province of Argentina/Brazil and Namibia) heralded the separation of Old World catarrhines from New World platyrrhines. In addition to the seas that developed along this intercontinental rift, Cretaceous seaways developed on *intra*-continental rifts in Africa that ultimately failed and filled with sediment. The linkage of spreading basins in the north and south "Atlantic Oceans" by Early Cretaceous transform faults produced the modern Atlantic and at the same time led to a major phase of basin development along the central

African rift system (Fairhead, 1986, 2003), notably in the Gulf of Guinea region. Before it eventually failed, this rifting produced the Benue trough in Nigeria. The trough is 5,000 m deep and has filled with sediment, although the Benue and lower Niger rivers still flow along it. The trough is delimited by large, subvertical faults, some with a throw of up to 1,000 m. As Africa and South America separated, the South Atlantic rift propagated north-ward into the Benue trough, and this developed as a complex, active, strike-slip basin (Guiraud and Plaziat, 1993; Basile et al., 2005). The trough is usually interpreted as the failed arm of three spreading ridges which meet at a triple junction in the Atlantic. It was the most prominent continen-tal structure formed in Africa as it separated from America (Keller et al., 1995). Transtension leading to extension and grabens with normal faults, together with wrench structures and strike-slip faults, were dominant dur-ing the formation of the trough. Strike-slip faulting continued from the lat-est Early Cretaceous (Albian, ~100 Ma) until the end of the Cretaceous.

The Benue trough is just one part of an extensive rift system which ramified on both sides of the Atlantic, in Brazil and in Africa. The African portion of this rift system has several arms, all Cretaceous basins >2,000 m deep (Fig. 5-2). The Early Cretaceous sea in the Benue trough connected with seas in the Sahara region, and so West Africa was completely separated from central Africa. In addition, the rifts extended east from Nigeria as far as southern Sudan and Kenya, forming a transcontinental rift system (Keller et al., 1995).

The Benue trough includes the great petroleum systems of the Niger delta, and most of the West and central Africa rift basins include signifi-cant quantities of hydrocarbons. The countries in sub-Saharan Africa currently producing the largest amounts of oil are Nigeria, Equatorial Guinea, and Angola. The Niger delta ranks among the world's most prolific hydrocarbon-producing deltas, along with the Mississippi, the Orinoco, and the Mahakam. Mann et al. (2003: 36) wrote:

> The rift history in west Africa consisted of a Berriasian to late Barremian–early Aptian period of continental separation between west Africa and Brazil, which encompassed a deforming zone as wide as 200–500 km. . . . [W]est African rifting progressed in several phases that became more focused with time as extension migrated to the west. Each rift phase produced deep, anoxic, lacustrine sections that are the source rocks for giants [oil fields] in the overlying passive margin section. After the cessation of rifting Aptian salt was deposited as a result of marine incursions. . . . The Benue rift con-tains a basal fill of early Cretaceous alluvial fan, braided river and lacus-trine deposits overlain by transgressive marine deposits of middle Albian

age. . . . Folding and inversion of the Benue rift occurred in the Santonian–early Campanian as a result of intraplate deformation of the African Plate, probably as a response to South Atlantic opening.

This sequence of events led to development of petroleum systems and may also have been involved in the production of unusual levels of biodiversity.

Keller et al. (1995) wrote that "The West and Central Africa Rift System is a large and complex feature whose elements have experienced distinct, usually multiphase geologic histories. . . . However, these elements all seem to have experienced a major phase of extension in the Cretaceous which is the basis for linking them." At this time, strike-slip movement, as seen in the Benue trough, occurred across the continent as it almost split apart. The amount of extension on these rifts was much greater (~70 km) than on the more well-known Cenozoic rifts of East Africa (~10 km) (Fairhead, 1986). Geologists regard the Benue trough and the Gulf of Guinea region as an ideal place to investigate the interaction of extension and shear during large-scale rifting. For ecologists and conservationists, the region is better known for its extraordinary species diversity, while phylogeneticists are interested in the coincident evolutionary breaks that occur there in many lineages. It is suggested here that these different phenomena are causally related.

The Benue trough abuts a Late Cretaceous–Cenozoic structure to its southeast, the Cameroon volcanic line with the associated Adamawa dome (Fig. 5-2; Njome and Suh, 2005). The volcanic line runs from the Gulf of Guinea islands (São Tomé, Bioko) toward Lake Chad and is an intraplate hotline, not a plate margin. It is unique in having volcanic centers in both continental and oceanic crust. The volcanoes along the line do not show a linear age sequence, and the line is probably controlled structurally rather than by plate movement over a hotspot. It has been active in episodes since the end of the Cretaceous, with alkaline intrusive magmatism from 65 to 30 Ma and volcanism from 35 Ma to the present. The Adamawa dome is the result of Cenozoic uplift. The Cameroon volcanic line and the Adamawa dome lie between the Benue trough to the northwest and the Sanaga fault system, an old dextral shear, to the southeast (Nnange et al., 2001). The Cameroon volcanic line runs parallel to the extensional structure of the Benue trough and meets the rift system in northern Cameroon, but the relationship between the tectonics and the magmatism is still debated (Marzoli et al., 2000). Magmatism, rifting, and uplift may all have contributed to the phylogenetic breaks here.

CONGO BASIN

African primates are most diverse between the Nigeria/Cameroon border area just discussed and the Congo/Uganda border, a region equivalent to the Congo basin. The richest forests in the world for primate species occur at Makoukou, Gabon, between the western margins of the Congo basin and the Gulf of Guinea area, and in Ituri on the eastern margin of the Congo basin, near the Congo/Uganda border (Fig. 5-2). Forests at these localities have 17 species (Chapman et al., 1999). The central Congo basin, the cuvette centrale, also has distinctive endemics such as the swamp monkey *Allenopithecus*. The direct ancestors of most of these taxa may have been stranded inland during regression of Cretaceous inland seas and lakes.

The Congo basin is unusual among the African basins as it is not located above an old rift but overlies rigid rather than stretched lithosphere (Hartley and Allen, 1994). The basin is a broad downwarp that includes Karoo-type sediments and above these a 1,000-m-thick sequence of mainly terrestrial sediments ranging in age from Triassic to Recent (Giresse, 2005). The Mesozoic sediments were deposited in swampy, lacustrine, or lagoonal basins close to the sea level, as shown by some intercalated strata with marine fossils (Fig. 5-2). These indicate a marine link between the eastern part of the basin and the Indian Ocean in the Late Jurassic. At the same time, a marine transgression is recorded in southern Congo (Kasai). At the beginning of the Cretaceous, there was a minor uplift of the basin and kimberlite intrusion along the southern border of the cuvette centrale. Late Cretaceous marine deposits near Bangui in the northern part of the basin suggest a connection with the trans-Saharan seas of that time. As in the central Africa rift system, the constantly changing shallow seas and coastlines in and around the basin could have had a major impact on the primate fauna.

The Congo basin was finally uplifted in the Late Cretaceous, and the regression of the seas may have stranded many groups. Cenozoic uplift of the basin's borders has prevented any subsequent marine incursion. In 2006, large reservoirs of petroleum were discovered at Lake Albert on the Congo/Uganda border, 70 km northeast of the Ituri site of maximum primate diversity.

Epicontinental seas in Cretaceous Africa were not simply barriers for the pre- and proto-primates, as the new coastlines also provided opportunities for expansion by any "weedy" inhabitants of mangrove and associated swamp and secondary forest. These taxa would have had large areas of new habitat to colonize. Weedy ecology and the ability to thrive in heavily disturbed habitat are retained in modern primates such as *Cercopithecus*, *Chlorocebus*, and *Papio*. But even in a phase of general mobilism, the mobilism has limits, and primates do not cross open sea. The incursions themselves may have caused differentiation, for example, between *Allenopithecus*, endemic in the Congo cuvette centrale, and the other cercopithecines. Differentiation around the boundary of the cuvette centrale (the area of marine fossils) led to *Allenopithecus* inside the boundary and the others outside. This was followed by differentiation further west (around the Benue trough/Gabon) and east, within *Chlorocebus* and *Miopithecus/Cercopithecus*, which was probably associated with activity on the central African rift system. As the Congo basin drained, endemics including *Allenopithecus* and many others were left around its margins and in the cuvette centrale, deposited like the rings in a bathtub.

LATE CRETACEOUS: CONGO-RUFIJI

The West and central Africa rifts were active from mid- to late Cretaceous (125–80 Ma), prior to elevation of the Congo basin to 50 m above sea level toward the end of the Cretaceous. The uplift was due to the rifting of the Atlantic that also caused rift-shoulder uplift in the west of the basin (Fig. 5-2). At this time, the Congo may have drained into the Indian Ocean at the Rufiji delta, an intriguing idea suggested by Stankiewicz and de Wit (2006). Later, with Eocene–Oligocene uplift of East Africa, drainage of the Congo shifted to the Atlantic. This history would explain why the current Rufiji River has a delta that is too large (500 km wide) for the present length of the river (800 km). The new idea is also relevant for biogeography, as it could explain the biodiversity and phylogeny of strepsirrhines in the Rufiji area; the diversity (not necessarily the modern species) could be relictual from the mangroves and associated forests of a large Cretaceous delta. Madagascar, with the bulk of strepsirrhine diversity, was formerly located next to mainland Africa in this same region.

CENOZOIC UPLIFT IN EAST AFRICA

Cenozoic uplift occurred in West Africa on the Cameroon volcanic line/ Adamawa plateau and also over much of southern and East Africa. In southwestern Uganda, rapid Neogene uplift of the Ruwenzori range has outpaced erosion. This has led to the preservation of relict land surfaces, including uplifted river terraces and extensive bogs stranded at high elevation (MacPhee, 2006). Along with the land surfaces, the resident fauna may also have been uplifted *in situ*. *Pan* extends from mangrove in West Africa up to 2,750 m in the Ruwenzori range. Further south along the Congo/Uganda border, *Gorilla* survives at even higher elevations, up to 4,000 m, in the Virunga Mountains.

Of 163 tree species in montane forest at Bwindi, southwestern Uganda, chimpanzees used only 38 for nesting, and 70% of the time used just four species, one each in *Chrysophyllum* (Sapotaceae), *Drypetes* (Euphorbiaceae), *Teclea* (Rutaceae), and *Cassipourea* (Rhizophoraceae) (Stanford and O'Malley, 2008). These families are typical components of back-mangrove worldwide rather than being characteristic montane families (Rhizophoraceae include the "true" mangroves), and this choice of sleeping trees by chimps may reflect old, pre-uplift associations.

Neogene volcanism may have had some minor biogeographic effects, but endemic taxa on the volcanoes of the Congo/Uganda/Rwanda border area, for example, have survived the volcanism more or less *in situ*. They have probably survived as metapopulations, colonizing forest as it forms on new lava flows from forest on older ones. In this way old taxa can survive on very young stratigraphy as the volcano builds up with successive eruptive phases. Endemics on young volcanoes are often dated as older or much older than the volcano they currently survive on; examples include insects on the East African rift volcanoes (Voje et al., 2009; see Chapter 2).

The development of African basins through Phanerozoic time has been related to the polyphase breakup of Gondwana. Rifting occurred from the Carboniferous onward, but "began in earnest during the Early-Middle Jurassic" (Bumby and Guiraud, 2005). Since then, the history of African basins has involved the following (cf. Hartley and Allen, 1994):

1. widespread rifting in the Late Jurassic to Early Cretaceous, fracturing primate populations;

2. extensive inundation under elevated Cretaceous sea levels, enabling mangrove and swamp taxa to expand inland while at the same time dividing other populations;

3. continued rifting through the Cretacous, further fracturing populations;

4. widespread epeirogenic uplift and retreat of the inland seas in the Late Cretaceous–Paleogene, stranding and redepositing coastal "weedy" taxa inland; and

5. off-rift domal uplift associated with alkaline igneous activity in the Cenozoic, especially the Neogene, leading to passive uplift of populations.

The two orders related to primates are Southeast Asian groups and, within primates, the African lorisids and colobines have Asian sister groups (Fig. 5-1). The sister of the sub-Saharan papionines is north African–Asian. In hominids there is a basal Asian grade (the gibbons, then *Pongo*), although as usual this does not imply a center of origin there, but an early center of differentiation.

The breaks between Africa and Asia, or north Africa–Asia, involve the Tethys region, a zone with complex, folded geology. Some aspects of primate evolution and tectonics in Asia are discussed next.

SULAWESI: A COMPLEX SUTURE ZONE

Madagascar has lemurs, America has the New World monkeys, and Sulawesi is famous for its tarsiers (*Tarsius*). Tarsiers extend to Borneo and Sumatra (*Cephalopachus*, one species; Groves and Shekelle, 2010) and the Philippines (*Carlito* of Greater Mindanao, one species; Groves and Shekelle, 2010), and are known from fossils in Thailand and China. But they are most diverse in Sulawesi, where there are eight or nine species of *Tarsius* (Shekelle et al., 2008; Merker et al., 2010; Groves and Shekelle, 2010). Tarsiers have some very distinctive features (Chapter 3), and the Sulawesi fauna as a whole is very unusual. The earlier idea that it represents a "transition" between Oriental and Australasian faunas does not account for the strange Sulawesi endemics or the conspicuous absences there. For example, in primates, only two genera are present, *Tarsius* and *Macaca*, macaque monkeys. In addition to the unusual presences and absences, the Sulawesi biota as a whole is characterized by much regional parallelism. In the butterflies there, unrelated taxa share distinctive aspects of their structure such as gigantism and "Celebes forewing." As Vane-Wright and de Jong (2003) concluded, while the origins and significance of the parallelism "remain mysterious, they appear to be further manifestations of the endemism and peculiarity of the Sulawesi fauna."

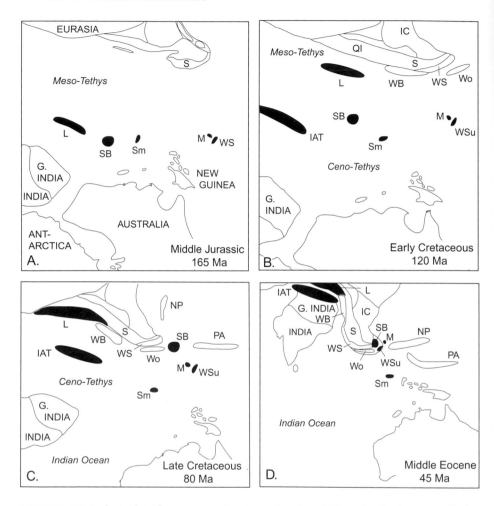

FIGURE 5-13A–D. Jurassic to Eocene reconstructions of eastern Tethys, showing terranes rifted from Gondwana rafting across Tethys and becoming incorporated into Asia (simplified from Metcalfe, 2006, 2008, 2010). Terranes: IAT = island arc terrane (= Incertus arc of Hall et al., in press), L = Lhasa, M = Mangkalihat, NP = North Palawan, PA = Philippine arc, SB = Southwest Borneo, Si = Sikuleh, Sm = Sumba, WB = West Burma, Wo = Woyla, WS = West Sumatra, WSu = West Sulawesi.

The anthropoid/tarsier split in haplorhines is the second node in the extant primate phylogeny (Chapter 3). The break occurred after the differentiation of haplorhines and strepsirrhines at the Lebombo monocline (180 Ma, Early Jurassic) but before differentiation of lemurs and lorises with the opening of the Mozambique Channel (160 Ma, Middle Jurassic). At about this time (165–155 Ma), an arc of terranes

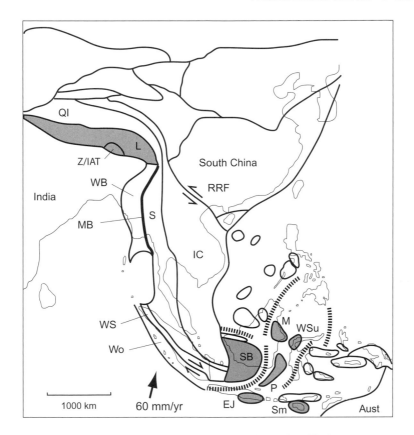

FIGURE 5-14. Asian terranes in their modern configuration. The last Gondwana-derived terranes that were incorporated into Asia before the arrival of India are shown in gray. Terranes: EJ = Eastern Java, IC = Indochina, L = Lhasa, M = Mangkalihat, MB = Mogok metamorphic belt, P = Paternoster, QI = Qiangtang, S = Sibumasu, Sm = Sumba, WB = West Burma, Wo = Woyla, WS = West Sumatra, WSu = West Sulawesi, Z/IAT = Zedong terrane, perhaps equivalent to the island arc terrane of Figure 5-13 (Hall et al., in press) (based on Metcalfe, 2006, 2008; Zhou et al., 2008). Broken lines in Borneo are the Borneo suture in the north, the Lupar suture in the west, and the Meratus suture in the southeast. RRF = Red River fault.

is proposed to have split off the margin of northern Gondwana (India–Australia–New Guinea), rafted across the Meso-Tethys basin and eventually accreted with Asia (Fig. 5-13; Metcalfe, 2006, 2010; Heine and Müller, 2005). One of these terranes ended up jammed in the middle of Sulawesi (Fig. 5-14).

Biogeographic breaks around Sulawesi include the boundary between Sulawesi and Borneo, at Makassar Strait (Wallace's line). This break is seen in tarsiers and in many other groups. For example, Schulte et al. (2003) discussed different clades of varanid and agamid lizards that show boundaries here between Australasian and Southeast Asian groups. Schulte et al. concluded that while late Tertiary dispersal events could be ruled out as an explanation, the Asian clades could have been introduced from Gondwana by earlier rafting on continental fragments. The authors also cited possible ancient divergence here in insects and passerine birds. In Osteoglossidae, a group of large freshwater fishes, there are also great differences between clades in Australasia and in Southeast Asia (Borneo and Sumatra to Vietnam). Kumazawa and Nishida (2000) proposed that the Asian osteoglossids had been introduced by arriving on India and/or smaller terranes in Indonesia as these accreted to Asia in the Cretaceous. (As the authors noted, this model "implicates the relative long absence of osteoglossiform fossil records from the Mesozoic"; p. 1869.)

The terranes that rifted off Gondwana would also have been occupied by eutherian mammals (Rich et al., 2001), and current distributions may reflect the rift and the accretion. In the same way, clades in marsupial and placental mammals in New Guinea have distribution breaks closely correlated with the boundary of the craton and the accreted terranes in the north of the island (Heads, 2001: Figs. 39–41). In addition, Asia (perhaps even including Taiwan and Japan) was probably occupied by primates ever since their origin, when they split from plesiadapiforms. One of the main questions here concerns the location of the original divide in Asia between strepsirrhines and haplorhines.

Terranes now located in Tibet, southwestern Borneo, eastern Borneo, and western Sulawesi were all parts of Gondwana in the Late Jurassic, parts of Asia by the Late Cretaceous (Fig. 5-13 and 5-14). The intervening period marks a critical phase of tectonic, phylogenetic, and biogeographic history in which the Earth's geography and the primates' morphology were completely recast. This was the last time such a radical modernization took place. Primates are essentially a lowland/coastal group, and while Cenozoic developments such as the final rise of the Andes and Himalayas involved uplift and survival of some primates, these events were of much less significance for phylogeny than the Mesozoic revolution.

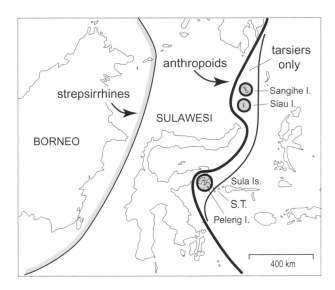

FIGURE 5-15. Primates at their eastern limit, with the eastern limits of the strepsirrhines, anthropoids, and tarsiers. The three tarsiers located east of strepsirrhines and anthropoids are: *Tarsius sangirensis* endemic to Sangihe Island, *T. tumpara* on Siau Island, and *T. pelengensis* on Peleng Island. S.T. = Salue Timpaus Strait.

In its biogeography, Sulawesi is unique as it lies near the boundary of the three main primate clades—strepsirrhines, anthropoids, and tarsiers (Fig. 5-15). In its geology, Sulawesi is also unique as it represents the most complex suture zone in eastern Indonesia (Hall and Wilson, 2000). There have been several phases of accretion from the south and east, involving ophiolite, arc, and continental-fragment terranes. Some of the thrust faults are shown in Figure 5-16. Most of the metamorphism, magmatism, and uplift is Cenozoic, but older, high-pressure metamorphic rocks are widespread in accretionary complexes found in Sulawesi (the Pompangeo Schist), southeastern Borneo, and Java. Metamorphism occurred here along a late Early Cretaceous (120–110 Ma) north-dipping subduction zone at the margin of the Sundaland craton and may have been caused by the collision of a Gondwanan continental fragment with the margin (Parkinson, 1998; Parkinson et al., 1998; Kadarusman, 2000).

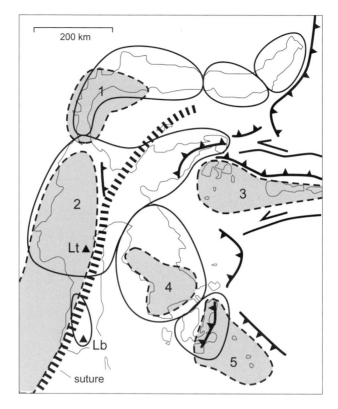

FIGURE 5-16. Sulawesi geology and centers of endemism.
Continuous lines = the seven species of *Macaca* on Sulawesi.
1–5 = continental fragment terranes. 1 = West Sulawesi
terrane, 2 = buried Gondwana fragment (Smyth et al., 2007),
3 = Banggai-Sula terrane, 4 = East Sulawesi terrane, 5 = Tukang
Besi-Buton block (Hall and Wilson, 2000). Thrust faults as lines
with barbs (barbs on the overriding plate), transcurrent faults
(North and South Sorong faults) as lines with arrows.
Lt = Latimojong Mountains; Lb = Mount Lompobatang (= Bonthain).

PHILIPPINES-SULAWESI: PRIMATES AT THEIR EASTERN LIMIT

Apart from humans and human introductions, primates reach their
eastern limit at the Philippines, Sulawesi, and Timor (Brandon-Jones
et al., 2004; Figs. 3-1 and 5-15). Many other taxa have a similar distribution, as they are widespread in the tropics but absent from suitable habitat in Australia, New Guinea, and the Pacific islands. For
example, several angiosperm families have a wide distribution in
warmer America, Africa, and Asia, but extend east only to Borneo

(Peraceae, Gelsemiaceae, Anisophylleaceae; cf. strepsirrhines) or to Sulawesi (Buxaceae, Erythropalaceae; cf. haplorhines) (Stevens, 2010). The current absence of these groups from temperate regions could be due to simple ecological factors, although their absence from tropical Australasia is probably not. This absence could reflect an ancestral absence or at least a low ancestral diversity in the latter region.

Primate clades such as strepsirrhines range east to Makassar Strait, while the eastern limit of primates and the main center of the tarsiers, one of the group's most distinctive clades, lie east of Sulawesi (Fig. 5-15). The reason for *any* limit in the Indonesian archipelago is not obvious; primates in the region, such as *Nasalis* and *Macaca*, are excellent swimmers in rivers and the sea (*Nasalis* individuals have been picked up by fishing boats some distance offshore). Despite this, *Nasalis* does not cross Makassar Strait. This boundary does not seem to be related to potential means of dispersal. The limit of all primates in eastern Sulawesi is even more striking as the Moluccan Islands (Maluku) are so close. While primates occur on the Banggai Islands (*Tarsius pelengensis* is endemic on Peleng Island; Fig. 5-15), none has ever been recorded across Salue Timpaus Strait (20 km wide) in the Sula Islands of the Moluccas. The strait is a minor feature in terms of current geography but despite this it is a biogeographic node of intercontinental significance. Primates and other widespread groups reaching their eastern limit there are juxtaposed with Australasian taxa reaching their western limit at the same strait, for example, the widespread passerine bird *Monarcha* (Heads, 2001: Fig. 16). These precise eastern and western biogeographic limits and the high levels of endemism on Sulawesi itself show no obvious relationship with current geography or ecology and could instead be due to the earlier tectonic history of the different component terranes.

SANGIHE ISLANDS, BETWEEN SULAWESI AND THE PHILIPPINES

The Sangihe (Sangir) Islands represent part of a double volcanic arc that extends from northern Sulawesi to the southern Philippines. Sangihe Island itself is about 50 km across and has an endemic tarsier, *Tarsius sangirensis* (Fig. 5-15). Another species, *Tarsius tumpara*, is endemic to nearby Siau Island, only 20 × 10 km in area and dominated by an active volcano. Shekelle et al. (2008) wrote that "How tarsiers ever came to these islands is a mystery," and the details of the individual islands and populations along the arc may never be known, as the small islands in an arc system appear and disappear so rapidly. Nevertheless, the general historical relationship between tarsiers and the arc as a whole could be elucidated with comparative biogeographic studies of the rich biota.

The Sangihe Islands are an important center of local endemism in many groups, including marsupials, squirrels, and birds. The distinctive passerine *Eutrichomyias* (Monarchidae) is endemic on Sangihe and shares morphological characters with both *Hypothymis* (India to the Philippines) and *Terpsiphone* (Africa to the Philippines) (Riley and Wardill, 2001). Sangihe is a young, active volcanic island, and the endemic bird on the island is probably older than the island itself, although it may be related to the age of the subduction zone that produces the arc. The endemism seems more related to the location than to simple isolation; there are thousands of other islands between Sangihe and Africa, and it is significant that *Eutrichomyias* occurs right on the margin of both its putative relatives. Whatever its true phylogenetic position, *Eutrichomyias*, like the tarsiers and the Western Sulawesi terrane, may represent the last, small fragment of a group which in the past was more widespread north of New Guinea.

The Siau and Sangihe tarsier species are putative sisters (Shekelle et al., 2008), and so in dispersal theory the question would be: How did tarsiers get from Siau to Sangihe? Instead, in a vicariance approach, the question is: How has this arc clade differentiated from the others on Sulawesi and the Philippines? As shown in Fig. 5-16 (from Hall and Wilson, 2000), the geological structure in eastern Sulawesi is dominated by a series of thrusts originating from the direction of New Guinea, and it was recognized early on that the Banggai–Sula block (labeled 3 in Fig. 5-16) is allochthonous—it did not form *in situ*. Likewise, Sangihe Island taxa are often outliers of Australia/New Guinea/Moluccas groups in the east (cf. *Eutrichomyias* with allies in the west). Examples include the following.

A genus of small flycatcher birds, *Colluricincla*, occurs in Australia and New Guinea and is disjunct with *C. sanghirensis* endemic on Sangihe. The closest ally of *C. sanghirensis* may be a *Colluricincla* clade on the eastern Papua New Guinea islands (D'Entrecasteaux, Trobriand, and Louisiade Archipelagos), 3,000 km to the east (Rozendaal and Lambert, 1999). This disjunction is unlikely to be due to a colonizing flight and is probably associated with large-scale transcurrent displacement on strike-slip faults. The parrot *Eos* occurs only between the islands of Geelvink Bay, northwestern New Guinea (not mainland New Guinea), and the Sangihe Islands, where populations of Sangihe, Siau, and the neighboring Talaud and Nenusa islands form a clade. Another parrot, *Loriculus amabilis*, is on Sangihe, Siau, and also Peleng and Sula Islands, but is absent on mainland Sulawesi. This distribution follows the disjunct strip occupied by tarsiers but no other primates (Fig. 5-15). As mentioned, the Australasian bird *Monarcha* extends west to the Sula Islands, and it is also on the southern Sangihe Islands (Siau, etc.) but, as in *Loriculus* and *Eos*, it is notably absent on the Sulawesi mainland. *Tarsius* is often recorded in mangrove and secondary, disturbed forest, and the Sangihe–Peleng strip is interpreted here as maintaining an accreted fauna of subduction zone mangrove weeds.

DIFFERENTIATION WITHIN SULAWESI: *MACACA*

The macaques (*Macaca*) range from the Atlas Mountains of northern Africa (living) and southern England (fossil, Miocene) to Japan, Sulawesi, and Timor, along the former Tethyan margins. This is a standard biogeographic pattern seen in many groups. *Macaca* species often have a weedy ecology and tolerate high levels of disturbance; several are known to raid human crops and rubbish dumps. *M. mulatta*, the rhesus monkey, is largely commensal with humans and ranges from mangrove up to 4,000 m elevation on the Tibet Plateau. *M. fascicularis*, the crab-eating monkey, is a characteristic species of Asian mangrove.

The primary phylogenetic split in *Macaca* is between the geographically isolated north African species and the Asian species. The latter comprise two well-supported groups, with one mainly in India, China, and Japan, and the other further south, mainly in India and Malesia (Chatterjee et al., 2009). In the first group, the Taiwan representative (*M. cyclopis*) is phylogenetically closer to the Japanese species (*M. fuscata*) than to the mainland Chinese species (*M. mulatta*). The Taiwan–Japan connection recurs in many groups, for example, in ranid frogs (Tanaka-Ueno et al., 1998) and in rhacophorid frogs, where *Buergeria* (Japan, Ryukyu Islands, Taiwan, Hainan) is sister to the other Asian and African genera (Wilkinson et al., 2002). Plants with the pattern include *Trochodendron* (Trochodendraceae) of Taiwan and Japan (with its sister *Tetracentron* in China; Stevens, 2010), the cypress *Chamaecyparis obtusa* (Taiwan and Japan), and the pair *Chamaecyparis formosensis* (Taiwan) + *C. pisifera* (Japan) (Li et al., 2003).

Macaca and *Tarsius* are the only non-human primates on Sulawesi. The locality is not a minor outlier for *Macaca* but an important center of diversity, with seven species out of 21 in the genus, all endemic

(Fig. 5-16). (India plus Sri Lanka, and China each have six species, but these are much larger areas and few of their species are endemic.) The widespread, weedy, mangrove species M. *fascicularis* is absent from Sulawesi, suggesting that its distribution is caused by tectonic and phylogenetic factors rather than an ecological ones. The Sulawesi–central Asia–Mediterranean sector marks the range of *Macaca* and was occupied by different "Tethyan" basins at different times. Most of the earlier ones have been destroyed by subduction beneath Asia (Fig. 5-13). Nevertheless, any primates that occupied the island arcs and intraplate volcanic islands in these basins could have survived subduction *in situ* and been accreted to Asia.

In Sulawesi, the seven endemic *Macaca* species have distributions that are, as Evans et al. (2003) emphasized, "virtually identical" with clades of toads. The authors concluded that the pattern indicates general vicariance in a whole fauna and is comparable to that found at other major biogeographic boundaries. In studies of the nymphalid butterfly *Cethosia* in Wallacea (i.e., Sulawesi and neighboring islands), Müller and Beheregaray (2010: 314) concluded that the high levels of island endemism and the essentially allopatric groupings are best explained by "vicariant processes linked to the history of formation of micro-continent and associated palaeo islands."

One of the biogeographic boundaries within Sulawesi occurs in the southwestern arm of the island, where the highest point in the island (Mount Rantemario, 3,440 m, in the Latimojong Mountains; Fig. 5-16) is located. There is an important biogeographic break between the biota of the Latimojong Mountains and that of Mount Lompobatang in the south. The break is known in plants, amphibians, primates, and many other groups and represents a classic problem in biogeography. Although biologists have challenged geologists to explain the pattern, until recently there was no known geological basis for it. Now there is good evidence from dated zircons in xenoliths that a fragment of Gondwana lies buried beneath one of the volcanic arcs of East Java. Smyth et al. (2007) proposed that this extends northeast to northern Sulawesi, beneath Latimojong but west of Lompobatang (Fig. 5-16). This Gondwana fragment could have collided with Sundaland during Late Cretaceous northwest-directed subduction. A mélange of arc and ophiolitic material was accreted first to the Sundaland margin, in Java and eastern Borneo, but the arrival and accretion of the Gondwana fragment would have terminated subduction here. In the Paleogene, subduction shifted to take place along the Java trench in a northeastward direction. The suture shown by Metcalfe (2008) running through central Sulawesi also separates Latimojong from Lompobatang (Fig. 5-16).

In the Sulawesi tarsiers, Merker et al. (2009) found strong evidence of sharp concordant genetic, acoustic, and morphologic breaks between *Tarsius lariang* (on terrane 2 in Fig. 5-16) and *T. dentatus*, north and east of there, at a point within an expanse of unbroken primary forest. The break occurs near a suture between two microplates, represented by the Palu–Koro fault (between 1 and 2 on Fig. 5-16). Shekelle et al. (2010) concluded that the numerous regions of endemism in Sulawesi are not merely products of Pleistocene vicariance, but also reflect evolution on the separate Sulawesi terranes before they fused to form the modern island.

PRIMATES IN THE PHILIPPINES

The only non-human primates in the Philippines, as in Sulawesi, are *tarsiers* and *Macaca*. As suggested in Chapter 1 (see "Basal Nodes Around the Philippine and Madagascar"), groups in the Philippines, including mammals, should probably not be regarded as recent derivatives from the north, west, and south, as traditional theory suggested. For example, in rodents, the 500 species of Murinae occur throughout the Old World and comprise the largest subfamily of mammals; the basal clade is endemic to the Philippines. This does not mean that the Murinae had a center of origin there, but instead suggests that an already widespread Old World ancestor had its primary phylogenetic differentiation there, or at least around the terranes now amalgamated there.

Until recently, most biogeographic work on the Philippines and Sulawesi explained the patterns with reference to changes in Pleistocene geography. But, as discussed in Chapter 2 (see "Case Study: the Philippines"), molecular analyses have now refuted the notion that Pleistocene events had a major role in causing regional patterns. Instead, authors working on groups such as lizards (Siler et al., 2010) and birds (Jones and Kennedy, 2008) have concluded that new hypotheses are needed, especially models that incorporate pre-Pleistocene geography.

The tectonics of the Philippines, as in Sulawesi, are complex and far from resolved, but ideas that have been proposed on the general movement of the crustal blocks and arcs are of great biogeographic interest. Details of emergent land are often not known to geologists, whose work is focused on the body of the rock rather than its surface. In contrast, biologists have a wealth of information on the topic due to the persistence of many taxa around subduction zones as metapopulations on small, individually ephemeral islands.

The Philippine archipelago is a composite, formed by the fusion of continental blocks in the south and west (the North Palawan and Zamboanga terranes) with the accreted island arc terranes of the Philippine mobile belt, which forms most of the country. In a clock study, Blackburn et al. (2010) suggested that the frog *Barbourula* of the southern Philippines (Palawan) and Borneo diverged from its sister taxon *Bombina*, of Europe and eastern Asia, in the Eocene (47 Ma) (Chapter 2). The clock in the study was fossil-calibrated, and so the Eocene date is only a minimum age, but a useful one. The North Palawan terrane rifted from the Indochina/South China blocks between Early and Late Cretaceous (Fig. 5-13; Metcalfe, 2010), and separation of the two frog genera could have resulted from this.

The arc terranes of the Philippine mobile belt were accreted from the east in the Late Miocene to Pliocene (Yumul et al., 2004). Thus, most Philippines terranes, like many in Taiwan, northern New Guinea, New Caledonia, and New Zealand, have docked with continental crust in the west and southwest of these countries after arriving from the Pacific basin to the east (Fig. 5-13).

Other Asian regions are discussed below, and as an introduction to this, the biogeography of some particular groups is considered.

SCANDENTIA (TREE SHREWS)

The Southeast Asian order Scandentia (Fig. 5-17) is closely related to primates (Fig. 5-1). In the main clade of Scandentia (Olson et al., 2005), *Anathana* (Indian plate) is allopatric with is sister group *Tupaia* (Eurasian plate) + *Urogale* (Philippine mobile belt). These breaks correlate with the India/Eurasia plate boundary and the Eurasia plate/ Philippine mobile belt boundary. Olson et al. (2005: 666) wrote that the circumstances leading to the disjunction between *Anathana* and the other genera across the Bay of Bengal "remain a mystery for which our results offer no clear explanation." The same break occurs in lorisid primates (Fig. 5-18) and could be the result of vicariance. As Olson et al. (2005: 668) noted, the absence of Scandentia from such proximate islands as the Andamans "suggests severe limitations to overwater dispersal [and] vicariance can almost certainly be assumed to have played

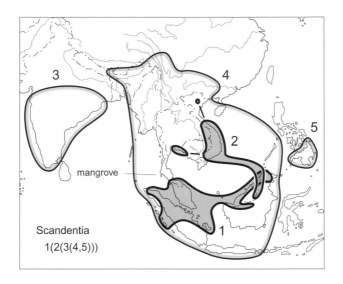

FIGURE 5-17. Distribution of Order Scandentia. 1 = *Ptilocercus*, 2 = *Dendrogale*, 3 = *Anathana*, 4 = *Tupaia*, 5 = *Urogale*. Phylogeny is indicated.

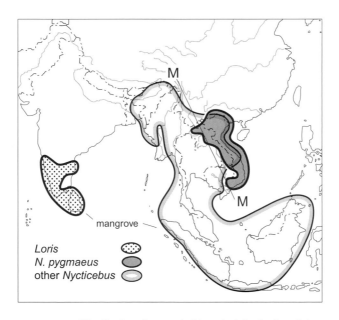

FIGURE 5-18. Distribution of strepsirrhines in Asia: *Loris* and the two primary clades of *Nycticebus*. Line M-M = "Mekong line."

a prominent role in the past diversification and resulting distribution of treeshrews." The same conclusion is reached here for primates.

The two basal clades in the Scandentia, *Ptilocercus and Dendrogale*, have different overall distributions but meet and overlap in Borneo. The main clades in the widespread *Tupaia* also have their main breaks in Borneo:

> *Tupaia* group 1: Borneo species basal to others which range to Nepal.
>
> *Tupaia* group 2: Borneo species basal to others which range to Thailand.
>
> *Tupaia* group 3 (*T. gracilis*): Borneo.

Although these phylogenetic breaks all involve Borneo, this does not necessarily imply a center of origin there. (Likewise, the oldest fossils of Scandentia are found in Thailand, but this does not necessarily mean the area was a center of origin.) Borneo is a geological composite, and several clades may have been juxtaposed there with terrane accretion or differentiated during accretion. Thus Borneo—or rather the terranes that became Borneo—could represent sites of differentiation in an already widespread proto-Scandentia. The distribution of *Ptilocercus* is centered on the Riau (Riouw) Islands region (off Singapore), while that of *Dendrogale* is based further north around the central South China Sea, and the two are almost completely allopatric. (The South China Sea basins opened with Late Cretaceous and Cenozoic rifting of continental crust.) The allopatry of the two genera suggests early zones of differentiation around the two regions before the opening of the South China Sea and before the plate boundary breaks in *Anathana/Tupaia/Urogale*. Within Borneo, *Ptilocercus* and *Dendrogale* are restricted to the north and northwest of the island (north of the Lupar line, cf. Heads, 2003, and west of the Mangkalihat terrane). Within this region they remain largely allopatric. *Tupaia* has complex diversity in different parts of Borneo, including the south and east, and so perhaps this was its original sector in "Borneo." When the terranes of Borneo were juxtaposed, so were the genera. *Tupaia* has expanded its range to overlap that of the two basal genera, and the two basal genera themselves show minor overlap, but apart from this, the biogeography of the Scandentia genera reflects their original evolution by simple vicariance.

The small order Dermoptera, or "flying lemurs," are the other primate relatives and comprise *Galeopterus* of Borneo, Sumatra, Peninsular Malaysia, and Indochina, and *Cynocephalus* of the Philippines. The pattern is a repeat of *Tupaia* + *Urogale*, and again indicates the independent history of the Philippines. If the Philippines clade was derived from an Asian group by dispersal, it would have a sister group at a localized center of origin in, say, Vietnam, and not be the sister of the entire Asian clade.

ASIAN STREPSIRRHINES

The only Asian strepsirrhines, a clade of lorisids, have a primary split between southern India/Sri Lanka (*Loris*) and Southeast Asia (*Nycticebus*) (Fig. 5-18). The node is located somewhere in the Bay of Bengal/Bangladesh area, as in Scandentia. The same break, near the plate margin, also occurs in both the *silenus* and *sinica* groups of *Macaca* (Tosi et al., 2003). In *Nycticebus*, the primary break is between *N. pygmaeus* of Vietnam/Laos and the other *Nycticebus* species, in western Indochina/Sundaland but also in parts of Vietnam (Chen et al., 2006). The overlap between the two clades is widespread but far from complete, and the original break may have taken place along a boundary near the Mekong River.

ASIAN COLOBINES

Colobines are diverse in Asia and comprise three widespread clades.

Trachypithecus and Semnopithecus

These two genera are allopatric (Fig. 5-19), although the boundary is complex, as discussed below. In contrast, *Presbytis* (Fig. 5-20) is overlapped almost entirely by *Trachypithecus*. The only locality where *Presbytis* is present (with distinctive endemism) and *Trachypithecus* is absent is the Mentawai group, off Sumatra. In the same way, the *fascicularis* and *silenus* species groups of *Macaca* show extensive overlap, but the former is absent from the Mentawai Islands and Sulawesi, where the *silenus* group is present with endemics in each (Tosi et al., 2003).

The arrangement of the langur group shown here is: *Presbytis* (*Trachypithecus* + *Semnopithecus*), a phylogeny based on nuclear (X-chromosome) DNA (Ting et al., 2008; cf. Osterholz et al., 2008). In contrast, a mitochondrial DNA phylogeny indicates the sequence: *Semnopithecus* (*Presbytis* + *Trachypithecus*) (Ting et al., 2008). The authors suggested this incongruence could be due to either ancient hybridization or differential lineage sorting.

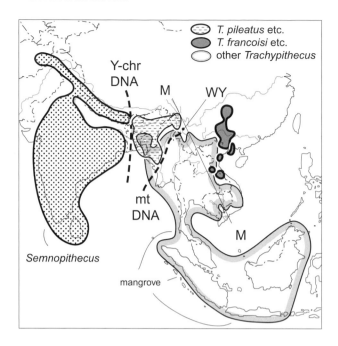

FIGURE 5-19. Distribution of *Semnopithecus* (incl. *Kasi*) and *Trachypithecus* (phylogeny from Ting et al., 2008, Osterholz et al., 2008). The *T. pileatus* group is closest to *Semnopithecus* according to mitochondrial DNA but to *Trachypithecus* according to nuclear DNA. WY = biogeographic node in western Yunnan. Line M–M = "Mekong line."

There are also ambiguities within *Semnopithecus* + *Trachypithecus*. In the most important case, the subclade from Bangladesh/Bhutan/ western Burma (*T. pileatus* and allies) clusters with the Southeast Asian *Trachypithecus* according to nuclear DNA (Y-chromosome) sequences, but with the Indian *Semnopithecus* based on mitochondrial DNA (Osterholz et al., 2008; Karanth et al., 2008). Again, these authors attributed the pattern to ancestral hybridization or incomplete lineage sorting. The latter is more compatible with the largely allopatric distribution of the *pileatus* group with respect to *Semnopithecus* and *Trachypithecus* s.str (cf. the Bangladesh–Burma distribution of *Hoolock* in gibbons, below). *Semnopithecus* and the *pileatus* group are allopatric, even at a local scale in Bhutan (Wangchuk et al., 2008).

Before the molecular studies, species of *Semnopithecus* from Sri Lanka (*S. vetulus*) and the Western Ghats of southwestern India (*S. johnii*) were united with the Southeast Asian *Trachypithecus* on

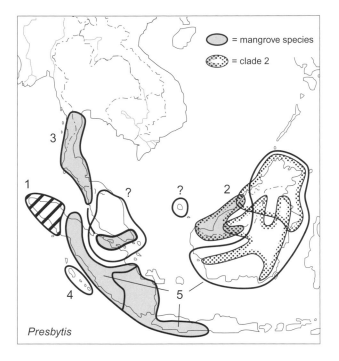

FIGURE 5-20. Distribution of *Presbytis*. The two species with queries have not been sequenced. The phylogeny of the sequenced species is: 1 (2 (3 (4 + 5))) (Meyer et al., 2011).

morphological characters (Brandon-Jones et al., 2004). This gives a disjunct affinity across the Bay of Bengal. The phylogenetic distribution of the morphological characters is incongruent with the molecular phylogeny as mapped here. It could represent simple convergence (Karanth et al., 2008), but the disjunction across the Bay of Bengal represents a standard biogeographic distribution and so it may be another case of incomplete lineage sorting. The morphological syndrome, involving cranial morphology, neonate pelage color, and sexually dichromatic pubic integument, could have been inherited from polymorphism already present in the ancestor and distributed *across* the Bay of Bengal before the modern genera evolved with their main *break* at the Bay of Bengal.

To summarize, the whole langur complex could have evolved by vicariant evolution and incomplete lineage sorting in a widespread, polymorphic ancestor. The primary geographic breaks include the mitochondrial boundary in Burma, which corresponds to the India plate/Eurasia plate boundary (Fig. 5-14), and may be related to activity along it. The

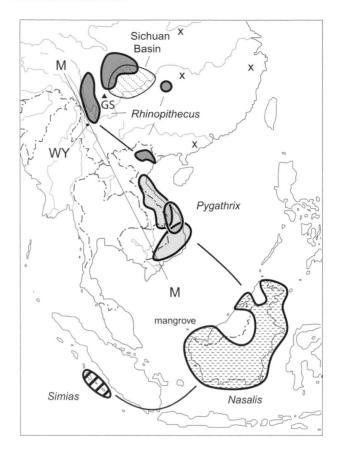

FIGURE 5-21. Distribution of the odd-nosed clade. GS = Gongga Shan (7,556 m). x = fossil localities of *Rhinopithecus*. WY = biogeographic node in western Yunnan. Line M–M = "Mekong line."

nuclear DNA boundary, near the Ganges–Brahmaputra delta, also occurs in Scandentia and lorisids. This may also relate to the plate margin, but the *pileatus* goup (and *Hoolock*) indicate the two breaks may each have distinct origins. The whole region between the Ganges–Brahmaputra delta and the Burma arc is a collision zone, and large earthquakes have occurred in the delta in historical times (Steckler et al., 2008).

In addition to the boundaries at Bangladesh and Burma, a third main break in *Trachypithecus* occurs along the line of the Mekong River (Fig. 5-19). The *T. francoisi* group is endemic east of the Mekong, mainly on limestone cliffs. The biogeographic "Mekong line" is also seen in lorisids (Fig. 5-18), other colobines (Fig. 5-21), and gibbons (Fig. 5-22). For

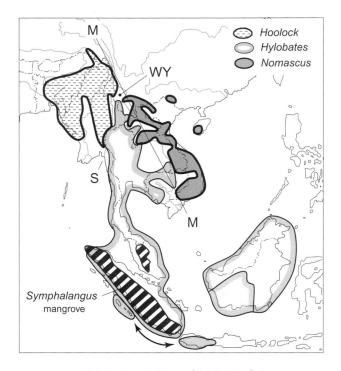

FIGURE 5-22. Distribution of gibbons (Hylobatidae). Arrows connect sister species of *Hylobates* on the Mentawai Islands and Java (Takacs et al., 2005; Whittaker et al., 2007). Fossils and early Chinese paintings indicate gibbons were formerly present in Sichuan and Hubei, around the gorges of the Yangtze (Groves, 1972). WY = biogeographic node in western Yunnan. Line M–M = "Mekong line."

differentiation at the Mekong line in another group, see the main break in the freshwater gastropod genus *Sulcospira* (Köhler and Dames, 2009).

The Mekong line meets the Burma boundary at a node in western Yunnan (WY in Figs. 5-19–5-22). Boundaries in primates occur at the Mekong and Salween valleys (Figs. 5-21, 5-22), but the modern valleys themselves probably did not cause the original differentiation. Instead, both the valleys and the distributions could have been caused by tectonics. This is evident in the region of southwestern China near the Burmese border, where the Yangtse, Mekong, and Salween almost meet. This region is about 300 km east of the plate margin itself, which is represented by the eastern Himalayan syntaxis (where the Brahmaputra/Tsangpo crosses the India/Tibet border).

The langurs and allies range from mangrove forest around the South China Sea via hill forest to 4,000 m elevation in the Himalayas (*Semnopithecus schistaceus*). This shows a trend opposite to the usual, ecologically determined pattern in which widespread taxa occur in temperate latitudes at *lower* elevations than in the tropics. This anomaly in the langurs can be accounted for by tectonic uplift and survival *in situ*.

Presbytis (Fig. 5-20)

This genus is almost completely overlapped by its sister group, *Semnopithecus* + *Trachypithecus*, and the region of sympatry (Sumatra/peninsular Malaysia/Borneo) might be assumed to indicate a center of origin of the whole clade. Yet distribution at the global level, with a simple break between Asian and African colobine clades (Fig. 5-1), indicates that the Asian colobines evolved as the already widespread Asian vicariant of the African colobines, not at a center of origin. Within the Asian colobines, the genera could have evolved by allopatric differentiation around the component terranes of Borneo and Sumatra from an ancestor already distributed between India and Borneo. Subsequently the genera overlapped, but traces of the original allopatry persist in northern Indochina and India, where *Presbytis* is absent, and on the Mentawai Islands, where it is the only genus present. The phylogeny of *Presbytis* (Meyer et al., 2011) is: (Aceh (Borneo (Malay Peninsula (Mentawai Islands (Sumatra, Java, Borneo)))). The significance of Aceh (northern Sumatra) in the biogeography of Southeast Asia and its independent connections with areas such as India and Australasia, excluding the rest of Indonesia and Malaysia, are discussed elsewhere (Heads, 2003).

It would be useful to have more details on *Presbytis* and its relatives in Borneo; for example, the distribution of *Trachypithecus* there is not well known. Although IUCN (2009) mapped the genus as widespread throughout Borneo, Harrison (2001) and Wangchuk et al. (2008) recorded it only in coastal areas of mangrove, peat-swamp forest, and riverine forest in the north and south, not from central areas.

The distribution of the *Trachypithecus/Semnopithecus/Presbytis* clade has probably undergone considerable "shortening" as the different terranes in Tethys crashed into Asia, ending with the docking of Tibet, Burma, and India. Taxa in the Himalayas and the *T. pileatus* group in and around Bangladesh, for example, may have had their ranges shortened as the India plate subducted beneath China and Burma.

The Odd-nosed Clade of Colobines

Nasalis of Borneo and *Simias* of the Mentawai Islands, together with *Pygathrix* of Cambodia, Laos, and Vietnam, and *Rhinopithecus* of northwestern Vietnam and China, make up the "odd-nosed clade" of colobines (Fig. 5-21). The colobine fossil *Mesopithecus* is known from Europe and southwestern Asia (including the Sivalik Hills) and may belong here. Its affinities have been much debated, and the genus has been allied with African colobines (e.g., Hohenegger and Zapfe, 1990) or with the Asian colobines, especially the odd-nosed clade (e.g., Pan et al., 2004).

The four extant genera in the odd-nosed clade are allopatric and range from the mangrove forests of Borneo (*Nasalis*) to the highest altitude attained by primates, 4,700 m (15,420') in China on the edge of the Tibetan plateau (*Rhinopithecus*) (Sterner et al., 2006; Whittaker et al., 2006). At the highest elevations, *Rhinopithecus* occupies conifer forest in which snow can lie for six months of the year and the monkeys live on lichen. The ecophylogenetic cline or track between the mangroves of Borneo and the mountains of central Asia is a repeat of that seen in the langurs, although here it takes place east of the Mekong line, not west of it. The high latitude (north of 30°N) and elevation of all these subalpine colobine populations are anomalous for primates, but can be compared with the *Macaca* species of Tibet and northern Japan (north of 40°N). The widespread, common ancestor of the four odd-nosed genera may have occupied mangrove, back-mangrove, monsoon, evergreen dipterocarp, edge, and secondary forest. Some populations have been trapped in the uplift of Borneo and stranded in the center of the island; others have been passively uplifted in central Asia following the arrival of India and the subsequent Himalayan orogeny.

In the proboscis monkey *Nasalis*, the nose is long and pendulous, whereas in the snub-nosed monkeys, *Rhinopithecus*, there is extreme reduction or even absence of the nasal bones. (Nasal bone reduction may also reach the point of disappearance in Hylobatidae and in *Pongo*; Hershkovitz, 1977: 127.) Jablonski (1998: 26) wrote that "The reasons for the dramatic broadening and shortening of the mid-face in *Rhinopithecus* species are unclear. . . . It is interesting to speculate that the driving force may have been natural selection." Instead, the nose of *Rhinopithecus* may reflect the general trend to reduction of the muzzle in haplorhines, which may in turn result from a trend to reduction and fusion in the skull of vertebrates.

Nasalis, endemic to Borneo, inhabits mangrove, *Nypa* palm swamp, riverine forest, the freshwater peat-swamp forest that is widespread in the south of the island, and mosaics of all these types of forest, up to about 200 m elevation (Salter et al., 1985; Meijaard and Nijman, 2000; Onuma, 2002). The favored roost tree is *Sonneratia* (Lythraceae: Myrtales), and the foliage is one of its main food items. *Sonneratia* is best known as a widespread Old World mangrove, but it also occurs inland in low hill forest where there has been uplift (e.g., in southern Thailand, pers. obs.). *Nasalis* is sister to *Simias*, which is endemic to the Mentawai Islands (Fig. 5-21) and is found in lowland, swamp, and montane forest.

The usual connection between Mentawai Islands/Sumatra on one hand and Borneo on the other involves records at the Riau Islands, Bangka Island, and Belitung (Billiton) Island, as seen in mangroves (*Sonneratia alba* and *Lumnitzera littorea*), Scandentia (*Ptilocercus*), and primates (*Nycticebus*, *Tarsius*, *Trachypithecus*, and *Presbytis*). Neverthless, the *Nasalis*/*Simias* clade, *Pongo*, and others are absent from these islands. This could be due to chance local extinction or hunting, or the absence could be part of a general, community-wide biogeographic pattern. Many trees that are otherwise widespread mangrove-associates (for example, several *Ficus* species) are also absent from this area, and botanists have termed it the "Riouw pocket" (Heads, 2003). Other primates that skirt the pocket include *Presbytis* (Fig. 5-20) and gibbons (Fig. 5-22), and the latter also show Mentawai–Java affinities. *Macaca*, well known in mangrove and many other habitat types, has been mapped as absent from the Riau Islands (IUCN, 2009).

On mainland Asia, the odd-nosed clade as a whole is absent in Indochina west of the Mekong. Again, Indochina is divided into eastern and western sectors. There is no obvious ecological break, and a general tectonic explanation seems more likely.

GIBBONS (HYLOBATIDAE)

Groves (1972) emphasized that: "5 of the 6 gibbons are vicarious in their distributions; only the siamang [*Symphalangus*] broadly overlaps any of the others." This is still an accurate summary (although four of the 1972 species are now regarded as genera). The genera are shown in Fig. 5-22; gibbons, both living and fossil, are absent west of Bangladesh. The phylogeny is controversial; in one phylogeny the first split is at the Salween River, between *Hoolock* and the rest (Takacs et al., 2005; cf. Matsudaira and Ishida, 2010); in another the first split is at the Mekong, between *Nomascus* and the rest (Thinh et al., 2010). (*Nomascus* has a distribution east of the Mekong line resembling that of *Nycticebus pygmaeus*, the *Trachypithecus francoisi* group, and the odd-nosed clade.) Israfil et al. (2011) suggested that the lack of phylogenetic resolution among the gibbon genera may reflect rapid vicariance in a widespread ancestor. Simple vicariance on its own around a node in western Yunnan (WY in Fig. 5-22) can explain the divergence of *Hoolock*, *Hylobates*, and *Nomascus*, and only one range expansion is required to account for the sympatry of *Symphalangus* and *Hylobates*. Even in the area of potential overlap, *Symphalangus* is usually found in lowland and submontane forest, while *Hylobates* is most abundant at mid-elevation (O'Brien et al., 2004). *Symphalangus* is recorded from mangrove (World Wide Fund for Nature, 2009). In contrast with the simple vicariance/range expansion model for gibbons proposed here, a dispersal–vicariance analysis that assumed local centers of origin and founder dispersal resulted in a complex scenario involving 23 dispersal events (Chatterjee, 2006). Thinh et al. (2010) also proposed a center of origin model but, again, this required a "complicated biogeographic pattern leading to the current distribution of gibbon taxa" (p.7).

We are now in a position to discuss some of the key areas in the region in terms of their tectonics and biogeography.

BORNEO

Borneo is a composite island occurring at the intersection of at least three major sutures represented by subduction/accretion belts (Fig. 5-14). Terranes in the north and west have accreted from parts of Laurasia (Cathaysia and Indochina blocks), while southern and eastern terranes have accreted from Gondwana. The Southwest Borneo terrane has been proposed as one of the blocks that broke off from northwestern Australia in the Jurassic (Barber and Crow, 2008a; Metcalfe, 2008), and the occurrence of diamonds in southwestern Borneo without any obvious local or regional source may support this (Metcalfe, 2010). The terrane sutures in Borneo mark former seas, represented by Late Cretaceous sequences in the southwest (Lupar and Boyan sutures), southeast (Meratus suture zone), and north (Adio–Palawan suture) (Metcalfe, 2006; Zhou et al., 2008). These are important biogeographic boundaries in many groups (Heads, 2003). The most species-rich area in Borneo for primates occurs in central-eastern Borneo (Meijaard and Nijman, 2003), around the Gondwanan Mangkalihat terrane. The great diversity in eastern Borneo and the phylogenetic breaks between here and Sulawesi are associated with giant oil fields in the Mahakam delta and Makassar Strait. The strait has either developed as a foreland basin caused by the crustal flexure or has an extensional origin with middle Eocene rifting caused by slab rollback (Puspita et al., 2005). In the latter process, a subducting slab and the trench sink backwards into the subducting plate. This was mentioned in Chapter 5 as a cause of back-arc basin formation in the pre-Andean landscape, and it may also be important in basin formation around the southwest Pacific and Asian margins (cf. Schellart et al., 2006). After the formation of Makassar Strait, very rapid uplift took place in Borneo. Mountains such as the granite batholith of Mount Kinabalu (4,094 m), the highest mountain on the island, have been forced up through overlying strata and have inherited their biota from these (including the three main clades of Scandentia). The uplift has also resulted in many coastal taxa being stranded inland.

SUMATRA AND THE MENTAWAI ISLANDS

The Mentawai Islands lie off the west coast of Sumatra in the Indian Ocean and, as Chivers (1986) pointed out, show an anomalous concentration of primate endemism. There are five endemic species, including two in *Macaca* (Roos et al., 2003) and an endemic genus, *Simias*, in an area of about 7,000 km^2 (Whittaker, 2006). This gives one endemic species per 1,400 km^2. In Madagascar, Groves (2005) recognized 60 species, but with changes in species concepts that figure soon climbed to 84 (Mittermeier et al., 2006). Madagascar has a total area of about 600,000 km^2, which gives one endemic species per 7,100 km^2. There are perhaps four "old species" in the Mentawai Islands and 48 "old species" in Madagascar (Tattersall, 2007), giving one endemic per 1,750 km^2 in the Mentawai Islands and one every 12,500 km^2 in Madagascar. The actual figures are only a rough estimate of biodiversity, as different species (and genetic differences) cannot always be compared. But the density of primate species endemism on the Mentawai Islands is very high—perhaps five times (new species) or seven times (old species) higher than that of Madagascar.

In addition to the high diversity, the Mentawai Islands have distinctive biogeographic affinities. In *Presbytis*, the Mentawai endemic (*P. potenziani*) is sister, not simply to the adjacent Sumatran species, but to a widespread clade in Sumatra, Java, and Borneo (Fig. 5-20; Meyer et al., 2011). In gibbons, the Mentawai Islands endemic *Hylobates klossii* is sister to the Javan species (*H. moloch*), not the Sumatran species, although Java is much further away (Fig. 5-22). This affinity, based on DNA sequences and vocalisations (Takacs et al., 2005; Whittaker et al., 2007), skirts the Sumatran mainland and so is incongruent with current geography and ecology. Nevertheless, the Mentawai–Java connection occurs in many groups, such as the tree genus *Citronella* (Cardiopteridaceae) (see discussion in Heads, 2003: 394). The biogeographic arc probably reflects a tectonic connection of some sort and may have been caused by dextral dislocation along strike-slip faults.

The West Sumatra block was emplaced against the Sibumasu block (Figs. 5-13, 5-14) in the Triassic by westward movement involving dextral transcurrent faulting along the Medial Sumatra tectonic zone (Barber et al., 2005; Barber and Crow, 2008a, 2008b). The last pre-Cenozoic tectonic unit to be accreted to Sumatra was the Woyla nappe (Wo in Fig. 5-14), a Jurassic–Early Cretaceous intraoceanic complex thrust westward over the West Sumatra block in the mid-Cretaceous. The terranes are formed from ocean floor and also incorporate collided seamounts, plateaus, and volcanic arc fragments. The Woyla group has been compared with the Luk Ulo complex of central Java, the Meratus complex of southeastern Borneo, and the Bantimala complex of central Sulawesi (Zhou et al., 2008). If its emplacement west of western Sumatra took place by dextral strike-slip (cf. Fig. 5-14), this might explain both the diversity anomaly on the Mentawai Islands and the dextral disjunct connections between the primates there and in Borneo. As the Indian–Australian plate moves northward, it collides with Java almost at right angles but hits Sumatra at an acute angle. In the past the dextral strike-slip component of the oblique convergence has been taken up along the Medial Sumatra zone and the Woyla suture. Since the late Cenozoic, it has been accommodated mainly on the Great Sumatran fault, a currently active fracture along western Sumatra that has an associated volcanic arc. Disjunction along the fault in plants (such as groups of *Rhododendron*; Heads, 2003: Fig. 16) may have been caused by the dextral strike-slip.

THE BREAK BETWEEN INDIA AND BURMA

A break here, in the Bangladesh/Bay of Bengal region, was cited above in Scandentia, lorisids, *Trachypithecus/Semnopithecus*, and gibbons. For Scandentia, Olson et al. (2005) wrote that the circumstances leading to the disjunction remain a "mystery." The break occurs in groups that show a wide range of ecologies, and so, as with other breaks in Burma, it may be related in some way to the plate boundary in the region (the Indian plate is subducting beneath Burma and Sundaland; Stork et al., 2008) (Fig. 5-14). In trees, for example, *Aglaia cucullata* (Meliaceae) ranges from the Ganges delta east to the Mekong delta and New Guinea, in mangrove, tidal estuaries, *Nypa* palm swamp, and riverine forest, from

0 to 20 m (Mabberley et al., 1995), and there is no ecological reason for its absence from the large areas of mangrove forest in India west of the Ganges delta. In other primates, the same limit occurs in groups of *Macaca* such as *M. fascicularis. Macaca* ranges between Morocco and Sulawesi/Japan, and other groups with similar Tethys-oriented distribution also show breaks here. The freshwater crab family Potamidae ranges between Morocco and Borneo and comprises two clades, one western and one eastern, with the major division in Bangladesh/Burma where the two groups overlap. Shih et al. (2009) attributed the break to orogeny in Burma caused by India–Asia collision, and whatever the details, it probably did involve some kind of activity on the plate boundary.

EAST/WEST DIVISION OF INDOCHINA—THE MEKONG LINE

Discussing mammalian biogeography in Southeast Asian, Meijaard and Groves (2006: 306) wrote: "The outstanding problem, which is especially notable in the case of the primates, is that of the Mekong." The fauna east of the Mekong River is rich and quite different from that to the west. In primates, the Mekong is the western limit of *Nycticebus pygmaeus, Nomascus,* the *Trachypithecus francoisi* group, and the entire odd-nosed clade. This indicates that the boundary developed early in primate phylogeny and, judging from the breaks in African and American genera, could be mid-Cretaceous. Meijaard and Groves (2006) found no evidence for an ecological break at the Mekong boundary and concluded that the reason for the boundary is "unclear." It is a general break that occurs in many groups with different ecology. For example, the dolphin *Orcaella* is wide-ranging in estuarine and freshwaters from India to Australia and New Guinea, but in Indochina it only occurs west of a line: upper Irrawaddy–southern Thailand–lower Mekong.

The most widespread of the primate groups east of the Mekong line (other than macaques) is the odd-nosed clade, with a distribution extending to localities in and around the Sichuan basin (Fig. 5-21). A well-known component of the fauna here is *Ailuropoda,* the giant panda. This is found with *Rhinopithecus* in mountains bordering the Sichuan basin to the west, the Longmen Shan (Longmen Mountains), and has fossils through southern China, northeastern Burma, and northern Thailand, but mainly east of the Mekong. The phylogeny of bears (Ursidae) given by Pagès et al. (2008) and their distribution (IUCN, 2009) is:

Ailuropoda: southern China (extralimital fossils in Thailand and Burma)

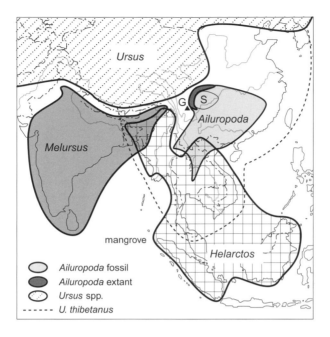

FIGURE 5-23. Distribution of the Ursidae in south and east Asia (Pagès et al., 2008; IUCN, 2009). G = Gongga Shan, S = Sichuan Basin.

Tremarctos: Bolivia to Venezuela (subfossil relatives in North America and Europe)

Melursus: India

Helarctos: Southeast Asia (fossils in Java). Tropical forest, including mangrove and freshwater swamp forest, up to 2,100 m elevation.

Ursus: Laurasia, with one species in Southeast Asia (south to southern Thailand).

The genera show a high degree of allopatry. There is some overlap but, apart from the single Southeast Asian species in the mainly northern *Ursus*, this is marginal. Thus the four Asian genera (Fig. 5-23) divide up Asia into four sectors:

· India,

· Indochina and Sundaland,

· China (south and east of Sichuan), and

· the far north.

The four meet at a node near Gongga Shan. The first two nodes indicate primary differentiation around the Pacific Ocean (west Pacific vs. the rest, then east Pacific vs. the rest). The first node, the break between *Ailuropoda* and the others, may have occurred around Gongga Shan region/Mekong line. At some time after this, there has been range expansion of *Helarctos* eastward across the line and expansion of *Ailuropoda* west of the line, followed by widespread extinction in this genus. *Melursus* and *Helarctos* show local overlap only in Bangladesh and adjacent parts of Assam, corresponding to similar breaks around Bangladesh in primates.

Another important boundary for primates, including gibbons, for example (Fig. 5-22), lies close to the Salween River and could be correlated with the Mogok metamorphic belt. This marks the junction of the West Burma and Sibumasu terranes (Fig. 5-14) and marks a major transcurrent fault zone that dates back to the Triassic. A Jurassic–Early Cretaceous subduction-related event resulted in emplacement of granodiorites and orthogneisses along the belt. Middle Jurassic augen gneisses in the belt are considered to be granitic intrusions into an active shear zone (Barber and Crow, 2008a). The zone is also intruded by undeformed granitoids of Cretaceous and younger ages (Searle et al., 2007).

STRIKE-SLIP ZONES IN VIETNAM/SOUTHERN CHINA

Strike-slip zones, whether plate boundaries (the transforms connecting divergent and convergent margins) or belts of deformation within plates, form some of the largest and most spectacular discontinuities found on Earth (Storti et al., 2003). Intraplate strike-slip belts may show great longevity and may be reactivated many times in successive phases of deformation and accretion. The belts have a profound influence on the location, architecture, and subsidence history of the associated sedimentary basins, which are often rich in hydrocarbons (Storti et al., 2003: 2). Good examples are seen in Southeast Asia, where long-lived strike-slip faults have repeatedly influenced the evolution of petroleum systems (Morley, 2002). In Burma, the oil wells at Yenangyaung on the Irrawaddy have been worked for centuries. Further east, many giant oil fields are located in the South China Sea (from the Gulf of Thailand to Brunei).

In much of Asia, significant Cenozoic deformation and displacement have occurred along a series of large, intraplate strike-slip deformation belts, notably the Red River fault (Fig. 5-14). Left-lateral ductile shear occurred along the belt, and Southeast Asia has been displaced 500 km southeastward with respect to China. The fault was active in the Oligocene–Miocene

(Jolivet et al., 1999) but is Mesozoic in origin. The shear may have brought about the divergence between *Pygathrix* and *Rhinopithecus* in the odd-nosed clade (Fig. 5-21). South of the Red River fault in Vietnam there are other left-lateral faults with a similar strike (Song Ca fault, etc.), and one or more of these may have been involved in producing the phylogenetic break along the "Mekong line" in the primates. These ideas would fit with the conclusion of Meijaard and Groves (2006) that differentiation here has not been due to the river itself or to ecological differences on either side.

In other groups, the spiny frogs *Nanorana* and *Quasipaa* form a clade ranging from the Himalayas through China and Southeast Asia. *Nanorana* is in the northwest (Tibetan plateau, etc.), *Quasipaa* is in the southeast (to Vietnam), and they have a mutual boundary at the Red River, near the China/Vietnam border (Che et al., 2010). Within *Quasipaa*, the two clades are separated neatly by the Red River fault. Che et al. attributed this break to a major tectonic event along the fault associated with Himalayan uplift and extrusion of the Indochina block from Asia.

The India–Asia collision caused complex Cenozoic deformation through Indochina and Sundaland. The rocks in the region can be divided into several fault-bounded tectonic domains (Fig. 5-14), each with a different rotation and/or translation history (Chi and Dorobek, 2004). Geologists are currently debating the relative importance of translation and rotation of terranes versus crustal shortening and thickening in accommodating the convergence of India and Eurasia. Paleomagnetic data indicate a large, clockwise rotation of the Shan Thai block (~Sibumasu terrane, S in Fig. 5-14) and a small amount of southward translation of the Indochina block, with Borneo and peninsular Malaysia undergoing a large, counterclockwise rotation (Chi and Dorobek, 2004). For the eastern Eurasian margin in general, Zhou et al. (2008) described a sudden and important change in the stress field in the early Late Cretaceous, from transpression to transtension, perhaps caused by oceanward retreat of the subduction zone (see "slab rollback," above). The transtension produced a series of large sedimentary basins from eastern Siberia to the Gulf of Thailand. Most of what is now the floor of the South China Sea was emergent, and rifting took place in the Late Cretaceous.

Deformation in eastern Asia, dominated by normal and strike-slip faults, has often been attributed to the "extrusion" of Indochina from Asia following the Eocene collision of India (Leloup et al., 1995). But there are problems with a model based on extrusion alone. As noted above, deformation started earlier than the collision, in the latest Cretaceous (Schellart and Lister, 2005). Before extrusion tectonics set in, deformation may have been caused by back-arc extension associated with the eastern Asian active margin. Back-arc extension here (with eastward slab rollback) may have continued from the Cretaceous through most of the Cenozoic (Schellart and Lister, 2005; cf. Morley, 2001).

NORTHERN INDIA AND SOUTHERN CHINA

The collision between Greater India and Asia (Fig. 5-13) produced a major geographical reorganization. Earlier studies portrayed the collision as a simple one between two land masses. In contrast, Aitchison et al. (2007) inferred the former presence of two intraoceanic island arc systems between pre-collision India and Asia and concluded that: "Neotethys was obviously more complex than originally envisaged" (p. 6; cf. Masters et al., 2006; Hall et al., 2009).

The Tibetan Plateau is the most extensive region of elevated topography on Earth. *Macaca* occurs around Lhasa and across the western part of the plateau, reaching 4,000 m. *Rhinopithecus* survives at 4,700 m on mountains near the eastern margin of the plateau. Tibet was formed when the Lhasa terrane collided with the Qiangtang terrane (northern Tibet) in the Early Cretaceous (Fig. 5-13). Much of the Lhasa terrane was inundated by the transgression of a shallow sea at the end of the Early Cretaceous (Albian–Aptian). The youngest marine strata of the Qiangtang terrane are Early Cretaceous, those of the Lhasa terrane are Late Cretaceous (Wang et al., 2008), but in some parts of south-central Tibet, Aitchison and Davis (2004) stressed that marine deposition continued until at least the end of the Eocene. The authors proposed that India collided with a mid-Tethyan, intraoceanic island arc around the Paleocene/Eocene boundary, and that the collision between India and Asia probably occurred in the Oligocene.

Primates could have occurred throughout the archipelagos formed by these arc terranes and continental fragments, as in the Caribbean islands with their diverse fossil taxa. Primates are absent from archipelagos in the Pacific, but that probably reflects a long-term absence from the Pacific and its margins that goes back to the origins of the group. In contrast, primates have been diverse around the Indian Ocean ever since they evolved, and their absence from the Mascarenes, Seychelles, and so on is probably due to the islands' small size. Larger islands or archipelagos could have maintained populations, and these would have been incorporated into and around Tibet during arc–continent collision. This colonization would have occurred by normal ecological processes as the arcs themselves were either accreted or slowly destroyed by subduction beneath Asia.

DeCelles et al. (2007) described a basin from the terrane suture zone in central Tibet that was a foreland at sea level in the mid-Cretaceous. During the Late Cretaceous, the Tibetan Songpan–Ganzi basin northwest of the Longmen Shan, and the mountains of the Longmen Shan themselves, were still close to sea level (Zhang, 2000). But there is a Late Cretaceous–Eocene depositional hiatus, which correlates with a phase of great crustal shortening and 5,000 m of uplift.

The interior basins of southwestern China provide a record of the transition from marine conditions in the Cretaceous to high-elevation conditions during the middle Tertiary. The southern Longmen Shan fold-thrust belt and the associated foreland basin, the Sichuan basin (Fig. 5-21), had its major deformation in the latest Cretaceous–early Cenozoic, and prospective petroleum traps have been described from here (Jia et al., 2006). On the western side of the Sichuan basin, crossing the Longmen Shan fault, the elevation changes in just 50 km from 600 m by the basin to 6,000 m on the Tibetan Plateau. Gongga Shan (7,556 m), the highest mountain in Asia east of the Himalayas, is located in this area. *Rhinopithecus roxellana* occurs around the Sichuan Basin and in the Longmen Shan up to 2,800 m. *R. bieti* occurs further east, in the mountains on the Yunnan–Tibet border (near northern Burma). The species is recorded in subalpine fir–larch and cypress forest from 3,000 to 4,700 m, and the lower and upper limits are the highest of any primate. The high elevation is significant as it occurs at latitude 30°N, well outside the tropics. In Ethiopia, *Theropithecus* is in grassland at 4,400 m, but the latitude here is only 10°N. Primates in montane China and Japan occur in cooler habitat than elsewhere, and this ecological anomaly may reflect an early presence of primates in the region. Fossil colobines from Japan (Iwamoto et al., 2005) indicate that the primate fauna there was more diverse in the past, and this is compatible with trans-northern Pacific connections in fossil primate clades (omomyiforms and adapiforms).

EVOLUTIONARY ECOLOGY OF OLD WORLD PRIMATES: FROM MANGROVE TO MONTANE FOREST

Many primates and primate relatives such as Dermoptera and Scandentia have been recorded in mangrove (Fig. 5-1). As discussed in Chapter 4, this is a distinctive form of tropical rainforest, a lowland swamp forest characterized by high levels of disturbance. The outer, seaward margin of mangrove has a simple floristic structure, and the striking zonation of the tree species is often depicted in textbooks. But in Asia, as in America

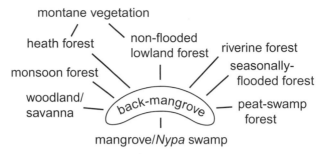

FIGURE 5-24. Geographical/ecological relationships among some forest types in tropical Asia.

and Africa, the inner edge of the mangrove is much more complex and can involve a mixture of characteristic taxa of all the main "types" of tropical forest; apart from mangrove and unflooded terra firma forest, the back of the mangrove also grades into different kinds of flooded forest (varying with sediment load and pH of the water), palm forest, monsoon forest, drier or open forest/savanna, and secondary forms of all of these (Fig. 5-24).

A typical example of a mosaic back-mangrove in Asia, on a floodplain in Borneo, was cited above; it is inhabited by a large population of orangutan and is covered with a patchwork of mangrove, *Nypa* palm forest, riverine forest, swamp forest, seasonally flooded forest, and, on areas of slightly higher ground, dry dipterocarp forest (Goosens et al., 2005). The modern forest types and their faunas can be derived from a similar ancestral "patchwork" forest by expansion of one or another of the "patches" when suitable environment becomes available.

Although the ecology of primates along the diverse land/sea ecotone of the back-mangrove is poorly understood, this may have been a key environment in the evolution of the group. Apart from providing a potential species pool for different kinds of forests, the back-mangrove itself has probably been an important site for phylogeny. Geological rifting will have led to great extensions of the back-mangrove and also the potential division of primate populations by new seaways. Groups such as the odd-nosed colobines occupy a wide-ranging ecological cline, with populations in mangrove at one end (Borneo) and in alpine woodland at the other (Yunnan, Tibet). If the phylogeny and chronology accepted here for primates and the tectonic reconstructions are correct, it is likely that the subalpine populations in China were originally mangrove-associates that were first landlocked and then uplifted with thrust faulting and orogenesis. A similar elevational/biogeographic cline occurs in other groups. For example, the *Antheraea frithi* complex of emperor moths occurs in mangrove forests of Sumatra, Belitung (Billiton) Island, western Borneo, and Palawan, while elsewhere in Indonesia and in mainland Asia, members of the complex inhabit hill and montane forest (Holloway and Hall, 1998).

Local aspects of ecology and behavior can often reflect processes operating at much larger scales of time and space. "Riverine refuging" is an interesting form of daily foraging dispersal, and *Macaca fascicularis*, the crab-eating monkey widespread in Asian mangroves, is a well-known example. Populations in Sumatra feed in terra firma forest by day and in the afternoon or evening return to sleep in trees at river edges on branches overhanging the water, while younger individuals swim and play in the water (Schaik et al., 1996). Riverine refuging is also recorded in *Nasalis* (Onuma, 2002; Matsuda et al., 2009; Bernard et al., 2011) and in the African genera *Allenopithecus*, *Miopithecus*, and *Cercopithecus* (Schaik et al., 1996; Maisels et al., 2006). It could be a retention of ancestral ecology in back-mangrove/riverine forest.

Comparative accounts of African primates in mangrove and non-mangrove are rare. Among populations of *Piliocolobus* on Zanzibar, group density and frequency of social play are both highest in mangrove populations (Nowak and Lee, 2011). The mangrove populations also visit vegetation on adjacent coral rubble, where other food species are available.

EVOLUTION IN THE OLD WORLD PRIMATES

For years, biogeographic structure in tropical rainforest taxa was interpreted as the result of evolution in Pleistocene refugia. The minimum ages given in fossil-calibrated clock studies now indicate that many primate species and even populations within species are much older than this, and Miocene ages are often proposed. If species are mid-Cenozoic, the subgeneric and intergeneric splits in primates could be Paleogene or Late Cretaceous. This would fit with an Early Cretaceous origin of Old World Monkeys (catarrhines) following the rifting of the Atlantic Ocean. It would also be compatible with the main nodes in African primates—at the Benue trough/central Africa rift system and around the Congo basin—being associated with Cretaceous rifting and uplift, respectively. The breaks around Indochina are harder to date because of reactivation of faults, but they may also involve Cretaceous deformation, perhaps caused by slab rollback before the collision of India with Asia.

American primates comprise eight widespread clades that all show considerable overlap with the others; African primates comprise ten widespread clades. Within these widespread clades (mostly genera or pairs of genera) there is a high to very high degree of allopatry among subclades (genera, species, and subspecies). In Asia there are nine widespread clades (the five mapped here plus tarsiers, two clades in *Macaca*, and orangutan). The sympatry among these parallels the situation in the American and African widespread clades. The origin of the widespread clades is unclear. It could be due to sympatric evolution, but the evolution of clades at both a higher level (catarrhines/platyrrhines, lemurs/lorisiforms) and a lower level (many genera, most species and subspecies) is notably allopatric, and so the overlap of the widespread clades is attributed here to secondary range expansion. The widespread clades evolved after the differentiation of catarrhines and platyrrhines, but perhaps soon after this or even during the same phase of Atlantic and intracontinental rifting. The mobilism and range expansion of the clades could also have taken place in Early Cretaceous time, before the groups settled down into another phase of immobilism and cladogenesis in the mid-Cretaceous.

In primates, the lorisids, colobines, papionines, and hominids all occur in both Asia and Africa. The location of the breaks between Asian and African clades is often difficult to specify, as no primates survive in the widespread deserts of the boundary area. In papionines, there is a Tethyan break between *Macaca*, from north Africa to Sulawesi, and the others in sub-Saharan Africa and (as fossils) India. Within Africa the main phylogenetic breaks in the widespread groups show fairly simple biogeographic patterns involving the Benue Trough–Sanaga River area and the Congo basin. This is a standard pattern seen in many other groups, such as *Osteolamus*, a small crocodile common in closed canopy forest. Its three main clades are allopatric in West Africa (west of Cameroon), Cameroon/Gabon, and the Congo basin (Congo-Brazzaville to Ituri) (Eaton et al., 2009).

In Asia, both the biogeography and the geology are more complex than in Africa and involve subduction and terrane accretion. There are major breaks around Bangladesh/western Burma, probably related to the subduction zone there, and in the Salween and Mekong regions, where the vicariance may relate to renewed Late Cretaceous activity on old rifts. This last process often leads to development of petroleum systems, and most areas with high primate diversity include oil belts. The generation of both may be induced by belts of low anticlines forming over old rifts at turning points in paleogeographic history, such as transitions from shallow marine/deltaic to fully terrestrial environments. Primate phylogeny in Asia may be related to the development of possible petroleum systems in the basins of southern China and the known giant fields of the South China Sea. Other main phylogenetic breaks in Southeast Asian primates occur along an arc, Mentawai Islands–Java–Makassar Strait, and these are also associated with giant oil fields.

Biogeographic breaks within Borneo (Fig. 5-14) include the Lupar line between Western Kalimantan and Sarawak, and this is seen in *Presbytis* and in gibbons. The Lupar line in Borneo and the boundaries in Sulawesi are correlated with zones of Cretaceous–Tertiary terrane accretion. Taxa occurring on both sides of a subduction zone will tend to have their range contracted due to subduction, and their total range may be shortened by hundreds or even thousands of kilometers. Many of the higher level clades in Southeast Asia may once have had extensive ranges on both arc terranes in the Tethys and Panthalassa seas (proto-Indian and Pacific Oceans) and on the Asian mainland.

Dramatic ruptures occur at the mid-ocean ridges with sea-floor spreading, in back-arc basins, and also at large-scale strike-slip faults, the transforms. A combination of sea-floor spreading and slippage on transforms has led to the Mesozoic separation of catarrhines from platyrrhines, lemurs from lorises, and tarsiers from anthropoids (before they were eventually juxtaposed again on the other side of Tethys). It is also likely that incomplete or failed rifting can cause biological differentiation, as at the Lebombo monocline, the central African rift system, and the Amazon. Transcurrent movement on strike-slip faults, as in the Benue trough, the equatorial Atlantic, the Amazon, and around the accreted terranes of Southeast Asia, seems to be especially important for biological evolution. Sea-floor spreading is well known as an agent of vicariance, and during this process the crust is extended. In contrast, in purely transcurrent movement, crustal blocks slide past each other and there is no crustal extension. Nevertheless, strike-slip movement can by itself cause disjunction and cladogenesis in biological communities, for example, along major faults in New Zealand, New Caledonia, and New Guinea (Heads and Craw, 2004; Heads, 2008a, 2008b). These are Cenozoic faults that generally separate subspecific variants or closely related species, making study of the disjunction straightforward.

Strike-slip faulting in the Benue trough lasted throughout the Late Cretaceous and could have had a much more profound effect on the mammals and other taxa in the region.

In the convergent margin/subduction context of northern Tethys, accreting terranes with their arcs and seamounts may have introduced primate populations northward. In addition, the collision process itself involved great transcurrent movement, disjunction, and evolution, affecting primates throughout Indochina and around the South China Sea. At the same time, subduction and thickening has shortened the crust along with the biological distributions. In areas such as Sulawesi and Borneo, belts of metamorphism are associated with terrane accretion, orogenesis, and phylogenesis.

East Africa—"high Africa"—has undergone regional, epeirogenic uplift, but the only orogenic (fold belt) mountains in Africa are in north Africa (the Atlas Mountains) and in South Africa. In contrast, Asian and American tectonics are dominated by orogenic folding and, especially in Asia, terrane accretion. Despite these differences, primate evolution has been similar in the three continents, with the formation of eight, nine, or ten groups, their later overlap, and finally the allopatric differentiation within the groups that produced the current, repeated patterns. This suggests that primate differentiation has been mediated by different kinds of tectonic activity: divergence, extension, and normal faulting have been important for primates in all areas, and in Asia and America the results of convergence, shortening, and reverse faulting are also evident.

The vertical component of tectonic movement is a primary factor determining aspects of ecology such as elevation, and primates illustrate many examples of "ecological lag" following uplift. Mangrove-associated taxa have been left at high elevation in the Andes, the central African mountains, and around Tibet. In lowland taxa, tectonic uplift and the retreat of inland seas may have led to the stranding of *Allenopithecus* in the cuvette centrale of the Congo, *Cacajao* in the Amazon basin, and *Nasalis* populations in central Borneo.

6

Biogeography of the Central Pacific

Endemism, Vicariance, and Plate Tectonics

East of the Philippines and Sulawesi there is abundant suitable habitat for non-human primates in the tropical rainforests of New Guinea and the Pacific islands. Despite this, they are absent (apart from an introduced *Macaca* in northwestern New Guinea), and the account of Pacific biogeography given in this chapter and the next two will rely instead on data from a range of other groups.

The islands of the Pacific, along with their plants and animals, provide the classic case of an island system, and the taxa are often cited as they seem to provide such remarkable evidence for the powers of long-distance dispersal (Whittaker, 1998: 2). Ever since the old idea of a former mid-Pacific continent was rejected by geologists, the prevailing paradigm for the biogeography of the islands has been dispersal from mainland sources. The central Pacific region is seen as a biogeographic sink, with the plant and animal taxa now found there having been derived from centers of origin in Asia, Australasia, and America—anywhere but the central Pacific. This view accepts no autochthonous land biota for the Pacific region, about a third of the Earth's surface. Many influential texts on Pacific biogeography, both terrestrial and marine, have adopted this model and assumed that the entire Pacific biota arose following long-distance dispersal from somewhere outside the basin, usually Southeast Asia (Darwin, 1859; Wallace, 1860; Mayr, 1942; Ekman, 1953; Darlington, 1957; see reviews in Heads, 2005b, 2006b). Mayr (1942), in a discussion of the avifauna, denied that a true Polynesian biota exists. Thus the Pacific has been dismissed as a biogeographic backwater having little significance for global studies apart from proving the efficacy of long-distance dispersal. Modern studies have continued in this tradition. Gillespie et al. (2008: 3337) wrote, "It is now clear that colonization of the islands occurred through transoceanic dispersal," and furthermore, "With the advent of molecular techniques, relatively precise estimations of timing and the source of colonization have become feasible" (p. 3335). Methods of timing were addressed in Chapter 2; sources of Pacific taxa and transoceanic dispersal are discussed here.

THE PACIFIC AS A CENTER OF ENDEMISM

Some interesting results from recent molecular work conflict with the traditional model, as they show that a number of Pacific taxa are not minor, unrelated, branches of mainland groups, as would be expected. Instead they form diverse, monophyletic clades endemic to the central Pacific and widespread there. These indicate that the central Pacific is a large, important center of endemism in its own right. In a similar way, taxa of southern regions that were formerly regarded as miscellaneous odds and ends or unrelated members of northern groups have turned out to be either large, basal grades (for example, Australasian groups in oscine passerines) or large, monophyletic groups (for example, African mammals in Afrotheria).

The molecular results showing widespread central Pacific endemics are not unexpected, as these were already well known from morphological studies. Examples from the latter include the following weevils (Curculionidae) from Kuschel (2008):

Platysimus: Micronesia (Kiribati), Solomon Islands, New Caledonia, Kermadec Islands, Austral Islands, and Society Islands. This gives a range ~4,000 km across.

Viticis: Japan (southern Ryukyu Islands: Ishigaki and Iriomote Islands), New Caledonia, Loyalty Islands, Vanuatu, Fiji, and one species in the Cook and Marquesas Islands. Total range: ~11,000 km.

Allorthorhinus: Philippines, Solomon Islands, Santa Cruz Islands, and Fiji. Total range: ~7,000 km.

While these groups have maintained their integrity over vast ranges, they all have a break around the western margin of the Pacific. Similar

Pacific plate distributions occur in many other groups, such as reef fishes (Springer, 1982).

Biogeographers were well aware of these patterns by the end of the 19th century and realized that the Pacific endemism, together with the correlated absence of many related groups in the central Pacific, is not easily explained by dispersal from outside the area (see, for example, Pilsbry, 1901). These workers described the local centers of endemism and the connections among them in detail, and sometimes suggested former land areas to account for disjunctions and centers of diversity. But geologists insisted there had never been continental crust in the central Pacific, and so all the early work on biogeography was rejected as "land-bridge building" in the 20th-century dispersalist accounts.

In fact, continental crust is not needed to maintain a metapopulation in an area indefinitely. All that is required are small islands appearing from time to time, and this occurs at subduction zones, spreading ridges, hotspots, and other tectonic cracks and fissures. De Queiroz (2005: 68) argued that "Geographical distributions of terrestrial or freshwater taxa that are broken up by oceans can be explained by either oceanic dispersal or vicariance in the form of fragmentation of a previously contiguous landmass." Yet this dichotomy overlooks the long-term survival of groups as metapopulations on ephemeral oceanic islands. A history of vicariance does not require an originally continuous landmass.

THE STANDARD MODEL OF PACIFIC BIOGEOGRAPHY AND ITS ASSUMPTIONS

The standard model of Pacific biogeography has been developed in the context of the center of origin/dispersal/adaptation (CODA) paradigm, as described in Chapter 1. The model is based on the following assumptions:

Assumption 1. Overall species diversity on the Pacific islands decreases moving east into the Pacific and away from Southeast Asia and Australia, suggesting that Asia/Australia is the major source (e.g., Keppel et al., 2009: 1035).

Instead, the decreasing diversity is probably due to decreasing island size eastward. East of Fiji, the islands are small and it is not surprising that the forests and reefs do not include continental or Indonesian levels of species diversity. Although the islands of southeastern Polynesia are small and have what appears to be low diversity, when total land area is taken into account they show levels of diversity comparable to those of the Hawaiian Islands (Gillespie et al., 2008). *Overall* species diversity decreases toward the east, but many individual groups show the opposite pattern, with most or even all species restricted to the central Pacific. Examples of these are discussed below.

While terrestrial diversity on many atolls is low, Pacific atolls reaching just a few meters above sea level include fossils of land snails that live only in montane cloud-forest (Solem, 1983). This is not surprising, as it has been known for over a century that atolls are former high islands that have eroded and subsided. This evidence indicates that a component of the Pacific atoll biota may be relictual.

Assumption 2. In phylogenies, island clades are nested in mainland clades and so have dispersed from the mainland.

This is not necessarily correct (see Chapter 1, especially Fig. 1-6). If island taxa survive as metapopulations, there may be little difference between the histories of island taxa and mainland taxa.

Assumption 3. Island endemics are younger than the islands they inhabit. The central Pacific islands are all young, and so the taxa must be even younger.

Many biogeographic and molecular studies have instead concluded that taxa endemic to young islands are older than the islands (see Chapter 2).

Assumption 4. There is no continental crust in the central Pacific, and so the islands must have been populated from the continents around the Pacific margins.

Nineteenth-century biologists such as T.H. Huxley recognized the biogeographic significance of Polynesia and proposed that larger areas of land existed there in the past—"either a great continent or an assemblage of subcontinental masses of land" (Huxley, 1896: 384). These land masses were rejected by geologists as they found no evidence for any continental crust. Many biogeographers have concluded that the lack of continental crust means there was no land in the Pacific before the current islands formed, and these are all young. But terrestrial organisms require land, not continental crust, and there were always islands present. Regional endemics in the Pacific may have survived *in situ* around areas of recurrent volcanism.

It is often emphasized that oceanic volcanic islands arose *de novo* and have never been connected with a continent. But this may not be relevant for their biogeography, as terrestrial life survives just as well on oceanic crust as on continental crust. The ephemeral islands on which populations "perch" until new islands appear do not develop at random localities; most form in areas of prior volcanism determined by tectonics, near other volcanic islands.

Factors of ecology and biogeography mean that some clades may survive as "volcano-weeds." This is especially evident along active margins. For example, the plant *Scaevola gracilis* (Goodeniaceae) is known only from north of New Zealand, on islands in the Kermadec and Tonga groups that are all recent or fairly recent volcanoes; Raoul and Tofua Islands are still active (Sykes, 1998). *S. gracilis* is most common on Raoul Island, where it forms dense stands on open pumice slopes in the crater. The distribution of this volcano-weed runs along the active Kermadec–Tonga Ridge, with the adjacent Kermadec–Tonga trench marking the subduction zone at the Australia/Pacific plate boundary. The species appears to be well adapted to life on a subduction zone and, within its sector of the margin, has probably been colonizing new volcanoes as they appeared for millions of years—much longer than the age of any individual volcanic island. Raoul Island endemics include many species and genera such as the terrestrial isopod *Okeaninoscia* (Schmalfuss, 2009).

The emplacement of large igneous plateaus in the central Pacific during the Early Cretaceous (~120 Ma) (Fig. 6-1) was the greatest magmatic event known in Earth history (Fitton et al., 2004). The Ontong Java plateau includes sedimentary strata with fossil wood, and the plateaus bear many seamounts, up to 24 km across on the Hikurangi plateau. As Neall and Trewick (2008) wrote, "The discovery in the last fifty years of a wide array of subsided seamounts throughout the Pacific that once formed above sea level provides insight into the potential origins and age of the island floras and faunas" (p. 3304). By the time the seamounts sank below sea level, part of their biota may have already colonized new islands in the region, maintaining the old geographic pattern. Thus Neall and Trewick (2008) suggested that biological evidence "in some cases reveals aspects of island geology not preserved in the stratigraphic record, providing new lines of evidence for island geological histories" (p. 3304).

Assumption 5. The central Pacific islands in Micronesia and Polynesia formed as the Pacific plate moved over hotspots, and so islands have been colonized from older islands downstream or by long-distance dispersal.

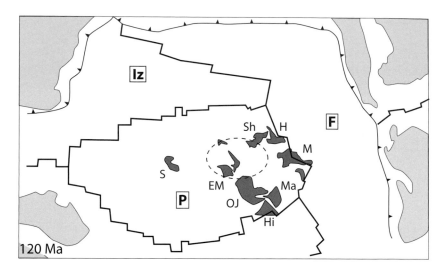

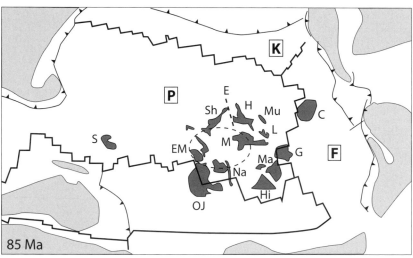

FIGURE 6-1. Reconstructions of Pacific paleogeography at 120 Ma (Early Cretaceous), 85 Ma (Late Cretaceous), 35 Ma (Eocene), and the present (based mainly on Utsunomiya et al., 2008; also Taylor, 2006; Ingle et al., 2007). The maps show the dispersal of the large igneous plateaus (dark gray) and the convergence of the Pacific/ Farallon Ridge (the East Pacific Rise) with America. The position of the South Pacific Superswell (ellipse) has remained the same while the crust has moved. Mid-ocean ridges as stepped lines, subduction zones as barbed lines (barbs on overriding plate). A = Antarctic plate, C = Caribbean plateau, Co = Cocos plate, E = Emperor seamount chain, EM = East Mariana basin, F = Farallon plate, G = Gorgona plateau, H = Hess Rise, Ha = Hawaiian Islands, Hi = Hikurangi plateau, Iz = Izanagi plate, L = Line Islands, M = Mid-Pacific Mountains, Mu = Musicians seamounts, N = Nazca plate, Na = Nauru basin, OJ = Ontong Java plateau, P = Pacific plate, S = East Sulawesi ophiolite, Sh = Shatsky Rise.

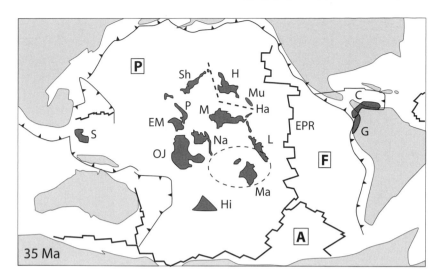

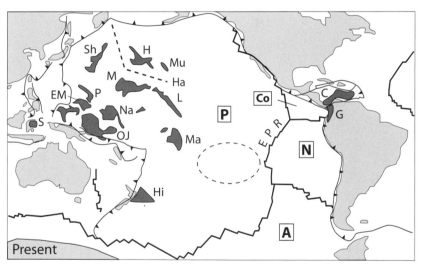

FIGURE 6-1. *Continued*

Mantle material is not normally able to burn its way through the crust or there would be volcanoes everywhere. Volcanoes develop at plate boundaries as the result of plate tectonics, but the causes of intra-plate (mid-plate) volcanism, far from the plate boundaries, are much less obvious. One explanation, the Wilson–Morgan model, proposed that central Pacific island chains have formed as the Pacific plate moved over a stationary hotspot in the mantle. The hotspot is caused by a narrow, cylindrical jet of abnormally hot material, or plume, in the mantle. Anomalies with the model have continued to accumulate, and the causes of the volcanism are the subject of current debate among geologists: Is intraplate volcanism caused by vertical mantle plumes or, instead, by flexing in the plate causing horizontal "hotlines" and propagating fissures? The second possibility is discussed in papers in Foulger et al. (2005) and Foulger and Jurdy (2007a), and especially Anderson (2005), Anderson and Natland (2005), and Foulger (2007).

Hotspot theory is one explanation for intraplate volcanism, and some biogeographic patterns may be caused by "integration" of hotspot traces. But the hotspot theory only accounts for one set of biogeographic tracks, those running parallel to plate movement. The new ideas on volcanism are an interesting alternative. They explain intraplate volcanism by stress- and crack-controlled propagation of fractures and fissures, not by plumes and hotspots. The stresses and cracks are caused by normal plate tectonic processes rather than movement over hotspots. In this model, changes in patterns of stress orientation (rather than shifts in plate motion over fixed hotspots) result in changes in directions of volcanic chains (Natland and Winterer, 2005). In the hotspot model, the youngest island lies nearest the hotspot; the oldest island is furthest away. This trend is also explicable by fissure propagation, which also accounts for the many islands arranged in linear chains that do not conform to age sequences.

Whether or not there is a plume under the Hawaiian Islands remains controversial. The red and yellow depictions of plumes in some papers may resemble photos of lava flows, but instead they are graphs based on complex interpretations of data. There is no direct way of measuring temperatures at this depth, and different tomographic studies infer plumes in different places (cf. Li et al., 2008; Wolfe et al., 2009).

The new model accounts for other geological "anomalies" such as the large igneous plateaus, and it may also explain the biogeographic patterns in the Pacific more simply, as these resemble a network rather than a simple pattern of parallel lines. Recurrent volcanism may have occurred on islands along orthogonal networks of hotlines instead of on the simple parallel lines of hotspot traces. The new model may also account for rejuvenated volcanism and complex patterns of phylogeny (non–progression-rule sequences) in island taxa.

The aim here is to integrate biogeography and tectonics, not to privilege any particular geological theory over biogeographic information, for example, by attributing patterns to chance if they do not fit with geological theory. With the molecular revolution, there is now a growing body of data on phylogeny and distribution available for comparative studies.

AN ALTERNATIVE TO THE LONG-DISTANCE DISPERSAL MODEL: METAPOPULATIONS, NORMAL DISPERSAL, AND VICARIANCE

Many authors imagine that young volcanic islands that are currently isolated must have acquired their plants and animals from far away. One study concluded: "Representatives of Sapotaceae are found on the islands of Fiji, Hawaii, New Caledonia, New Guinea, the Solomon Islands, Tahiti and Vanuatu. Some of these are geologically young volcanic islands that are a considerable distance from any continental land mass. Propagules of Sapotaceae are therefore clearly capable of dispersing significant distances across oceans" (Bartish et al., 2011). Again, this line of reasoning does not allow for volcanism and subsidence persisting within an area for tens of millions of years, or for the associated persistence of metapopulations.

Authors adopting a dispersalist model have also argued that individual volcanic islands were never joined to each other by continuous land, and so a vicariance history is impossible (e.g., van Balgooy et al., 1996, discussing Pacific islands). But vicariance is possible in a metapopulation model that accepts local dispersal ("normal ecological dispersal"; see Chapter 1). A metapopulation occupying islands along an arc and dispersing among the islands as part of its normal ecology may be sundered if the arc is rifted apart due to tectonics, with part of the arc moving away from the other. Thus the dispersalist model rejects ordinary, local dispersal among populations and *in situ* survival of metapopulations, but supports extraordinary, long-distance dispersal (founder speciation). In contrast, the model used here accepts normal, local dispersal among populations and vicariance between metapopulations, but rejects founder speciation.

REGIONAL GEOLOGY: THE PACIFIC PLATE AND THE LARGE IGNEOUS PLATEAUS

Mayr (1940) felt that the entire bird biogeography of the Pacific could be explained without any significant change in the configuration of land or sea and that the phylogenetic breaks and patterns were caused by chance dispersal. In contrast, the model presented here suggests that the patterns have been caused by geological change.

This chapter focuses on island taxa on the Pacific plate, east of Fiji and the plate boundary. The boundary was recognized early on as the "andesite line," as it is the eastern limit of andesite islands. Andesite (named after the Andes) is a typical feature of subduction zones and the volcanic arcs that they produce. On the Pacific plate itself, the volcanic islands often form linear chains and are composed of alkali basalts, not andesite. Normal seafloor, produced at mid-ocean ridges, is tholeiitic basalt.

The Pacific plate formed in the mid-Jurassic at a triple junction of three spreading ridges that developed in what is now the Cook Islands region. From here the new plate grew by outward migration of the ridges, and Figure 6-1 shows the later stages of the process. Sea-floor spreading at mid-ocean ridges is one of the most cited causes of vicariance, and growth of the plate would have had a major effect on the region's biota.

The other great geological revolution in the central Pacific was the Early Cretaceous eruption of large igneous plateaus, the largest of which is the Ontong Java plateau, lying under water next to the Solomon Islands. Following their emplacement, the plateaus dispersed with plate movement (Fig. 6-1), and this is compatible with many biogeographic patterns and anomalies. For example, disjunct western outliers of central Pacific clades occur in Sulawesi (e.g., the *Cycas rumphii* group of cycads of Micronesia to Tonga, plus Sulawesi; Keppel et al., 2008), and this may be explained by the tectonic origin of the Eastern Sulawesi ophiolite in the central Pacific (Fig. 6-1).

The connection between oceanic plateau formation and hotspot/hotline volcanism is not clear and is a key topic in the current debate about intraplate volcanism. In any case, the Ontong Java, Manihiki, and Hikurangi plateaus were emplaced in the Early Cretaceous, possibly as a single, very large plateau that broke up later (Taylor, 2006; Ingle et al., 2007). Some geologists have also suggested that the Ontong Java plateau formed in the central Pacific near the Gorgona plateau, now accreted in western Ecuador, western Colombia, and the Caribbean igneous plateau. The Ontong Java plateau moved westward until it eventually accreted with the Solomon Islands (Fig. 6-1), while the Gorgona and Caribbean plateaus moved east and collided with South America (Chicangana, 2005; Kerr and Tarney, 2005; Utsunomiya et al., 2008).

The formation of the large Ontong Java plateau and the associated 80° change in movement of the Pacific plate had far-reaching effects.

For example, in eastern China, large-scale orogenic lode gold mineralization occurred at this time, indicating the change from extension to compression (Sun et al., 2007). The emplacement of the Ontong Java plateau must have also had a significant impact on the life of the Pacific region. Many populations and taxa in the area would have been extirpated during the eruptions, but the new volcanic strata would have soon been invaded by plants and animals of marginal habitats—weedy taxa from the islands of the young Pacific plate and the adjacent Phoenix, Farallon, and Izanagi plates. These last three plates were later destroyed by subduction and the growth of the Pacific plate, but the descendants of their flora and fauna may survive on the islands of the new plate.

The South Pacific isotopic and thermal anomaly (SOPITA) is an important geological feature located around a topographic structure, the South Pacific superswell (Fig. 6-1). This is a bulge in the seafloor of French Polynesia and the southern Line Islands ~4,000 km across and 680 m high (Adam and Bonneville, 2005). As a mantle feature, it has maintained its position while the crust has moved over it (Fig. 6-1). The superswell is the last remnant of a mid-Cretaceous (120 Ma) upwelling that also produced the Ontong Java, Manihiki, and Hikurangi plateaus. It has been a major center of volcanism at all scales since the Cretaceous, and the West Pacific seamount province (north of the Ontong Java plateau) developed at the superswell before being translated westward with plate movement.

To summarize, there is no real evidence from either biology or geology that requires all the Pacific island biota to have dispersed into the region. The "total dispersal" theory remains popular largely for historical reasons. For some groups, new data from molecular phylogeny are more compatible with a vicariance interpretation, and examples are discussed below.

BIRDS AND BIOGEOGRAPHY

Most of the widespread Pacific endemics considered in this chapter are from case studies of birds and plants. Of all plant and animal groups, birds are the best known in terms of their distribution, ecology, and phylogeny. This is due to the massive collections and intensive study made by ornithologists and birdwatchers all over the world. Their diurnal habits and conspicuous plumage and song make birds much easier to study than, for example, many mammals. Detailed accounts of geographic variation among bird populations in the Pacific, including those of the smallest, most remote islets, were produced in the 1930s and 1940s. This work was based on 40,000 bird specimens collected by the American Museum of Natural History in the decade-long Whitney Expedition (1921–1932). Ideas on phylogeny and distribution were summarized as subspecies classifications, notably in publications by Mayr and other authors associated with the museum (available at http://digitallibrary.amnh.org/dspace/). Some of the subspecies represent points on clines, while others are more clear-cut and may themselves include well-documented geographic variation. This detailed distributional information can now be combined with results from molecular studies.

Research on birds has continued, and all the world's birds have now been illustrated and mapped in the magnificent set of volumes by del Hoyo et al. (1992–2011). This work indicates the great amount of information available for each individual species. It is unfortunate that the detailed distributional and phylogenetic data for birds is often overlooked because of preconceptions about their great powers of dispersal. Although most birds can fly, most are local or regional endemics, and even seabirds that roam for thousands of kilometers return to the same, localized breeding sites. It might seem unlikely that bird populations would remain in the same area for as long as suggested here. But as Agassiz (1850: 193) observed (in his critique of single centers of origin), "Nothing can be more striking to the observer than the fact that animals, though endowed with the power of locomotion, remain within fixed bounds of their geographical distribution." This also applies to birds, and many ornithologists have questioned the significance of birds' means of dispersal in establishing their distributions, as the following quotations indicate:

> Birds offer us one of the best means of determining the law of distribution; for though at first sight it would appear that the watery boundaries which keep out the land quadrupeds could be easily passed over by birds, yet practically it is not so. (Wallace, 1869: 24)

> To the person who is impressed by the bird's potential mobility, the occurrence of the same or representative species at widely separated localities is simply a matter of flight from one station to the other. . . . But the avian geographer is not so easily answered. He knows that most birds are closely confined to their own ranges. (Chapman, 1926)

> Most species of birds, especially on tropical islands, are extraordinarily sedentary. (Mayr, 1940: 198)

That birds can fly across barriers is one of those apparently simple facts that are not simple. (Darlington, 1957: 240)

Despite powers of dispersal through the air, we have the paradox that ornithologists constantly stress the value of birds as objects of distributional and dispersal studies derived from the fact that known cases of such dispersal are so *rare*. (Deignan, 1963: 263)

It may be argued that geographic patterns in birds are not to be expected due to their high ability to disperse, but this has been shown empirically to be wrong. Indeed, the geographic history can be reconstructed for most groups. (Ericson, 2008: 114)

Tropical birds can be very sedentary and have more limited dispersal than is often appreciated. (Burns and Racicot, 2009: 645)

To summarize, although birds can disperse, in most cases they do not, and so a study of their biogeography may be worthwhile. Many of the groups that Mayr and others worked on 70 years ago were not revised again until the molecular revolution. With the rise of the modern synthesis, work on regional biogeography declined as very localized ecological studies became popular. Perhaps researchers felt there was little point investigating phenomena that were due to chance processes, and by the time the molecular revolution occurred the basic geometry of the areas of endemism in the Pacific had become a neglected field.

In contrast, over the same period, structural geology made remarkable progress. Geologists discovered a convergent plate margin in the West Pacific (formerly recognized as the "andesite line"), a divergent plate margin in the East Pacific (the East Pacific Rise), major structures in southeastern Polynesia (the South Pacific superswell and the associated isotopic and thermal anomaly), the large igneous plateaus now dispersed in the west Pacific and the Colombia/Caribbean region, and many other features. Activity on all of these is reflected in the distributions of the new molecular clades.

Most of the geographic patterns shown in molecular phylogenies and highlighted here had already been suggested in morphological studies. The extensive collections of birds and Mayr's acute observations revealed patterns that are now being confirmed. A new generation of systematists is reexamining the biogeographic patterns that led directly, through Mayr's work, to the Modern Synthesis.

Dispersal theory predicts that Pacific members of a group will comprise unrelated waifs or will form complex stepping-stone series of clades in which groups found further into the Pacific are ever more nested phylogenetically. This could be interpreted as the result of taxa migrating into the Pacific and spreading there or as a sequence of differentiation. Repeated invasions of the Pacific and occasional back-colonizations of Asia might also be predicted. But, in fact, the phylogenies of many taxa show a simpler pattern in which Pacific members are diverse and form a single, widespread endemic clade. These are well documented in traditional treatments, and recent molecular studies have confirmed some of these, refined others, and also found intriguing new examples.

PARTULID LAND SNAILS

Land snails include a group of families, Orthurethra, in which the diversity is concentrated in the Pacific. Three of the families are endemic there: Amastridae in Hawaii, Draparnaudiidae in New Caledonia, and Partulidae, widespread in the central Pacific from Micronesia to south-eastern Polynesia (Fig. 6-2; data from Goodacre and Wade, 2001). Many other taxa have a range very similar to that of the Partulidae, for example, the plant *Lepinia* (Apocynaceae): Micronesia (Ponape), Woodlark Island off New Guinea, the Solomon Islands, and in the east on the Society and Marquesas Islands. In addition to its "anomalous" distribution (Lorence and Wagner, 1997), *Lepinia* has "without doubt, the most bizarre fruits in the Apocynaceae," a large, pantropical family (Endress et al., 1997). Thus, Micronesia–southeastern Polynesia is an important center of endemism for terrestrial plant and animal life and is ~10,000 km across.

Partulids as such are absent from New Caledonia and Hawaii, although the family has close relatives in both localities (Fig. 6-2) and so its absences there are probably due to phylogeny rather than lack of suitable habitat or inadequate means of dispersal. Partulids were paired with the Draparnaudiidae by Bouchet et al. (2005), although the exact relationship was unresolved in the study by Wade et al. (2006).

Within Partulidae, the massive disjunction in the range of *Samoana* suggests a former range including what are now the low islands of the Line, Phoenix, and Marshall groups and general allopatry north of *Partula*. Land snail taxa living in humid, montane forest on high islands also occur as fossils on low, dry atolls that have formed by the erosion and subsidence of high islands. For example, Endodontidae is another land snail family (not a member of Orthurethra) endemic to the Pacific. The members range from Micronesia (Palau) to

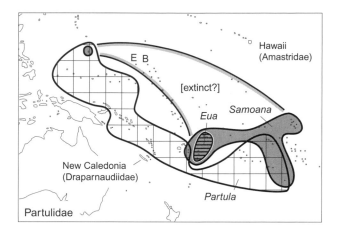

FIGURE 6-2. Distribution of Partulidae (Gastropoda) (Goodacre and Wade, 2001). E = Enewetak Atoll, B = Bikini Atoll; other forest land snails are known from these islands as fossils.

southeastern Polynesia (Henderson Island) and Hawaii, but on Midway Island (an atoll in the northwestern Hawaiian Islands) and Bikini Atoll (Marshall Islands; see Fig. 6-2) they are only known from fossil material. Likewise, on Enewetak (= Eniwetok, Marshall Islands; Fig. 6-2) the related Charopidae are only known as fossils (Solem, 1983: 275).

The break in partulids between *Eua* on Tonga and *Partula* on the Lau group, eastern Fiji (Fig. 6-2) corresponds to the opening of a well-documented back-arc basin between the Tonga and Lau ridges. As this developed it split the former, single arc in two.

MONARCH FLYCATCHERS (MONARCHIDAE)

Monarcha and related genera (Fig. 6-3) are small passerine birds of Australasia and the Pacific in which the phylogeny "does not follow the expected stepping-stone pattern from inshore to remote islands. Instead the most basal division . . . separates eastern and western clades" (Filardi and Moyle, 2005: 217). The eastern clade, *Chasiempis* and four other genera, defines a large central Pacific area of endemism that differs from that of the partulid land snails, as it encompasses New Caledonia and Hawaii but not Micronesia. Despite these differences, the Pacific forms again comprise a single, major clade, not a diverse,

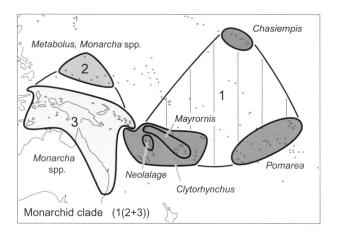

FIGURE 6-3. Distribution of *Monarcha* and allies (Monarchidae) (Filardi and Moyle, 2005). Reproduced from *Journal of Biogeography* (Heads, 2010b), with permission from John Wiley and Sons.

unrelated assemblage of waifs and strays as predicted in dispersal theory and traditional taxonomy. The authors discussed their important results in the context of prior studies on Pacific biogeography, in which they noted the "pervasive assumption" of continent to island dispersal. Their findings contradict the idea of "gradual accretion of independent lineages from continental sources" (p. 216).

The distribution of the central Pacific *Chasiempis* group could have developed in one of two ways. First, it could be the result of dispersal from an initial center of origin somewhere within the present distribution. Many methods have been used to locate this center. For example, it could be the southwestern Solomon Islands (as this is nearest locality to the putative source), in Hawaii (as *Chasiempis* is the basal genus in the group), or in the Vanuatu/Santa Cruz Islands area (as this is the center of diversity, with three genera present). The second possibility is vicariance, and in this case there is simple allopatry between the *Chasiempis* group and its sister, *Monarcha* + *Metabolus*. This suggests that the *Chasiempis* group did not evolve by spreading out from a center of origin within its present range but by simple vicariance of an ancestor that occupied both Australasia and the Pacific. If the location of the split signifies a major regional break in the ancestor, not a single, chance flight, it should be associated with a large-scale tectonic or climatic discontinuity and a similar break in other taxa. The split occurs

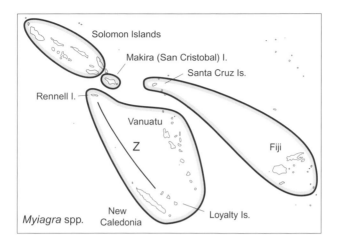

FIGURE 6-4. Distribution of *Myiagra* (Monarchidae) in eastern Melanesia (*M. ferrocyanea* in the Solomon Islands; *M. cervinicauda* on Makira; *M. caledonica* in New Caledonia, Vanuatu, and Rennell Island; and *M. vanikorensis* in Santa Cruz Islands and Fiji; Mayr and Diamond, 2001). The line Z indicates possible sister species of *Zosterops* in Rennell Island (*Z. rennellianus*) and the Loyalty Islands (*Z. inornatus*) (Moyle et al., 2009).

within the southern Solomon Islands, near the margin of the Ontong Java plateau and the area where this has collided with the Solomon Islands (the Australian plate) while drifting west.

Ongoing collision of the Ontong Java Plateau with the Solomon arc over the last 20 m.y. has resulted in uplift of the plateau's southern margin and created on-land exposures of the plateau in the southern Solomon Islands. Bird distributions here led to Makira (San Cristobal) Island and Rennell Island being described as centers of "mystery" and "paradox" in dispersal analyses (Mayr and Diamond, 2001: 249, 254). The boundary between the two main monarch clades runs between Rennell and Makira Islands, and this area is also an important break point in many other groups. For example, in *Myiagra* species (Monarchidae; Fig. 6-4) the boundaries occur around Rennell, Makira, and the Santa Cruz Islands, and the distributions do not conform with the current geographic boundaries of the archipelagos (the Solomon Islands, Vanuatu, New Caledonia, and Fiji).

Filardi and Moyle (2005) observed that the attenuation of bird species numbers in the Pacific with increasing distance from the continental source has led to the assumption of a flow of "colonists." "The assumption is evident in Mayr's seminal interpretations of Pacific bird diversity [Mayr, 1942, etc.], which have influenced general theory. . . . Mayr concluded that there is no Polynesian bird fauna [i.e., no autochthonous fauna], there are only colonists from New Guinea." The monarchid example is important as it tests the theory with new molecular information on a classic exemplar group and contradicts it.

Some biogeographers might accept that the whole *Monarcha* + *Chasiempis* group could have differentiated *in situ* by vicariance in a widespread common ancestor of the two, but would suggest that this range itself must have been achieved by dispersal in the ancestor. This may or may not be true and depends on the location of the sister group. If this is in central Australia or Southeast Asia, for example, where *Monarcha* + *Chasiempis* and so on do not occur, then, again, no dispersal is required. Eventually, of course, clades will be found to overlap, and this indicates simple range expansion at that level in the phylogeny (not necessarily in the modern genera). Overlap, not allopatry, is explained by range expansion, and so movement among the current monarchid genera is only needed to account for potential overlap of *Clytorhynchus* with *Neolalage* and *Mayrornis* (Fig. 6-3).

In fact, Filardi and Moyle (2005) found that the sister group of the clade in Figure 6-3 may be *Grallina*, widespread throughout Australia (including the southwest, the northwest, and Tasmania) and New Guinea, implying vicariance followed by secondary overlap along the eastern Australian seaboard and in New Guinea.

In their new interpretation of the Pacific monarchids, Filardi and Moyle (2005) simply reversed the traditional sense of dispersal and supported a center of origin in Polynesia followed by dispersal to the west and colonization of Australia. (This is because the two continental species are nested in the pan-Pacific radiation.) In fact, though, the precise, large-scale allopatry evident between the two main clades and among the subclades within them undermines the whole chance dispersal paradigm, and there is no need for dispersal in any direction (cf. Figs. 1-6, 1-7). In this model, passerines have been in the Pacific ever since the group's origin, which is consistent with the basal passerines (Acanthisittidae) being endemic to New Zealand.

The three subclades in the central Pacific group illustrate three standard sectors of the region:

- northern Polynesia (Hawaiian Islands: *Chasiempis*, the basal group);
- southeastern Polynesia (French Polynesia and the Cook Islands: *Pomarea*);

- southwestern Polynesia/eastern Melanesia (*Neolalage, Clytorhynchus, Mayrornis*). The islands occupied by this group (except for Samoa) lie west of the present Pacific/Australia plate boundary.

Birds such as *Neolalage* and plants such as the palm genus *Carpoxylon* are endemic to Vanuatu. The biota of this archipelago is sometimes suggested to be depauperate for the area of the islands, but this may be a myth that has developed because the islands are young. The idea is shown to be incorrect, at least for the lizard fauna, by Hamilton et al. (2009). Also, there are many Vanuatu–Fiji endemics, and these do not show up on lists of Vanuatu or Fijian endemics. Simply because the areas of endemism in the region do not coincide with the geographic boundaries of the archipelagos does not mean these rainforest biotas are depauperate or low in endemism.

EMOIA (SCINCIDAE)

The skink *Emoia samoensis* and its relatives make up a central Pacific clade that ranges from Vanuatu to the Cook Islands (Hamilton et al., 2010). The boundary at northern Vanuatu is near the southern Solomon Islands break in the monarchs. Hamilton et al. tested the idea that the skink group had dispersed eastward to the Cook Islands, either by waif dispersal or by human-mediated transport, but found that the Cook Islands species was basal to the rest of the whole complex, in Vanuatu, Fiji, Samoa, and Tonga. As with the monarchs, this phylogeny contradicts the traditional model of dispersal into the Pacific, and while Hamilton et al. (2010) introduced the Cook Islands species as either "Biogeographic Anomaly or Human Introduction" (in the title of their paper), it may be neither. This and other endemism in the Cook Islands–southeastern Polynesia region is consistent with long-term survival of clades in the area. The break between southeastern Polynesia (including the Cook Islands) and southwestern Polynesia/Vanuatu is a repetition of the break between *Pomarea* and *Neolalage/Clytorhynchus/Mayrornis* (Fig. 6-3).

SUCCINEIDAE

In the worldwide land snail family Succineidae, the basal clade is *Succinea manuana* (Holland and Cowie, 2009), recorded from southeastern Polynesia (Samoa, Cook, Austral, and Marquesas Islands; B. Holland, pers. comm.). As usual, this does not necessarily mean the area was a center of origin, but the pattern cannot be explained by simple dispersal to southeastern Polynesia.

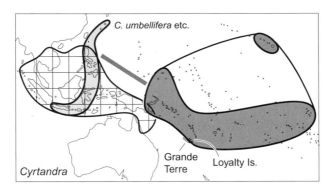

FIGURE 6-5. Distribution of *Cyrtandra* (Gesneriacerae) (Clark et al., 2008, 2009). Dark gray = Pacific clade, light gray = its sister group (*C. umbellifera* etc.), grid pattern = all other *Cyrtandra* species. Reproduced from *Journal of Biogeography* (Heads, 2010b), with permission from John Wiley and Sons.

CYRTANDRA (GESNERIACEAE)

Cyrtandra (Gesneriaceae) comprises 500–600 species, mostly shrubs, distributed through Southeast Asia and the Pacific (Fig. 6-5). The 250 species present in the central Pacific form a well-supported clade (Cronk et al., 2005; Clark et al. 2008, 2009). Most of the Pacific species are restricted to the understory of dense rainforest at higher elevation, especially in shaded gullies by streams and waterfalls. They are well known for their localized distributions; most are single-island endemics and many are rare and endangered. The exact western boundary of the Pacific group is not clear, and there may be minor overlap with other *Cyrtandra* clades in northern New Guinea, a zone of terrane accretion.

The sister of the Pacific clade comprises species of Java, Sulawesi, and the Philippines, and also *C. umbellifera* of small islands north of there: Batan Islands (between the Philippines and Taiwan), Lanyu Island (off Taiwan), and southern Ryukyu Islands (southern Japan) (Weber and Skog, 2007; Flora of China at www.efloras.org). This arc follows the plate margin which *C. umbellifera* occupies as a subduction zone weed. The phrase might not be quite accurate with respect to the narrow ecological sense of "weed," as *C. umbellifera* is an understory species of closed forest. Yet in a broader sense the concept may be useful, as over geological time the species and its ancestors have survived and persisted along the subduction zone by constantly invading new volcanic islands there. In this sense the whole forest, including taxa of the most shaded gullies and deepest forest, is a weedy, pioneer community (Heads, 2006b).

The widespread Pacific clade of *Cyrtandra* occurs only on small islands, and the question is: Why does it occur only on *these* small islands and not those of Indonesia or the Philippines? The small-island ecology of the clade may have been determined by its location in the central Pacific, where small islands are the only available habitat. The distribution was brought about by a break in a widespread ancestor, somewhere north of New Guinea (gray bar in Fig. 6-5) between the Pacific group and its sister in the Philippines–Sulawesi region. Thus the biogeography and the phylogeny, along with the morphology and physiology, have determined the ecology, not the other way around.

In their study of *Cyrtandra*, Cronk et al. (2005) predicted that "Given such a large and widely distributed genus, it would not be surprising if the oceanic islands of the Pacific had been colonized several times independently from the species rich source region in Asia" (p. 1019). Nevertheless, the authors concluded that the Pacific group is monophyletic and, as they noted, this raises three key questions:

- Why did the genus only colonize the Pacific once?
- Why is there no evidence of back-colonization to Southeast Asia?
- How exactly did the genus disperse to all the high islands in Melanesia and Polynesia?

Answering the third question, Cronk et al. (2005) suggested inter-island dispersal by birds, but this makes the first two questions even more difficult and they were left unanswered. Cronk et al. concluded that Pacific *Cyrtandra* has "extraordinary dispersability" and that it is a "supertramp clade" which occurs on small and remote islands because of its "extremely high vagility." Yet there is no actual evidence for any special vagility apart from the central Pacific distribution itself, and instead this could be the result of vicariance at the plate boundary (Ryukyu Islands, Taiwan, Philippines). The "extremely high vagility" assumed for the plants is contradicted by their ecology; most species are rare, local endemics in shady gullies of closed montane rainforest and could hardly be less vagile.

The basal grade in Asia does not necessarily indicate a center of origin for the Pacific clade (Fig. 1-6). The Pacific group is neatly allopatric with its sister in the Philippines region and there is no evidence that the Pacific was ever "colonized" by the Pacific clade. This clade may have developed by evolving in the Pacific region as a simple vicariant. A vicariance origin would also explain why the group only "invaded" the Pacific once: It evolved there. Its ancestors already occupied Polynesian and Micronesian sectors of the Pacific region long before the current islands or the modern taxa existed. The Pacific populations began as generalized Gesneriaceae, evolved into generalized *Cyrtandra*, and finally became individualized as the central Pacific clade, formed by the break at the plate margin.

An intriguing anomaly occurs in the distribution of *Cyrtandra* in the New Caledonia archipelago. Here the genus is present on the low, flat Loyalty Islands (Fig. 6-5) but has never been found on the main island, Grande Terre, although the steep mountains there are covered with rainforest and provide much more typical habitat. This difference between Grande Terre and the Loyalty Islands is a common pattern and can be attributed to the independent history of the Loyalty Ridge with respect to Grande Terre, rather than to chance dispersal. The Loyalty Islands lie along the Loyalty Ridge, an old, sunken arc that is independent in its origin from Grande Terre (part of Norfolk Ridge), and the two ridges were only juxtaposed later. The Loyalty Ridge may have been part of the Vitiaz arc, which formed possibly in the Cretaceous and was later rifted apart to form the precursors of the Solomon Islands, Vanuatu, Fiji, Tonga, and the Three Kings ridge (northeastern New Zealand) (Crawford et al., 2003).

The two remaining questions posed by Cronk et al. (2005)—why did *Cyrtandra* colonize the Pacific only once? and why did it not back-disperse to Asia?—are critical. No answers were returned because none are evident, and this may be because the fundamental assumption of colonization of the Pacific is incorrect. The precise vicariance of "Pacific *Cyrtandra*" with its sister group, the absence of any *Cyrtandra* from areas such as the New Caledonian mainland where there is abundant suitable habitat, the very high rates of local endemism and allopatry, and its closed forest ecology all suggest that vagility in the group is limited and that the distribution has been determined by historical, tectonic factors.

In their detailed study of *Cyrtandra* Clark et al. (2008) based analyses of the Pacific clade on several explicit assumptions.

In their assumption 1, the authors used ages of Pacific islands to give maximum ages of lineages endemic to the islands. This assumes that a lineage endemic to an island "most probably post-dates the origin of that area" (p. 697) but, as discussed in Chapter 2, this assumption is not valid.

In assumption 2, Clark et al. (2008) accepted that the ancestor of the Pacific clade was a species. In the Mayr–Hennig synthesis (see Chapter 1) the ancestor of a clade is always a single individual, parent pair, or homogeneous species and, in particular, it is monomorphic. In this model, all new clades originate by the development of new characters. But this is not necessary if the ancestor was already polymorphic before the descendant clades evolved. This has been proposed both in biogeography and in genetics, where the widespread incongruence of gene trees is often attributed to incomplete lineage sorting. This involves ancestral polymorphism that already existed before differentiation of the modern groups began and that is retained in the modern groups. Both the ancestral morphological variation and its biogeography may be retained (Fig. 1-11).

In assumption 3 (their "prior expectations"), Clark et al. (2008) reasoned that because the ancestor of the Pacific clade was a species and because the ranges of extant species are narrow, the range of the ancestor would also have been narrow. This seems doubtful and it eliminates the possibility of vicariance *a priori* (in vicariance the ancestor always has a wider distribution than its descendants). Contradicting this is the widespread central Pacific clade and its allopatry with the sister groups. Following extensive vicariance in a widespread ancestor, the many allopatric descendants may be very localized, as in *Cyrtandra*.

Clark et al. (2008) assumed a center of origin/dispersal model for Pacific *Cyrtandra* and used four different programs to find the center of origin (ancestral area). Analysis using the program DIVA, which sometimes accepts vicariance as a possibility, concluded that ancestors were widespread. Clark et al. wrote that this "conflicts" with their prior expectations and implies that lineages occurred in "areas that did not exist." For these reasons the results were rejected. Nevertheless, it is clear that there were islands in the Pacific before the current islands, as indicated by the numerous flat-topped (formerly subaerial) seamounts there.

Other programs for finding the center of origin did fit the authors' preconceptions of stepping-stone dispersal and so were favored: SM and DEC programs gave "plausible hypotheses" indicating a "major dispersal route into the Pacific through the islands of Fiji and Samoa" (p. 693).

Clark et al. (2008) rejected the obvious solution—simple vicariance— not because of their results, which provide new, strong evidence for it, but because of their "prior expectations"; they found a center of origin reconstruction "yields more intuitive results" (p. 705). This still leaves the three critical questions that Cronk et al. (2005) raised (see above) unanswered. In addition, although the two dispersal studies of *Cyrtandra* both assumed the same ultimate source in Asia and also agreed on the phylogeny, they reached different conclusions: Clark et al. (2008, 2009) supported a center of origin for the Pacific clade in Fiji/Samoa, whereas Cronk et al. (2005) proposed a center of origin in Hawaii. Dispersal analysis often concludes with a controversy about the location of the center of origin, and this would be expected if the center of origin is imaginary. Clark et al. (2009) themselves concluded that the center of origin hypothesis they proposed "needs to be explored in detail" (p. 993), but perhaps it is the underlying principle of "center of origin" itself that needs to be explored.

APLONIS (STURNIDAE)

This starling ranges from India to southeastern Polynesia (extinct on Society Islands; Steadman, 2006) (Fig. 6-6). Zuccon et al. (2006) described the genus as "remarkably speciose," with "great morphological diversity" and wrote that it "seems to include surviving species of an old radiation." A central Pacific clade, the "*cinerascens* group," was recognized on the basis of morphology (Mayr and Diamond, 2001), and it is starting to find molecular support. Lovette and Rubenstein (2007) sampled ten out of 22 *Aplonis* species and retrieved a clade comprising *cinerascens* (Cook Islands), *tabuensis* (Santa Cruz–Samoa), and *insularis* (Rennell Island) (dark gray in Fig. 6-6). Other Pacific species have been linked with members of this group on the basis of morphology (light gray in Fig. 6-6), but none from New Guinea, the main Solomon Islands (except Rennell), or Australia. *A. feadensis* (not yet sequenced) skirts the Bismarck Archipelago and Solomon Islands on small atolls in the Ninigo (3 km^2), Hermit (8 km^2), Tench ($<$1 km^2), Nissan (37 km^2), Nguria (5 km^2), and Ontong Java groups (10 km^2) (IUCN, 2009). Morphologists have linked it with *A. insularis* of Rennell Island. Because of its small islet habitat, Mayr and Diamond (2001) regarded *A. feadensis* as a "supertramp," the same conclusion that Cronk et al. (2005) reached for Pacific *Cyrtandra*. While Mayr and Diamond (2001) interpreted the distribution of *A. feadensis* as a function of its ecology, its atoll ecology does not explain why it is on these particular atolls and none of the thousands of others in the Pacific. Nor does the ecology explain the linear, arcuate shape of the distribution. Competition between this species and, for example, *A. cantoroides* on the large islands of Papua New Guinea and the Solomon Islands may explain why the two do not invade each other's territory, although it does not explain why the distributional break developed where it has. Simberloff and Collins (2010: 251) concluded that "Why *A. feadensis* is a supertramp and is not found on other islands is uncertain; it is highly vagile." On the other hand, IUCN (2009) concluded, "Its dispersal abilities are not well-known beyond flying between islands within a group, but inter-island morphological variation suggests limited gene flow."

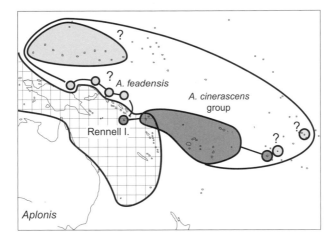

FIGURE 6-6. *Aplonis.* Dark gray = *cinerascens* group (Lovette and Rubenstein, 2007); light gray with queries = possibly related, unsequenced species; grid pattern = other species of *Aplonis.* Species that are possibly related to the *cinerascens* group include *A. opaca* of Micronesia, *A. feadensis* on islets off Papua New Guinea and the Solomon Islands, *A. mavornata* of the Cook Islands (Mauke, one old specimen only), and a fossil species on the Society Islands (Huahine).

Instead of its ecology determining its biogeography, *A. feadensis* may be a regional representative whose habitat (small islands) has been determined by the geological history of the region (subsidence of former high islands). The linear distribution may reflect the origin of the species as a metapopulation on a former island arc or arcs, and the geographic relationship with the Ontong Java plateau needs to be explained. The bird's ecology—for example, its ability to colonize new sandbars, live in disturbed forest and plantations as well as intact forest, and so on— explains its survival within its region but not why it was there in the first place. The ecology probably reflects "weedy" tendencies already present in the ancestor that were preadaptive in populations redeposited back down to sea level and atoll habitat.

PTILINOPUS (COLUMBIDAE)

A group of fruit doves once termed the *Ptilinopus superbus* species group (now including *Chrysoenas*, *Drepanoptila*, and *Alectroenas*) forms a diverse Indo-Pacific clade (Fig. 6-7; Gibb and Penny, 2010).

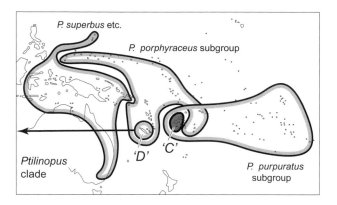

FIGURE 6-7. Distribution of a clade in *Ptilinopus* (Baptista et al., 1997; Gibb and Penny, 2010). Light gray = *Ptilinopus purpuratus* species group (Columbidae): 'C' = "Chrysoenas," sister to the *purpuratus* species group. 'D' = "Drepanoptila," sister to "*Alectroenas*" of Madagascar, Seychelles, and Mauritius (indicated by arrow).

The Fijian "*Chrysoenas*" (*P. victor*, etc.) is the basal group, ranging in the western Fijian Islands (absent from the Lau group in eastern Fiji). The next basal group is the disjunct pair "*Drepanoptila*" of New Caledonia and "*Alectroenas*" of Madagascar, Mauritius, and the Seychelles. A detailed molecular phylogeny for the last, main clade is not yet available, but morphologists have proposed three allopatric subgroups. Two of these, the *porphyraceus* and *purpuratus* subgroups, are widespread central Pacific groups that include 12 species (Ripley and Birckhead, 1942; Baptista et al., 1997). They have a western limit at Rennell Island, as in the Pacific monarchs (Fig. 6-3) and starlings (Fig. 6-6). Gibb and Penny (2010) sampled 13 of the 49 *Ptilinopus* species, but further work is needed to test the central Pacific subgroups.

The *P. porphyraceus* subgroup occurs in Tonga and the Lau Group (eastern Fiji), although not in western Fiji. The pattern can be related to the history of the Tonga and Lau arcs, which originally formed a single arc that was split along its length by back-arc basin formation (cf. partulid land snails, Fig. 6-2). The *P. porphyraceus* subgroup is replaced in the rest of Fiji by *Chrysoenas* and the *P. purpuratus* subgroup. Thus the archipelago of the Fiji Islands is a biogeographic composite, reflecting its structure as a tectonic composite of several arcs (Heads, 2006b).

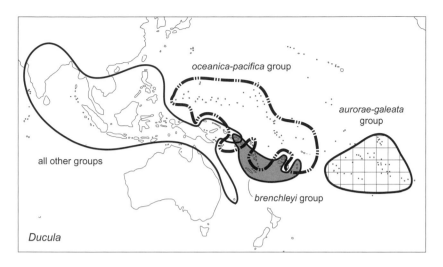

FIGURE 6-8. Distribution of *Ducula* (Columbidae), including a fossil record on Henderson Island. The groups shown are based on morphology (Baptista et al., 1997; Mayr and Diamond, 2001).

DUCULA (COLUMBIDAE)

Ducula, like *Ptilinopus*, is a diverse genus of Indo-Pacific fruit doves. In morphological studies, Amadon (1943) recognized three species groups endemic in the central Pacific (Fig. 6-8). (In their molecular study, Gibb and Penny, 2010, sampled nine of the 35 *Ducula* species, including one central Pacific representative.) Two of the morphological groups in the central Pacific, the *oceanica–pacifica* group and the *aurorae–galatea* group, may be linked (Baptista et al., 1997; Mayr and Diamond, 2001), and this would give a clade with a distribution similar to that of partulid land snails (Fig. 6-2). Both groups extend from Micronesia to southeastern Polynesia, although they are absent from New Caledonia and Hawaii. In *Ducula*, two of the Pacific species groups (the *oceanica–pacifica* and *brenchleyi* groups) and two of the western groups all overlap in the south and central Solomon Islands, indicating the usual complex break here followed by local overlap.

The Pacific region occupied by *Ducula* and *Ptilinopus* includes many atolls and low islets that represent former high, volcanic islands and, as in *Aplonis feadensis*, the current ecology of the atoll taxa may have developed as populations were lowered to sea level. A similar process explains the fossil land snails on atolls cited above and may also explain anomalous, low-elevation plant populations on subsiding islands in eastern Fiji (Heads, 2006b). Unlike the forest land snails that went extinct on the

Marshall Islands, a few columbids have survived the lowering from high elevation rainforest to harsh atoll conditions. These include *Ducula oceanica* in Micronesia and *Ptilinopus coralensis* endemic to atolls in southeastern Polynesia. Because *Ducula* and *Ptilinopus* can survive on small islands and even atolls, they may conserve old patterns even after subsidence and erosion of high islands, and so the many species and subspecies are valuable biogeographic indicators, even on a local scale. Further molecular work on these Pacific groups will be of great interest.

In *Ducula*, as in *Ptilinopus*, representatives on the Lau Group (eastern Fiji) belong to the *oceanica–pacifica* group and are linked with Tonga, not with the main Fijian islands to the west.

ALOPECOENAS (COLUMBIDAE)

In this genus, formerly treated under *Gallicolumba*, the only clade in the Pacific is subgenus *Terricolumba*. It is widespread and endemic from Palau and the Marianas islands, through Micronesia and Melanesia (but not Australia), to southeastern Polynesia (Jønsson et al., 2011). *Terricolumba* has nine extant species which have all been sequenced, and seven extinct species in the same region have also been attributed to the clade. The subgenus has its relatives in the Lesser Sunda Islands (Timor and Wetar: *G. hoedtii*), eastern Australia (*Leucosarcia*), Burma to southern New Guinea and all Australia (*Geopelia*, *Phaps*, *Ocyphaps*, *Geophaps*), the Philippines–Sulawesi (*Gallicolumba* in part), and the Philippines–New Guinea (*Gallicolumba* in part).

AERODRAMUS (*COLLOCALIA* OF SOME AUTHORS) (APODIDAE)

Swiftlets are known as a taxonomically difficult group, and so it is not surprising that molecular studies have led to revised ideas on genera and species. As mentioned in Chapter 2, *Aerodramus* ranges from the Seychelles and Mascarenes (not Madagascar) east to Polynesia. Sampling is still limited, but preliminary results indicate a well-supported new clade endemic to the central Pacific, including *A. bartschi* (Micronesia), *A. vanikorensis lugubris* (Solomon Islands), and *A. sawtelli* (Cook Islands) (Fig. 6-9; Dickinson, 2003; Price et al., 2005). Unsampled species that have been allied with members of this group by morphologists are indicated in Figure 6-9 (with queries). Within *Aerodramus*, the sister of the Pacific group ranges from the Seychelles to Samoa and includes *A. spodiopygius* (Fig. 6-9), highlighted here as it defines another important

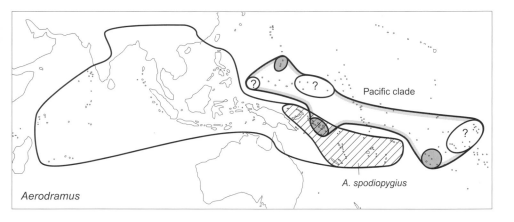

FIGURE 6-9. Distribution of *Aerodramus* (Apodidae) (Dickinson, 2003; Price et al., 2005). Dark gray = Pacific clade (*A. bartschi*, *A. sawtelli*, and *A. vanokorensis lugubris*), queries = unsampled species often allied with members of the Pacific clade (*A. pelewensis*: Palau; *A. inquietus*: Caroline Islands; *A. leucophaeus*: southeastern Polynesia). The only Marquesas record is a fossil species.

center of endemism, the Lapita region. This comprises the islands from New Britain to Samoa, including New Caledonia but not mainland New Guinea.

The Pacific group is allopatric with its congeners except at the Solomon Islands, the usual point of overlap.

TODIRAMPHUS S.STR. (ALCEDINIDAE)

Early studies of kingfishers recognized a clade of flat-billed species in southeastern Polynesia that was named *Todiramphus* (Fig. 6-10; distributions from Woodall, 2001). Members of the group were often included under *Halcyon*, and Mayr (1941) considered they were "nothing but aberrant offshoots" of *Halcyon chloris*, widespread from the Red Sea to New Zealand. Nevertheless, Mayr and Diamond (2001) continued to refer to "the *Todiramphus* group," and in molecular studies the sample of *Todiramphus* species emerged as basal not only to the widespread *H. chloris* but also *H. sanctus*, found through Australasia and Indonesia (Moyle, 2006). (Moyle used the name *Todiramphus* in a broad sense to cover this whole group.) The main clade of *Todiramphus* s.lat. ranges over 12,000 km between the Red Sea and New Zealand without a major break, and so the differentiation of the central Pacific *Todiramphus* s.str. (six species; Woodall, 2001) at the basal node and the precision of the allopatry between it and its sister are remarkable.

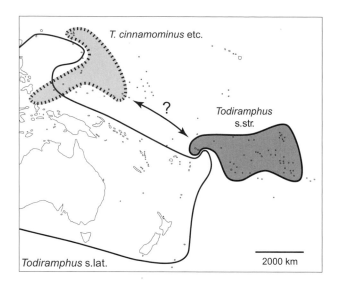

FIGURE 6-10. Distribution of *Todiramphus* s.lat. (Alcedinidae) in the Pacific (Moyle, 2006).

The Micronesian *Todiramphus cinnamominus* group (Fig. 6-10) has not yet been sequenced; affinities either with taxa in New Guinea (as in *Monarcha*) or with the central Pacific group (as in *Aerodramus*) would conform to standard biogeographic alternatives, and the latter possibility is indicated in Figure 6-10. Even if *Todiramphus* s.str. does not include the Micronesian species, it still occupies most of southeastern Polynesia (~10 million km²) and may have evolved there *in situ* over geological time. The break with the rest of *Todiramphus* s.lat. occurs in Samoa, near the Pacific plate boundary.

TWO CLADES OF PARROTS

The central Pacific parrots (Fig. 6-11) belong to two clades that are regional representatives of the Platycercini (in a new, broad sense) and Loriinae (also in a broad sense) (Wright et al., 2008). In the Platycercini, *Cyanoramphus* and allies occur throughout the New Zealand region, New Caledonia, Fiji, and southeastern Polynesia. *Charmosyna* and allies, in the Loriinae, range north of there, from New Guinea (most taxa) to southeastern Polynesia. (*Vini* is often assumed to have been introduced to the Line Islands by ancient Polynesians, although this needs to be confirmed. *Prosopeia* is sometimes thought to have been

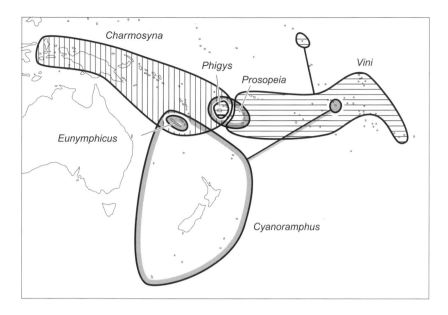

FIGURE 6-11. Distribution of parrots in the central Pacific. Line fill = Loriinae clade; gray = Platycercini clade (Wright et al., 2008).

introduced to Tonga from Fiji.) Within the Pacific Loriinae, the main break occurs within the Fiji archipelago. Here, the western *Charmosyna* and *Phigys* do not occur in the Lau group, while the eastern Pacific *Vini* is only known in the Lau group.

Platycercini and Loriinae were not regarded as related until the molecular work showed they were sisters. Platycercini are distributed from Africa and Madagascar though tropical Asia to the Philippines, New Guinea (one species only), Australia (most genera), and the Pacific (*Cyanoramphus* and allies). Loriinae are in Sulawesi, the Philippines, New Guinea (most genera), Australia, and the Pacific (the *Charmosyna* group). Thus there is overlap in Australia, Sulawesi, and New Guinea, but as in the Pacific, Platycercini dominate in the south, Loriinae in the north. In dispersal theory, the pattern (Fig. 6-11) would imply a double invasion of the Pacific by parrots, with one invasion by each group. Instead, the overall allopatry between the two clades is interpreted here as the remains of an old vicariance pattern that developed before the opening of the Coral Sea and the Tasman Sea. The break led to populations north of the New Caledonia–Society Islands line differentiating as Loriinae, while populations south of the line became Platycercini. A similar break occurs in orchids.

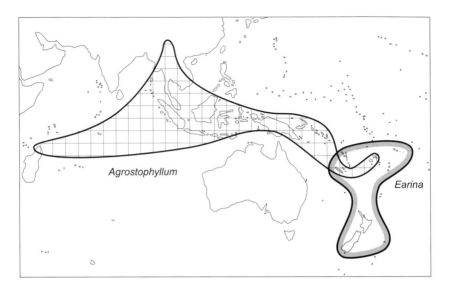

FIGURE 6-12. Distribution of *Agrostophyllum* and its sister group *Earina* (Orchidaceae) (St. George, 1993; Conran et al., 2009).

EARINA (ORCHIDACEAE)

The orchid *Earina* occupies a large sector of the southwest Pacific and overlaps with its sister group *Agrostophyllum* only in New Caledonia and Fiji (Fig. 6-12; Conran et al., 2009), and this can be compared with the parrots (Fig. 6-11). Conran et al. (2009) wrote that the extant distribution of *Earina* (and a similar pattern in *Dendrobium* sect. *Macrocladium*) "indicates an early radiation into Zealandia [the New Zealand plateau, including New Caledonia] prior to this subcontinent becoming largely submerged during the middle Cenozoic" (p. 471). This supports suggestions that vicariance has been important in the family (Brieger, 1981; Chase, 2005), despite the minute, wind-blown seeds. Because *Earina* itself is a vicariant of its sister (Fig. 6-12), there is no need for a radiation *into* the southwest Pacific, only differentiation in the southwest Pacific, out of a widespread Indo-Pacific group.

MERYTA (ARALIACEAE)

Trees in the genus *Meryta* are distributed from Micronesia (Yap) to southeastern Polynesia (Lowry, 1988; Fig. 6-13). The two main clades in the genus (not mapped here) are allopatric; one is only in Fiji and New Zealand,

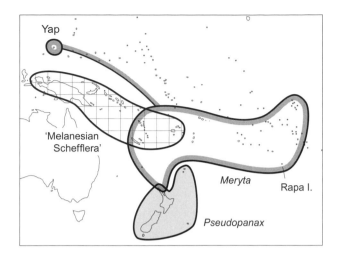

FIGURE 6-13. Distribution of "Melanesian *Schefflera*," *Meryta*, and *Pseudopanax* (Araliaceae) (Plunkett and Lowry, 2010).

the other widespread in the central Pacific (Micronesia, New Caledonia, Vanuatu, and southeastern Polynesia). Tronchet et al. (2005) accepted that the break between the two clades represents "ancient vicariance."

Possible relatives of *Meryta* include a clade termed "Melanesian *Schefflera*" (including *Plerandra*, *Dizygotheca*, Gabriellae group, etc.) (Plunkett et al., 2005; Plunkett and Lowry, 2007; Fig. 6-13). The two clades are largely vicariant, with a region of overlap: New Caledonia, Vanuatu, and Fiji. Both clades have most of their diversity in New Caledonia, which could be a result of the composite tectonic structure of the island. *Pseudopanax* of New Zealand is another relative, and the three groups may form a southwest Pacific clade (G. Plunkett, pers. comm.). Another possibility is that "Melanesian *Schefflera*," *Meryta*, and then *Pseudopanax* are the three basal branches in a widespread Indo-Pacific group, with the main, widespread clade being *Polyscias* s.lat. (not shown) (Plunkett and Lowry, 2010). In any case, *Meryta* is a distinctive genus and a typical central Pacific group.

THE *TETRAPLASANDRA* GROUP (A CLADE IN *POLYSCIAS* SUBG. *TETRAPLASANDRA*) (ARALIACEAE)

The clade known until recently as the *Tetraplasandra* group (including *Reynoldsia* and *Munroidendron*) is distributed in northern Polynesia (Hawaii) and southern Polynesia (Samoa, Society and Marquesas

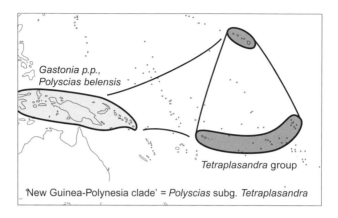

FIGURE 6-14. Distribution of *Tetraplasandra* group and *Gastonia p.p.* (Araliaceae) (Plunkett and Lowry, 2010).

Islands; Fig. 6-14; Costello and Motley, 2007; Plunkett and Lowry, 2010). Its tentative sister is *Polyscias belensis* of New Guinea–Solomon Islands, with the break located between the Solomon Islands and Samoa. This may have involved the Vitiaz trench (the former plate margin), which runs between the two areas, and disjunctions in this region also occur in the "volcanic islands clade" of *Meryta* (Tronchet et al., 2005). The closest relatives of the *Tetraplasandra* group also include *Gastonia spectabilis* and probably the unsampled *G. serratifolia*, of Sulawesi and New Guinea. All these plants form a Pacific basin group (the "New Guinea-Polynesia clade" of Plunkett and Lowry, 2010) (Fig. 6-14). (Lowry and Plunkett, 2010, placed six genera under *Polyscias* and treated the Pacific basin group as *Polyscias* subg. *Tetraplasandra*.) The clade has a notable absence in the New Caledonia–Vanuatu–Fiji region, where related groups have their centers of diversity (e.g., 25 of the 27 species in *Polyscias* subg. *Tieghemopanax*; see below). A similar affinity connecting the Solomon Islands and Samoa, but absent in Vanuatu–Fiji, occurs in groups of Diptera (Bickel, 2009).

The sister group of the Pacific basin group is the "Indian Ocean basin clade" (Plunkett and Lowry, 2010), present through Africa, Madagascar, and the Mascarene Islands. Thus a widespread Indo-Pacific group ranging from West Africa to the Hawaiian Islands (strangely absent from the Seychelles) is divided into Indian Ocean and Pacific Ocean clades.

The Indian + Pacific group itself may in turn have developed its large range by simple vicariance with its sister, endemic to the Seychelles (three species, now *Polyscias* subg. *Indokingia*). Lowry and

Plunkett (2010: 79) suggested that the Seychelles clade "colonized the archipelago from the east, most likely from somewhere in Australasia" and represents a radiation "completely independent" of that of the Indian + Pacific group (Plunkett and Lowry, 2010: 48). Yet this does not explain the neat allopatry between the two and the strange absence of the Indian + Pacific clade from the Seychelles. Both phenomena are consistent with a vicariance history. The initial differentiation in the Seychelles + Indian + Pacific ancestor may have been caused by Gondwanan rifting around the Seychelles.

The sister group of the whole Indo-Pacific + Seychelles group is *Polyscias* subg. *Tieghemopanax*, which fills the conspicuous gap in its sister clade in New Caledonia, Vanuatu, and Fiji. Lowry and Plunkett (2010) treated all these taxa, with others, under *Polyscias* s.lat. This broad *Polyscias* occurs in the Marianas and Hawaii, but not in the New Zealand region or the Rapa Island region. Conversely, its immediate relatives *Meryta* and *Pseudopanax* are absent from the Marianas and Hawaii, but are present in the New Zealand region and Rapa. This difference could reflect initial allopatry, while the overlap of *Polyscias* s.lat. and *Meryta* along the central strip: New Caledonia–eastern Polynesia could be a secondary development. This is the same zone of overlap seen in the Pacific parrots (Fig. 6-11).

The primary phylogenetic and geographic breaks in these araliads occur in the southwest Pacific (New Caledonia region) and the southwest Indian Ocean (Seychelles region) and, as usual, these nodes may represent early fracture zones in an already widespread ancestor.

FITCHIA AND *OPARANTHUS* (ASTERACEAE TRIBE COREOPSIDEAE)

This clade (Fig. 6-15) exemplifies central Pacific groups with Caribbean affinities. *Fitchia* + *Oparanthus* is endemic to southeastern Polynesian and is sister to a Caribbean plate clade, *Selleophytum* + *Narvalina* (Motley et al., 2008). This connection follows the tectonic affinity between large igneous plateaus in the two areas (Fig. 6-1). *Fitchia* is a small tree of montane forest distinguished by its stilt roots, the largest fruits in the family (5 cm long, not including the awns) and, at least occasionally, viviparous germination (pers. observ., December 2000, on Te Manga, Rarotonga). The plants may have survived as a metapopulation on the volcanic islands and atolls which have come and gone around the Manihiki plateau (Cook Islands/Tokelau) and other localities in the region since the Cretaceous. *Fitchia* can be interpreted as

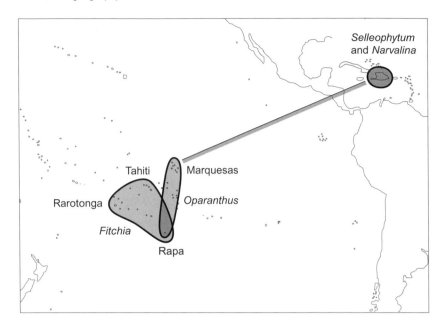

FIGURE 6-15. Distribution of *Fitchia* + *Oparanthus* and their sister group *Selleophyton* + *Narvalina* (Asteraceae) (Motley et al., 2008).

an uplifted derivative of weedy mangroves which colonized subaerial parts of the plateaus and their seamounts from other islands in the area. Some of their descendants have survived *in situ*, around the Manihiki plateau and the South Pacific superswell. As Sharma and Giribet (2009: 289) concluded for New Caledonia–American affinities in Opiliones, survival in the central Pacific by "Short-range dispersals among a group of ephemeral islands is a far more parsimonious hypothesis than to postulate a single 'jump' dispersal event across the expanse of the Pacific."

CUCURBITACEAE TRIBE SICYEAE

This group of 16 genera (Schaefer et al., 2008) has a concentration of diversity in western Mexico and, by virtue of *Sicyos*, a trans-Pacific distribution (Fig. 6-16; distributions from www.gbif.org). The disjunctions have probably involved extinction in central and eastern Polynesia. The basal member of the tribe is *Linnaeosicyos* in Hispaniola, and the only other member of the tribe in the West Indies is a single species of *Sicyos* on the same island. This arrangement is compatible with an early break in a Pacific–Caribbean ancestor around Hispaniola, or at least its

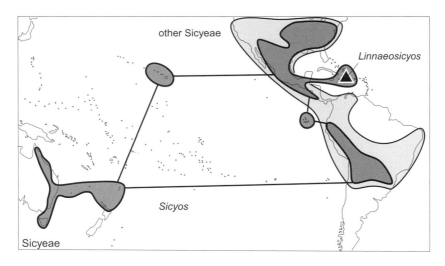

FIGURE 6-16. Distribution of Cucurbitaceae tribe Sicyeae (Schaefer et al., 2008).

precursor terranes. The diversity of the tribe in western Mexico (and the Ecuador–Galapagos–western Panama disjunction in *Sicyos*) may be related to Cretaceous terrane accretion in the region.

REEF FISHES

Finally, apart from widespread, central Pacific endemics, molecular studies are now showing groups in which the sequence of differentiation (nodes) moves from the central Pacific westward to the continents. The example given here is from reef fishes. These depend on shallow water, and this is usually associated with land.

Drew et al. (2008) studied molecular phylogeny in five species of reef fishes that are widespread in the Indo-Pacific region and show regional color variation in the southwest Pacific. In all cases, populations from Fiji and Tonga (with or without Samoa) were sister to clades that ranged through the rest of the Indo-Pacific. This repeats the pattern in fruit doves, where the Fijian *Chrysoenas* is basal in the *Ptilinopus superbus* group (Fig. 6-7). As usual with biogeographic patterns, the Fiji–Tonga–Samoa fish groups show a wide-ranging degree of divergence from their sisters, from shallow to deep (sequence divergence 2–17%). Drew et al. wrote that the marked regional endemism in the patterns "suggests that larval dispersal may be much more limited than previously thought" (p. 973). The repetition of the geographic pattern in the five fish species suggests a community-wide tectonic mechanism for the basal node. This may involve the Cenozoic breakup of the Vitiaz arc, the precursor of the Melanesian archipelagos.

One of the five fish species surveyed, *Pomacentrus moluccensis*, was studied in detail by Drew and Barber (2009) who found a phylogeny: (Fiji (Vanuatu (Solomon Islands/Papua New Guinea (eastern Indonesia + Sumatra)))). Given its pelagic larvae, the authors observed that "this species does not live up to its dispersal potential" (Drew and Barber, 2009: 338). The phylogeny is incompatible with the usual scenario of migration from Asia/Australasia into the Pacific, as with the widespread central Pacific endemics in land snails, passerines, and *Cyrtandra*. Instead, Drew and Barber (2009) inferred migration in the opposite direction, westward from a center of origin around Fiji. No dispersal is necessary, though (see Chapter 1), and the phylogeny may instead represent a westward sequence of differentiation in a widespread ancestor. This wave of differentiation moving out of the Pacific may relate to the complex regional tectonics, which have been dominated by accretion of terranes from the Pacific and back-arc basin formation.

Thomas (2006) discussed marine biodiversity hotspots and suggested that "areas of exceptional diversity may be the result of accumulation into areas rather than dispersal out of supposed centers of origin. Biogeographic patterns in reefs appear more congruent with geotectonic assembly processes that accumulate species into 'composite' areas of lineage-based diversity." Thus accumulation of terranes and their biota by "geological dispersal" and accretion at subduction zones can lead to local hyperdiversity. The normal faulting and rifting open of extensional back-arc basins that has accompanied the accretion (perhaps generated by slab rollback) has also led to biological diversification by normal vicariance.

Many subregions are evident within the central Pacific and differentiation among southwestern, southeastern, and northern sectors is indicated in groups such as the monarchs (Fig. 6-3).

SOUTHWEST PACIFIC

Most of the islands here have been formed by volcanism associated with active plate margins. The volcanic islands of the Solomon Islands–Vanuatu–Fiji–Tonga belt developed along a former plate margin as a continuous arc (the Vitiaz arc) that was later pulled apart by sea-floor spreading. This also caused the rapid, anticlockwise rotation of the Fiji

Islands (Taylor et al., 2000). Differentiation in insects, reef fish, and lizards has been attributed to vicariance caused by the breakup of the Vitiaz arc (Gillespie and Roderick, 2002; Heads, 2006b). *Melonycteris*, endemic to the Solomon Islands, is a basal fruitbat, and Pulvers and Colgan (2007) suggested suitable habitat may have been "permanently available in the Melanesian Arc for many millions of years, either as continuously emergent land or as successively emergent, geographically proximal islands." In another bat genus, *Emballonura*, one clade comprises three allopatric groups (Colgan and Soheili, 2008):

- Micronesia (Palau, Marianas, Caroline Islands), Vanuatu, Rotuma, Fiji, Samoa, and Tonga (*E. semicaudata*). The disjunction between Micronesia and Vanuatu/Fiji/Polynesia, with a gap in the Bismarck Archipelago/Solomon Islands, is illustrated here in other groups (Figs. 6-8, 6-9, 6-12).
- New Ireland and the Solomon Islands (*E. raffrayana*).
- New Guinea mainland (Sepik), Alcester Island (near Woodlark Island off the eastern tip of the Papuan Peninsula), and New Ireland (*E. serii* and *E. beccarii*).

Colgan and Soheili (2008) noted that the fossil record of emballonurids and the extent of molecular divergence among the taxa suggest an early date of origin and "that the influence of the southwestern Pacific's tectonic history on evolution within the family may have been substantial. There have been major changes in the relative position of the island arcs . . . in the last 5 to 20 million years" (Colgan and Soheili, 2008: 227).

Other authors have interpreted the island arcs as routes of dispersal into the Pacific. But if there were always terrestrial taxa in the Pacific, as suggested here, the arcs have probably been important not so much as "land bridges" or "stepping stones" for long-distance migration, but simply because they provide habitat within the region of the metapopulation where it can survive until the next arc appears.

SOUTHEASTERN POLYNESIA

This large region, including French Polynesia and the Pitcairn group, is associated with the South Pacific superswell (Fig. 6-1). Its biogeography is far from resolved.

Southeastern Polynesia is a center of endemism, as already shown in several groups, and is also the site of important boundaries. For example, in spiders, a clade of *Dicaea* (Thomisidae) occurs in the western Pacific (New Caledonia, New Zealand, Fiji, Tonga, and the Austral Islands, including Rapa) and vicariates with *Mimusenops* further east (in the Society, Marquesas, and Hawaiian Islands and the Americas), with a neat break in southeastern Polynesia between the Austral and Society Islands (Garb and Gillespie, 2006). The authors suggested that the pattern is due to island-hopping, but this does not account for the precision of the east Pacific/west Pacific break. Garb and Gillespie suggested the boundary might have arisen because the two clades invaded from the east and west and met in the middle, although they admitted that the lack of any overlap at the boundary zone is "surprising." It is suggested here that the division between the two clades is so precise because it represents a simple phylogenetic/biogeographic break in a Pacific-wide ancestor, not a zone of convergent chance dispersal.

Rapa Island in the Austral Islands has one of the most interesting biotas in southeastern Polynesia. The island is only 5 m.y. old and 40 km² in area, but the biogeographic connections and the high biodiversity suggest a much longer history for the flora and fauna currently preserved there. For example, there are more than 100 species of land snails and an "astonishing radiation" in *Miocalles* weevils (67 species with "striking morphological and ecological differentiation"; Gillespie et al., 2008: 3341). Instead of representing a radiation on Rapa, the diversity there may be a temporary juxtaposition of ancient regional elements on an ephemeral fragment. Paulay (1984) suggested that the speciation of the Rapa weevils has taken place on Rapa itself, but close relatives occur on nearby Marotiri, an almost sunken island, and on the other Austral Islands. Gillespie et al. (2008) admitted that the diversity is "enigmatic," and this indicates that alternative hypotheses need to be developed. The biota of Rapa is only "astonishing" and "enigmatic" if the age of the currently exposed strata is assumed to be the maximum age of the biota.

Rapa taxa show connections east and west. The Rapa endemic *Apostates* (Asteraceae) is sister to southwestern North American members of *Bahia* (Baldwin et al., 2008), and New Zealand–Rapa connections are documented in the plant genus *Hebe* (Plantaginaceae; Heenan et al., 2010).

The plant genus *Fuchsia* (Onagraceae) has a trans-Pacific pattern, with a significant representation in southeastern Polynesia. It comprises one clade in New Zealand and Tahiti (Society Islands), with the remainder through the Andes and Central America. In northern New

Zealand, *Fuchsia* occurs in lowland forest and sandy, gravelly, or rocky places only a little above high-tide mark, where it is sometimes submerged by the highest tides (Allan, 1961). In Tahiti, the genus occurs in montane rainforest and can be interpreted as an uplifted mangrove associate, as suggested above for *Fitchia*. In the mangrove family itself (Rhizophoraceae), *Crossostylis* occurs in the montane forest of the Marquesas and other Pacific islands, where it retains well-developed stilt-roots (Hallé, 1978).

Southeastern Polynesia is not currently the site of a plate boundary, but the region coincides with the South Pacific superswell and is characterized by intraplate volcanism at all scales. Phases of volcanism have often recurred in the same locality. For example, the Marquesas plateau has been interpreted as "an ancient edifice overprinted by recent volcanism" (Gutscher et al., 1999). It may have formed adjacent to the Inca plateau, now completely subducted beneath Peru. Likewise, the Nazca Ridge off Peru may have formed with the Tuamotu Plateau (Rosenbaum et al., 2005). Gillespie et al. (2008) cited repeated episodes of volcanism in the Marquesas, Austral, and Cook Islands, and the biological patterns suggest that this has been widespread through the region. In the blackfly genus *Simulium*, Tahiti has 29 species and "As Tahiti ages and erodes away, most of these species will no doubt become extinct" (Gillespie et al., 2008: 3328). This will be the outcome if there is no "secondary," "anomalous" volcanism in or near Tahiti and if there is no dispersal. If there is renewed volcanism and suitable habitat develops, it is possible that *Simulium* species may occupy these by normal ecological colonization (not founder dispersal with speciation).

NORTHERN POLYNESIA

This region only includes one group of high islands, the Hawaiian Islands. The biogeography here is of special interest and is treated separately in the next chapter.

MEANS OF DISPERSAL

Although dispersal has been the dominant paradigm in Pacific island biogeography for many years, several authors have suggested that distributions of Pacific taxa are not explained by their means of dispersal. Whitmore (1973) observed: "There is no predominance in the Pacific floras of water- or wind-dispersed genera. Many genera have similar distribution patterns, and there is not the tendency to random distribution which would follow from the chances of long-distance dispersal. Many botanists . . . have agreed that this present pattern of distribution of flowering plants in the Pacific could not have arisen from the present-day distribution of land and sea." Springer (1982), writing on reef fishes, also argued that evolution at tectonic boundaries such as the Pacific plate margin were more important than means of dispersal in determining biogeography.

Terrestrial mammals are diverse in the Pacific islands—for example, in the Solomon Islands—and they range east to the Cook Islands. Carvajal and Adler (2005: 1561) found that "levels of endemism did not differ between volant and non-volant species." They also wrote, "It is perhaps surprising that levels of endemism in skinks, birds and mammals are similar in the tropical Pacific region, despite major differences in vagility. Birds and bats actively disperse by flight, whereas skinks and non-volant mammals can disperse only passively, by rafting" (p. 1565).

Many dispersalist authors accept that long-distance chance dispersal cannot rely on the normal, observed means of dispersal, as these will either be too efficient (as in birds) or not efficient enough (as in plants whose fruits are neither wind-blown nor eaten by birds). This leads to the idea that taxa must have dispersed using mysterious means that are, as yet, unknown.

IS DISTRIBUTION THE RESULT OF MIRACULOUS, ONE-OFF EVENTS?

Mayr and Phelps (1967: 299) accepted that the eastern Pacific islands were populated by animals through long-distance, over-water dispersal, and wrote that "the distances involved in some of these colonizations are truly miraculous." Instead of accepting miraculous dispersal, the model adopted here requires only normal, local dispersal, recurrent volcanism, and evolution by vicariance. This implies that similar patterns will recur in other groups. Trewick and Wallis (2001) and McDowall (2007) have criticized panbiogeography for being concerned with general patterns and not accepting the lawless, one-off events of long-distance founder dispersal. Likewise in the 17th century, good theologians such as Bossuet criticized the early scientists and sympathetic philosophers for their dangerous belief in "general laws" of nature, as this was incompatible with the existence of miracles (Kenny, 2006). In the Middle Ages, as in modern dispersalism, "There was proof by miracle as well as proof by authority. . . . What made medieval minds agree to believe in something was not what could be observed and proved by a natural law or by a regularly repeated mechanism. On the contrary, it was the extraordinary, the supernatural or at least the abnormal. Science itself was more willing to take as its subject the exceptional, the *mirabilia*, and prodigies" (Le Goff, 2001: 329).

THE STANDARD PARADIGM OF PACIFIC BIOGEOGRAPHY: THE CENTER OF ORIGIN–DISPERSAL MODEL OF MATTHEW, DARLINGTON, AND MACARTHUR AND WILSON

The equilibrium theory of island biogeography (MacArthur and Wilson, 1967) is the most cited work in biogeography and has influenced many studies on Pacific groups. The theory was an attempt to unify ecology and biogeography (Lomolino and Brown, 2009: 378) and proposed that an island biota develops by dispersal from a center of origin on the mainland. (In contrast, in metapopulation theory there is no mainland center of origin.) The equilibrium theory interprets a fauna as a function of immigration and extinction, and so the key attributes of an island are its area and isolation. Equilibrium theory was not concerned with the prior history of volcanism and island formation in a region or geological history in general. Instead, it focused on the process of normal ecological movement (e.g., colonization of new volcanoes appearing within a group's range, as at Krakatoa) and saw this as the mechanism causing patterns of diversity. The theory also assumed that speciation was unimportant, except for the largest and most isolated archipelagos, and was more or less restricted to phenomena occurring within ecological timescales (Lomolino and Brown, 2009).

Lomolino and Brown (2009) traced the development of equilibrium theory and wrote that when E.O. Wilson began writing, the "normal science" in biogeography was that being advanced by W.D. Matthew, G.G. Simpson, and P.J. Darlington. These authors postulated that species attained their distributions by expanding from centers of origin and dispersing into new regions as they adapted. Lomolino and Brown (2009: 366) recognized this as a "20th-century articulation of the Center of Origin-Dispersal-Adaptation (CODA) tradition (developed largely by Darwin and Wallace)."

One of E.O. Wilson's most influential mentors at Harvard was Philip Darlington, who sent Wilson to New Guinea. Lomolino and Brown (2009) described how Wilson was "enchanted" by the ideas of

Matthew and Darlington and how he worked to apply the "normal science of the CODA tradition" to ants in New Guinea and the Pacific. (For Wilson's reliance on Matthew's, 1915, concepts and synthesis, see also MacArthur and Wilson, 1967; Hölldobler and Wilson, 1994: 214; Wilson, 2001a: viii, 2010: 2). Wilson (1959: Fig. 1) surmised that the ponerine ants invaded New Guinea and the other Pacific islands from centers of origin in Southeast Asia and Australia. "From New Guinea a fraction of the invading stocks presses on to outer Melanesia. An ever-diminishing number reaches the Bismarck Archipelago, Solomon Islands, New Hebrides [Vanuatu] and the Fiji islands. Progress along this route follows the classical 'filter' effect that applies generally to the faunas of archipelagic chains with permanent water gaps" (Wilson 1959: 123). Wilson thus concluded that colonization across the Pacific archipelagos follows the "immigrant pattern" described by Darlington (1957: 485), with attenuation of species composition across islands driven by differences in immigration abilities among species (Lomolino and Brown, 2009). (Instead, the eastward decrease in species numbers on Pacific islands could be due to eastward decrease in land area of the islands.)

Lomolino and Brown (2009: 380) wrote that although biogeography in the mid-20th century "had experienced a succession of paradigms, ranging from that of unique to multiple periods and sites of creation, as well as those of the extensionists [who favored previous land in oceanic areas] versus the Center of Origin-Dispersal-Adaptation tradition, none of these seemed particularly well-suited unifying paradigms for island biogeography." By the time E.O. Wilson was writing, historical biogeography "had progressed to the point that a burgeoning morass of anomalies and novelties could no longer be accounted for by the reigning paradigm" (p. 361). "The burgeoning but disarticulated morass of information on the diversities, distributions, and dynamics of insular biotas had accumulated well beyond the tensile strength of the CODA tradition" (p. 381). Lomolino and Brown argued (p. 381) that MacArthur and Wilson's (1967) theory of biogeography was a "bold and visionary" response to this crisis and a "new paradigm," while Hubbell (2001) called it a "radical" approach. Despite these claims, equilibrium theory is just a version of traditional CODA theory, as developed by Matthew and passed on to Simpson, Darlington, and Wilson.

The two main methods used to analyze island biotas—equilibrium theory and ancestral area analysis—are both based on the same assumptions. Equilibrium theory is a formalization of classic dispersal theory and uses processes taking place at an ecological, local community scale to explain biogeography, instead of using regional biogeography, phylogeny, and tectonics to explain local ecology. Authors such as Whittaker (1998) have commented that the theory failed to live up to its promises, and a journal issue devoted to the theory concluded that there were "deep conceptual and empirical problems" with it (Brown and Lomolino, 2000: 89). The second method, ancestral area analysis (using programs such as DIVA and Lagrange), is also based on center of origin/dispersal theory (as shown by Santos, 2007) and, in particular, endorses the key concept of speciation by dispersal.

Character optimization of morphological characters on a phylogeny aims to find the ancestral character within the *current* range of morphological variation, as in the medieval argument about whether Adam and Eve spoke French or German. It assumes that the ancestral form of two characters is either one or the other, and so it is almost guaranteed to fail. Likewise, ancestral area reconstruction uses character optimization algorithms to find the original center of origin among the current areas of the range, instead of at a break or margin with a sister group, and without considering former geography. Ancestral area analyses of groups in Hawaii and southeastern Polynesia, for example, will find a center in either of the two areas, or in both, but the Line Islands will not be mentioned, as the model does not incorporate geological change. Nevertheless, these islands represent a massive phase of non-hotspot volcanism that developed in the Cretaceous over a vast swathe 4,000 km long, stretching from Johnston Atoll southeast to Caroline Island. Whatever the precise ecology of these former high islands was, it does not seem realistic to discuss the biogeography of Hawaii or southeastern Polynesia without referring to them.

7

Biogeography of the Hawaiian Islands

The Global Context

The Hawaiian Islands have one of the most distinctive biotas of all the Pacific islands, and they have been the subject of many in-depth studies. Most authors now interpret the flora and fauna as classic cases of long-distance dispersal. For these reasons the biogeography of the group is discussed here in more detail.

As explained in Chapter 6, the most influential theory of island biogeography, MacArthur and Wilson's (1967) equilibrium model, applied Matthew's (1915) center of origin concepts to island taxa. Thus Wilson (2001b: 56) described the process whereby "The Hawaiian Islands are colonized by birds, crickets, wasps, damselflies, beetles, snails, flowering plants and other kinds of organisms arriving as occasional wind-blown waifs. As the first colonists multiply and spread, they evolve in response to the distinctive environment of the islands and hence diverge from the ancestral populations left behind on the mainlands of North America and Asia." But did the groups really come from North America and Asia? What about the former high islands in the Pacific that are now seamounts and atolls—surely they had a diverse fauna and flora?

Many authors have stressed the *current* isolation of the Hawaiian Islands; they lie 3,900 km from the nearest continent (North America)

Some of the material included in this chapter has appeared previously in *Systematic Biology* (Heads, 2011) and is reproduced here with permission from Oxford University Press.

and about the same distance from the nearest high islands, the Marquesas group in southeastern Polynesia. The Hawaiian Islands biota differs in many ways from others, and most authors have attributed its evolution to the "distinctive environment" (Wilson, 2001b) and the isolation of the islands. Vargas et al. (1999: 235) wrote: "The volcanic history, extreme geographic isolation, and disharmonic biota of the Hawaiian archipelago demonstrate that terrestrial life in the islands must have arrived by long-distance dispersal." Craddock (2000: 2) concluded, "It is incontrovertible that speciation in Hawaii is somehow tied to founder events." Eggens et al. (2007) agreed: "Because the islands are isolated and volcanic in origin, and have never been attached to the continental mainland, plant colonists could only have arrived by long-distance dispersal."

Although the consensus is now almost universal, some experienced biologists have been more cautious. Springer and Williams (1994: 183), discussing reef fishes, concluded that "The causal factors contributing to Hawaiian Islands' endemism are undoubtedly complex and remain to be elucidated." For example, the Hawaiian–Emperor chain has long been interpreted as a hotspot, "conveyor-belt" system and so could provide an example of an area where taxa have survived and evolved as a metapopulation throughout the Cenozoic, constantly colonizing new islands from older ones (Menard, 1986: 214). This concept undermines the center of origin/dispersal model, as presented by Carson and Clague (1995), for example. The metapopulation model has been rebutted by the suggestion that there was a period between 33 and 29 Ma in which no islands (Clague et al., 2010: 1022) or no high islands (Clague et al., 2010: 1031) were emergent. This is discussed further below. There are many problems in reconstructing the paleogeography of the region, and in particular it is difficult to eliminate the possibility of small islands, although these may have held considerable biodiversity. The other main evidence used to support the dispersal model is the young age suggested for many clades in molecular clock analyses, although there are problems with the calibrations in many of these. Some of the key assumptions that the standard model of Hawaiian biogeography is based on are examined next.

THE STANDARD MODEL OF HAWAIIAN BIOGEOGRAPHY

Assumption 1. The Hawaiian Islands have never been connected to other land masses and so must have received all their biota by long-distance dispersal from the continents (mainly Asia and America).

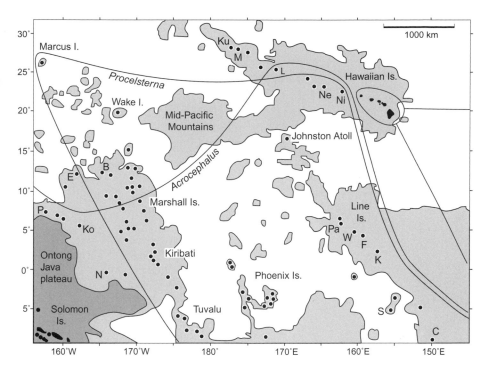

FIGURE 7-1. The northern central Pacific, showing atolls (black dots, not to scale), the 5,000-m isobaths, and the 4,000-m isobath (more or less delimiting the Ontong Java plateau). Also indicated are the range of the passerine *Acrocephalus* (Acrocephalidae) and the seabird *Procelsterna* (Sternidae) in the region, and other biogeographic connections of the main Hawaiian Islands with the Marquesas Islands and California. B = Bikini, C = Caroline Island, E = Enewetak, F = Fanning (Tabuaeran), K = Kiritimati (Christmas), Ko = Kosrae (Caroline Islands), Ku = Kure, L = Laysan, M = Midway Atoll, N = Nauru, Ne = Necker, Ni = Nihoa, P = Ponape, Pa = Palmyra, S = Starbuck, W = Washington (Teraina).

This assumes that the clades on the present Hawaiian Islands have not survived on former land in the vicinity and overlooks the evidence for former high islands around Hawaii. Apart from the atolls in the northwestern part of the chain itself (the Northwestern or Leeward Islands), there are many atolls south of Hawaii, such as Johnston Atoll and the Line Islands (Fig. 7-1). In recent years, many authors have described an affinity between the Hawaiian Islands and the Marquesas Islands (see below) and attributed this to dispersal between the two, but without mentioning the Line Islands. Seamounts (Fig. 7-2) are also relevant, especially guyots—islands that have been planed to sea level by wave erosion. The Musicians seamounts north of Hawaii have not been referred to by terrestrial biologists even though there is now good evidence for guyots there (Kopp et al., 2003).

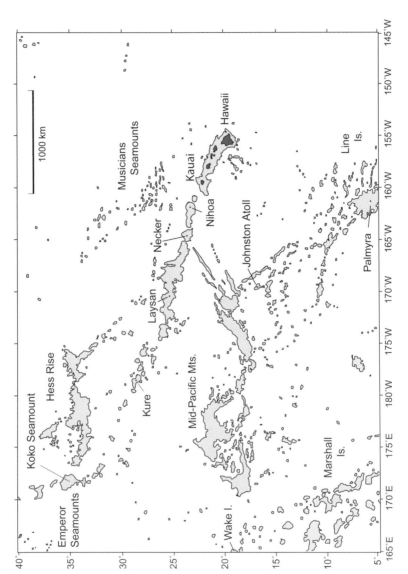

FIGURE 7-2. The Hawaiian Islands region showing the 2,000-m isobath that indicates many seamounts.

Assumption 2. The intraplate volcanism on the Hawaiian Islands is well understood and is caused by fixed, mantle-plume hotspots.

Volcanoes develop at plate boundaries as the result of plate tectonics but, as discussed in Chapter 6, the causes of intraplate volcanism, far from any plate boundaries, are less obvious. One theory, the Wilson–Morgan model, proposes that central Pacific island chains such as the Hawaiian Islands each formed as the Pacific plate moved over a stationary hotspot in the mantle, caused by a mantle plume. Plate movement would explain the linear age sequence of islands, with older ones further away from the hot spot. The Hawaiian–Emperor chain is 5,800 km long and extends from the active submarine volcano Loihi Seamount, 35 km southeast of Hawaii; to Hawaii Island (dated at 0.6 Ma), the oldest main island, Kauai (5.1 Ma); the atolls of the Northwestern Islands such as Midway Atoll (27 Ma); and the Emperor Seamounts, the oldest of which is Meiji seamount (85 Ma) at the edge of the Aleutian trench (Koppers, 2009). Older islands have probably been subducted at the trench. In the Wilson–Morgan model of intraplate volcanism, successive islands formed individually as separate islands, and so new islands have been colonized from older islands downstream of the hotspot (see papers in Wagner and Funk, 1995).

But is the hotspot really fixed? Is the mantle below Hawaii in fact hotter than normal? Do mantle plumes even exist? Most biologists have accepted the standard geological model, arguing that the geology of Hawaii is "clearly understood" (Cowie and Holland, 2006) and fitting biological patterns to the geological scenario. In fact, the whole subject of intraplate volcanism and its causes is currently being debated among geologists. The American Geological Society has devoted two large volumes to the discussion (Foulger et al., 2005; Foulger and Jurdy, 2007a), and this research may have the potential to bring about "the most significant paradigm shift in Earth science since the advent of plate tectonics" (Foulger and Jurdy, 2007b: vii). Several key papers have discussed alternative interpretations of the Hawaiian Islands (Anderson, 2005; Natland and Winterer, 2005; Stuart et al., 2007; A.D. Smith, 2007; Norton, 2007; Sager, 2007).

Instead of being caused by narrow, vertical mantle plumes, as in the Wilson–Morgan model, intraplate volcanism could be the result of narrow, horizontal "hotlines" and propagating fissures that are caused by plates flexing of plates and setting up stress fields. The Hawaiian chain may have originated "by a propagating fracture controlled by the direction of regional stress, the fabric of the seafloor or stresses caused by previously erupted volcanoes" (Neall and Trewick, 2008: 3304). The distinctive linear age sequence of volcanism could have developed along a propagating fissure instead of at a hotspot. If volcanism follows prior lines of stress rather than being the result of a mantle plume, recurrent volcanism could occur (see below) and bury an island under younger rock.

The present review attempts to integrate the biological evidence with ideas on tectonics, rather than basing the biogeographic interpretation on any particular geological theory.

Assumption 3. At one time in the Oligocene, none of the Hawaiian Islands were emergent.

Using different assumptions for the Hawaiian Islands (a hot spot origin, estimates of prior island area and slope, rates of erosion and subsidence, etc.), Clague (1996) calculated that there was a period in the Oligocene, between 34 and 30 Ma, when no islands were emergent along the chain; "30 Ma marks a time when colonizers from distant continents had to start over completely" (p. 45). Clague emphasized that estimating the longevity of each island "is far more complex and, therefore, far more uncertain than estimating either the age or size of the volcanoes" (p. 40). But despite this caveat, his estimates and similar ones by Price and Clague (2002) have been accepted without question by many biologists. For example, Geiger et al. (2007) wrote that Price and Clague (2002) "provide compelling evidence that the ancestors of extant Hawaiian biota could certainly not have colonized the Hawaiian chain prior to about 23 Ma." In fact, the method that Clague (1996) used to estimate the heights (and therefore the ages) of the former islands underestimated the height of the extant volcanoes on Maui and Hawaii by over 1,000 m (Table 7-1), and so the method might have also underestimated the maximum height that the seamounts reached in the past.

Price and Clague (2002) were explicit about the "assumptions and uncertainties" in their model. For example, they accepted a constant rate of erosion and admitted that while this "is probably inaccurate (especially considering the occurrence of massive [submarine] landslides; Moore et al., 1994a), there are too few data to determine how this rate may change over time" (p. 2431). Some of the landslides are more than 200 km long and are among the largest landslides on Earth (Moore et al., 1994b). Price and Clague did not refer to the new interpretations

TABLE 7-1   ACTUAL HEIGHTS OF VOLCANOES ON MAUI AND HAWAII
AND MAXIMUM HEIGHTS PREDICTED BY CLAGUE (1996)

Volcano	Actual Height (m)	Clague 1996 Estimate (m)
E. Maui	3,055	2,180
Kohala	1,670	1,740
Hualalai	2,521	1,040
Mauna Kea	4,205	3,050
Mauna Loa	4,170	3,050
Kilauea	1,277	1,040

of intraplate volcanism, and these also affect the assumptions that their reconstruction is based on. This sort of modeling is interesting but does not seem reliable enough to serve as the sole foundation for analyzing Hawaiian biogeography.

One seamount in the southern Emperor chain, Koko (Fig. 7-2), is formed from three large coalesced volcanoes and has a flat summit about 100 km across (Clague et al., 2010: Fig. 1c). Clague et al. (2010) showed that subsidence caused the cessation of growth in shallow-water coral there "probably around 33 Ma." This was before the eruption of Kure Atoll (the oldest emergent island) at ~29 Ma. Based on this evidence, Clague et al. (2010) wrote in their Abstract (p. 1022): "There was a period between at least 33 and 29 Ma in which no islands existed." Yet apart from the uncertainty in the dates, there are 16 other seamounts between Koko and Kure (Clague, 1996) and in the body of their paper Clague et al. (2010: 1031) concluded instead: "There was a time period from about 33–30 Ma in which no high islands, and only transient low islands, existed." They also wrote that "Koko Seamount was quite distant from these ephemeral islands and was itself a flat, low-lying coral island with limited terrestrial biodiversity" (p. 1030), although there is no direct evidence for the biodiversity there. ("The Deep Sea Drilling Project and later Ocean Drilling Program drilled there several times, but always drilled through the carbonate cap and did not start coring until they hit basement rocks, thus passing by any chance of recovering sediments containing pollen or spores"; D. Clague, pers. comm.). Some low, flat coral islands are biologically depauperate, while others are rich; the rainforests on the Loyalty Islands, for example, are very diverse.

Assumption 4. All lineages in the central Pacific have been derived from elsewhere.

The idea that large, continental landmasses formerly existed in the mid-Pacific was rejected by geologists. Following this, biogeographers have jumped to the conclusion that the islands have been populated by dispersal from mainland sources, either in the east or the west (see Chapter 6). Terrestrial groups require land, not continental crust, and a central Pacific group could exist indefinitely in the region as a metapopulation surviving on ephemeral islands around centers of volcanism. There is no mainland source in a metapopulation model, and this distinguishes it from MacArthur and Wilson's (1967) theory of island biogeography.

Assumption 5. Island clades nested in otherwise mainland clades have a center of origin on the mainland.

If a clade comprises groups *a* and *b* in areas A and B, and group *a* is shown to be paraphyletic with *b* nested in it, authors take this as evidence indicating dispersal from A to B. Yet this pattern could be due to primary vicariance—not between areas A and B, but within B (Fig. 1-6). Bellemain and Ricklefs (2008: 463) suggested that "In general, when an island lineage is imbedded within a clade otherwise restricted to a continental region, one can infer that colonization occurred from continent to island." But this argument is not valid. The pattern will also result if a widespread (mainland + island) group first differentiated around boundaries within the mainland. In a similar example, a group in the Hawaiian Islands with the phylogeny: (Hawaii (Hawaii (Hawaii, Maui))) might be assumed to have lived on Hawaii before it dispersed to Maui and speciated there, yet this cannot be deduced from the phylogeny alone. If the ancestor was already on both islands, two divergences in Hawaii followed by divergence between Hawaii and Maui (and possibly overlap within Hawaii) would produce the same pattern.

The phylogeny: (mainland (mainland + island)) is a common pattern, as island biotas are small and relictual. Despite this, O'Grady and DeSalle (2008) found the reverse arrangement in Drosophilidae, as non-Hawaiian members of the widespread drosophilid *Scaptomyza* were nested among the Hawaiian species. Again, instead of indicating colonization by the group from Hawaii, this pattern could be due to basal breaks in a widespread ancestor developing in metapopulations around the Hawaiian region.

Assumption 6. The degree of differentiation of a clade is related to its age (evolutionary clock theory).

The evolutionary clock idea (Matthew, 1915; Mayr, 1944) predicts that different degrees of differentiation in the taxa of an area indicate colonization at different times. The theory assumes that evolution is more or less continuous, and so the degree of divergence of a group is proportional to the time since its origin (see Chapter 2). In this model an endemic genus, for example, will be older than an endemic subgenus, at least within the same family.

Nevertheless, degree of differentiation could instead be due to prior aspects of genome architecture that determine the "evolvability" of a group, its intrinsic propensity to diversify (Lovette et al., 2002). This explanation for branch length accounts for Avise's (1992) paradox: All biogeographic patterns show significant geographic concordance among groups, indicating vicariance, but all patterns are shared by different taxa showing different degrees of differentiation, indicating dispersal (see Chapter 2). The paradox only arises if degree of differentiation is assumed to be related to time since divergence. It is suggested here that the evolutionary clock is extremely relaxed and very local, and it cannot be assumed that a species, for example, is younger than a related genus.

Authors sometimes suggest that if a genus is known from fossils back to, say, the Miocene, the proposal that its actual age is Mesozoic is unacceptable, as it would imply that its family is impossibly old. But this is true only if the rate of molecular evolution is constant. With an elastic evolutionary clock, family and genus may have both evolved in a period of rapid evolution, with subsequent evolution occurring much more slowly. Instead of calibrating a phylogeny with a series of fossils, it can be calibrated with multiple biogeographic events.

Assumption 7. The age of an island establishes the maximum age of its endemic taxa.

If this were true, island ages could be used to calibrate the time course of molecular phylogenies. Yet the minimum ages given in many molecular studies show that island endemics can be much older than their islands (Chapter 2). Whether the endemic taxa survived on former islands nearby or on a mainland, later going extinct there, using the age of islands to date the endemic taxa there will often give unreliable results with unpredictable and sometimes massive errors (Heads, 2011).

In the Hawaiian Islands, the endemic genus *Hillebrandia* (Begoniaceae) has been dated at ~50 Ma, predating the oldest emergent island (Kure, 29 Ma) by ~20 million years (Clement et al., 2004). Goodall-Copestake et al. (2009) found that *Begonia*, sister of *Hillebrandia*, could be as old as Late Cretaceous. This study assumed that eudicots were no older than their oldest fossil, and so the actual age of *Begonia* (and thus *Hillebrandia*) could be older than this. In the mid-Cretaceous the Hess Ridge/Mid-Pacific Mountains area that became the site of the Hawaiian Islands was located at a major spreading ridge, probably the East Pacific Rise (Fig. 6-1), that could have caused the vicariance.

Kim et al. (1998) dated the Hawaiian endemic *Hesperomannia* (Asteraceae) at 17–26 Ma, much older than the oldest of the islands it is endemic to (Kauai, 5.1 Ma). They suggested its progenitor arrived on one of the low, Northwestern Islands when these were higher islands. The time course of the phylogeny was calibrated using a rate from *Dendroseris* (Asteraceae), endemic to Juan Fernández Islands. This was derived by assuming that *Dendroseris* could be no older than the current Juan Fernández Islands (4 Ma) (Sang et al., 1994). But other Juan Fernández endemics include Lactoridaceae, a group dated to more than 125 Ma using a molecular clock (Wikström et al., 2001). If Lactoridaceae are so much older than the current islands, *Dendroseris* could be too. Thus the *Dendroseris* dates and the *Hesperomannia* dates based on them could be greatly underestimated. *Hesperomannia* is placed near the base of the Vernonieae, a large pantropical clade (Funk and Bonifacino, 2009). Thomas and Hunt (1991) dated Hawaiian *Drosophila* at 10 Ma, 5 m.y. older than the main Hawaiian Islands where they are endemic. But their calibration (Rowan and Hunt, 1991) was based on the assumption that a species could not be older than the island it is endemic, to and so the dates could, again, be underestimates.

Assumption 8. The oldest fossils of a group give a maximum age for the group.

The oldest fossil of a group gives a minimum age for the group, not a maximum or absolute age. Mummenhoff et al. (2001) studied the worldwide genus *Lepidium* (Brassicaceae) (the Hawaiian species are discussed below) and calibrated the phylogeny with fossil *Rorippa* fruit dated at 2.5–5 Ma, a minimum date. Nevertheless, the authors concluded that *Lepidium* radiated "in [not in or before] the Pliocene/Pleistocene." Only by transmogrifying the fossil-calibrated (minimum) dates into maximum dates in this way were they able to rule out earlier vicariance and argue that long-distance dispersal was "unequivocal."

Assumption 9. Hawaiian clades cannot be older than the (inferred) age of their worldwide group.

The global age of groups is often estimated using calibrations made with island ages and oldest fossil ages. The calculated ages are then treated as maximum dates for clades. Both these methods will give underestimates of clade age (Chapter 2). For example, angiosperm families such as Asteraceae (Compositae) are always assumed to be Cenozoic, based on the treatment of oldest fossil records and fossil-calibrated ages—minimum ages—as maximum ages. In the 1960s, literal readings of the fossil record led to the idea that angiosperms themselves originated only in the Cretaceous, but molecular phylogenies now confirm the earlier view that they are much older than this; Magallón (2010) suggested they originated in the Permian. Many authors are now assuming vicariance and correlating the biogeography of molecular clades with dated events from plate tectonics. This avoids using oldest fossils to give anything except minimum ages for clades.

REGIONAL GEOLOGY

The main geological structures in the Pacific basin are the tectonic plates (labeled with letters in squares on Fig. 6-1). These are being produced by a mid-ocean ridge (a divergent plate margin or spreading center), the East Pacific Rise. The Pacific plate is being produced to the west of the ridge, while the Antarctic, Nazca, Cocos, and Juan de Fuca plates are produced to the east (the latter three are derived from the original Farallon plate). At the same time, older parts of the plates with their seamounts are being destroyed by subduction around the Pacific rim. Much of the oceanic crust that formed east of the mid-ocean ridge, with its seamounts, has been subducted beneath western America; north of Mexico, nearly all of it has been.

The ridge and the other plate margins have themselves shifted over time. The Pacific plate has undergone tremendous growth since its mid-Jurassic origin at a ridge–ridge–ridge triple junction in the Cook Islands region. A broad swathe of mid-ocean ridge-type volcanics extends for 7,000 km from Easter Island in the southeast to the Tuamotu Plateau and Austral Islands, across the equator to the Line Islands, the Mid-Pacific Mountains, and the Shatsky and Hess Rises in the northwest (Watts et al., 2006). This belt of Cretaceous on-ridge volcanism (surrounding the off-ridge, intraplate volcanism of the Hawaiian chain) may mark the former position of the East Pacific Rise (Fig. 6-1; Hillier, 2006; A.D. Smith, 2007; Utsunomiya et al., 2008). Samples from the next-to-oldest seamount in the Hawaiian–Emperor chain, the Detroit seamount, show an isotopic signature indistinguishable from mid-ocean ridge basalt (Keller et al., 2000). This is consistent with the interpretation that the chain was located close to a mid-ocean ridge in the Late Cretaceous, at about 80 Ma.

Rocks from the Hawaiian Ridge itself that have been dated as Cretaceous include samples collected from the northern slope of Necker Island and from Wentworth seamount, 80 km northwest of Midway Island (Clague and Dalrymple, 1989). The Necker sample may have been from an erratic, but Necker Ridge itself (extending southwestward from Necker Island, see Fig. 7-2) seems to be well established as Cretaceous.

Other large-scale geological structures in the central Pacific include large oceanic plateaus, areas of anomalously thickened oceanic crust that were emplaced in the Cretaceous (Chapter 6). The plateaus bear numerous guyots, and fossil wood is recorded from intercalated sedimentary strata.

In the Wilson–Morgan model of intraplate volcanism, the orientation of central Pacific seamount trails is attributed to the direction of Pacific plate over hotspots (Chapter 6). Nevertheless, many seamounts are not in simple trails, and different trails show differing orientations. For example, the alignment of the Marquesas chain differs by 20–30° from that of Pacific plate motion. This deviation is "quite odd" (Bonneville, 2009), and the Marquesas line of volcanoes could be controlled by weaknesses in the plate rather than a hotspot (Koppers, 2009). If plate tectonics effects rather than mantle plumes could produce these islands, they may have also produced other islands.

There is an abrupt 60° change in orientation between the Hawaiian Islands and the Emperor seamounts to the northwest. This Hawaii–Emperor bend (HEB) has been attributed to a change in the direction of Pacific plate movement, but while similar bends occur in other seamount trails, such as the Tokelau and Gilbert chains, these developed at different times. "The remarkable differences observed in these colinear seamount trails fundamentally question the existence of HEB-type bends in the formation of Pacific plate volcanic lineaments" (Koppers et al., 2007: 1). Hamilton (2007: 19) concluded that:

> Powerful evidence contradicts the notion of fixed hot spots. . . . Geophysics of the Hawaiian region misfits plume predictions. . . . Pacific spreading patterns . . . , paleomagnetism of Emperor seamounts . . . and paleomagnetic latitudes of cores from the floor of the Pacific plate . . . show independently that the Pacific plate did not change direction by 60° above a fixed hot spot

at the time of the Emperor elbow, 50 Ma, as required by fixed Hawaiian plume speculation. Other island and seamount chains once conjectured in the absence of data to fit a Hawaiian trajectory in fact misfit it badly in chronology, trends, and geometry. . . . Hawaii and other chains are properly explained as responses to within-plate stresses.

Although it is often argued that the islands of Micronesia and Polynesia have all formed at mantle-plume hotspots, a non-plume origin for the volcanism opens up many other possibilities for recurrent volcanism and the survival of terrestrial life.

Atolls

Zimmerman (1948: 125) wrote that "Many of the peculiar endemic groups of the Hawaiian and southeastern Polynesian islands owe their existence, if not their very origin, to ancient high islands of the one time splendid archipelagos now marked by clusters of coral reefs [or submerged seamounts]. . . . Atolls have been overlooked, generally." Sixty years later, the significance of the atolls and seamounts that surround the Hawaiian Islands is still neglected by biogeographers. Where they are acknowledged, the former islands that they represent are regarded merely as stepping stones that may have facilitated dispersal from the Pacific margins to the Hawaiian Islands. But former islands may have had a more fundamental role than this, as the different generations of islands could have allowed taxa to persist in the region indefinitely. Atolls around the Hawaiian Islands are shown in Figure 7-1.

Land snail taxa typical of humid, montane forest on high islands also occur as fossils on low, dry atolls that have formed by the erosion and subsidence of high islands. For example, the family Endodontidae was cited in Chapter 6 as a central Pacific endemic. It occurs from Micronesia to southeastern Polynesia and Hawaii, but on the low Midway Island (Northwestern Hawaiian Islands) and Bikini Atoll (Marshall Islands), the family is known only from fossil material (Fig. 7-1; Solem, 1983). On Enewetak (= Eniwetok, Marshall Islands), the related Charopidae are only known as fossils. Solem (1983: 275) wrote: "The extinction of both endodontids and charopids on these islands was the result of natural processes. As the high islands degraded into atolls, the more frequently interrupted moisture supplies in the latter wiped out the moisture-dependent endodontoid land snails."

Seamounts

With respect to seamounts (Fig. 7-2), maps of the ocean floor are far from complete; we have better topographic maps of the Moon and Mars (Koppers, 2009). Only ~15,000 out of possibly ~200,000 seamounts more than 1 km in height have been mapped (Wessell, 2009), and only a few hundred have been sampled. In 2005, the nuclear submarine U.S.S. *San Francisco* collided with an uncharted seamount south of Guam. Koppers (2009: 713) emphasized that "Compared to our knowledge about the volcanic islands in the Pacific Region, we know very little about seamounts and guyots and how they formed." The Musicians seamounts north of the Hawaiian archipelago formed mainly in the Late Cretacous (105–62 Ma; Clouard and Bonneville, 2005) (possibly through the interaction of a hotspot and a spreading center). They include guyots such as the Rossini seamount, with a flat top 11 km across (Kopp et al., 2003).

In Chapter 1, a distinction was drawn between normal, ecological dispersal that does not involve phylogeny, and biogeographic founder dispersal that leads to speciation. If the Hawaiian biota formerly survived on what are now atolls and seamounts, the current Hawaiian Islands may have been colonized by the first process, not the second. Many taxa that are now Hawaiian endemics could have evolved not on the Hawaiian chain but on other islands that no longer exist.

The Line Islands

These islands south of Hawaii include the world's largest atoll (Kiritimati or Christmas Atoll) and the most pristine coral reefs (Koppers, 2009). All the islands are low, flat atolls but were originally high volcanic islands erupted in the Cretaceous (86–81 Ma and 73–68 Ma) and the Eocene (50–35 Ma) (Clouard and Bonneville, 2005; Charles and Sandin, 2009). The volcanism extended for more than 4,000 km, from Johnston Atoll southeast to Caroline Island. The Tuamotu Plateau may be the southern continuation of the Line Islands, although the tectonic origin of the two remains unresolved (Bonneville, 2009). The Line Islands do not show an age progression, and so the chain was probably not hotspot related. Instead, the volcanism may have been associated with lithospheric extension along preexisting zones of weakness. These could have formed in response to the changing stress field of the Pacific plate and broad upwarping of the South Pacific superswell region (Fig. 6-1; Davis et al.,

2002; Natland and Winterer, 2005; Charles and Sandin, 2009). The activity may have formed the precursor of the East Pacific Rise. The Line Islands are now all reduced to low, flat limestone islands, but in the past they would have formed "splendid archipelagos" (Zimmerman, 1948) thousands of kilometers long that provided habitat for many lowland and montane lineages. As newly formed oceanic crust moves away from the ridge that formed it, it cools and subsides. As with plateau breakup, regional subsidence following the original volcanism could have led to the vicariance of metapopulations.

The biogeography of the remnant terrestrial biota of the Line Islands provides tantalizing glimpses of early affinities. For example, in jungle leeches (Haemadipsidae), *Abessebdella* of the Line Islands (Palmyra) is sister to *Nesophilaemon* of Juan Fernández (Borda and Sidall, 2010). The authors regarded the idea of long-distance dispersal in these leeches as 'problematic', as they do not swim (they are hydrophobic) and drop off any host once they are engorged with blood. Recent collections on Juan Fernández found no evidence for an association with birds. The Line Islands–Juan Fernández clade is sister to a clade from the Madagascar, Seychelles, the Philippines, and New Guinea, giving a primary break in the widespread Indo-Pacific group between the Line Islands and New Guinea.

AFFINITIES OF HAWAIIAN TAXA OUTSIDE HAWAII

Biogeographic connections of the Hawaiian biota include the following patterns, supported by representative examples. Well-supported clades retrieved in molecular analyses are cited first, followed by a selection of examples from morphological studies. Unless otherwise stated, plant distributions are from Wagner et al. (1990) and the website http://botany.si.edu/pacificislandbiodiversity/hawaiianflora/index.htm.

GLOBALLY BASAL GROUPS IN THE HAWAIIAN ISLANDS

Several cases are now known in which Hawaiian clades have globally distributed sister groups. A dispersal interpretation seems unlikely, as it would suggest that a group has only colonized Hawaii once, early in the group's history and before differentiation anywhere else. Instead, the pattern is compatible with normal vicariance in metapopulations at boundaries around the Hawaiian region.

Some authors in the dispersalist tradition have argued that new, advanced species of a group occupy the group's center of origin and force out the primitive species there by competition. In contrast, a second group of authors has proposed that the primitive or basal clade in a group occurs at the center of origin (Chapter 1). Neither view is accepted in a vicariance model; a "basal" group (a small sister group) is not necessarily ancestral and does not necessarily occupy a center of origin. The existence of a global group that has its basal clade in Hawaii implies the early differentiation of a widespread ancestor in the region, not a center of origin there. Examples include the following.

Hillebrandia (Begoniaceae) is endemic to the Hawaiian Islands (Kauai to Maui), and the single species is restricted to wet ravines in montane rainforest (900 to 1,800 m elevation). Its sister is *Begonia*, the only other member of the family, which is more or less pantropical, with 1,500 species. It is significant that indigenous species of *Begonia* are absent on the Hawaiian Islands, although introduced species have naturalized. The allopatry of the two genera implies simple vicariance with no subsequent overlap of the two clades. Goodall-Copestake (2010) gave a dispersal analysis of *Begonia*, invoking transoceanic long-distance dispersal. But they did not account for the strange absence of the genus from Hawaii. The sister group of the tropical Begoniaceae is Datiscaceae of central Asia, eastern Mediterranean, and California, neatly allopatric to the north of Begoniaceae (Stevens, 2010).

Nototrichium (Amaranthaceae) is endemic to the Hawaiian Islands and is widespread there, with three species in mesic and dry forest and shrubland. It is sister to the pantropical genus *Achyranthes* (Müller and Borsch, 2005), also widespread in the Hawaiian Islands and with endemic species there. The pattern implies a break between the two genera in the region, followed by range expansion in one or both leading to overlap.

In *Dodonaea viscosa* (Sapindaceae), the Hawaiian clade is sister to one from Africa, Australia, Micronesia, and the Caribbean (Harrington and Gadek, 2009). A related species, *D. triquetra*, has a distinctive pollen type well documented in Australia back to the late Eocene (~35 Ma) (Martin, 1994), giving a useful minimum age for the group.

The moa-nalos are a clade of ducks known from subfossils on the main Hawaiian Islands. They are among the most extraordinary of the birds that went extinct on the islands following human settlement about 1,600 years ago, presumably due to hunting. The moa-nalos are very large, with heavy lower bodies, hypertrophied pelvises, and deep beaks, but minute wings (Sorenson et al., 1999). There are three genera, *Thambetochen*, *Ptaiochen*, and *Chelychelynechen*, the last with a massive rostrum and mandible suggestive of those of a tortoise (Olson and James, 1991: 85). Analyses of ancient DNA showed that the moa-nalos are a monophyletic group sister to the more or less cosmopolitan Anatini (*Anas*, *Speculanas*, *Amazonetta*, *Lophonetta*, and *Tachyeres*). *Anas* occurs through the Hawaiian Islands, including the Northwestern Islands (Laysan, etc.). This indicates that range expansion has occurred after the differentiation of a worldwide ancestor into moa-nalos and the other Anatini; for example, in the Hawaiian region, the Anatini may originally have been restricted to the northwestern part.

AFFINITIES WITH PACIFIC MARGIN GROUPS

Many Hawaiian taxa show affinities with taxa from different areas of land bordering the Pacific. For example, Baldwin (1998) listed the clades of Hawaiian Asteraceae with the location of their nearest relatives. These occur in western North America, tropical America, Micronesia, Australasia, and Africa. This information is of great interest, although its interpretation is not straightforward. Baldwin treated the sister areas as the "source of the founder" (p. 51), but the taxa and their ancestors may have previously occurred on former Pacific islands that have been reduced to atolls or seamounts, or been subducted. If this was the case, there may have been no founder event, just normal ecological colonization and extinction on the individually ephemeral islands. Monophyletic Hawaian clades are often assumed to be the result of *in situ* adaptive radiation there, following a single colonization event. It is suggested here that many of these clades may have existed in the north Pacific region for long periods before they colonized Hawaii.

HAWAIIAN ISLANDS–EASTERN ASIA

In the fern *Dryopteris*, the Hawaiian "exindusiate group" (five species) is sister to a group of Europe, China, and Japan (Geiger and Ranker, 2005).

Broussaisia (Hydrangeaceae) of the Hawaiian Islands is sister to a clade of southern China–New Guinea (*Dichroa) and eastern Asia (Hydrangea hirta* and *H. macrophylla) (Hufford et al., 2001).*

Nothocestrum (Solanaceae) of the Hawaiian Islands is sister to *Tubocapsicum*, with one species of China (Olmstead et al., 2008).

The petrel *Pterodroma hypoleuca* breeds on the Bonin Islands south of Japan and in the Hawaiian Islands (Brooke, 2004).

HAWAIIAN ISLANDS–TEMPERATE AND ARCTIC NORTH PACIFIC

Viola (Violaceae) is a subcosmopolitan angiosperm genus. Most species are herbs with inflorescences reduced to solitary flowers, but the Hawaiian species form a clade of subshrubs or small trees. In addition, the flowers are grouped in inflorescences, "strengthening the hypothesis of a very ancient origin" (Ballard and Sytsma, 2000). Yet Ballard and Sytsma wrote that ITS sequences revealed a different origin for the group. The Hawaiian clade of *Viola* is sister to *V. langsdorfii*, a herbaceous species that ranges from Japan via Beringia to California, and possibly to Beringian populations of the species. Thus Ballard and Sytsma inferred herbaceous ancestry, an arctic origin, and long-distance dispersal from there for the Hawaiian group. Instead, this pattern could reflect simple vicariance within a widespread north Pacific ancestor that occupied former islands now subducted beneath North America or northern Asia. The geologic blocks that comprise Beringia reached their modern configuration sometime in the Cretaceous, probably at ~100 Ma (Fiorillo, 2008). Following this, the Kula plate and the Pacific–Kula ridge have been completely destroyed by subduction at the Aleutian trench as they moved northward (Fig. 6-1). This geological history may also explain the Hawaii–Arctic affinities and central Pacific–Arctic migratory patterns seen in many birds (Ballard and Sytsma, 2000). Beringia has often been interpreted simply as a Pleistocene land bridge enabling dispersal between Asia to America, but instead it may be important as a biogeographic center in its own right.

Schiedea (Caryophyllaceae) is endemic to the Hawaiian Islands and includes 34 species of herbs, shrubs, and vines. Discerning its relationships has been complicated by the extensive morphological divergence of the group from the other members of its family (Baldwin and Wagner, 2010). Molecular work has shown that it is sister to two circum-boreal genera, *Honkenya* and *Wilhelmsia* (Harbaugh, Nepokroeff et al., 2010), that both occur in the Bering Strait region.

HAWAIIAN ISLANDS–WESTERN NORTH AMERICA

These affinities have been treated in a detailed, beautifully illustrated review by Baldwin and Wagner (2010).

Carex kauaiensis and *C. alligata* of Hawaii are sister to *C. obnupta* of western North America (Dragon and Barrington, 2009).

In *Lepidium* (Brassicaceae), the Hawaiian species form a clade of shrubs and subshrubs. This is sister to a group of herbaceous species in the south-central and southwestern U.S. and Mexico (one species is widespread in the western U.S. and Central America) (Mummenhoff et al., 2001). Two of these herbaceous species (*L. lasiocarpum* and *L. oblongum*) have varieties restricted to Baja California and the Channel Islands. Mummenhoff et al. (2001) regarded *Lepidium* as an "unequivocal" example of dispersal from western North America to Hawaii, and they calculated an age of ~350,000 years for the Hawaiian clade, although this is a minimum date based on a fossil calibration. The Hawaiian–American clade probably occupied former islands now subducted beneath western America and is estimated here as more than two orders of magnitude older than 350,000 BP.

The Hawaiian endemic *Rubus hawaiensis* (Rosaceae) is sister to *R. spectabilis* (Howarth and Gardner, 1997) of western North America (San Francisco to Alaska Peninsula, inland to Idaho; http://data.gbif.org).

The Hawaiian mints (*Haplostachys*, *Phyllostegia*, and *Stenogyne*—Lamiaceae; 78 species) form a monophyletic clade sister to either "*Stachys*" *chamissonis* from the western U.S. (Channel Islands to northwestern Washington) or *S. quercetorum*, with a similar range (Lindqvist and Albert, 2002). The authors wrote that the Hawaiian group "obviously arrived in the island chain via long-distance dispersal from temperate North America," but did not mention the subduction of the northeast Pacific seafloor beneath western North America.

Hawaiian members of *Euphorbia* subg. *Chamaesyce* (Euphorbiaceae) are a monophyletic group. Baldwin and Wagner (2010) described the morphological diversity of Hawaiian *Chamaesyce* as spectacular; the group includes trees, shrubs, and subshrubs. Their sister is a clade of four species of annuals (*E. leucantha*, etc.) known only from the Chihuahuan Desert in Texas and northern Mexico (Yang et al., 2009; Baldwin and Wagner, 2010).

The Hawaiian species of *Geranium* make up a clade that has its closest relatives in the western United States and Mexico (*G. vulcanicola*, *G. subulatostipulatum*, and *G. richardsonii*) (Pax et al., 1997).

The Hawaiian *Plantago* species (Plantaginaceae) form a monophyletic clade. Nuclear DNA sequences show it as sister to *P. macrocarpa*, a seashore species from the west coast of North America (Oregon to the Aleutian Islands, 178°W; http://data.gbif.org) (Dunbar-Co et al., 2008). Chloroplast DNA sequences instead have it sister to a clade of Rapa Island (*P. rupicola* and *P. rapensis*). Both are standard connections. Dunbar-Co et al. (2008) discussed possible dispersal pathways from New Zealand via Rapa to Hawaii, and from North America to Hawaii, but did not cite the Line Islands, the westward migration of the East Pacific Rise, or the dispersal of the large igneous plateaus.

Nama sandwicensis (Boraginaceae s.l.) has its closest relatives among annuals of sect. *Nama* in southwestern North America (S. Taylor and B. Simpson, pers. comm. in Baldwin and Wagner, 2010).

Great white sharks, *Carcharodon carcharias*, are widespread in temperate seas and have three main populations. These occur off South Africa, around Australia/New Zealand, and in the Hawaiian Islands–Mexico/California sector of the northeastern Pacific. The South Africa and Australasian groups are reciprocally monophyletic, while the northeastern Pacific clade is nested in the diverse Australasian group. Jorgensen et al. (2010) inferred dispersal from Australasia to the northeastern Pacific, but instead, the pattern could be the result of vicariance in a widespread ancestor (as in Fig. 1-6). In the northeastern Pacific clade, populations from central California migrate every winter to the Hawaiian Islands and intermediate points (Jorgensen et al., 2010), and this might be assumed to be for ecological reasons. However, populations from Guadalupe Island, west of Baja California, also migrate to the Hawaiian Islands (Domeier and Nasby-Lucas, 2008). If the eastward plate movement has gradually separated foraging areas (Hawaii) from breeding areas (now accreted in California), the westward annual migrations of the populations may have been extended a few centimeters every year in step with the geological movement.

The Hawaiian clade of the goose genus *Branta* is sister to *B. canadensis* s.str. (southwestern Alaska–Canada), and so Paxinos et al. (2002) proposed a recent colonization of Hawaii from Canada. Again, this whole clade's former range may have included islands, now subducted, between Canada and Hawaii.

MORPHOLOGICAL EXAMPLE:

Hesperocnide (Urticaceae) comprises one species in the Hawaiian Islands (Hawaii only), and one in California, including the Channel Islands.

HAWAIIAN ISLANDS–WESTERN NORTH AMERICA–TEMPERATE SOUTH AMERICA

The Hawaiian silversword alliance (28 species of *Argyroxiphium*, *Dubautia*, and *Wilkesia*: Asteraceae) belongs to the subtribe Madiinae (tarweeds), a group that is otherwise restricted to mainland America and

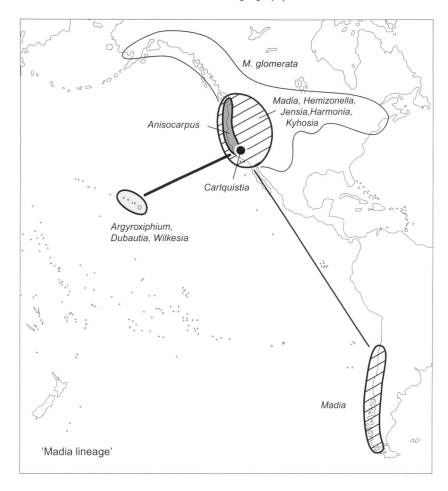

FIGURE 7-3. The distribution of the "*Madia* lineage," including the "silversword alliance" (*Argyroxiphium*, *Dubautia*, and *Wilkesia*), this group's immediate allies (*Anisocarpus* and *Carlquistia*), and the remaining members (*Madia*, etc.) (Asteraceae; phylogeny from Barrier et al., 1999; distributions from GBIF at http://data.gbif.org, Flora of North America at www.efloras.org, and Moreira-Muñoz, 2007).

the Channel Islands of California. Unlike the mainland Madiinae, which are nearly all herbs, the plants in the silversword alliance are all at least semi-woody and include trees, shrubs, and lianes. Within the Madiinae, the silversword genera belong to the "*Madia* lineage," along with seven mainland genera, *Madia*, *Jensia*, *Harmonia*, *Hemizonella*, *Kyhosia*, *Anisocarpus*, and *Carlquistia* (Fig. 7-3; Baldwin, 1996, 1999, 2009).

Baldwin (1996) used a paleoecological method to date the Madiinae. In the mainland members, eight of the 14 genera and 73 of the 85 species are endemic to the California Floristic Province (California state and immediately adjacent areas). The other 12 species range more widely in western North America, with *Madia sativa* also in Chile and adjacent Argentina and *M. chilensis* endemic there. Baldwin and Sanderson (1998) suggested that because the "vast majority" of Madiinae are restricted to the California Floristic Province, the group must have originated after the present, summer-dry climate developed there in the Miocene (~15 Ma). This is not necessary, though, and species of Madiinae such as *Kyhosia* (formerly *Madia*) *bolanderi* and *Raillardella pringlei* grow in areas with summer-wet climates (Baldwin, 1996). Madiinae may have occurred in California before the Miocene, despite the different climate then, by surviving in refugia, or their ecological preferences may have changed over time, or they may have occurred elsewhere before the Miocene.

There are many endemics in the California Floristic Province, including 52 plant genera (www.biodiversityhotspot.org), of which at least one (*California*: Geraniaceae) is basal to a worldwide clade (Fiz et al., 2006). Endemic animals include basal stick insect, *Timema*, and the salamander *Batrachoseps* that differentiated "possibly in the late Mesozoic" (Jockusch and Wake, 2002: 385). It is unlikely that all the Californian endemics are younger than 15 Ma. Using the postulated 15 Ma age of Madiinae as a calibration gave a very young age (5.2 Ma) for the Hawaiian silversword alliance (Baldwin and Sanderson, 1998). Baldwin has argued strongly for the calibration (writing that a "maximum age" of 13–15 Ma for the Madiinae "dictates" that the silversword alliance is less than ~6 Ma; Baldwin, 1997; and referring to "The upper limits of conceivable age for the Californian group, 15 Ma"; Baldwin and Sanderson, 1998), but while the dating is often cited, it is not strongly supported.

Baldwin and Sanderson (1998) argued that the mainland tarweed group "gave rise to the silversword alliance," although the fact that the silverswords are nested phylogenetically in the mainland tarweeds does not mean they are derived from them; instead, the groups could have had a common ancestor (cf. Fig. 1-6). Barrier et al. (1999) found that the closest mainland relatives of the silverswords are *Raillardiopsis muirii* (now *Carlquistia*) and *R. scabrida* (now *Anisopappus*) of California, with both included in different parts of the silversword alliance. The authors proposed that the Hawaiian silverswords were allopolyploids derived from a hybridization event. The hybridism involved members of lineages that include *Carlquistia* and *Anisocarpus*, but not necessarily the modern genera or groups with their distribution. Barrier et al. (1999) suggested that the hybrid has dispersed to the Hawaiian Islands and established there. Nevertheless, the hybrid has not diversified or even survived in America. Another possibility is that a widespread eastern Pacific ancestor has differentiated into diploid (possibly hybrid) groups on the mainland and polyploid, hybrid clades on the islands.

Vargas et al. (1999) regarded the silversword alliance as an "unequivocal example" of dispersal from America to the Hawaiian archipelago and suggested (p. 238) that "no geological evidence exists for now-extinct islands that could have served as 'stepping-stones' for dispersal from North America to the Hawaiian archipelago." Baldwin (2009) also emphasized that no land bridge or intervening islands exist, and so he concluded that the ancestor of the *Argyroxiphium* group must have dispersed across more than 3,500 km of open ocean to reach the Hawaiian archipelago. While it is true that there are no present-day islands between California and Hawaii, there are many seamounts (mapped by Etnoyer et al., 2010). In addition, since the Cretaceous, when the precursor of the Hawaii–Emperor chain already existed, a vast amount of the seafloor that existed between Hawaii and Californa (the Farallon plate and derivatives), along with its seamounts, has been destroyed by subduction. Some surface material of the seafloor, and also the Caribbean plateau, have been accreted rather than subducted. The suggestion that there are currently no islands in the region overlooks the subduction, the movement of the spreading ridge from the central Pacific to California (obliterating even the youngest parts of the Farallon plate north of Mexico), and extant topographic features such as the seamounts.

The eastern Pacific subduction is recorded in sequences such as the Franciscan Complex of central California, one of the world's best-known subduction complexes. It comprises rock material that has been scraped off during subduction from 160 Ma to <20 Ma, including oceanic-island, mid-oceanic ridge, and island-arc type basalts (Saha et al., 2005). The Coastal Belt of the Franciscan Complex was emplaced from 70 to 25 Ma. Within the Coastal Belt, Late Cretaceous basalts may have formed near the Pacific–Farallon ridge and been translated northeastward to reach California in the Paleocene–Middle Eocene (56–40 Ma). In contrast, another unit in the Franciscan Complex, the oceanic Wheatfield Fork terrane, formed in the Eocene. It has been equated with the Siletz terrane (Siletzia) and may have originated on the southeastern Kula plate, accreting by the Miocene (McGlaughlin et al., 2009).

The Siletzia terrane forms the coast of most of Washington and Oregon. It is an oceanic seamount terrane that formed during Paleocene–Eocene time "at some unknown distance offshore" and was accreted early in Eocene time (Dickinson, 2004: 33). The submarine and subaerial basaltic lavas form a large igneous province possibly >30 km thick. Wells (2007) suggested that the terrane represents an accreted oceanic plateau that has been partially obducted onto the continent in the same way that the Ontong Java plateau has been obducted onto the Australian plate.

In other cases, seamounts and other asperities on the seafloor are subducted along with the crust, and occasionally there is direct evidence for this. For example, tomographic analysis showed that a major earthquake in western Costa Rica in 1990 was caused by a seamount currently subducting at 30 km below the Earth's surface (Husen et al., 2002).

In *Deinandra*, another genus of tarweeds (Madiinae), five shrubby or facultatively perennial taxa occur only on the Channel Islands and Guadalupe Island, further south off Baja California; the other 19 taxa in the genus are mostly annuals that all occur in mainland California, with three species also on the Channel Islands. The Guadalupe endemics form a monophyletic clade and, as Baldwin (2007) indicated, this parallels the diversification of the silversword alliance in the Hawaiian Islands. Baldwin used the same calibration as cited above and interpreted the Guadalupe clade as an adaptive radiation on the current island. Annual habit and a California center of origin were assumed to be ancestral for the group, based on its immediate relatives, but this may not be necessary (Fig. 1-6). The Guadalupe–Channel islands region is an important center of endemism for groups also represented in Hawaii and is located in a zone of tectonic extension (the concept of a Baja hotspot is controversial; cf. Favela and Anderson, 2000). The groups could be much older than the present islands and could have existed previously on other islands that have been subducted.

In the widespread *Sanicula* (Apiaceae) a Hawaiian–American group, *Sanicula* "clade 1" (Vargas et al., 1999; Fig. 7-4), "provides a remarkable parallel" to the silverswords (Vargas et al., 1999: 239). In "clade 1," the Hawaiian group is sister to a group (*S. arctopoides*, *S. arguta*, and *S. laciniata*) that ranges along the coast from southern California to northwestern Washington (all three species are present on the Channel Islands). This whole subclade is sister to a second subclade (*S. bipinnnata* and *S. crassicaulis*) that ranges between southern California,

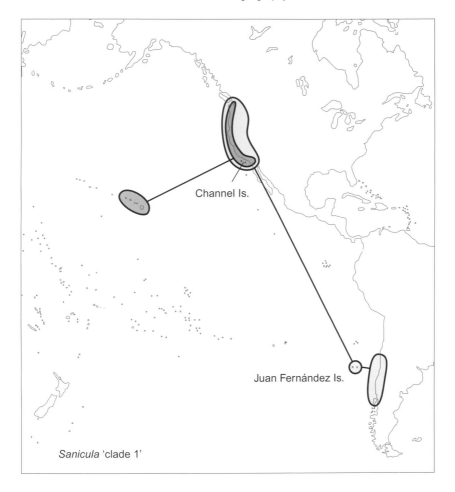

FIGURE 7-4. Distribution of *Sanicula* "clade 1" (Apiacae) (phylogeny from Vargas et al., 1999; distribution data from GBIF at http://data.gbif.org and Flora of Argentina at www.darwin.edu.ar/Proyectos/FloraArgentina/ Generos.asp).

inland Oregon, and northwestern Washington, with *S. crassicaulis* also in Juan Fernández Islands, central Chile (~30°–41°), and adjacent Argentina. (Both species are also on the Channel Islands.) The only overlap between the two subclades occurs along the west coast of the U.S. and is attributed here to local range expansion, possibly following accretion of the Hawaiian subclade onto the mainland; this may be the only dispersal that has occurred in the group. Another *Sanicula* species, *S. graveolens*, has a similar North–South America disjunct distribution: southern California (including Channel Islands) to southernmost British Columbia, also in Chile (30°–44°S) and adjacent Argentina.

Vargas et al. (1999) proposed that the distribution pattern resulted from long-distance dispersal and stressed that these *Sanicula* species have fruits with hooked prickles. But these do not explain why dispersal stopped, and they are absent in groups that have a similar distribution, such as the silversword alliance and the next example.

Silene (Caryophyllaceae) is a subcosmopolitan, mainly herbaceous genus, although Hawaiian species are woody or semi-woody. The Hawaiian members form a clade that has its sister group, *S. antirrhina*, in the western and eastern U.S., including the Channel Islands, and also South America around Paraguay (http://data.gbif.org; Eggens et al., 2007). The pattern would usually be taken as evidence for an American origin, but Eggens et al. pointed out "the difficulties . . . in unambiguously inferring the direction of dispersal": "It is possible that the ancestor of the endemic Hawaiian *Silene* arrived to the Hawaiian Islands from an unknown [Old World] source area and that a representative from this lineage was dispersed to the American continents. . . . This scenario does, however, require two long-distance dispersals over the Pacific Ocean" (pp. 215–216). As a third possibility, the group and its ancestors could have evolved as a metapopulation on mainland North America and in the northeast Pacific on former islands.

MORPHOLOGICAL EXAMPLE:

> *Fragaria chiloensis* (Rosaceae): Hawaiian Islands (Maui and Hawaii); Alaska to California; Juan Fernández, central Chile/adjacent Argentina (Wagner et al., 1990).

The examples just discussed show conspicuous similarities in distributions in groups with different ecology and means of dispersal. The disjunction correlates with subduction of the East Pacific seafloor and eastward translation of the East Pacific Rise. This has been overlooked in biogeographic accounts but may be a key to understanding these and related patterns.

HAWAIIAN ISLANDS–MEXICO/CARIBBEAN

In ferns, the Hawaiian *Lellingeria saffordii* (Grammitidaceae) is sister to *L. hellwigii* of southern Mexico (Oaxaca: Guerrero terrane) (Labiak et al., 2010).

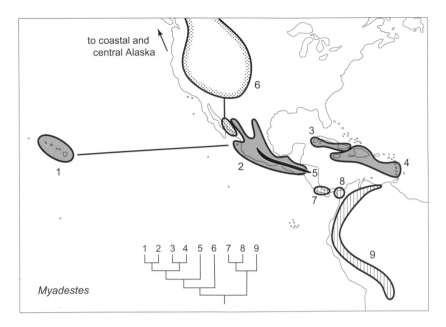

FIGURE 7-5. Distribution of *Myadestes* (Turdidae) (phylogeny from Miller et al., 2007; distributions from Collar, 2005).

The Hawaiian endemic *Gossypium tomentosum* (Malvaceae) is sister to *G. hirsutum* of Mexico, Central America, and the Carribean (Baldwin and Wagner, 2010).

Argemone glauca (Papaveraceae) of Hawaii is sister to a clade of Mexico and Central and South America (Schwarzbach and Kadereit, 1999).

Hawaiian species of *Sicyos* (Cucurbitaceae) are nested in a clade from Mexico (Sebastian et al., 2009; S. Renner, pers. comm. in Baldwin and Wagner, 2010).

Jacquemontia sandwicensis (Convolvulaceae) of Hawaii is sister to *J. obcordata* of the eastern Mexico coast and the Greater and Lesser Antilles. The pair is sister to *J. ovalifolia* s.str of West and East Africa (Namoff et al., 2010). As the authors noted (p. 57), the group is another example of a biogeographic link between the Caribbean Basin and Polynesia, along with groups such as *Siemensia* and allies (Rubiaceae) and *Fitchia* and allies (Asteraceae; Fig. 6-15).

The passerine *Myadestes* (Turdidae) includes a clade: ((Hawaiian Islands + Mexico) (Jamaica and Lesser Antilles + Cuba)) (Miller et al., 2007). This distribution and the rest of the genus range is shown in Figure 7-5 and can be related to the Pacific origin of the Caribbean plate (Fig. 6-1).

The passerine family Mohoidae of the Hawaiian Islands, extinct within the last century, is sister to the Ptilogonatidae ranging in the southwestern U.S. to western Guatemala and, disjunct, in Costa Rica and western Panama. The pair are sister to *Dulus* of Hispaniola (Fleischer et al., 2008). The whole clade is sister to the trans-Atlantic family Bombycillidae of Europe, northern Asia, and North America south to Colombia.

HAWAIIAN ISLANDS–GALAPAGOS

MORPHOLOGICAL EXAMPLES:

The fly *Thinophilus hardyi* (Dolichopodidae) is recorded only from the Hawaiian Islands and the Galapagos (Bickel and Dyte, 2007). There may be slight differences in the genitalia of the two populations but the two are "certainly one unit" (D. Bickel, pers. comm.).

The petrels formerly treated as *Pterodroma phaeopygia* s.lat. comprise *P. phaeopygia* s.str. of the Galapagos and *P. sandwichensis* on the main Hawaiian Islands (Brooke, 2004).

HAWAIIAN ISLANDS–SOUTHEASTERN POLYNESIA

Adamson (1939: 72) discussed the Polynesian fauna and wrote that "Positive affinities between the Marquesas and Hawaii, not shared with other islands, are few but well defined and significant." Endemism defining this disjunct, transequatorial sector is not predicted in a dispersal model, given the direction of the prevailing winds and currents, and it has been neglected by biogeographers. The connection is indicated in a growing number of molecular studies and involves groups, both terrestrial and marine, that have quite different ecology and means of dispersal. The distribution pattern is correlated with a major geological feature, a belt of Cretaceous on-ridge volcanism. The only reason vicariance has not been accepted for the pattern is that the groups concerned are thought to be too young, although there is no real evidence for this.

Charpentiera (Amaranthaceae) is widespread in the Hawaiian Islands (five species and several varieties) and southeastern Polynesia (Austral Islands: Tubuai and Raivavae Islands, and Cook Islands: Rarotonga; Florence, 2004). It is basal to a global clade comprising all Amaranthaceae s.str. except *Bosea* (Müller and Borsch, 2005).

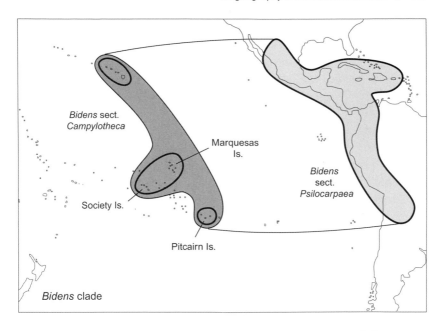

FIGURE 7-6. Distribution of *Bidens* sect. *Campylotheca* and its sister sect. *Psilocarpaea* (minus the pantropical weed *B. pilosa*) (Asteraceae) (Kim et al., 1999; Ganders et al., 2000; Kimball and Crawford, 2004).

A clade in *Metrosideros* (Myrtaceae) is endemic to the Hawaiian and Marquesas Islands (Wright et al., 2001).

The diverse *Melicope* sect. *Pelea* (Rutaceae) is endemic to the Hawaiian islands (48 spp.) and Marquesas Islands (four spp.) (Harbaugh Wagner, Allan and Zimmer et al., 2009).

A clade ("*Nesoluma*") in *Sideroxylon* (Sapotaceae) is on the Hawaiian Islands and southeastern Polynesia (Rapa, Raivavae, Tahiti, and Henderson Islands) (Smedmark and Anderberg, 2007).

Ilex anomala (Aquifoliaceae) is endemic to the Hawaiian and Society Islands. It forms a clade with American species (Cuénoud et al., 2000).

In the widespread genus *Bidens* (Asteraceae), sect. *Campylotheca* comprises 19 species in the Hawaiian Islands and about the same number in southeastern Polynesia (Marquesas, Society, Henderson, and Pitcairn Islands; Kim et al., 1999; Ganders et al., 2000; Kimball and Crawford, 2004; Fig. 7-6). Members of this section have woody, indeterminate growth, whereas the group's sister from western tropical America and the Caribbean (sect. *Psilocarpaea*) has herbaceous, determinate growth. The Pacific group includes many local endemics;

in contrast, the American clade includes one species, *B. pilosa*, that has become a weed throughout the tropics (not shown in the figure). Kim et al. (1999) suggested that the wide distribution of *Bidens* in Polynesia, America, and Africa "attests to the dispersal ability of these plants." This is questionable as there are no indigenous members of the genus in Southeast Asia, Australasia, or the islands of the southwest Pacific. The two clades mapped here are separated by the East Pacific Rise, and the simple allopatry can be explained by activity on this spreading ridge.

Kadua (Rubiaceae; formerly in *Hedyotis*) is on the Hawaiian Islands (21 endemic species), the Marquesas (3 endemics), the Society Islands (2 endemics), and Rapa Island (one endemic) (species from these last three localities comprise sect. *Austrogouldia*). Finally in the genus, there is one widespread species, *K. romanzoffiensis*, on the Line, Ellice, Tokelau, Cook, Austral, Society, Tuamotu, Gambier, and Pitcairn Islands (Terrell et al., 2005; Kårehed et al., 2008). Thus the group is a typical widespread central Pacific endemic (Chapter 6), but apart from the populations of the last species on the Ellice Islands in Tuvalu and on Tokelau, the genus is distributed along the Hawaiian Islands–Line Islands–southeastern Polynesia sector. Kårehed et al. (2008: 855), citing Motley (in press), concluded that "the Hawaiian *Kadua* species are paraphyletic with respect to the French Polynesian species and provide strong evidence for migration of a plant lineage out of the archipelago," but this reasoning is not accepted here (cf. Fig. 1-6).

In land snails, a clade in the genus *Succinea* (Succineidae) occurs in the Hawaiian Islands and Tahiti (Cowie and Holland, 2008).

The weevil *Rhyncogonus* (Curculionidae) is endemic to the Hawaiian Islands, Wake Island (near the Mid-Pacific Mountains; Fig. 7-1), the Line Islands (Fanning and Christmas Atolls; Fig. 7-1), and southeastern Polynesia (Cook, Austral, Society, Tuamotu, and Marquesas Islands) (Zimmerman, 1948; Claridge, 2006). A molecular study showed two main clades, one in Hawaii, the Society Islands, and Rarotonga (Cook Islands), the other throughout the Austral and Marquesas Islands (Claridge, 2006). The genus was found to be allied with the western Pacific Elytrurini.

Many reef taxa show the pattern. In the sea urchin *Diadema paucispinum*, Lessios et al. (2001) recorded Hawaii–southeastern Polynesia (Pitcairn/Easter) clades. They interpreted these as the result of "chance arrival of larvae" and inferred "high rates of gene flow" between the localities. Nevertheless, they noted that the Hawaii–southeastern Polynesia connection runs perpendicular to both the North and the South Equatorial Currents, and so the affinity is "remarkable," as it indicates a "tremendous capacity for dispersal." Instead, the affinity could be due to vicariance.

The hermit crab *Calcinus laurentae* and its immediate relatives are endemic to the Hawaiian Islands and southeastern Polynesia (Malay and Paulay, 2010).

In the surgeonfish *Acanthurus triostegus*, Marquesas Islands populations are geographically closest to those in the Tuamotu and Society Archipelagos, although genetically they are closest to Hawaiian populations (Planes and Fauvelot, 2002). The authors concluded: "These observations favor the hypothesis of biogeographic vicariance as an evolutionary process leading to the differentiation of the *A. triostegus* populations in the Hawaiian and Marquesas Archipelagos" (p. 391). They also cited congruent biogeographic patterns in other Hawaiian and Marquesan coral reef fishes (see Heads, 2005b, for further discussion).

Gaither et al. (2010) studied two widespread Indo-Pacific snappers in the genus *Lutjanus*. *L. kasmira* showed no population structure across the Indian Ocean and (most of the) Pacific Ocean, while *L. fulvus* was highly structured. Despite these differences, Gaither et al. found that "both fishes demonstrate a remarkably strong phylogeographic break at the Marquesas Islands." The authors attributed the isolation of Marquesan populations to contrarian ocean currents and an unusual local environment. But this does not account for the Hawaii–Marquesas connection in other groups. In the coral reef fishes of the *Dascyllus trimaculatus* species complex, a clade of the Hawaiian and Marquesas Islands is sister to the rest of the group, found throughout the Indo-Pacific west of the Hawaii–Marquesas clade (Leray et al., 2010). Populations on the Line and Phoenix Islands are intermediate.

In the passerine *Acrocephalus* (Acrocephalidae), one clade has the distribution: Hawaii, Marquesas, Tuamotu, Society, Gambier, Pitcairn, Austral, and Cook Islands (Cibois et al., 2011).

MORPHOLOGICAL EXAMPLES:

Cheirodendron (Araliaceae): Hawaiian and Marquesas Islands.

Phyllostegia (Lamiaceae): Hawaiian Islands (27 species) and Society Islands (*P. tahitensis* of Tahiti).

Vaccinium sect. *Macropelma* (Ericaceae): Hawaiian Islands and Marquesas, Society, Rapa, and Cook Islands (Vander Kloet, 1996).

Oreobolus furcatus (Cyperaceae): Hawaiian and Society Islands (Tahiti) (Meyer and Salvat, 2009).

Leptecophylla (formerly *Styphelia*) *tameiameiae* (Ericaceae): Hawaiian and Marquesas Islands.

Korthalsella complanata (Santalaceae): Hawaiian Islands plus Henderson Island (Pitcairn group) (Molvray et al., 1999, sampled *Korthalsella* populations from Hawaii and New Zealand/Australia but not those from southeastern Polynesia).

Pisonia wagneriana (Nyctaginaceae): Hawaiian Islands; "seems to be more similar to some Society Islands members of this affinity than to the variable Hawaiian populations" (Wagner et al., 1990: 988).

Campsicnemus (Diptera: Dolichopodidae) is primarily a north temperate genus but has 300+ species in the Hawaiian Islands (138 described) and three in the Marquesas Islands, the only southern hemisphere locality for the genus (Evenhuis, 1999). The Hawaiian and Marquesas species appear to form a distinct clade (D. Bickel, pers. comm.).

Shallow-water moray eels of the "*Anarchias cantonensis* Group" comprise three species (Reece et al., 2010):

- *A. exulatus*: eastern Pacific (Johnston Atoll, Hawaiian Islands, southeastern Polynesia: Rapa, Pitcairn, Henderson, and Gambier Islands) and Tonga.
- *A. schultzi*: southwest Pacific (Caroline Islands, Solomon Islands, New Caledonia) and Tonga.
- *A. cantonensis*: Indian Ocean and central west Pacific, to the Marshall, Phoenix, and Line Islands, and Tonga.

The authors pointed out that all three species occur only in Tonga, at the Pacific plate margin. *A. exulatus* is disjunct between Hawaii and southeastern Polynesia, skirting *A. cantonensis* in the Line and Phoenix Islands.

HAWAIIAN ISLANDS–COOK ISLANDS

In *Tetramolopium* (Asteraceae), the Hawaiian clade is sister to a species of the Cook Islands (Mitiaro) (Lowrey et al., 2001).

MORPHOLOGICAL EXAMPLES:

Myoporum sandwicense (Myoporaceae): Hawaiian Islands and Cook Islands (Mangaia).

In *Lepidium* (Brassicaceae), an undescribed species of the Cook Islands (Mitiaro) is thought to be related to Hawaiian species (Lowrey et al., 2005).

HAWAIIAN ISLANDS–WEST AND SOUTHWEST PACIFIC ISLANDS

In the fern *Grammitis*, a clade of three Hawaiian species (*G. forbesiana*, *G. hookeri*, and *G. baldwinii*) is sister to *G. knutsfordiana* of Borneo to Fiji and *G. padangensis* of Malesia to New Guinea (Geiger et al., 2007).

In *Planchonella* (Sapotaceae), Swenson et al. (2007) found a well-supported clade in Hawaii (*P. sandwicensis*) and Fiji (*P. umbonata*).

In a sample of spiny solanums (*Solanum* sect. *Leptostemonum*; Solanaceae), two Hawaiian species are sister to the New Caledonian *S. pancheri*. This is sister to and vicariant with Australian species (Levin et al., 2006).

In a sample of Pacific *Pittosporum* species (Pittosporaceae), Hawaiian species formed a clade with species of Fiji, Tonga, and New Caledonia (Gemmill et al., 2002).

The Hawaiian *Diospyros sandwicensis* (Ebenaceae) is sister to a large clade of New Caledonian species (Duangjai et al., 2009).

Nestegis (Oleaceae) of the Hawaiian Islands is related to "*Nestegis*" of New Zealand and Norfolk Island, *Osmanthus* sect. *Notosmanthus* of New Caledonia, and *Notelaea* of northern and eastern Australia and Tasmania (Wallander and Albert, 2000).

In land snails, a clade of *Succinea* is in the Hawaiian Islands and Samoa (Cowie and Holland, 2008).

MORPHOLOGICAL EXAMPLES:

Keysseria (Asteraceae) comprises one section in Borneo, Sulawesi, and New Guinea, and the other in the Hawaiian Islands (Nesom, 2001).

Gahnia vitiensis (Cyperaceae): Hawaiian Islands and Fiji.

Carex meyenii (Cyperaceae): Hawaiian Islands and Caroline Islands.

Sophora chrysophylla (Fabaceae) of the Hawaiian Islands "appears to be closely related to the *S. tetraptera* complex of New Zealand" (Wagner et al., 1990: 706).

In *Dianella* (Hemerocallidaceae), the Hawaiian species shares distinctive inflorescence characters with the New Caledonian species (Wagner et al., 1990).

In isopods, the genus *Myrmecodillo* occurs in South Africa (KwaZulu-Natal), Madagascar, Mascarene Islands, New Britain, Queensland, Tonga, and Hawaii, with one endemic species in each locality. *Ligia hawaiensis* is recorded only in Hawaii and Fiji (Schmalfuss, 2009).

The squid *Pterygioteuthis microlampas* is known from the Hawaiian Islands and New Zealand (Lindgren, 2010).

TRANS-TROPICAL PACIFIC DISTRIBUTION

Schleinitzia (Fabaceae) ranges from the Philippines to New Caledonia and the Society Islands (including Guam and the Marianas but not Australia). This widespread central Pacific endemic is in a well-supported clade with *Kanaloa* of Hawaii (on Kahoolawe and, fossil, on Kauai; Burney et al., 2001) and *Desmanthus*, in warm America and the Caribbean (Lewis et al., 2005). The three genera divide up a vast Pacific–America range without any overlap.

MORPHOLOGICAL EXAMPLES:

Perrottetia (Celastraceae): southern China through Malesia to northeastern Australia and the Solomon Islands; Hawaiian Islands; Mexico to Venezuela and Bolivia (van Balgooy and Ding Hou, 1966; http://data.gbif.org). *Perrottetia* and *Mortonia* (Mexico and the southwestern U.S.) form a trans-tropical Pacific clade that is basal in a worldwide group, the Celastraceae (Simmons et al., 2001).

Osteomeles (Rosaceae): southern China, islands south of Japan; Tonga, Cook, and Pitcairn Islands; Hawaiian Islands; Costa Rica to Venezuela and Bolivia (van Balgooy, 1966; http://data.gbif.org).

Cibotium (Cyatheaceae): southern China to Indonesia, Philippines, and New Guinea; Hawaiian Islands; southern Mexico and Guatemala (Large and Braggins, 2004).

HAWAIIAN ISLANDS AS AN EAST/WEST BOUNDARY

The Hawaiian Islands can represent a boundary between western and eastern Pacific groups. This is seen in *Hyles* (Sphingidae), a widespread genus of hawkmoths. There are four main clades, and these are largely allopatric (Hundsdoerfer et al., 2009; distributions from www.cate-sphingidae.org and Roque-Albelo and Landry, 2002):

1. (*H. lineata*). Widespread in the Americas (in South America mainly in the west and not in Brazil; south to Concepción in Chile); also in the Galapagos and Hawaii.

2. Southern South America and eastern Brazil.

3. Madagascar and Australia.

4. Madeira and through Eurasia to Mongolia and Hawaii.

The clades have the phylogeny (1 (2 (3 (4)))). The distribution could, as always, reflect chance processes, but the overall allopatry and the distinctive Madagascar–Australia affinity is compatible with differentiation in an already global ancestor. This implies that the global range has been sundered by the opening of the Indian Ocean and Atlantic Ocean basins and that the group is Mesozoic. Here it is a question of weighing up, on one hand, the fossil record of Lepidoptera and insects in general, and, on the other, global patterns of clear-cut allopatry.

The only overlap among the four clades seems to occur in parts of South America (this is fairly minor and probably due to secondary range expansion) and in the Hawaiian Islands. Here one clade, *H. lineata*, possibly vagrant, has populations further east, in the U.S., while the other (*H. perkinsi* and *H. calida*) has relatives to the west, in Asia. This may reflect a former occurrence of both groups in the north-central Pacific on older, now submerged or subducted islands east and west of Hawaii. In the Hawaiian Islands, *H. lineata* is lowland and widespread (in drier areas, to Midway Island), while the *perkinsi/calida* clade is at high elevation and on the main islands only; the two clades are not found together; D. Rubinoff, pers. comm.). *H. lineata* is well known for its great dispersal capabilities and is said to land on ships in mid-ocean (A. Hundsdoerfer, pers. comm.). But this does not explain why it is absent in areas such as eastern Brazil, southern South America, and the western Pacific, where it is replaced by other clades. *H. lineata* may have been dispersing around the eastern Pacific since it differentiated there in the Cretaceous, repeatedly dying out on some islands and reinvading others as a weedy metapopulation.

WIDESPREAD CENTRAL PACIFIC ENDEMICS

Many groups are widespread in the Pacific islands and are endemic to the region, as discussed in Chapter 6. *Lycium* (Solanaceae) comprises two clades, one in the Old World and one in the Americas and the Pacific (Fig. 7-7; Fukuda et al., 2001; Levin and Miller, 2005; Miller et al., 2011). The only Pacific representative of the genus, *L. sandwicense*, is endemic to the Pacific basin and widespread there. It is nested in the New World clade but, as shown in Figure 1-6, this does not necessarily mean it originated there. It ranges from the Daitou (Ryukyu) and Ogasawara

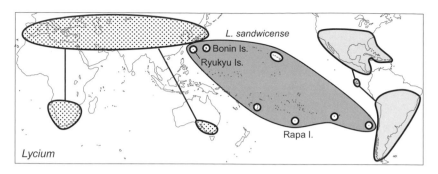

FIGURE 7-7. Distribution of *Lycium* (Solanaceae). There are two clades, the Old World group and the New World/Pacific clade, in which *L. sandwicense* is the only Pacific island representative (Fukuda et al., 2001; Levin and Miller, 2005).

(Bonin) Islands (Fig. 7-7) to the Hawaiian Islands, Rapa, and the Juan Fernández Islands. The margins of the species in the northwest and southeast both occur at plate boundaries. The neat allopatry of the range with the Old World clade of *Lycium* suggests the break was formed as the group evolved, not by a secondary meeting following dispersal across the Pacific. Fukuda et al. (2001) cited six other plants of the Ogasawara Islands that have their closest relatives in the Hawaiian Islands. Miller et al. (2011) explained the distribution of *Lycium* as the result of "dispersal from the Americas to Africa approximately 3.64 Ma . . . followed by subsequent dispersal to eastern Asia approximately 1.21 Ma," but this does not explain the precise global allopatry of the *Lycium* groups or the location of the breaks at plate margins.

Metrosideros s.str. (Myrtaceae) is another central Pacific group; its Hawaii–Marquesas branch was cited above. The trees and shrubs in this genus are a characteristic feature of many forests on the high islands in the Pacific (Fig. 7-8; based on Wright et al., 2000, 2001; Heads, 2009d). Percy et al. (2008) wrote that "The presence of *Metrosideros* throughout the Pacific region has long been considered the result of highly effective dispersal via small, wind-dispersed seed." Yet in their study of Hawaiian and southeastern Polynesian populations, they found "a number of monophyletic [allopatric] island lineages, which does not support a pattern of numerous repeated colonization" (Percy et al., 2008: 1486). The five main lineages in the genus as a whole (Fig. 7-8) maintain more or less complete allopatry over large areas of the Pacific. The main central Pacific clade (dark gray in Fig. 7-8)

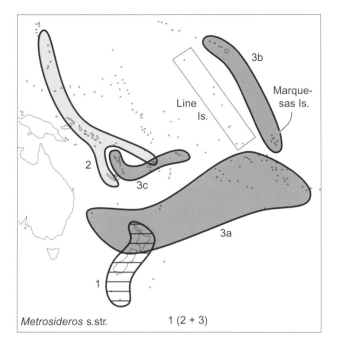

FIGURE 7-8. Distribution of *Metrosideros* s.str. (Myrtaceae), showing the three main clades. Possible former habitat includes the Line Islands (Wright et al., 2000, 2001; Heads, 2009d).

is divided into three subclades that more or less surround the Line Islands. The genus is probably absent in the Line Islands due to the lack of suitable habitat on these low atolls. *Metrosideros* has not colonized the Pacific margins, where it is replaced by its closest relatives *Mearnsia* (New Zealand, New Caledonia, New Guinea, Philippines) and *Tepualia* (southern South America). This clade (Metrosidereae) in turn is notably absent from Australia, where it is replaced by its sister group Tristanieae (Australia, New Caledonia, New Guinea, and eastern Malesia) (Biffin et al., 2010). Many other Australian Myrtaceae, including *Eucalyptus*, do not occur naturally in the Pacific islands, despite having small, wind-dispersed seed.

As Percy et al. (2008) suggested, the evident "geographic structure" in *Metrosideros* indicates a phase of colonization followed by a phase of "limited" colonization. The key question is: What caused the change? Vicariant breakup has occurred throughout the ancestral Metrosidereae + Tristanieae (at tribe, genus, and subgenus level), but there is overlap with the sister group, Backhousieae + Syzygieae, throughout the Pacific. Thus there has been overlap between the two clades, and this has occurred early in the evolution of the family, not among the present genera or even tribes.

Widespread geological revolutions that may have contributed to this phase of mobilism and range expansion include the global sea-level maximum in the Cretaceous, the Jurassic–Cretaceous rifting of Gondwana, and the Cretaceous emplacement of large igneous provinces and volcanic island chains in the Pacific. The mid-Cretaceous plateaus and seamounts would have been colonized from earlier Cretaceous islands already in the region. Subsequent dismembering and dispersal of the plateaus and subsidence of the island chains may have led to allopatry.

Although Percy et al. (2008) wrote that known volcanic activity in the Austral Islands extends back to 58 Ma, they did not use this date for clock calibrations of *Metrosideros* as it would imply "extremely slow" rates of evolution in the genus and would suggest that it has been in Hawaii since 13 Ma. Instead, they used the age of the youngest Australs hotspot.

Metrosideros species show vicariance even at a local level. For example, in *M. polymorpha* on eastern Hawaii Island, analysis of population structure revealed dispersal limitation and significant differentiation of extreme-habitat varieties (Stacy et al., 2010). Several species of *Metrosideros* are pioneers in primary successions on recent lava flows. On the Bonin Islands, *M. boninensis* is also a pioneer species and regenerates in disturbed open spaces, such as abandoned fields, canopy gaps, and steep slopes after landslides. Kaneko et al. (2008) found that *M. boninensis* comprises genetically distinctive populations and subpopulations that reflect geographical distribution. They concluded that "Populations of *M. boninensis* showed significant genetic differentiation and isolation by distance over a small geographical scale, despite the fact that this species should have extensive gene dispersal ability" (Kaneko et al., 2008: 119).

In their sample of *Melicope* (Rutaceae), Harbaugh, Wagner, Allan, and Zimmer (2009) recovered three main clades: (Lord Howe Island + New Zealand + Tahiti) ((Vietnam/Taiwan + Australia + New Guinea) (Hawaii + Marquesas Islands)). The authors interpreted the phylogeny as "unequivocal" evidence for long-distance dispersal to Hawaii. This is debatable and in any case does not explain the fact that the same break between a Lord Howe–New Zealand–Society Islands clade and one in the Marquesas–Hawaiian Islands occurs in both *Melicope* and *Metrosideros* (see above). Nor does it account for the perfect allopatry of the three *Melicope* clades or the break between Australia and Lord Howe Island, attributed here to the Cretaceous opening of the Tasman Sea.

In the central Pacific, *Tetramolopium* (Asteraceae) is represented not by unrelated waifs but by a single, widespread endemic clade with a phylogeny: (New Guinea (Cook Islands, Hawaii)) (Lowrey et al., 2001, 2005). According to Carlquist (1995), the records on Hawaii and the Cook Islands "demonstrate clearly" long-distance dispersal from New Guinea and show it is "clear" that evolutionary diversification in Hawaii has been recent, but there is no actual evidence for these proposals. Lowrey et al. (2005) also argued that the phylogeny "clearly supports the hypothesis . . . that *Tetramolopium* evolved in the New Guinea highlands and was subsequently dispersed to the eastern Pacific" (p. 451). In fact, the phylogeny in itself only supports a split between east Pacific (Cook Islands–Hawaii) and west Pacific (New Guinea) and sheds little light on the mechanism for the allopatry. Although Lowrey et al. (2001, 2005) proposed long-distance dispersal from New Guinea to Hawaii and from there to the Cook Islands, they only referred to these current islands and did not mention atolls, seamounts, or any tectonic change. Instead, they suggested that the migrant plover *Pluvialis fulva* could have transported the seeds from Hawaii to the Cook Islands. These birds breed in Alaska and Siberia and migrate in the winter to southern Asia and many Pacific islands, south to Australia and New Zealand. This migration pattern is a poor match with the standard central Pacific distribution of the *Tetramolopium* clade and, in particular, does not explain the clade's allopatry with its sister group (*Vittadinia* and *Peripleura*) in Australia and New Zealand.

In the Hawaiian Islands, *Tetramolopium* is confined to dry, open habitats, and this is "highly consistent with the compelling hypothesis that *Tetramolopium* is a recent arrival in the Hawaiian Islands compared to the silversword alliance [i.e., post-Miocene]" (Baldwin, 1998: 65). Instead, it is suggested here that the central Pacific endemism of the *Tetramolopium* clade is compatible with a much older origin, as discussed above for *Metrosideros*. An older origin is also compatible with the biogeography of *Tetramolopium* in the mountains of New Guinea, where at least five species have distributional limits at the major tectonic boundary there, the margin of the Australian craton and the accreted terranes (the species are mapped in Heads, 2001: Fig. 35).

The *Tetraplasandra* group (Araliaceae) was cited in Chapter 6 as a central Pacific endemic group (Fig. 6-13). It comprises one clade in

Hawaii and a second in the Marquesas, the Society Islands, and Samoa (Costello and Motley, 2007).

In a sample of *Ochrosia* (Apocynaceae), a clade of (Hawaii (Fiji + Tonga)) appeared as sister to one from New Caledonia, Guam, and the Moluccas (Hendrian and Kondo, 2007). Species from the Marquesas and others were not included, but the break probably occurs between a widespread clade endemic to the central Pacific and one in New Caledonia–Asia.

MORPHOLOGICAL EXAMPLES:

The southwest Pacific–southeast Polynesia–Hawaiian Islands triangle occurs in many groups.

Pritchardia (Arecaceae): Fiji/Samoa/Tonga, Tuamotu, and Hawaiian Islands (Laysan–Hawaii).

Phyllostegia (Lamiaceae): Tonga, Tahiti, and the Hawaiian Islands (Lindqvist et al., 2003).

Machaerina angustifolia (Cyperaceae): New Guinea, Society Islands, and Hawaiian Islands.

M. mariscoides: Marianas and Caroline Islands, New Guinea, Solomon, Marquesas, and Hawaiian Islands.

Freycinetia arborea (Pandanaceae): New Caledonia, Samoa, Cook, Austral, Society, Marquesas, and Hawaiian Islands.

Hyalopeplus (Hemiptera: Miridae): Fiji, Samoa, Marquesas, and Hawaiian Islands (Asquith, 1997).

The weevil *Proterhinus* (Belidae) comprises 167 species, with 159 endemic to the Hawaiian Islands, and the remainder in Fiji, Samoa, and the Phoenix, Austral, Society, and Marquesas Islands. (The Samoan species has been widely dispersed throughout Oceania with human transport of coconuts) (Marvaldi et al. 2006). The sister genus is *Aralius* of New Zealand and New Caledonia (Kuschel, 2003; Marvaldi et al., 2006) (cf. *Tetramolopium*).

HAWAIIAN ISLANDS, JOHNSTON ATOLL, AND WAKE ISLAND

The terrestrial biota of the northern Pacific atolls (Fig. 7-1) is relictual, with only a fraction of its original diversity. Nevertheless, the few species that do remain along with the reef taxa provide valuable information on earlier biogeographic affinities. Johnston Atoll (85 Ma) is connected with the main chain of the Line Islands to the south by the

Christmas Ridge, where drilling has revealed Late Cretaceous reef fossils (Kosaki et al., 1991). These indicate that the ridge was near the surface. Despite this, the Johnston Atoll fauna is more closely related to the Hawaiian Islands fauna than to the Line Islands fauna. Thirty-nine fishes are only known from Johnston Atoll and the Hawaiian Islands (Randall et al., 1985). Slight divergences between the populations suggest restricted gene flow (Kosaki et al., 1991). Four other species are restricted to southern Japan, the Hawaiian Islands, and Johnston Atoll. In addition, many species of the Marshall, Phoenix, and Line Islands are absent from Johnston Atoll and the Hawaiian Islands (Randall et al., 1985). These authors also listed ten pairs of closely related fish taxa in which one member occurs on Hawaii and Johnston, and the other on the Line Islands.

In contrast with Johnston, the atoll of Wake Island shows fewer links with Hawaii. The insect fauna of Wake Island is "markedly different" from that of the Hawaiian Islands and Johnston Atoll; as with the flora, it closely resembles that of atolls further south, from the Caroline Islands to the Tuamotu Islands (Bryan, 1926). For example, the dolichopodid fly *Chrysosoma complicatum* (Dolichopodidae) is on Wake, Micronesia (eastern Caroline, Marshall, Gilbert, Tuvalu, Nauru Islands), Fiji, Samoa, Tonga, and the Line Islands (Palmyra) (Bickel and Dyte, 2007). Giant clams (Cardiidae: Tridacninae) have a similar pattern, as they are widespread through the Indian and Pacific Oceans to Wake Atoll and the central Line Islands but are not in Hawaii (Newman and Gomez, 2002; http://data.gbif.org). The reef fishes at Wake Island are most similar to those in the Marshall and Marianas Islands. For example, in the widespread Pacific surgeonfish *Acanthurus triostegus*, cited above, the population at Wake Island has two forms, one very similar to that of the Marshall and Marianas Islands, the other resembling a variation seen in the Phoenix and Line Islands. The Hawaiian form is distinctive.

What is the reason for the repeated break between Hawaii/Johnston Atoll on one hand and Wake Island/South Pacific on the other? It seems more likely to be due to vicariance than to dispersal. A current flows between Wake and Hawaii, at least periodically, and "based on this evidence, one would expect an overall close [faunistic] similarity between Wake and Hawaii; however, this does not seem to be the case" (Lobel and Lobel, 2004: 83).

This chapter has analyzed the connections of the Hawaiian biota with other areas. The next chapter reviews patterns within the archipelago.

8

Distribution within the Hawaiian Islands

The plants and animals of the Hawaiian Islands survive in a landscape that has been dominated, at least at some stage in the past, by volcanism. The geology of the islands is summarized below before looking in more detail at distribution in the archipelago.

GEOLOGY OF THE HAWAIIAN ISLANDS

The Hawaiian Islands form a chain with the youngest island in the southeast and the oldest in the northwest. Although the archipelago comprises one chain of islands, it has been formed by two parallel chains of volcanoes, termed the Loa and Kea trends, that show geochemical and isotopic differences (Fig. 8-1; Tanaka et al., 2008). These distinctions imply differences between the two trends in magma supplies and plumbing systems (Eiler et al., 1996). Tanaka et al. (2008) suggested that a single original volcanic lineament has split along its length at 3 Ma, with the pivot at Oahu. If the Hawaiian chain is the result of a propagating crack, the two trends may relate to lateral extension taking place within the chain at the rate of ∼1 mm/yr (Foulger, 2007).

If islands form as a plate moves over a hotspot, it is possible to date the age of each island by dating the exposed rock. On the other hand, if the chain has developed along lines of prior stress, there may be recurrent volcanism and the age of the island might not be the same age

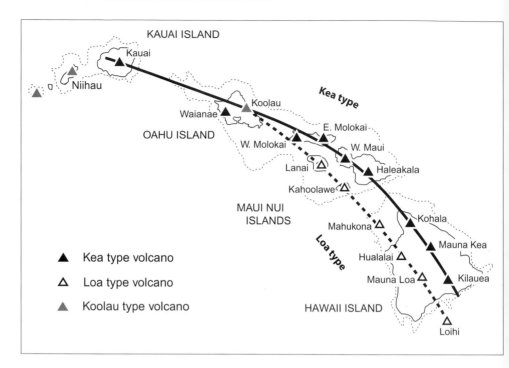

FIGURE 8-1. Hawaiian volcanoes, showing the Kea and Loa trends. Coastline as solid line, 1,500-m isobath as dotted line (from Tanaka et al., 2008).

as the exposed rock. Apart from the linear age sequence of Hawaiian volcanism, rejuvenated-stage volcanism occurs on Niihau (dated at 2.2–0.4 Ma) and Kauai (ongoing since 3 Ma), and the causal mechanisms of this "remain difficult to explain" (Sherrod, 2009). If rejuvenated volcanism continued, it could eventually bury an older island under younger rock, leading to an underestimation of the island's age.

The mantle plume/hotspot theory assumes that individual islands in chains such as the Hawaiian Islands formed as separate islands, although this is debatable. While the modern bathymetry around Hawaii is well known (Figs. 7-2 and 8-1), it has changed in many ways over time due to regional tectonics (including the giant landslides cited under assumption 3), thus "rendering bathymetry around the modern islands a dubious basis" for inferring past geology (Nelson, 2006: 2154). Holland and Cowie (2006) criticized Nelson's "portrayal of the bathymetry around the current islands as 'dubious,'" but Nelson did not suggest that the bathymetry is dubious. His point was that in such a dynamic region the modern geography and bathymetry, no matter how well known, may not be a suitable basis for interpreting past events.

The original, maximum extent of the individual islands and the location of former shorelines have been estimated by examining the change in slope of the sea bed off the islands. Lava extruded underwater cools quickly and forms a steep slope, while lava extruded subaerially forms more gentle slopes. The breaks-in-slope do not have a constant depth as they have been affected by later uplift and subsidence; in particular, there is a pronounced tilt of the seafloor platforms toward Hawaii caused by the volcanic loading there.

Maui, Kahoolawe, Lanai, and Molokai originally made up a single island referred to as "Maui Nui" (i.e., "greater Maui"). The breaks-in-slope indicate that the oldest volcano in Maui Nui was initially connected by land to Oahu, forming a large island, "Oahu Nui" (Price and Elliot-Fisk, 2004). The breaks-in-slope off Oahu and Kauai (see map in Coombs et al., 2004) indicate that the original channel between these islands may have been ~50 km wide, less than half its present width (120 km).

Breaks-in-slope off Maui and Hawaii are only ~8 km apart, suggesting this was the original distance between the two islands, now separated by a channel ~50 km wide. Price and Elliot-Fisk (2004) plotted one of the breaks-in-slope (their "H terrace") for 125 km from Molokai, where it is about 500 m deep, to east of Maui, where it is over 2,000 m deep (Price and Elliot-Fisk, 2004: Fig. 5). This means that the originally horizontal feature has tilted as it subsided; the tilt has resulted from higher rates of subsidence closer to the current zone of volcanic loading, the volcanoes on Hawaii. The profile of the H terrace, as far as it was mapped, shows direct evidence for at least ~1,500 m of subsidence and, as the sea between Maui and Hawaii is currently only between 1,500 and 2,000 m deep, there may have been land between the two islands.

In vicariance theory, biological change develops with geological change, and Nelson (2006) pointed out that growth of the channels may have led to vicariance among populations on different islands. For example, a channel originally 8 km wide between Maui and Hawaii may have been within the normal dispersal range of an ancestral taxon on both islands, whereas the current channel, 50 km wide, may be beyond its capabilities. This geological change could have caused vicariance in the ancestral metapopulation. There is no need for the islands to have been connected by land for vicariance to occur; the only requirement is for a significant *change* in the channel width.

DISTRIBUTIONS WITHIN THE HAWAIIAN ISLANDS: THE PROGRESSION RULE AND OTHER PATTERNS

In the following discussion, plant distributions are from the Smithsonian website "Flora of the Hawaiian Islands" (http://botany.si.edu/pacificislandbiodiversity/hawaiianflora/), unless otherwise stated. Island names are abbreviated as follows: Ni = Nihoa, Nii = Niihau, K = Kauai, O = Oahu, Mo = Molokai, L = Lanai, M = Maui, H = Hawaii.

NORTHWESTERN ISLANDS VS. THE MAIN ISLANDS (KAUAI TO HAWAII)

The islands in the northwestern part of the Hawaiian chain are older, smaller, and lower (and uninhabited by humans), while the main islands in the southeast are younger, larger, and higher. One common biogeographic break in the Hawaiian Islands occurs between these two groups of islands.

In *Banza* katydids (Tettigoniidae), the four clades have a phylogeny: Ni (O (K, O, L) (H, M)) (Shapiro et al., 2006). The main break occurs between the Northwestern Islands (Nihoa) and the main islands.

In the cone-cased clade of *Hyposmocoma* moths, the phylogeny is: (Necker (K (O (Mo + M)))) (Rubinoff, 2008).

MORPHOLOGICAL EXAMPLES:

Cenchrus agrimonioides (Poaceae) comprises var. *laysanensis*: Northwestern Islands (Kure, Midway, and Laysan), and var. *agrimonioides*: main islands (O, Mo, L, M, H).

Achyranthes atollensis (Amaranthaceae): Northwestern Islands (Kure, Midway, Pearl and Hermes, and Laysan); other Hawaiian species of *Achyranthes*: all main islands.

The tern genus *Procelsterna* is endemic to the central Pacific; the northern part of its range is shown in Fig. 7-1. Although the species are seabirds, they do not roam far from their breeding grounds. In Hawaii the genus is restricted to the Northwestern Islands.

The passerine bird *Acrocephalus* (Acrocephalidae) is a widespread genus that occurs on many Pacific islands, including Micronesia, the Line Islands, and elsewhere, but in the Hawaiian Islands it has only been recorded in the Northwestern Islands of Laysan (extinct there by about 1920) and Nihoa (Fig. 7-1).

Neither *Procelsterna* nor *Acrocephalus* is known from the main Hawaiian Islands east of Nihoa, either in the extant fauna or among the many bird fossils described from there (Olson and James, 1982; H. James, pers. comm.). In southeastern Polynesia *Acrocephalus* occurs on low and high islands where its habitat includes montane forest (Cibois et al., 2007), and so its absence from the main Hawaiian Islands may be due to geological and phylogenetic causes rather than ecological ones. Similar factors may have also determined the boundary of *Procelsterna* and other groups.

A diverse range of birds has been wiped out on the main Hawaiian Islands by humans and introduced predators, and many of the taxa extirpated there are now restricted to the Northwestern Islands. A second factor, namely the absence of high islands in the Northwestern group (Niihau, elevation 273 m, is the highest), means that many taxa cannot survive there for ecological reasons. A third factor that could have determined differences between the biotas of the Northwestern Islands and the main islands involves large-scale regional evolution. The boundaries in groups such as *Acrocephalus* and *Procelsterna* may reflect prior aspects of geology and biogeography that predate the modern arrangement of the north Pacific islands (cf. Fig. 6-1).

DISTRIBUTION IN THE MAIN HAWAIIAN ISLANDS

Many accounts of biogeography in the Hawaiian Islands, such as those in Wagner and Funk (1995), refer to a "progression rule" pattern in which the basal species of a group occurs on an older island in the northwest and progressively less basal species inhabit younger islands further southeast. This sequence is a particular case of the prior, more general meaning of the phrase, which refers to a linear sequence of phylogeny (e.g., Fig. 1-3). (A group with the basal species on the *youngest* island, next basal species on the next youngest, and so on would follow a progression rule in the broad sense but would not conform to the "Hawaiian progression rule"). The progression rule has been explained by dispersal from older islands to new ones with accompanying founder speciation.

The progression rule in its purest form was illustrated by Funk and Wagner (1995b: Fig. 17.1) and Cowie and Holland (2006: Fig. 1). These figures indicated populations on the main Hawaiian islands and their 12 volcanoes, with a phylogenetic sequence from most basal taxa on the oldest volcano to least basal on the youngest volcano. Cowie and

Holland (2006: 195) described this as "an example of a non-stochastic dispersal pattern observed for many plant and animal lineages," although the figure represented an ideal, hypothetical case and does not seem to apply to any real example. The closest actual pattern to this "ideal" form is the simpler sequence: (Kauai (Oahu (Maui, Hawaii))).

Carson and Clague (1995) suggested that "high species endemism on the present [individual] islands speaks against the vicariance view" (i.e., that taxa could be older than their islands), and they accepted the progression rule as further evidence for dispersal. They argued that the progression rule sequence shows that island endemic species have each evolved on their present island and are not older relics. They cited the picture-winged *Drosophila* group (112 spp.; Bonacum et al., 2005) and the silverswords (*Argyroxiphium*: Asteraceae). Yet two of the picture-winged clades (the *grimshawi* complex and the *planitibia* group) have subsequently been sequenced and neither conforms to the progression rule; nor do the silverswords (the distributions are given below). Chromosomal and morphological studies of members of the picture-winged group also show non–progression rule patterns, for example, in the *Drosophila ochracea* group and the *Drosophila assita* group (Kaneshiro et al., 1995; distributions given below).

The progression rule has come to be seen as the fundamental reality underlying distribution in the Hawaiian Islands. Biogeographic patterns "generally conform" to it, while "more complex" patterns are "superimposed on the basic progression rule pattern" (Cowie and Holland, 2006). But the number of clades that are exceptions to the rule is growing. Cowie and Holland cited "back colonizations" from younger to older islands; radiations within, rather than among, islands; and "island skipping" patterns in which dispersal "passes over" intermediate islands. Molecular work is turning up many examples of these non–progression rule patterns, and in a recent review Cowie and Holland (2008: 3366) suggested that "there may be considerable stochastic dispersal." While the rule provides a "predictive framework . . . a number of biogeographic patterns in terrestrial Hawaiian radiations are complex," and in some cases "neither the original island colonized nor the subsequent pattern of diversification is clear" (p. 3369). It seems that the progression rule has become a conceptual straitjacket. Patterns that do not conform to it are simply attributed to "stochastic dispersal" rather than being analyzed and compared with others in order to find general patterns. Leaving aside the theory, what are the actual biogeographic patterns?

POLARITY BETWEEN KAUAI AND HAWAII

The distribution of groups in the Hawaiian Islands often shows conspic-uous polarity with respect to the two ends of the chain. For example, the grasses *Poa* and *Trisetum* are both diverse and widespread globally, but in the Hawaiian Islands *Poa* is only on Kauai (three species, all endemic), while *Trisetum* is only on Hawaii, Maui, and Lanai (two spe-cies, both endemic). Groups represented in the Hawaiian Islands only on Kauai include *Drosera* (Droseraceae); groups only on Hawaii include *Vicia* (Fabaceae) and *Hesperocnide* (Urticaceae). Many taxa are on all the main islands except Hawaii (e.g., *Gossypium*: Malvaceae); others are on all the main islands except Kauai (e.g., *Portulaca*: Portulacaceae, and *Cuscuta*: Convolvulaceae).

THE PROGRESSION RULE PATTERN IN THE MAIN ISLANDS: (K (O (M, H)))

As already indicated, the progression rule in the strict sense, with a sequence from oldest volcano to youngest volcano (Cowie and Holland, 2006: Fig. 1), is not documented. A simpler pattern, with the phylogeny (K (O (M + H))) or minor variants, occur in some groups. For example, the cone-cased clade in *Hyposmocoma* moths, cited above, has the phy-logeny: (Necker (K (O (Mo + M)))) (Rubinoff, 2008), given as: Laysan (K (O (O (Mo + M)))) in Rubinoff and Schmitz (2010). In these stud-ies, many populations were sampled and, with the exception of two clades on Oahu, the island clades are reciprocally monophyletic.

Other groups that show variants of the progression rule include the spider *Orsonwelles*, with a phylogeny: (K (K (O (O (Mo (M + H)))))) (Hormiga et al., 2003). The cricket genus *Laupala* has a phylogeny: K ((O (Mo (M (M + H)))) (O (O (M (M + H))))) (Mendelson and Shaw, 2005). Six of these monophyletic, single-island clades include more than one species, and multiple populations were sampled. A dis-persal origin implies that one or a few individuals from a diverse island group colonized another, new island. This model would predict that all island clades would be deeply nested in diverse groups from the next older island: (K (K (K (K (K (K . . . (O (O (O (O (O . . . , etc. In fact, well-sampled molecular studies do not show deeply nested pat-terns but instead a simple sister relationship between K and the rest, and O and the rest. Even in *Orsonwelles* and *Laupala*, there is no deep nesting. Cowie and Holland (2008) hinted at this in observing that the phylogeny in *Orsonwelles* resembles a progression rule, "although the pattern is predominantly within-island speciation."

In addition to the lack of deep nesting, the clades on one island may be more or less allopatric. Within Hawaiian *Plantago* (Plantaginaceae), Dunbar-Co et al. (2008) recovered the phylogeny: (K (K (O (O (Mo, M (M, H)))))) and wrote that the data "unequivocally indicate a Kauai ancestor for the group," although they did not consider the possibility of vicariance. The first Kauai clade comprises samples of *P. pachyphylla* from Sincock Bog, Waialeale, Wainiha, and North Bog, whereas the second Kauai clade is *P. princeps* var. *anomala* from Kalalau (also at Upper Hanapepe; Wagner et al., 1990). Thus the phylogeny is compatible with a widespread ancestor dividing first into one clade on central and northern Kauai (Sincock Bog, etc.) and its sister on western Kauai (Kalalau, Hanapepe) and the other islands. Any overlap between the two Kauai clades may be secondary following initial vicariance (cf. Fig. 1-6). On its own, the phylogeny is just as compatible with a sequence of differentiation in a widespread ancestor as with a center of origin on Kauai and a series of founder speciation events.

A similar case is seen in a group of Hawaiian *Viola*, the "wet clade," known from wetter habitats (Havran et al., 2009). There are three subclades, and these show perfect allopatry, as follows:

1. Kauai: Waihiawa Stream. Sister to the next two.
2. Kauai: one species from Waineki to Alakai Swamp; one species endemic at Alakai Swamp.
3. Oahu to Hawaii.

The three clades could have differentiated by simple vicariance. The other main group in Hawaiian *Viola*, the "dry clade," is distributed from Oahu to Maui and so overlaps with its sister, although it is allopatric at an ecological scale and lives in drier habitat.

The progression rule has been seen as compatible with the standard models in biology (dispersal/founder speciation) and geology (the hotspot model), and so it has been supported as the basic distribution pattern. Nevertheless, several other patterns are emerging in molecular work and, as noted, founder dispersal would lead to clades on younger islands being deeply nested in grades on older islands, not the simple K (O (M, H)) pattern. This pattern and the non–progression rule patterns may have other origins, and the progression rule may not be a result of dispersal matching plate movement over a hotspot. The progression rule sequence may be caused by a sequence of differentiation in an ancestor that was already widespread in the main islands or in the region the current islands now occupy, with the first division being between northwest (Kauai) and southeast (the other islands). Patterns that do not conform to the progression rule include the following.

KAUAI + MOLOKAI BASAL

In Hawaiian *Metrosideros* (Myrtaceae), one of the two main clades is on Kauai and Molokai, missing Oahu, while the other is on Oahu, Molokai, Maui, and Hawaii (Percy et al., 2008). Although the statistical support for this is not strong, the two clades are largely allopatric, and the only overlap is on Molokai. Percy et al. (2008) did not refer to the basic allopatry but acknowledged that the pattern is an "anomaly," as it does not follow the progression rule. Despite this, they accepted a "stepping-stone pattern of colonization from older to younger islands." Harbaugh, Wagner, Percy et al. (2009) agreed, writing that the study by Percy et al. (2008) showed "a strong biogeographic pattern where ancient dispersal events likely proceeded down the island chain from the older to younger islands." The problem in *Metrosideros*, as Percy et al. (2008) implied in citing the distributional "anomaly," is the strange, non–progression rule allopatry within the Hawaiian Islands. The origin of the pattern is a key question for the biogeography of the genus as a whole, not just the Hawaiian clades. As indicated in Chapter 7 (Fig. 7-8), allopatry occurs throughout *Metrosideros* (found on most high islands in the Pacific) and its relatives.

OAHU BASAL

The *Stenogyne* clade (Lamiaceae) has a phylogeny (O) (Mo) (K, O, M, H) (Lindqvist et al., 2003). The authors suggested that "the most-parsimonious hypothesis" was an origin on Oahu or Maui Nui, "followed by numerous dispersal events" (p. 488).

The isopod genus *Ligia* is represented in the Hawaiian Islands by *L. perkinsi* on K, O, and H, and *L. hawaiense* on all the main islands. In a sample, Taiti et al. (2003) retrieved the phylogeny: (Oahu *perkinsi* (Kauai *perkinsi* (Oahu *hawaiensis* + Kauai *hawaiensis*))). The authors wrote that Oahu is basal, "in contrast to what we might expect."

As cited above, *Banza* katydids (Tettigoniidae) show the phylogeny: Ni (O (K, O, L) (H, M)) (Shapiro et al., 2006), and so in the main islands one of the Oahu clades is basal to a group on all the others.

MOLOKAI BASAL

The isopod *Hawaiioscia* comprises four Hawaiian species with the phylogeny: (Mo (O, K)) (Rivera et al., 2002). (The Maui species was not sampled.) The authors found the phylogeny "unexpected," as it is not congruent with the progression rule.

MAUI BASAL

In the "windswept clade" of the moth *Thyrocapa*, a species from Maui, *T. apatela*, is basal to the rest of the clade, which ranges from Necker to Hawaii (skipping Oahu) (Medeiros and Gillespie, 2011).

HAWAII + MAUI BASAL

The silversword alliance (Asteraceae: Madiinae) comprises *Argyroxiphium*: H and M, sister to *Dubautia*, and *Wilkesia*: on K, O, Mo, L, M, and H. The three genera do not constitute a progression rule sequence, so the absence of *Argyroxiphium* from islands other than Hawaii and Maui is "enigmatic" (Baldwin, 1997: 116) and a "paradox" (Carlquist, 1995). Funk and Wagner (1995a) proposed that the pattern may be due to extinction and suggested that *Argyroxiphium* is "confined to habitats that do not exist [outside Hawaii and Maui]" (p. 403). Baldwin (2009) agreed that the genus is absent from Kauai and Oahu due to loss of high-elevation habitat and cited the "predominant occurrence of most species" of *Argyroxiphium* on Maui and Hawaii at higher elevations than are present on Kauai and Oahu. Nevertheless, two of the five species occur on Maui at lower elevations than the high point of Kauai, 1,598 m (*A. caliginis* down to 1,350 m and *A. grayanum* down to 1,200 m), and Baldwin (1997: 116) noted that the genus is absent from high-elevation bogs on Kauai "that appear similar to bogs occupied by *A. caliginis* and *A. grayanum* on West Maui." Thus the pattern, with Hawaii and Maui basal, cannot be easily "massaged" into a progression rule sequence. Nevertheless, the same pattern is repeated in several groups.

As shown above, *Silene* (Caryophyllaceae) has an east Pacific distribution similar to that of the silversword subtribe Madiinae. The phylogeny of the Hawaiian *Silene* clade is: (H, M) (H, M, L, Mo, O, K) (Eggens et al., 2007), again repeating the silversword pattern.

The six Hawaiian species of *Geranium* form a clade that has its closest relatives in the western United States and Mexico (Pax et al., 1997),

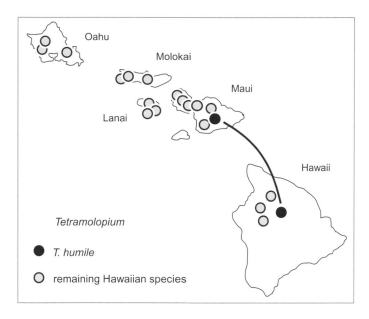

FIGURE 8-2. *Tetramolopium* (Asteraceae) in the Hawaiian Islands, showing the basal species, *T. humile*, and the other Hawaiian species (Lowrey, 1995).

as in *Silene* and the silverswords, and the distribution within the islands is also similar. The phylogeny of the Hawaiian members is: H (*G. cuneatum* subsp. *hypoleucum*) + M (*G. arboreum*) basal to all the others on H, M, K (skipping O) (Kidd, 2005).

Tetramolopium (Asteraceae) has 17 species and subspecies in the Hawaiian Islands; these occur on all the main islands except Kauai (Fig. 8-2). Lowrey (1995) and Lowrey et al. (2001) found the Hawaii–Maui species *T. humile* to be basal to the remaining Hawaiian species. Baldwin (1998: 65) wrote that "The area cladogram for *Tetramolopium* is, in fact, highly unusual among those of Hawaiian organisms in general for showing unequivocal evidence of origin on a younger island group (Maui Nui) followed by dispersal to younger and higher islands." The location of a basal group (a small sister group) does not necessarily indicate a center of origin, and in *Tetramolopium* the distribution of the basal species, *T. humile*, and the others is allopatric (Fig. 8-2).

In Hawaiian *Dicranomyia* (Tipulidae, craneflies), one of the two clades has a phylogeny: (H, M) (H, M, Mo, O, K) (Nitta and O'Grady, 2008).

MORPHOLOGICAL EXAMPLE:

The basal clade in the beetle *Blackburnia* subg. *Blackburnia* (Carabidae) comprises nine species with a phylogeny: (H, M) (H, Maui Nui, O, K) (Liebherr and Short, 2006).

HAWAII BASAL

Clermontia (Campanulaceae) is made up of three components (Givnish, 2003): a basal species on Kauai and Oahu, a basal grade from Oahu to Hawaii, and a large clade with the phylogeny H (H, M (H, M, L (H (H (all islands))))). The base of the phylogeny indicates breaks around Oahu; the large clade indicates a phase of main breaks in or around Hawaii.

The Hawaiian endemic land snail *Succinea caduca* (Succineidae) occurs on all the main Hawaiian Islands, and the primary division in the species is between one clade ranging from Kauai to Maui Nui and one restricted to Hawaii (Holland and Cowie, 2007). *S. caduca* belongs to "clade B" of *Succinea*, which is otherwise restricted to the island of Hawaii, and so neither the species nor "clade B" follows the progression rule sequence. A center of origin on Hawaii, the youngest island, implies a very rapid evolution rate, and Holland and Cowie (2007: 2432) found the pattern "challenging to explain." They suggested that the species might have had a center of origin on one of the older Hawaiian islands and that populations there with basal haplotypes have either gone extinct or were not sampled (cf. *Argyroxiphium*, above). These are ad hoc explanations for a simple pattern that is repeated in many other taxa.

The native bee fauna of Hawaiian Islands consists of 60 species that form a single clade in the cosmopolitan genus *Hylaeus*. A DIVA analysis gave a center of origin in Hawaii (Magnacca and Danforth, 2006). This is the opposite of the progression rule, and the authors found it "surprising." Individual clades of *Hylaeus* in which Hawaii is basal include:

The *Hylaeus inquilina* group: (H) (H, M, L, Mo, O, K).

H. sphecodoides + *H. volatilis*: (H) (M, L, Mo, O).

H. specularis: (H) (Mo, O, K)

Medeiros et al. (2009) found a similar pattern in *Schrankia* moths (Noctuidae), with the "most ancestral" (i.e., basal) forms on Hawaii.

Phylogeny in the damselfly *Megalagrion hawaiiense* shows an "upchain" pattern: (H) (M, Mo, O) (Jordan et al., 2003).

Paxinos et al. (2002) studied the Hawaiian goose *Branta sandvicensis* and its fossil allies using ancient DNA. "Unexpectedly," they found that the extinct giant goose of Hawaii Island, an undescribed, flightless species, is basal to *B. hylobadistes* of Maui (fossil) + *B. sandvicensis* (extant) of Kauai, Maui Nui, and Hawaii.

MORPHOLOGICAL EXAMPLE:

Coprosma foliosa (Rubiaceae) is on M, L, Mo, O, K. *C. menziesii* of H is "closely related" to *C. foliosa* and is "difficult to separate from it" (Wagner et al., 1990: 1128.). *C. cymosa* of H is "closely related" to these two species (p. 1124).

KAHOOLAWE BASAL

Medeiros and Gillespie (2011) sampled 23 of 31 species in the endemic moth *Thyrocopa*. The only group showing a progression rule is the pair: *T. albonubila* + *T. indecora*, with a phylogeny: K (O (M + H)), but the sister of this clade is *T. kanaloa* from Kahoolawe (where the first two species are absent). This is the arrangement in the Bayesian tree. In the parsimony tree, *T. kanaloa* from Kahoolawe is supported as sister to all other *Thyrocopa* species (Necker to Hawaii).

"LEAPFROG" PATTERNS SKIPPING KAUAI, OAHU, OR MAUI

So-called "leapfrog" distribution patterns, in which clades skip central areas in a range, have been described in many places, such as the Andes, New Zealand, and New Guinea, and are also conspicuous in the Hawaiian Islands. They do not conform to a simple progression rule.

DISTRIBUTION SKIPPING KAUAI

Reynoldsia sandwicensis (Araliaceae) is on Nii, O, Mo, L, M, H. It is replaced on Kauai by its sister group, *Munroidendron racemosum*, endemic there (Costello and Motley, 2007). The authors suggested that *R. sandwicensis* may have been on Kauai and has died out there due to loss of suitable habitat, but they did not mention the pattern of simple allopatry there with its sister group or that the same pattern also occurs in another araliad, *Cheirodendron* (next).

MORPHOLOGICAL EXAMPLES:

Cheirodendron trigynum subsp. *trigynum* (Araliaceae) is on Nii, O, Mo, L, M, and H. It is replaced on K by the only other subspecies, subsp. *helleri*.

Cenchrus (Poaceae): Kure, Midway, Laysan, O, Mo, L, M, H.

Portulaca (Portulacaceae): Northwestern Islands and all main islands except K.

Portulaca villosa: Ni, Kaula, and all main islands except Nii and K.

Portulaca lutea: Pacific islands (widespread), Hawaii: Northwestern Islands (except Kure and Pearl and Hermes) plus O, Mo, L, M.

Lipochaeta lobata (Asteraceae): Nii, O, M.

Vigna o-wahuensis (Fabaceae): Nii (extinct), O (extinct), Mo, L, Ka, M, H.

Artemisia australis forma *skottsbergi*: Nii, Mo, L (skipping both K and O).

Cyperus laevigatus (Cyperaceae): Laysan, Ni, O, Mo, M, H.

Eleocharis calva (Cyperaceae): mainland U.S. plus Nii, O, Kahoolawe (possibly not indigenous on Kahoolawe).

Eragrostis paupera (Poaceae): "Pacific equatorial region," plus Kure, Midway, Pearl and Hermes, French Frigate Shoals, and Oahu.

In the beetle genus *Plagithmysus* (Cerambycidae), *P. nihoae* of Nihoa is related to the *usingeri, funebris, immundus, atricolor*, and *kraussi* groups, all widespread in the main islands but absent from Kauai, and so the suggested affinity skips that island (Gressitt, 1978).

In the Hawaiian honeycreepers, the related pair *Telespiza cantans* and *T. ultima* are on Laysan, Nihoa, and, as fossils, O and Mo. Their sister, the fossil *T. persecutrix* fills the gap on K and O (Olson and James, 1991b; James, 2004; Burney et al., 2001).

DISTRIBUTION SKIPPING OAHU

An example was given in *Metrosideros*, above. Some examples of the pattern may be due to anthropogenic habitat loss on Oahu, but Maui has also suffered in this respect, and few clades appear to have skipped that island. In *Metrosideros*, the gap in Oahu in one clade is replaced with the other clade, suggesting vicariance rather than extinction as the reason for the absence.

In Grammitidaceae ferns, *Adenophorus epigaeus* (K) is sister to *A. montanus* (Mo, M, H) (Ranker et al., 2003). Five other species of *Adenophorus* do occur on Oahu, including the endemic *A. oahuensis*. The genus does not follow a progression rule; the basal species is *A. periens*, widespread through the main islands.

In the highly variable *Metrosideros polymorpha* (Myrtaceae), one genetic grouping was only recorded on Kauai and Maui (the "orange group" in Harbaugh, Wagner, Percy et al., 2009).

In *Cyanea* (Lobeliaceae), a phylogeny of a sample (24 species) retrieved a "*hirtella* clade" of Kauai, sister to a "*solanacea* clade" of Maui Nui (Givnish et al., 1995).

In *Dicranomyia* (Tipulidae), the basal clade is on Kauai, Molokai, Maui, and Hawaii, but not Oahu (Nitta and O'Grady, 2008)

In the *Drosophila haleakale* group, O'Grady and Zilversmit (2004) sampled 18 of the 54 species and found that "one pattern that is not observed" is the progression rule. Both the *fungiperda* complex and the *iki* complex show disjunctions between Kauai and Maui/Hawaii, skipping Oahu, where related clades such as the *polita* subgroup have most of their species.

In the moth *Thyrocopa*, the "windswept clade" (five species) has a phylogeny: (Necker + Nihoa) (K, M, La, Ka, H). Alternatively, the species on Kahoolawe may be basal to all the rest (Medeiros, 2009; Medeiros and Gillespie, 2011). In any case, the genus is absent from Oahu.

MORPHOLOGICAL EXAMPLES:

Keysseria (Asteraceae) (3 spp., formerly treated in *Lagenifera*): K, Mo, M.

Brighamia (Campanulaceae) (2 spp.): Nii, K, Mo, L, M.

Remya (Asteraceae) (3 spp.): K, M.

Geranium (Geraniaceae) (6 spp.): K, M, H.

Haplostachys (Lamiaceae) (5 spp.): K, Mo, L, M, H.

Hibiscadelphus (Malvaceae) (6 spp.): K, L, M, H.

Acaena (Rosaceae) (1 sp.): K, M.

Rubus (Rosaceae) (2 spp.): K, Mo, M. H.

Deschampsia (Poaceae) (1 sp.): K, Mo, M, H.

Uncinia (Cyperaceae) (2 spp.): K, Mo, M, H.

The presence of several otherwise southern elements in the list, such as *Keysseria, Acaena, Uncinia*, and, probably, *Remya*, is interesting and may be significant. Perhaps Oahu has been colonized by a flora that never had the South Pacific connections held by the other floras preserved in the Hawaiian Islands.

The following examples are absent from Oahu but have congeneric species there:

Pritchardia remota complex (Arecaceae): Laysan (extinct), Ni, Nii, K, M. Other *Pritchardia* species are on K, O, Mo, L, M, H.

Deparia kaalaana (Athyriaceae): K, M, H.

Deparia marginalis: K, Mo, L, M, H.

Grammitis forbesiana (Grammitidaceae): K, Mo, M.

Huperzia mannii (Lycopodiaceae): K, M, H.

Microsorum spectrum var. *pentadactylum* (Polypodiaceae): K, M. (The typical variety is widespread in the Hawaiian Islands.)

Polypodium pellucidum var. *acuminatum* (Polypodiaceae): K, M. (The typical variety is widespread in the Hawaiian Islands.)

Lipochaeta connata (Asteraceae): Nii, K, M.

Canavalia pubescens (Fabaceae): Nii, K, L, M.

Peperomia cookiana (Piperaceae): K, Mo, M, H.

Lysimachia mauritiana (Primulaceae): Old World, Micronesia, New Caledonia, Nii, K, Mo, M, H.

Psychotria mauiensis (Rubiaceae): K, Mo, L, M, H.

Zanthoxylum hawaiiense (Rutaceae): K, Mo, L, M, H.

Solanum incompletum (Solanaceae): K, Mo, L, M, H.

Carex echinata (Cyperaceae): K, M, H.

Carex montis-eeka: K, Mo, M.

Dichanthelium cynodon (Poaceae): K, Mo, M

Dichanthelium hillebrandianum: K, Mo, M, H.

Dichanthelium isachnoides: K, Mo, M. (The only other *Dichanthelium* in Hawaii is endemic to Oahu.)

Panicum pellitum (Poaceae): Nii, K, L, M, H. Wagner et al. (1990: 1571) commented: "Surprisingly, this species has not been collected on Oahu." Wagner et al. regarded it as closely related to *P. konaense*, next.

Panicum konaense: K, Mo, M, H.

Scaevola procera (Goodeniaceae): K, Mo.

Sicyos cucumerinus (Cucurbitaceae): K, Mo, M, H.

The land snail *Newcombia*: K (fossil), Mo, M (Cowie and Holland, 2008).

In carabid beetles, the *Blackburnia sulcipennis* clade (9 species): K, Mo, M (Liebherr and Short, 2006).

The *Blackburnia bryophila* clade (four species): K, Mo, M (Liebherr and Short, 2006).

The *Blackburnia pauma* clade (three species): K, M (Liebherr, 2006).

The *Blackburnia kavanaughi* clade (three species): K, Mo, M (Liebherr, 2006).

Aeletes molokaiae (Coleoptera: Histeridae): K, Mo (20 other species occur on O) (Yélamos, 1998).

The *Drosophila ochracea* group: K, M, H (Kaneshiro et al., 1995).

DISTRIBUTION SKIPPING LANAI

In many cases, this pattern may be due to extinction on Lanai, as the island is smaller and lower than the other main islands and so most examples are not cited here. Nevertheless, in one case there is a vicariant on the island, and so the pattern may not be due simply to extinction there.

MORPHOLOGICAL EXAMPLE:

Kadua cordata (= *Hedyotis schlechtendahliana*; Rubiaceae) comprises three subspecies (nomenclature follows Terrell et al., 2005): *K. cordata* subsp. *cordata*: K, O, Mo, M; *K. cordata* subsp. *remyi*: L; and *K. cordata* subsp. *waimeae*: K.

DISTRIBUTION SKIPPING MAUI

Drosophila setiger subgroup (in the "Modified Mouthparts group") (4 spp.): K, O, Mo, L, H (Magnacca and O'Grady, 2009).

MORPHOLOGICAL EXAMPLES:

Kokia (Malvaceae) (4 spp.): K, O, Mo, H.

DISTRIBUTION SKIPPING MAUI NUI (ON OAHU AND HAWAII, OR KAUAI, OAHU, AND HAWAII)

The isopod *Ligia perkinsi*: K, O, and H (Taiti et al., 2003).

An eastern Oahu – western Hawaii clade in the shrimp *Halocaridina* (Atyidae) (Craft et al., 2008).

The *Drosophila ateledrophila* species group: O, H (Magnacca and O'Grady, 2008).

The *Drosophila kahania* subgroup of the *nudidrosophila* group: O, H (Magnacca and O'Grady, 2008). Both these groups have related clades on Maui Nui.

The *Drosophila grimshawi* complex comprises two clades, one on K, O, and H, skipping Maui Nui, the other endemic to Maui Nui (Cowie and Holland, 2008).

A passerine bird, the Hawaiian honeycreeper *Loxioides* (Fringillidae: Drepanidini) has a similar distribution: K (fossil), O (fossil), H (extant) (Olson and James, 1991b, Burney et al., 2001).

The passerine *Chasiempis* (Monarchidae) was discussed in Chapter 6 as part of a widespread central Pacific endemic clade (Fig. 6-3). In the Hawaiian Islands it is known from Kauai, Oahu, and Hawaii, with living and fossil records on each, but it has never been recorded on the central islands of Maui, Molokai, and Lanai (Olson and James, 1991b). The absence of the genus there (not shown in Fig. 6-3) has been described as "perplexing" (Olson and James, 1982), an "intriguing problem" (Berger, 1972), and one that is "peculiar given the ordered geologic history of the Hawaiian Islands" (VanderWerf, 2007: 326). The lack of any records of *Chasiempis* in the extensive fossil collections from Maui Nui (Olson and James, 1991b; H. James, pers. comm.) may indicate that the genus never occurred there, as VanderWerf (2007) argued. Although *Chasiempis* is quite sedentary, with natal dispersal distance usually less than a kilometer and breeding dispersal less than 400 m (VanderWerf et al., 2010), the authors suggested that the bird dispersed directly from Oahu to Hawaii and "inadvertently bypassed the Maui Nui stepping stone." Yet the molecular study by VanderWerf et al. (2010) included multiple samples from each of the three islands and gave a clear-cut phylogeny: K (O + H); the Hawaii species was sister to the Oahu species, not nested in it. The study indicated substantial morphological and molecular variation among and within islands but, as with the progression rule studies cited above, the three island species were reciprocally monophyletic. VanderWerf et al. (2010) suggested that the phylogeny reveals the Kauai species as ancestral, but in fact the phylogeny shows that it is monophyletic and the other two species are sister to it, not nested in it. Thus, the assumption of its ancestral status cannot be justified and the idea of jump dispersal inadvertently missing Maui Nui by chance seems implausible in a clade that divides up the central and west Pacific so neatly (Fig. 6-3). Perhaps it is absent from Maui Nui *because* its clade is in the central and southern parts of the Pacific rather than being an eastern group. It would then resemble the southern plant groups listed above as skipping Oahu.

With respect to dating, VanderWerf (2007) suggested that *Chasiempis* only occupied Oahu after the island was separated from Molokai; if it was on Oahu before this, the genus "presumably would have occupied both islands." This assumes that a taxon will always occur throughout the landmass it occupies, but this is not valid as a generalization even at a local level.

VanderWerf et al. (2010) found that using the age of Kauai to calibrate the phylogeny of *Chasiempis* was "not appropriate" as it gave a divergence time for the Hawaii species substantially older than the age of the island, contradicting assumption 7 (Chapter 7): "A species cannot be older than its island." An alternative published rate was used instead as this gave an "appropriate" age for the Hawaii species (cf. the study of Percy et al., 2008, on *Metrosideros*).

MORPHOLOGICAL EXAMPLES:

Ophioglossum nudicaule (Ophioglossaceae): K, O, H.

The histerid beetle *Aeletes eutretus*: K, O, H (Yélamos, 1998). (There are 14 species on the intervening Maui Nui islands.)

The *Drosophila assita* group: O, H (Kaneshiro et al., 1995).

The xylorictid moth *Thyrocopa usitata*: K, O, H (Medeiros, 2009). (The species is "very similar" to *T. subahenea* of M and Mo.)

DISTRIBUTION SKIPPING OAHU AND MAUI NUI (ON KAUAI AND HAWAII)

In *Havaika* jumping spiders, a weakly supported clade is on K and H, and is sister to one on Maui Nui (Arnedo and Gillespie, 2006).

A clade in the spider *Argyrodes* (Theridiidae) is on K and H, with relatives on O, Mo, and M (Gillespie et al., 1998: Fig. 1).

Drosophila ceratostoma subgroup (Modified Mouthparts group) (four spp.): K and H only (Magnacca and O'Grady, 2009).

The *Drosophila planitibia* group comprises a basal clade, the *picticornis* subgroup, on Kauai and Hawaii only, and a diverse clade from Oahu to Hawaii (not on Kauai) (Bonacum et al., 2005; O'Grady et al., 2009). The authors attributed the absence of the *picticornis* subgroup on all the central islands to extinction there, but this does not explain the allopatry of the group and its sister outside a possible area of overlap on Hawaii.

In *Hylaeus* bees, the clades *H. rugulosus* (H) + *H. solaris* (K), and *H. hula* (H) + *H. kokeensis* (K) are weakly supported (Magnacca and Danforth, 2006).

MORPHOLOGICAL EXAMPLES:

Melanthera faurei (Asteraceae) of K is "closely related" to *M. subcordata* of H (Wagner et al., 1990). (These species were previously treated in *Lipochaeta*; see Wagner and Robinson, 2001, for the new taxonomy.)

Schoenoplectiella juncoides (Cyperaceae): K, H.

Achyranthes mutica (Amaranthaceae): K, H.

Kadua cookiana (Rubiaceae): K, H. (The species was formerly treated in *Hedyotis*.) Wagner et al. (1990) wrote that it is "presumably relictual, and was probably formerly much more widely spread" but did not cite any evidence for this. Other *Kadua* species are endemic on Oahu, Maui, etc.

In *Kokia* (Malvaceae), *K. kauaiensis* of K appears as sister to *K. drynarioides* of H in Funk and Wagner (1995b) (the other species are on Oahu and Molokai).

In insects, *Sarona* (Heteroptera: Miridae) is endemic to the Hawaiian Islands, with 40 species. One of the five main clades is endemic to K (*S. mokihana*) and H (*S. alani*). Many other species occur on the intervening islands (Asquith, 1995).

A group in the beetle *Plagithmysus* (Cerambycidae) is recorded only from K (*P. permundus*, *P. polystictus*) and H (*P. swezeyi*) (Gressitt, 1978).

PARALLEL ARC DISTRIBUTIONS THAT DIVIDE THE HAWAIIAN CHAIN ALONG ITS AXIS

Distributions in the Hawaiian Islands are often listed in terms of current geographic areas—the islands—and so clades that are allopatric *within* individual islands, such as those listed next, will appear to be sympatric. One pattern that may be quite common, but will not show up in distribution records listed only by islands, involves taxa arranged in parallel arcs along the axis of the archipelago. While these distributions could be related to rainfall—it is wetter in the northeast than in the southwest—there may be more to it than this. The parallel arc distribution patterns often involve leapfrog distributions, and the two phenomena may be related. The skipping pattern is not accounted for by rainfall, and so perhaps both it and the parallel arc pattern are associated with other factors. These may include the former islands to the north (Musicians seamounts), south (Line Islands), east (unnamed seamounts), and west (Wake Island, etc.) of Hawaii that may have supplied the archipelago with its endemic taxa. Other possible factors include the two trends in Hawaiian volcanism noted above (Fig. 8-1) and possible extension between the arcs.

In *Tetramolopium* (Asteraceae) (Fig. 8-2), the two Hawaiian clades are *T. humile* on Hawaii and Maui, and the other species, which are widespread from Hawaii to Oahu, but on Hawaii and Maui all lie west of *T. humile* (Lowrey, 1995; Lowrey et al., 2001).

On the island of Hawaii, *Drosophila silvestris* has two clades, eastern and western (DeSalle and Templeton, 1992), that correspond closely to the two volcanic arcs.

Kadua fosbergi (Rubiaceae) of Oahu and Lanai, skipping Molokai, is sister to *K. axillaris* on Molokai, Maui, and Hawaii (Terrell et al., 2005; Kårehed et al., 2008). This distribution illustrates how a skipping pattern may be associated with an underlying pattern of parallel arcs.

In Hawaiian *Mecaphesa* spiders, the phylogeny showed "contradictory tendencies" (Garb and Gillespie, 2009); instead of following a progression rule sequence, there are many widespread species and dramatic ecological diversity. The most widespread group, the *insulana–cavata* clade, ranges from Necker to Hawaii, but is not on Kauai, Lanai, or Maui (Fig. 8-3). It is sister to *discreta* + *juncta*, on Kauai and Molokai. The other species occupy all the main islands, from Kauai to Hawaii. The break between Nihoa and Kauai is a common one, but the *Mecaphesa* pattern ssuggests that it may not be a simple transverse break across the line of the chain.

As in *Mecaphesa*, the fan-palm genus *Pritchardia* has a Northwestern Islands clade (the *remota* complex; Wagner et al., 1990) that is also present on the main islands, where it shows a skipping pattern (present on western Kauai and western Maui) (Fig. 8-4). The other Hawaiian species occupy all the main islands.

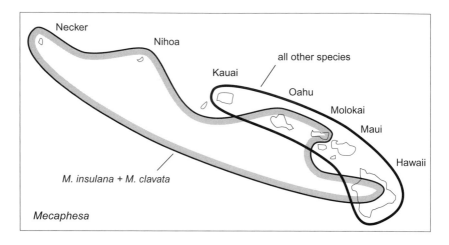

FIGURE 8-3. Distribution of *Mecaphesa* (Thomisidae) spiders in the Hawaiian Islands (Garb and Gillespie, 2009). Gray: the *insulana* + *clavata* clade. Thick line: all other species, including two, distributed from Kauai to Hawaii.

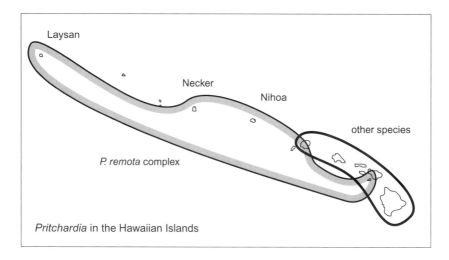

FIGURE 8-4. Distibution of *Pritchardia* (Arecaceae) in the Hawaiian Islands (Wagner et al., 1990), based on morphology. Gray: the *Pritchardia remota* complex (Arecaceae): Laysan (extinct by 1896; no specimen, photo and fossil pollen only; Athens et al., 2007), Necker, and Nihoa (*P. remota*), Niihau (*P. aylmer-robinsonii*), Kauai (*P. napali*), Maui (Iao valley; *P. glabrata*). Thick line: all other Hawaiian species.

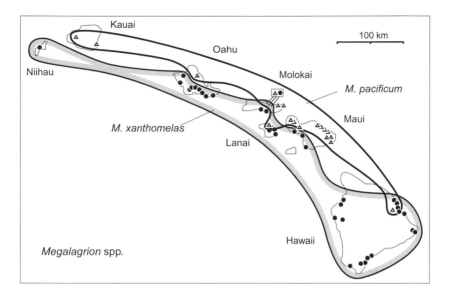

FIGURE 8-5. Distribution of two sister species of *Megalagrion* (Odonata), *M. pacificum* and *M. xanthomelas* (Jordan et al., 2005).

The damselflies *Megalagrion xanthomelas* and *M. pacificum* are widespread on the Hawaiian Islands and are sister species (Fig. 8-5). Jordan et al. (2005) suggested that the distribution is the result of dispersal, writing that "our results point to the importance of Pleistocene land bridges [between the islands]." In fact, the pattern shows a neatly allopatric division between the southwestern *M. xanthomelas* (present at lower elevation, lower rainfall sites; abundant) and the northeastern *M. pacificum* (present at higher elevation, higher rainfall sites; rare), although the authors did not mention the allopatry. There is minor geographic overlap in central northern Molokai, interpreted here as secondary range expansion, and this is the only dispersal required. In their molecular clock study Jordan et al. (2003) inferred that *Megalagrion* arrived in Hawaii at 10 Ma, but the clock was calibrated using island ages.

In *Lipochaeta* (Asteraceae), two of the Hawaiian Islands species have skipping patterns, cited above, and the group as a whole can be resolved into a series of parallel arcs with local secondary overlap (Fig. 8-6; within-island distributions generalized). Five of the six species occur on Maui, but rather than being a center of origin, the island may represent a zone of juxtaposition.

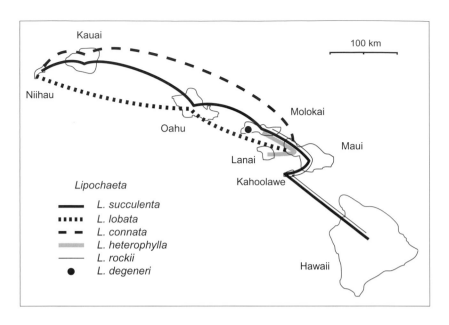

FIGURE 8-6. Distribution of *Lipochaeta* (Asteraceae) in the Hawaiian Islands (Wagner et al., 1990). (Distribution within each island is diagrammatic only.)

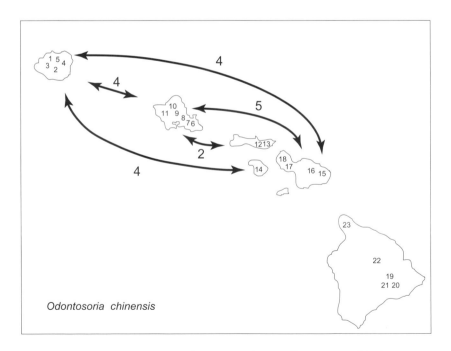

FIGURE 8-7. Distribution of the fern *Odontosoria chinensis* (Lindsaeaceae) in the Hawaiian Islands, showing affinities among populations (Ranker et al., 2000). Small numbers = sampled populations, lines with arrows = "inferred interisland dispersal patterns," numbers by arrows = "inferred minimum number of dispersal events."

In Hawaiian populations of the fern *Odontosoria chinensis*, patterns of allozyme differentiation indicated repeated connections skipping Oahu, Molokai, and Oahu plus Molokai (Fig. 8-7; Ranker et al., 2000). The authors interpreted the patterns as the result of interisland dispersal; another possibility is that the parallel arcs were caused by vicariance in a widespread ancestor.

These parallel arc patterns and many of the "leapfrog" patterns in the current Hawaiian groups may reflect differentiation between old clades to the southwest (in the central Pacific) and northeast (in the eastern Pacific), a division that lies athwart the progression rule sequence. Other non–progression rule patterns may also be related to this history.

BIOGEOGRAPHIC PATTERNS WITHIN THE HAWAIIAN ISLANDS IN MARINE TAXA

Many coastal marine taxa are restricted to areas of shallow water and can provide valuable information on the biogeography and history of islands. Four examples are cited here to indicate the possibility of integrating marine and terrestrial distribution patterns (cf. Heads, 2005b).

The isopod *Ligia hawaiensis* is widespread in the littoral zone of all the main Hawaiian Islands. Taiti et al. (2003) found deep genetic divergences within the species, "well above the mitochondrial divergence reported between species, or even genera, of other Hawaiian invertebrates." The variation shows "Remarkable geographic structure" and this "seems to contradict the hypothesis that littoral species have a great facility for dispersal as commonly supposed" (p. 100).

Bird et al. (2007) agreed, writing that "The marine environment offers few obvious barriers to dispersal for broadcast-spawning species, yet population genetic structure can occur on a scale much smaller than the theoretical limits of larval dispersal. . . . Population partitioning in the three species of Hawaiian *Cellana* [limpets] was observed at finer spatial scales (< 200 km) than was expected for broadcast-spawning invertebrates with a pelagic larval phase" (Bird et al., 2007: 3173, 3183). For example, *C. talcosa* showed strong population structure with a major restriction between Kauai and the other main Hawaiian islands, a pattern seen in many terrestrial taxa.

In near-shore waters around the Hawaiian Islands, a number of cetacean species show evidence that populations are associated with particular islands. Baird et al. (2009) photographed 336 individuals of the dolphin *Tursiops truncatus* and found "Their generally shallow-water distribution, and numerous within-year and between-year resightings within island areas suggest that individuals are resident to the islands, rather than part of an offshore population moving through the area" (p. 251). Comparisons of identifications obtained from Niihau/Kauai, Oahu, Maui Nui, and Hawaii showed no evidence of movements among these four island groups, although movements from Kauai to Niihau and within the Maui Nui group were documented. In California, *T. truncatus* individuals have been documented moving 670 km, and an individual in the western Atlantic moved over 2,000 km. Baird et al. (2009: 270) wrote: "Given such known dispersal abilities, the lack of movements among areas documented in our study is surprising." There may be as many as four discrete populations.

Andrews et al. (2010) sampled widely from the Hawaiian populations of spinner dolphin, *Stenella longirostris*, and found that with few exceptions, dolphins at every island were significantly different from dolphins at every other island. As exceptions, adjacent Midway and Kure populations were not distinct for all *F*-statistics, and French Frigate Shoals, Niihau, Kauai, and Oahu populations were not distinct from each other for most *F*-statistics.

DISTRIBUTION OF HAWAIIAN CLADES OUTSIDE THE ARCHIPELAGO AND WITHIN THE ARCHIPELAGO

Due to sampling limitations, few molecular studies permit the integration of large-scale distributions of clades outside Hawaii with their small-scale patterns within the chain, and so those that do are of special interest. The examples discussed above suggest a general pattern with southern and western Pacific groups in the Hawaiian Islands on Kauai or Kauai plus leapfrog disjunctions, but not on Oahu, Maui, or both, and eastern Pacific groups in the Hawaiian Islands on Hawaii, Maui, and Oahu.

The silverswords and their relatives are in temperate South America, California, and the Hawaiian Islands, where the basal clade (*Argyroxiphium*) is in Hawaii and Maui (Baldwin, 1997). A clade in *Silene* is in temperate South America, the U.S., and the Hawaiian Islands, where the basal clade is also in Hawaii and Maui (Eggens et al., 2007). A clade in *Geranium* is in the Americas and in the Hawaiian Islands, where the basal clade is again in Hawaii and Maui (Pax et al., 1997; Kidd, 2005).

A red-flowered clade in *Santalum* (Santalaceae) (Fig. 8-8) comprises one group on the Bonin Islands (*S. boninense*), southeastern Polynesia (Cook, Society, Austral, and Marquesas Islands; *S. insulare*), and Kauai (*S. freycinetianum* var. *pyrularium*), and others on Oahu, Molokai,

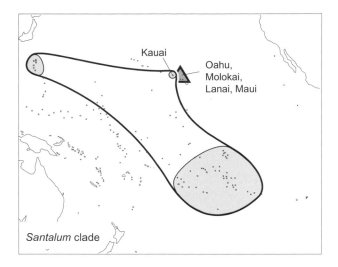

FIGURE 8-8. Distribution of a clade in the sandalwood genus, *Santalum* (Santalaceae) and its two subclades (Harbaugh and Baldwin, 2007). Gray = *S. insulare*, etc. Triangle = *S. haleakalae*, etc.

Lanai, and Maui (*S. freycinetianum* s.str., *S. haleakalae*) (Harbaugh and Baldwin, 2007; Harbaugh, Oppenheimer et al., 2010).

In the cosmopolitan land snail genus *Succinea* there are two clades in the Hawaiian Islands. One occurs elsewhere only in Tahiti, the other only in Samoa (Cowie and Holland, 2008; Fig. 8-9). If an ancestor that was already widespread through the Samoa–Tahiti–Hawaiian Islands region and other areas underwent a major phase of differentiation with the breaks centered on the Hawaiian Islands area, species from Samoa and from Tahiti might each end up having their closest affinities with different groups in the Hawaiian Islands. In vicariance analysis, the question is not: How did the species disperse to Samoa or Tahiti? (it is assumed they evolved there), but: How and why did populations in the Samoa–Hawaii sector differentiate from those in the Tahiti–Hawaii sector? If intraplate volcanism is caused by flexing plates and propagating fissures, there are many possible scenarios. As with subgeneric groups in *Metrosideros* and others, the break may have involved the volcanism and subsequent subsidence of the Line Islands.

In the land snail *Succinea*, the Hawaiian Islands–Samoa clade has a basal group in Kauai and Oahu and follows "the" progression rule. In contrast, the Hawaiian Islands–Tahiti clade has its basal group in Hawaii, the youngest island, and follows another progression rule

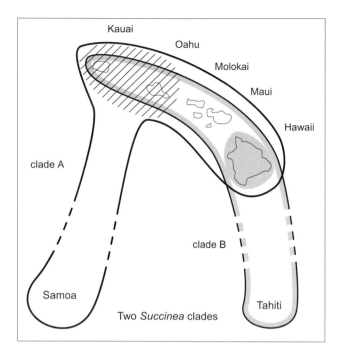

FIGURE 8-9. Distribution of the land snail genus *Succinea*
(Succineidae) in the Hawaiian Islands (Cowie and Holland, 2008).
Oblique lines = basal clade in clade A. Gray = basal clade in clade B.

(Fig. 8-9). The different senses of the phylogenetic sequence could be
due to chance dispersal, or the pattern might indicate that waves of differentiation proceeded through two widespread ancestors in different
directions (cf. the northwest–southeast polarity of distributions in the
chain cited above). This could be related to different genomic architecture in the two ancestors, different times of the waves, or different prior
distributions of the two clades. For example, the Hawaiian components
of the two clades may have been allopatric within the Hawaiian Islands
region on two parallel arcs, and subsequently overlapped on the present islands. This possibility could be investigated with detailed study of
local distribution and phylogeny.

In another example, the bird *Chasiempis* of Kauai, Oahu, and Hawaii,
skipping Maui Nui, has its relatives in the south Pacific. As emphasized
in Chapter 6, birds are often remarkably sessile, despite the fact that
they can fly. This means their distributions can represent inherited information and be passed on more or less intact for millions of years. In
the same way, even birds with much greater mobility than passerines,

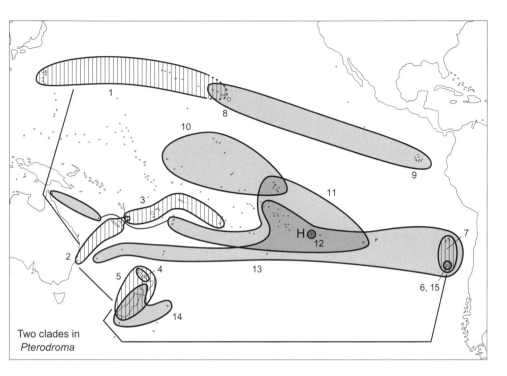

Two clades in
Pterodroma

FIGURE 8-10. Distributions of the petrel *Pterodroma* subg. *Cookilaria* (incl. *P. hypoleuca*) (1–7) and subg. *Hallstroma* (incl. *P. inexpectata*) (8–15). Breeding localities only are shown except for *P. hypoleuca* (phylogeny from Kennedy and Page, 2002; distributions from Brooke, 2004). H = Henderson Island (Pitcairn group). 1 = *P. hypoleuca* (fossil records as dotted line), 2 = *P. leucoptera*, 3 = *P. brevipes*, 4 = *P. pycrofti*, 5 = *P. cooki*, 6 = *P. longirostris*, 7 = *P. defilippiana*, 8 = *P. sandwichensis*, 9 = *P. phaeopygia*, 10 = *P. alba*, 11 = *P. heraldica*, 12 = *P. atrata*, 13 = *P. neglecta*, 14 = *P. inexpectata*, 15 = *P. externa*.

the seabirds, can have distributions that show intricate, detailed patterns of allopatry, disjunction, and overlap. As an island eventually crumbles to become an atoll or rock stack, the last surviving terrestrial taxa will be groups such as lichens, flies, and seabirds, and so these have special value in establishing biogeographic connections.

Members of the genus *Pterodroma* (Procellariidae), the gadfly petrels, are widespread in the Atlantic, Indian, Pacific, and Southern Oceans (Brooke, 2004). The birds are strong fliers and range for thousands of miles when feeding. Nevertheless, they show extreme natal philopatry, always returning to breed at the same, localized sites. Two species breed on the Hawaiian Islands: *Pterodroma sandwichensis*, related to birds of the Galapagos, and *P. hypoleuca*, which has other breeding populations on the Bonin Islands south of Japan (Fig. 8-10).

The "dark-rumped petrels," formerly treated as *P. phaeopygia* s.lat., comprise *P. phaeopygia* s.str. of the Galapagos and *P. sandwichensis* on the main Hawaiian Islands from Hawaii to Niihau (living and fossil), but with no records from the Northwestern Islands (Browne et al., 1997). In contrast, *P. hypoleuca* is on the Bonin Islands (Ogasawara), the Volcano Islands (= Kazan-retto, south of the Bonin group) and the Hawaiian Islands, where it is extant on the Northwestern Islands and fossil on Kauai, Oahu, and Molokai. There are no records, either living or fossil, from Hawaii (Olson and James, 1982).

The Galapagos–Hawaiian archipelago connection in the dark-rumped species would usually be interpreted as the result of chance, long-distance dispersal as these seabirds are such strong flyers. While the dark-rumped species favor tropical island ecology, this does not explain why they breed on these particular tropical islands and not countless others which they could reach easily. Nor does it explain why the dark-rumped pair vicariates to the north of the South Pacific species, to the east of *P. hypoleuca*, and to the west of the Caribbean species. Finally, founder dispersal does not account for the different distributions within the Hawaiian Islands, with *P. sandwichensis* absent from the Northwestern Islands and *P. hypoleuca* absent from Hawaii. These local differences may seem minor, but they reflect the regional vicariance of the subgenera that the two species belong to.

Mitochondrial DNA trees (Kennedy and Page, 2002) indicate that *Pterodroma hypoleuca* of the Bonin and Hawaiian Islands is a member of subgenus *Cookilaria*. This group occurs mainly in the western Pacific and also has two outlying species on Juan Fernández and the Desventuradas Islands off Chile (Fig. 8-10). (For the center of endemism: Juan Fernández + Desventuradas, see, for example, the clade *Dendroseris* + *Thamnoseris* in the Asteraceae; Baldwin, pers. comm., in Kim et al., 2007, and 22 species of coastal fishes; Dyer and Westneat, 2010.) The affinity in *Cookilaria* between west Pacific and far east Pacific may have involved former Antarctic connections (as suggested by the connecting lines in Fig. 8-10). In contrast, the dark-rumped petrels of the Hawaiian and Galapagos Islands (*P. sandwichensis* and *P. phaeopygia*) belong to subgenus *Hallstroma*. This ranges widely in the Pacific but is concentrated in the central region (there are three species breeding on Henderson Island in the Pitcairn group), where subg. *Cookilaria* is absent (Fig. 8-10). The mitochondrial data suggest that subgenera *Cookilaria* and *Hallstroma* are sister clades in *Pterodroma*, although as yet there is little statistical support for this, and seabird phylogeny remains controversial (cf. Rheindt and Austin, 2005). The point stressed here is that the difference in species distribution within the Hawaiian Islands is a function of the different overall distributions of the broader clades.

MORPHOLOGICAL EXAMPLES:

Hesperocnide (Urticaceae): California and Hawaiian Islands: Hawaii.

Eleocharis calva (Cyperaceae): mainland U.S. and Hawaiian Islands: Nii, O, Kahoolawe (possibly not indigenous on Kahoolawe).

Lepechinia (Lamiaceae): western Americas from California to central Chile (41°S, near Valdivia) (http://data.gbif.org; Moreira-Muñoz, 2007) and in the Hawaiian Islands: Maui only.

Portulaca lutea (Portulacaceae): Micronesia to New Caledonia, Pitcairn, and the Hawaiian Islands: Northwestern Islands (except Kure and Pearl and Hermes) plus O, Mo, L, M. The species is related to *P. molokiniensis*: Kahoolawe and Molokini (off Kahoolawe).

Keysseria (Asteraceae): Borneo, Sulawesi, New Guinea, Hawaiian Islands: K, Mo, M (skipping O).

Gahnia vitiensis (Cyperaceae): Fiji and Hawaiian Islands: Kauai.

Oreobolus furcatus (Cyperaceae): Society Islands (Tahiti) and Hawaiian Islands (Kauai to Maui; not Hawaii) (Meyer and Salvat, 2009).

Pisonia wagneriana (Nyctaginaceae): Hawaiian Islands (Kauai only); "seems to be more similar to some Society Island members of this affinity [the widespread Indo-Pacific *P. umbellifera*] than to the variable Hawaiian populations" (all main islands) (Wagner et al., 1990: 988).

Pseudoscorpions include endemics on Hawaii and even on atolls such as Bikini in the Marshall Islands (*Garypus ornatus*; vander Velde, 2009). Pseudoscorpion fossils were known back to 45 Ma, and in the standard model this would have precluded a Cretacous origin for Hawaiian members. But in 1989 a fossil dated as 380 Ma was discovered that is closest to ("perhaps even a member of") the extant superfamily Chthonioidea (Schawaller et al., 1991). Three chthonioid genera are represented on the Hawaiian Islands (Muchmore, 2000):

- *Lechytia*: genus cosmopolitan. There is one Hawaiian species, *L. sakagamii*, and this has a typical northwestern Pacific distribution: Marcus Island, western Caroline Islands (Ulithi), Marshall Islands (Enewetak and Taka), and the Hawaiian Islands: Midway, Kauai, and Oahu.

- *Tyrannochthonius*: genus pantropical. There are four species on the Hawaiian Islands, on Kauai, Oahu, and Maui. Muchmore (2000: 130) pointed out, "It is especially interesting to note that none has yet been found on Hawaii, where three species of the related genus *Volcanochthonius* occur."

- *Volcanochthonius*: genus endemic to Hawaii Island, with three species. Muchmore (2000: 133) noted, "it is remarkable that so much change has been accomplished in the short span of time available on Hawaii Island, less than one million years."

NON-PROGRESSION RULE PATTERNS AND THEIR SIGNIFICANCE

Molecular studies on Hawaiian groups are turning up more and more examples of phylogenies that do not conform to the progression rule sequence. The rule describes an important pattern, but it is only one of several. Different mechanisms, such as back-colonization, dispersal "passing over" intermediate islands, extinction, and stochastic processes, have all been invoked to account for these distributions, but only in an ad hoc way.

In *Dicranomyia* (Tipulidae), Nitta and O'Grady (2008: 1188) found that "patterns do not strictly adhere to the progression rule or any other simple pattern of dispersal observed in a number of endemic Hawaiian plant and insect lineages." They were unable to resolve the "ancestral island" for the majority of nodes in their phylogeny; in parsimony analyses, the basal node is unresolved between the oldest island, Kauai, and the youngest, Hawaii. The authors attributed this lack of a progression rule pattern to the (inferred) ease with which crane flies, though large, fragile insects, are blown from island to island. The phylogeny shows three clades in Hawaiian *Dicranomyia*. The basal one is on Kauai, Molokai, Maui, and Hawaii, but not Oahu—a standard pattern discussed above. The second clade is on Oahu (with the basal species) and all the other islands—also a standard pattern. The third clade is on Hawaii and Maui, and is basal to clade 2, giving a phylogeny: (H + M) (all islands), another standard pattern. Although none of the three clades follows a progression rule, each has a standard distribution pattern shared with many groups with different means of ecology and dispersal. There is a strong polarity in the overall distribution, with clade 1 having its basal member on Kauai, and clade 2 + 3 having its basal clade on Hawaii and Maui. This polarity between the two ends of the chain is itself a standard pattern (cf. *Succinea*; Fig. 8-9).

In *Hylaeus* bees, there is no progression rule, and Magnacca and Danforth (2006) reached the "surprising" conclusion that the ancestral area was on Hawaii, the youngest island. Several groups cited above have their basal clade on Hawaii. The authors explained "the almost complete lack of any pattern in terms of species distribution" (p. 405) by inferring a high dispersal ability of *Hylaeus* bees, yet as Cowie and Holland (2008: 3371) pointed out for Hawaiian taxa, "in most cases, the level of vagility is assumed based on the distribution and phylogeographic structure, introducing a certain circularity."

Non–progression rule patterns also occur in other archipelagos with volcanic islands arranged in linear age sequences. In the blackfly genus *Simulium* (Simuliidae) on the Society Islands, "all the species appear to have arisen on the youngest island, Tahiti," with back-dispersal to older islands (Moorea, etc.) in the northwest (Gillespie et al., 2008)—the reverse of the sequence seen in the Hawaiian progression rule.

The examples of non–progression rule distributions cited above may indicate general patterns, although these are complicated by anthropogenic extinctions. Patterns such as the two progression rule sequences in the Hawaiian Islands (with basal groups in Kauai or in Hawaii), the "skipping" of islands, the parallel arcs patterns, and the involvement of these with broader Pacific connections are not predicted in the simple dispersal model, and an alternative is needed to the single hypothesis available at present.

"Leapfrog" patterns are important in taxa of New Guinea, New Caledonia, and New Zealand, and these can be explained by large-scale strike-slip movement on faults (Heads, 2008). In the Hawaiian Islands, leapfrog patterns and parallel arc distributions may instead be due to tectonic extension normal to the chain direction in a propagating lithospheric crack model (Stuart et al., 2007) or to colonization of the islands from former high islands in the south (such as the Line Islands), the north (such as the Musicians seamounts), and other parts of the broader Hawaiian region.

THE EXTRAORDINARY HAWAIIAN BIOTA

Many Hawaiian taxa do show high levels of endemism and an unusual range of morphological diversity for such small islands. Adopting the standard model of Hawaiian biogeography has rendered these aspects of the biota even more surprising. It has also led to the idea that Hawaiian taxa have extraordinary means of dispersal and exceptional rates of evolution, but these are inferences rather than observations.

Extraordinary Endemism

Many taxa in Hawaii show extraordinary levels of endemism (Gillespie et al., 1998).

The current geographical isolation of Hawaii may have led to the endemism there not because of long-distance dispersal/founder speciation, but because of the lack of it. The extinction of populations on all other former high islands in the north-central Pacific region and the consequent isolation has meant that Hawaii has not received the steady stream of widespread, weedy immigrants over millions of years that many areas have. There has not been the fraying around the margins and gradual interdigitation that normally take place between adjacent centers of endemism. For example, if all of North America was flooded except for one central area the size of the Hawaiian Islands, and this was preserved and isolated from other continents, after 70 million years it would probably have a very high level of endemism, assuming that there was little or no colonization of the community from outside areas.

The Hawaiian biota is often described as "disharmonic" or "unbalanced" (Gressitt, 1978; Coyne, 2009), as some groups show high diversity while others are absent. Nevertheless, all continent-sized regions (including the north-central Pacific) have biotas that are "unbalanced" with respect to each other, and their absences and presences are strange or even astonishing to a visitor from another region. None can be regarded as "normal" or "complete."

Extraordinary Morphological Diversity

Many Hawaiian groups do show great morphological diversity or structures that appear to be anomalous and bizarre in their current, reduced geography. Only a few examples are cited here.

In plants, the Hawaiian lobeliads comprise 126 species of trees, shrubs, succulents, epiphytes, vine-like species, and rosettes (Givnish et al., 2009) an "extraordinary" range of morphological diversity (Gillespie, 2009).

Most plants have stomata either on the lower surface of the leaf or the lower and upper surfaces. The only plants with stomata restricted to the upper leaf surface were thought to be various aquatic plants, grasses, and some high-elevation herbs. But Sporck and Sack (2008) reported that Hawaiian members of the cosmopolitan *Chamaesyce* (Euphorbiaceae) have stomata on lower, lower and upper, or only upper surfaces, diversity unknown in any other lineage.

The endemic fern genus *Diellia* shows "remarkable morphological variation" (Schneider et al., 2005). The authors suggested it was a "palaeo-endemic," "although it is equally possible that the lineage evolved its distinctive morphology before dispersing to Hawaii, with apparent endemism resulting from extinction of extra-Hawaiian populations." This proposal is accepted here for the biota in general.

Many Hawaiian plants are woody but belong to clades that are otherwise herbaceous or mainly herbaceous. Examples include *Viola* (Violaceae), *Chamaesyce* (Euphorbiaceae), *Schiedea* (Caryophyllaceae), *Silene* (Caryophyllaceae), *Chenopodium* (Amaranthaceae), *Geranium* (Geraniaceae), the silversword alliance (Asteraceae), and *Bidens* (Asteraceae).

Hawaiian *Drosophila* is well known for its high species numbers (see below). In addition, the species show "bizarre" modifications in their legs, mouthparts, heads, and wings, and also their courtship behavior (Bonacum et al., 2005).

In birds, the Hawaiian honeycreepers (Fringillidae: Drepanidini) may show the "most extreme radiation within a single subfamily or family of birds, with forms that have bills of almost any type known from songbirds" (Fleischer, 2009).

Groups such as the fossil moa-nalo ducks and a giant subfossil goose show impressive gigantism. The goose, not yet named, is a member of the genus *Branta*, known with extant species through the north temperate zone and one, *B. sandvicensis*, on the Hawaiian Islands. The giant fossil species is only known from Hawaii Island. It is highly modified and flightless, suggesting that "extreme morphological change occurred on the island of Hawaii" (Paxinos et al., 2002: 1404). The authors used an oldest-fossil calibration (fossils from 4–5 Ma) and pointed out that this gives minimum ages only. They calculated that "the estimated time of divergence of *B. sandvicensis* and the giant Hawaii goose is 0.566 Myr," and this is consistent with an origin of the Hawaii species on that island, which is dated at ~0.5 Ma. As a minimum date, the fossil-based calculation is also consistent with a much older origin and there is no reason to assume that the species on Hawaii evolved there, other than the fact that it occurs there now.

Iwaniuk et al. (2009) described a fossil duck from Hawaii, *Talpanas*, that has extraordinary cranial morphology. The optic foramen, which transmits the optic nerve, is remarkably small, whereas the maxillo-mandibular foramen, the exit point of the trigeminal nerve (mainly supplying sensory receptors in the face), is grossly enlarged, being an order of magnitude larger in cross-sectional area than that of extant Anseriformes. The bird had very small eyes and probably poor vision, and combined with enhanced somatosensory abilities, this indicates "a dramatic adaptation for tactile foraging that appears to have been better developed than in any other known bird" (Iwaniuk et al., 2009: 47). The authors suggested that it may have been nocturnal. Another fossil bird from Hawaii, *Hemignathus vorpalis* (Fringillidae: Drepanidini), has an extraordinary bill with a long, scimitar-like maxillary rostrum and a much shorter mandibular rostrum, "giving the bird a most unusual appearance" (James and Olson, 2003).

Flightlessness occurs in many Pacific clades of birds and insects. Flightlessness in other Pacific island insects has been attributed to selection, as it prevents the insects being blown off their island. Zimmerman (1948) and others have dismissed this idea, pointing out that the Hawaiian Islands are too large for this to be likely. In hydrophiline beetles, Short and Liebherr (2007) reported that "The majority of the Hawaiian species exhibit vestigial wings, a very unusual condition in aquatic beetles. No other island-endemic members of Hydrophilinae are known to be flightless, suggesting insularity *per se* is not responsible for the condition."

All these features may be older than the current islands and could represent relictual aspects of what was a much larger biota in the north-central Pacific region, rather than adaptive radiations in the modern Hawaiian environment.

Anomalous Ecology

Insects are the most diverse form of multicellular life on the planet, dominating both terrestrial and freshwater ecosystems. Yet the only species with a life stage able to breath, feed, and develop either continually submerged or without access to water are moth larvae in mountain streams across the Hawaiian Islands. These truly amphibious caterpillars belong to certain groups in the endemic moth *Hyposmocoma* (Rubinoff and

Schmitz, 2010). A molecular phylogeny of 89 *Hyposmocoma* species indicated that the amphibious lifestyle is an example of parallel evolution. It has arisen from strictly terrestrial clades at least three separate times in the genus, starting more than 6 million years ago, before the current high islands existed. Rubinoff and Schmitz (2010) concluded: "Why and how *Hyposmocoma*, an overwhelmingly terrestrial group, repeatedly evolved unprecedented aquatic species is unclear, although there are many other evolutionary anomalies across the Hawaiian archipelago."

An area of endemism or an ecological habitat or niche that appears as "distal" in a phylogeny does not necessarily mean that it was colonized after the group itself formed (cf. Fig. 1-6). An ancestor of *Hyposmocoma* (not the genus itself) that was ecologically and geographically widespread may have already occupied both amphibious and terrestrial ecospace in the broad Hawaii–north Pacific region before differentiating into the modern genus and its clades.

Exceptional Means of Dispersal

Most modern authors would agree with Price and Wagner (2004) that "The biota of the Hawaiian Islands is derived entirely from long distance dispersal." All taxa in the Hawaiian biota are thought to be there because of their "exceptional dispersal capabilities" (e.g., Garb and Gillespie, 2009: 1746). There is no actual evidence for this beyond the presence of the groups in Hawaii, and as Cowie and Holland (2008: 3371) observed, there is a certain circularity in this argument. The very high level of endemism in the biota and the absence or low diversity of groups with outstanding means of dispersal, such as mangroves and orchids, argue against the importance of means of dispersal in the establishment of the biota. The alternative model discussed here suggests that a component of the biota was not derived by extraordinary long-distance dispersal events from present-day land areas, but by normal, shorter-distance colonization from other land areas, high islands that no longer exist.

Elevated Rates of Evolution

In the cricket genus *Laupala*, Mendelson and Shaw (2005) calibrated the time scale by assuming that the clades were no older than the islands they are endemic to. This gave a rate of 4.2 species per million years, a speciation rate that is an order of magnitude greater than the average for arthropods and exceeded in all taxa only by African cichlids. In fact, the time course of the cichlid phylogeny that is usually cited may be incorrect (Chapter 2). Analyses using calibrations based on biogeography and plate tectonics instead of oldest fossils (Genner et al., 2007, Azuma et al., 2008) give much older dates than the traditional ones, and so the postulated *Laupala* rates may be even more anomalous than was thought.

Craft et al. (2008) calibrated a clock for the Hawaiian shrimp *Halocaridina rubra* (Atyidae) by assuming that anchialine clades were the same age as the volcanic strata that they occupied. This implied a rate of 20% divergence per million years – probably the fastest rate yet reported for an invertebrate.

In the Hawaiian members of the bee *Hylaeus*, Magnacca and Danforth (2006) inferred a center of origin on the island of Hawaii. They assumed that the biota of Hawaii could be no older than the current island, and so they proposed that the 60 Hawaiian Islands species of *Hylaeus* have all evolved in the last 700,000 years, "an unusually short time for such a large radiation" (p. 393). They argued that their results "demonstrate unequivocally that rapid genetic change is taking place" (p. 406). Later studies showed "low" nuclear DNA variation (Magnacca and Danforth, 2007) and confirmed that the bees are "clearly evolving at an elevated rate" (p. 914). Gillespie (2009) wrote that the pattern "awaits explanation." Other groups with basal clades on Hawaii are cited above and would all be assumed to be less than the age of the island, originating at ~0.5 Ma rather than, say, 50 Ma as suggested here.

PRESENCE/ABSENCE AND HIGH/LOW DIVERSITY ANOMALIES IN HAWAIIAN GROUPS

Some Hawaiian groups show anomalous, high species diversity. For example, in flies, the high diversity of Drosophilidae on the Hawaiian Islands comprises one-sixth of the world diversity of the family and is "extraordinary" (O'Grady et al., 2009). There are an estimated 1,000 species on the Hawaiian Islands, all in the genera *Drosophila* and *Scaptomyza*. On the other hand, some groups are present but notably species poor. In the dispersal model, these discrepancies are regarded as "great mysteries" (H.L. Carson, pers. comm. in Hoch, 2006).

Discrepancies in diversity level may have a origin similar to patterns in which some groups are present while related clades are absent. In a vicariance interpretation, presence/absence and high/low diversity do not reflect the means of dispersal of a group but prior centers of diversity, vicariance in global patterns, and aspects of prior genomic architecture ("evolutionary potential"). If a lineage never had a north-central Pacific center of diversity it will probably not be on Hawaii, no matter how effective its means of dispersal are.

In passerines, the endemic Hawaiian honeycreepers (Fringillidae: Drepanidini) have 55 species, including many now extinct. In contrast with the endemism and diversity in this group, Perkins (1901) noted that "whole families of birds far better adapted to cross wide extents of ocean are quite unrepresented in the Hawaiian Islands, although we know that some of them thrive exceedingly when imported." What is the reason for this anomaly?

Lovette et al. (2002) compared the Hawaiian honeycreepers (with 55 species including fossils) and Hawaiian thrushes (a clade of *Myadestes*, Turdidae, with five species, including fossils). In a molecular clock study, the two groups showed similar degrees of differentiation from their mainland relatives and so were assumed to have been on the archipelago for about the same amount of time. Thus, the marked difference of diversity between the two was not attributed to a difference in the time available for evolution in each group. Instead, Lovette et al. (2002) suggested that the contrasting diversity is due to "an intrinsic, clade-specific trait" that fostered the striking "radiation" in the honeycreepers. Their diversity represents "an extreme manifestation of a general clade-specific ability to evolve novel morphologies."

Complementing the high-diversity taxa such as lobeliads, drosophilids, and honeycreepers, the Hawaiian biota has many conspicuous absences. In the flora, diverse, pantropical groups such as mangroves are absent in the indigenous flora despite having very efficient means of dispersal. Introduced mangrove species are flourishing. Other plants that characterize tropical shorelines and small islands either globally (such as *Calophyllum*, *Terminalia*, *Hernandia*, *Salicornia*, and *Atriplex*) or through the Old World and Pacific (such as *Barringtonia*) are also absent from Hawaii, at least in the indigenous biota. Nevertheless, species of most of these genera thrive in Hawaii as introduced weeds. Transport in mud on birds' feet is often proposed for Hawaiian groups, but characteristic marsh plants such as Eriocaulaceae and *Juncus* are absent from the indigenous flora (although several introduced *Juncus* species are now widespread). Diverse groups typical of rainforest throughout the tropics, such as Zingiberales (Zingiberaceae, Marantaceae, etc.), Meliaceae, Melastomataceae, Araceae s.str., Cyatheaceae, *Ficus* (figs), and *Piper*, are all absent from the indigenous flora, although many are naturalized.

Other groups show anomalous, low diversity. Orchids have very efficient dispersal due to the vast quantities of dust-like seeds they produce and are usually among the first plants to colonize islands (Roberts and Bateman, 2009). Yet despite this ecology, there are only three indigenous orchid species in the Hawaiian islands (endemic species of *Liparis*, *Anoectochilus*, and *Platanthera*) compared with 164 in the Fiji Islands, volcanic islands with a similar land area (18,330 km^2 versus 16,640 km^2 in Hawaii). This paucity is probably not due to ecological factors, and introduced species of orchids in genera such as *Spathoglottis*, *Phaius*, *Arundina*, *Epidendrum*, *Polystachya*, and *Zeuxine* have established in Hawaii.

Of all oceanic islands, the Hawaiian archipelago presents "the most aberrant and anomalous insect fauna, and the one most difficult of interpretation" (Zimmerman, 1940: 273). Although Hawaii has 5,818 described insect species (Price, 2009), only about half of all the insect orders are represented. There are many conspicuous absences, for example, there are no mayflies, no cockroaches or termites, no chrysomelid, scarabaeid, or buprestid beetles, no ants, and no papilionid butterflies. Complementing these absences in the Hawaiian biota, beetles, flies, moths, and bugs are proportionately better represented there than elsewhere (Howarth, 1990). In Lepidoptera, there are over a thousand moths but only two endemic butterflies, and in Diptera a thousand drosophilids but only 13 crane-flies (Cowie and Holland, 2008). Whatever the reason for these and other, similar cases, it cannot be related to normal means of dispersal, and so most authors have appealed to the extraordinary, unpredictable events of founder dispersal.

The family Miridae (Heteroptera) has ~10,000 species in eight subfamilies. They occur worldwide and inhabit nearly all oceanic island groups in the tropics. In the Hawaiian Islands, one subfamily, the Orthotylinae, is predominant, and the islands have about 73% of the world's genera (Asquith, 1997). Distributions of the other subfamilies of Miridae are very different; none has most of its genera in Hawaii and four are entirely absent. Asquith (1997: 357) reasoned that "The preponderance of Orthotylineae in Hawaii is unlikely to be the result of greater dispersal capabilities. In fact other remote Pacific island groups such as Samoa and the Marquesas are depauperate of Orthotylinae, suggesting that they are actually poor dispersers compared with other subfamilies." The pattern may reflect allopatric centers of diversity in the Miridae subfamilies related to their original vicariance.

Because Hawaiian land snails are exceptionally diverse (1,243 species; Price, 2009), they are thought to have "extraordinary" dispersal ability (Holland, 2009: 540) and be "surprisingly adept at dispersing across vast stretches of open ocean" (Holland, 2009: 537). Nevertheless, land snail groups such as the family Partulidae, which is endemic, diverse, and widespread in the South Pacific from Micronesia to southeastern Polynesia (Fig. 6-2), have never been recorded from Hawaii, either living or fossil.

Other anomalies difficult to explain in the long-distance dispersal model include the fact that many of the diverse Hawaiian groups comprise monophyletic, endemic clades. Many were thought to be polyphyletic and the result of several founder dispersal events, but in molecular studies an increasing number have been shown to be monophyletic. The Hawaiian lobeliads (six genera, 126 species) and *Cyrtandra* (Gesneriaceae, 58 species) are among the most remarkable examples. Because the Hawaiian lobeliads are monophyletic, Givnish et al. (2009) assumed that the group resulted from a single colonization event. Nevertheless, they admitted that this origin is a "paradox," given that the seeds are minute and that the group's relatives "have repeatedly dispersed across large areas of ocean."

Cronk et al. (2005) raised the same question in relation to the single, widespread clade of *Cyrtandra* in the Pacific discussed in Chapter 6 (Fig. 6-5). If dispersal occurred, Cronk et al. asked, why did *Cyrtandra* only colonize the Pacific once, despite its great ecological and phylogenetic diversity? Why did this happen so early in the history of the genus and never again? Why does the Pacific clade show simple, large-scale allopatry with its sister group rather than following a linear sequence and a "progression rule" sequence along a series of islands into the Pacific?

Not only are the Hawaiian clades monophyletic, they are often the basal member in pan-Pacific groups and not nested in them. This is seen in plants such as the *Tetraplasandra* group (Fig. 6-13) and animals such as the passerine *Chasiempis* (Fig. 6-3) (Costello and Motley, 2007; Filardi and Moyle, 2005). Continuing studies on the Pacific clade of *Cyrtandra* (Fig. 6-5) indicate that it has three subclades, with one in the Hawaiian Islands, one in the South Pacific islands, and one in Fiji (Clark et al., 2008, 2009). Why did *Cyrtandra*—a "supertramp" (Cronk et al., 2005)—colonize Hawaii only once? This pattern is not strictly incompatible with dispersal theory, as it could be the result of a single, early dispersal event before any other differentiation in the group. Yet it is not expected in a model that stresses the extraordinary means of dispersal of all the groups present and the Hawaiian clade would be more likely to be deeply nested in other Pacific taxa. Carlquist (1995: 5) suggested that "There are no spectacular relics on the present Hawaiian chain, and calling any of them ancient is a misnomer if one compares Hawaiian plant or animal groups with those on continental islands." This was written before molecular studies showed that Hawaiian endemics such as *Hillebrandia* (Begoniaceae), *Nototrichium* (Amaranthaceae), and the moa-nalo ducks are all basal in global groups (see Chapter 7).

While the monophyly and basal position of the Hawaiian groups is not expected in dispersal theory, it is of special interest, as it is the standard signature of a normal vicariance event. When it is a repeated pattern in groups with very different ecology and means of dispersal, vicariance is even more likely.

ADAPTIVE RADIATION OR NONADAPTIVE JUXTAPOSITION?

The concept of adaptive radiation is derived from center of origin theory. It assumes a center of origin at a point in morphological and ecological space, with evolution and dispersal radiating out from there into other kinds of morphology and habitat. This is the CODA model of evolution (center of origin/dispersal/adaptation) discussed in Chapters 1 and 6. Groups such as the 1,000 Hawaiian drosophilid species are often regarded as classic examples of adaptive radiation (O'Grady et al., 2009). Instead, much of their phylogenetic, biogeographic, and ecological diversity may have been inherited from widespread neighboring archipelagoes (where it is now extinct due to subsidence), rather than from a center of origin in the Hawaiian Islands.

For example, the Hawaiian endemic *Broussaisia* is the only dioecious genus in Hydrangeaceae. As mentioned in Chapter 7, it is sister to *Dichroa* of southern China to New Guinea, plus *Hydrangea macrophylla* and *H. hirta* of eastern Asia including Japan (Hufford et al., 2001). Hufford wrote that *Broussaisia* "may represent a long distance dispersal from Asia" and that "The shifts of *Broussaisia* and *Dichroa* into wet tropical forests are clearly derived in the family" (Hufford, 2004: 207). This assumes an ecological center of origin for the group, but the ancestor of the central Pacific–east Asia clade may have been widespread geographically and ecologically at the time of its differentiation.

In another example, the silversword alliance (*Argyroxiphium*, etc., Asteraceae) comprises 28 species of trees, lianes, shrubs, cushion plants, and mat plants. These inhabit very wet to very dry areas from 75 to 3,750 m elevation. The group has been described as "one of the most remarkable examples of adaptive radiation in plants" (Baldwin, 2007: 237).

In an alternative hypothesis, the ancestor may have already been widespread in the eastern Pacific and western America before the modern genera differentiated. If the ancestor was so widespread, it may have also already occupied diverse habitats.

In *Tetramolopium* (Asteraceae), Lowrey (1995) showed that the Hawaiian species are all neatly allopatric and occupy different habitat types. "Therefore," he concluded, "speciation appears to have resulted from local radiation after dispersal to a suitable habitat" (p. 217). Instead, the simple allopatry of all the species suggests that there has been no dispersal in the modern genus, only geographic division and subdivision. Lowrey (1995) argued that *T. consanguineum* "may represent a very recent speciation," and in support of this he noted that the species favors lava flows only a few hundred to 1,000 years old. Yet this could also indicate that it is a very old species that has always occurred on very young flows. Its direct ancestors may have had a similar ecology.

Within Hawaiian *Plantago*, Dunbar-Co et al. (2008) recovered the phylogeny: (K (K (O (O (Mo, M (M, H))))))), and while the authors supported a Kauai ancestry for the group, this is not necessary, as suggested above. In addition to the progression rule geographic sequence, the phylogeny also shows an "ecological progression rule." The basal clade on central and northern Kauai consists of minute herbaceous rosettes that grow in bogs, whereas its sister clade in western Kauai and the other islands includes larger herbaceous species from bogs and also woody plants—shrubs and small trees—from woodlands. "The data suggest . . . there was movement out of the bogs and into neighboring woodland habitats" (Dunbar-Co et al., 2008: 1182), but as in *Tetramolopium* and the silverswords, the ancestor may have already had a wide ecological range as well as a wide geographical range. Dunbar-Co et al. suggested that shifts in growth form from herbaceous to woody and back to herbaceous "may have been mediated by the availability of bog habitats." Instead, the small-leaved, herbaceous habit may determine the bog habitat and the woody habit may determine the woodland habitat. Dunbar-Co et al. (2008) wrote that "Habitat shifts appear to be a major driver of speciation," but speciation may instead drive the habitat shift; alternatively, woody/herbaceous differentiation may have already occurred as a cline in a diverse ancestor.

The three species of Hawaiian geese (one living and two fossil species of *Branta*) provide a comparable example. Paxinos et al. (2002: 1404) wrote that "niche shift from mainly wetland to mainly terrestrial habitats . . . may underlie two morphological traits . . . reduction in wing length (presumably because of loss of migration) and increase in the depth of skull and bill (probably because of dietary shift)." Instead, the dietary shift could have been determined by the skull modifications and loss of migration could be the result of wing reduction. Long-term phylogenetic and morphogenetic trends in development, or "laws of growth" (Darwin, 1859), underlie and constrain particular morphologies and adaptations (Heads, 2009c). In a similar way, the inherent propensity for evolution in a group's genome determines the diversity of the group (Lovette et al., 2002; see above).

Megalagrion (Odonata: Zygoptera) is a genus of damselflies endemic to the Hawaiian Islands. The 23 species occupy such a wide variety of habitats that their ecological range equals that of all the other damselflies of the world put together, and Jordan et al. (2003) interpreted the pattern as an adaptive radiation. They wrote, "The influence of ecology in *Megalagrion* speciation is illustrated by the fact that these damselflies breed in an astonishing array of habitats" (Jordan et al., 2003: 105). The authors cited the shift of *Megalagrion* taxa into terrestrial habitats such as the leaf axils of plants and litter beneath ferns. Nevertheless, an ancestral complex could have already occupied these habitat types before differentiation within the genus or even before differentiation of *Megalagrion* and its sister genera.

Hawaiian hydrophilid beetles show a habitat shift "nearly identical" to that of *Megalagrion*, as four derived Hawaiian taxa "have moved away" from true aquatic habitats (Short and Liebherr, 2007). But, as in the study on *Megalagrion*, these arguments rely on optimization of features on phylogenies and overlook the possibility that the ancestor may have already been ecologically diverse (or geographically widespread; cf. Fig. 1-6) before the modern taxa evolved. A critique of adaptive radiation as the default explanation does not mean that no new habitat type is ever entered, any more than a critique of founder dispersal means that there is no range expansion. But as with geographic differentiation, the ecological differentiation may have occurred at a much earlier stage in phylogeny than is often assumed, long before the origin of the species it is currently seen in.

Aspects of Elevational Range

Adaptive radiation is thought to account for the wide elevational range of Hawaiian taxa such as the silversword alliance. This overlooks the survival of taxa *in situ* on growing volcanoes and subsiding islands. In the dispersal model, a montane habitat develops and is then colonized by its taxa. Yet as mountains rise, they are never devoid of life. Habitats can change while a taxon remains *in situ*, even when the population is in theory quite mobile. This is best seen in rapidly rising mountains such as the Andes, the Himalayas, and the central ranges of New Guinea. Volcanoes grow in altitude with each eruption of lava and other material. Populations repeatedly colonize new flows from adjacent older ones and gradually gain elevation. This explains how taxa endemic on mainland volcanoes can be older than their volcano. For example, the volcanic edifice of Mount Kilimanjaro in Tanzania has built up since ~2.5 Ma, while insects endemic there have been dated at ~7–8 Ma (Chapter 2).

Artemisia (Asteraceae) is represented in the Hawaiian Islands with a clade of three species; *A. australis* occupies littoral sites, while *A. kauaiensis* and *A. mauiensis* are in subalpine shrubland. The clade is sister to *A. chinensis* of littoral habitats in China, Taiwan, the Ryukyu Islands, and the Philippines (Hobbs and Baldwin, 2008), and the authors suggested that the littoral habitat is ancestral in the group as a whole. They proposed that the Hawaiian clade has "colonized" inland areas; an alternative possibility is that they gradually occupied montane areas as the elevation of the islands increased with eruptions and the clade persisted *in situ* as a metapopulation.

The isopod genus *Ligia* occurs worldwide and has most of its species in the supralittoral zone. It is represented in Hawaii by two endemic species, *L. hawaiensis*, widespread in the littoral zone, and *L. perkinsi* (probably polyphyletic), in montane forest to above 600 m. Taiti et al. (2003) suggested "independent colonization" of the upland habitats by *L. perkinsi*. Instead, any movement may have been local and only within the prior range of the species; there is no need to propose unidirectional migrations into the mountains. Littoral and inland species pairs in *Ligia* also occur in Taiwan and the Austral Islands.

After its active phase, a volcano becomes dormant, and with subsidence, landslides, and other erosion, the surface of the volcano is lowered along with its living community. The plant *Portulaca sclerocarpa* is recorded from 1,030–1,630 m on Hawaii and also on Poʻopoʻo Island, off Lanai, at 30 m. The latter population "apparently represents a colonization from Hawaii" (Wagner et al., 1990: 1074), although it could also be due to subsidence. This could be tested with comparative studies of other groups to establish whether there is a general pattern of low-elevation anomalies in the area.

In *Tetramolopium* (Asteraceae), the central Pacific clade of New Guinea, Hawaii, and the Cook Islands (Chapter 7) is mainly a high-elevation group, and many species are common in the mountains of New Guinea and Hawaii. But one Hawaiian species, *T. rockii*, only grows at 10–200 m on lithified coral sand dunes in western Molokai. This represents the lowest elevation of the genus in Hawaii, and *T. rockii* or its immediate ancestors may have survived more or less *in situ* as the area was lowered by subsidence and erosion. In the Cook Islands, *Tetramolopium* is only known from one locality on the low, flat coral island of Mitiaro. It occurs at less than 10 m elevation on coral limestone. This population may also be relictual and the result of subsidence of the original high island with its biota.

RETENTION OF ANCESTRAL POLYMORPHISM AND ITS GEOGRAPHY

The weevil *Rhyncogonus* was cited above for its Hawaiian–southeastern Polynesian distribution. It is related to the western Pacific Elytrurini (Claridge, 2006). The affinities of a new weevil genus *Coconotus* from Cocos Island (between the Galapagos and Costa Rica) are unclear, but some features suggest a relationship with *Rhyncogonus* (Anderson and Lanteri, 2000). While these features may prove to be symplesiomorphies, that do not reflect the phylogeny, their geographic distribution would still be of interest and may predate the emergence of the modern genera. In other words, the ancestor may have been polymorphic before the modern groups began to differentiate, with some characters having had a central and east-Pacific distribution. The new groups could have differentiated along geographic lines different from those of the prior polymorphism. Incomplete lineage sorting of the ancestral polymorphism may have led to *Rhyncogonus* having affinities with the western Pacific Elytrurini, but at the same time inheriting characters shared with the eastern Pacific *Coconotus*. Precedents for a Hawaii–Galapagos connection in birds and insects were cited above.

The genus *Cuscuta* (Convolvulaceae) is represented in the Hawaiian Islands by a single endemic species, *C. sandwichiana*. Chloroplast sequences show it to be in "clade B," which is cosmopolitan but mainly in North America, while nuclear sequences place it in "clade H" with three species of Mexico and the southern U.S., and one in Australia, Southeast Asia, and Africa (Stefanović and Costea, 2008). The differences among individuals of *C. sandwichiana* and also between the species and its relatives are consistent with a "relatively ancient" hybridization event. The authors suggested that this may have occurred in the southwestern U.S., followed by dispersal to Hawaii. This would also require extinction on the mainland, no dispersal of either parent (or any other *Cuscuta* species) to Hawaii, and no further hybridization on the mainland. This combination of events seems unlikely, and the hybridization may instead reflect the early presence of east Pacific and west Pacific lineages in the central Pacific around the East Pacific Rise. The authors discussed incomplete lineage sorting as another possibility for the discordance between nuclear and chloroplast phylogenies, and although they preferred hybridization as an explanation, ancestral polymorphism would explain the biogeography simply.

In the diverse Hawaiian honeycreepers (Fringillidae: Drepanidini), Pratt (2001) accepted the genus *Oreomystis* for *O. bairdi* in Kauai and *O. mana* in Hawaii, disjunct at each end of the chain. Pratt mentioned that if the great morphological, behavioral, and ecological similarities between the species did not define a monophyletic group, they would present the most dramatic example of convergent evolution yet discovered in birds. In fact, subsequent osteological and molecular studies indicated that the two species were not sisters (James, 2004; Reding et al., 2009), indicating a level of incongruence between phylogeny and morphology that is "surprising among birds generally" (Reding et al., 2009: 222). While Reding et al. regarded the similarities as the result of convergence due to natural selection, they admitted that "It is . . . surprising that a behavioural trait not clearly related to foraging, such as juvenile begging calls, would be convergent" (p. 222). These calls are "nearly identical and distinct from those known for other honeycreepers" (p. 221). The surprising similarities between the Kauai and Hawaii "creepers" may be due to retention of ancestral polymorphism rather than selection and convergence, and the distribution conforms to a standard pattern.

Until the molecular studies of Fleischer et al. (2008), the endemic Hawaiian passerine family Mohoidae (*Moho, Chaetoptila*, etc.) was always placed with the Australasian/southwest Pacific Meliphagidae. DNA sequences instead showed that Mohoidae are much closer to Ptilogonatidae of southwestern U.S. to Panama, Dulidae of Hispaniola, and Bombycillidae of North America, South America, and Eurasia. Fleischer et al. (2008) suggested that the similarities between Mohoidae and Meliphagidae were due to convergence. They wrote that what are now seen as "the closest relatives [of Mohoidae], and presumably their common ancestor, look nothing like melaphagids," but there is no direct evidence for the morphology of the ancestor, and it may have been polymorphic. If this was the case, the typical "meliphagid" appearance of Mohoidae could be due to the retention of characters that were already present in the central and southwest Pacific ancestor long before the modern passerine clades evolved. In Mohoidae, the "parallelisms" shown with the southwest Pacific Meliphagidae and the "true affinities" shown with American birds may both reflect the standard biogeographic connections of the central Pacific.

In the silversword alliance, relationships of *Dubautia* based on chloroplast DNA conflict strongly with those based on nuclear DNA (Baldwin, 1998). Incomplete lineage sorting predicts ancestral polymorphism, but Baldwin concluded, "The lineage sorting scenario is difficult to reconcile with expectations of genetic bottlenecks during insular speciation [i.e., the standard model of Hawaiian biogeography] . . . particularly in light of the extensive ancestral cpDNA diversity that would need to be postulated in this example" (p. 68). Because of this conflict between ancestral polymorphism and a dispersal model of Hawaiian biogeography, Baldwin (1998: 68) suggested that multiple instances of hybridization "appear to be the only tenable explanations" for the cpDNA/nrDNA incongruence. Hybridization may well have occurred, but rejecting incomplete lineage sorting because it conflicts with the standard model is not supported here.

THE HAWAIIAN BIOTA AS A RELIC OF A NORTH-CENTRAL PACIFIC REGIONAL BIOTA

While most studies on the origins of the Hawaiian biota refer to the geographic isolation of Hawaii from the continents and other high islands, references to the neighboring atoll groups such as the Line Islands are much less common, and seamounts and guyots in the vicinity are seldom mentioned. It is usually assumed that Hawaii has been invaded from the continents bordering the Pacific, either directly (by wind or water) or by island-hopping. Instead, there may have always been a regional biota in the north-central Pacific, and in this metapopulation model Hawaii was colonized by normal ecological dispersal within the region, not by extraordinary, one-off founder events. The north-central Pacific biota may have been distinct and widespread following the Cretaceous volcanism and dispersal of the oceanic plateaus, although only a small fraction of its terrestrial diversity now remains, preserved on the Hawaiian Islands. The immediate sources of the Hawaiian biota may not be Asia, America, or the high islands of southern Polynesia and Melanesia, but former high islands represented by atolls (Phoenix and Line Islands, Johnston Island, Wake Island), seamounts (the Mid-Pacific Mountains, Musician seamounts), and the islands that have been destroyed by subduction beneath America and Asia.

Character optimization algorithms examine the distribution of extant characters on a phylogeny and calculate the arrangement that requires the fewest character transformations in a pure, monomorphic ancestor. This leads to ideas on which extant character was primitive. But in a character with, say, two extant states, neither state is likely to be ancestral, and instead of one being derived from the other, both are usually derived from a prior, ancestral condition, or the ancestor may have already been polymorphic (cf. widespread distribution). This is also a major theme of vicariance biogeography, where distribution is treated as a character. In the worldwide plant genus *Geranium*, woody growth forms occur only in the Hawaiian taxa (*G. arboreum* is a shrub or small tree), and Baldwin and Wagner (2010: 861) suggested this "is consistent with a hypothesis of secondary woodiness evolving in the islands from an herbaceous founding lineage." Yet it is also consistent with a widespread ancestor in which the central Pacific populations were already woody. Baldwin and Wagner (2010: 858) also suggested that woodiness in the Hawaiian endemic *Schiedea* (Caryophyllaceae) "appears to be secondarily derived from an herbaceous habit," but this was only because the herbaceous habit is common in its closest relatives (*Honckenya*, *Wilhelmsia*, etc.) and secondary derivation may not be necessary.

Modern methods of ancestral area reconstruction, such as DIVA, employ character optimization algorithms to find the original center of origin among the current areas of the range. These methods, as in traditional center of origin analyses, consider only the present-day geographic areas occupied by the taxa and overlook former geography and the significance of localities such as the atolls and seamounts. Ancestral area analyses of groups in Hawaii and southeastern Polynesia, for example, will find a center in either of the two areas, or in both, but the Line Islands will not be mentioned as the model does not incorporate geological change. Nevertheless, these islands represent the remains of a vast archipelago 4,000 km long that developed in the Cretaceous. The precise details of the geography, biogeography, and ecology of these former high islands may never be known, but it does not seem realistic to discuss the biogeography of Hawaii or southeastern Polynesia without referring to them.

HAWAIIAN BIOGEOGRAPHY AND THE METHOD OF MULTIPLE
WORKING HYPOTHESES

Holland and Cowie (2006: 2155) wrote: "In our view it is time for island biogeography to move on, finally leaving unbalanced, vicariance-only thinking behind." But no one has ever supported "vicariance only," as this would be incompatible with any overlap of distributions. Nor has any vicariance biogeographer ever denied that many taxa have dispersed into the Pacific; the introduced tree weeds present even on the summits of islands such as Rarotonga attest to that. It is modern phylogeography that has adopted a narrow, dispersal-only position, denying that there is any autochthonous Pacific land biota at all. Keppel et al. (2009) reviewed 29 molecular studies on central Pacific groups and found that all of these (apart from six in which the mode was untested or unclear) supported recent dispersal into the Pacific. While the molecular work thus reaches a remarkable consensus, it is all based on the same method: the use of calibrations based on fossils (minimum ages) or island ages (minimum ages) to give maximum ages for clades, which are then used to rule out older events (i.e., vicariance) as relevant. Transmogrifying minimum ages into maximum ages in this way will always give dates that are too young and so will always provide "evidence" for dispersal.

Apart from the fact that it is not logical, dispersal theory leaves many questions unanswered. For example, if taxa such as the *Chasiempis* monarchids (Fig. 6-3) and the others really do have extraordinary, cryptic means of dispersal and could invade so efficiently, colonizing the entire Pacific basin, why did they only invade once and why are most of the species local endemics? Why do the distributions of the clades within the Pacific show such precise allopatry? Why is the Hawaiian clade of the *Chasiempis* group sister to the other Pacific clade in the south and west, not nested in it? Why are there so many non–progression rule taxa in the Hawaiian Islands? Whittaker et al. (2010: 109) wrote that in their General Dynamic Model of island biogeography (based on MacArthur and Wilson, 1967), the progression rule "should be expected to be a dominant pattern" in chains such as the Hawaiian Islands. Instead, the progression rule is only one of several patterns observed.

The consensus in current work on the Hawaiian biota is all the more remarkable given the many paradoxes and enigmas admitted within that consensus. This indicates a need for the method of "multiple working hypotheses" often cited by geologists (see Chapter 1). The method proposes that it is never desirable to have just one working hypothesis to explain a given phenomenon, and that accepting a single interpretation as definitive can hold up progress for decades (Chamberlin, 1890, reprinted 1965). The usual interpretation of Hawaiian biogeography—chance long-distance dispersal with founder speciation—is repeated without question in almost all studies on the topic. At the same time, guyots such as those in the Musicians seamounts are overlooked and the large area of east Pacific crust subducted with its seamounts beneath western America is never mentioned. It is strange that the one paper in modern times to have even considered an alternative model for the Hawaiian biota (Nelson, 2006) was accused of trying to fit biogeography into a straitjacket (Holland and Cowie, 2006).

A METAPOPULATION MODEL FOR CENTRAL PACIFIC BIOGEOGRAPHY

Oceanic island systems and their biota have been seen as unique and invaluable "natural experiments" for evolutionary studies. Authors have dismissed the idea of autochthonous Pacific groups because there is no continental crust in the region, but this rejection may have been too hasty. Terrestrial groups require land, not continental crust, and the geological evidence, let alone the biological evidence, suggests there has always been land available on Pacific oceanic crust in the form of islands. Diverse rainforest can exist on very small islands, and endemic taxa can survive as metapopulations on old systems of individually ephemeral islands. The metapopulation model differs from the equilibrium theory of island biogeography in that there is no migration from a continental source—there is no mainland where the species is constantly present (Hjerman, 2009). In the metapopulation model, the colonization of islands is by normal ecological dispersal, not the extraordinary, singular events of long-distance dispersal theory. Dispersal takes place continuously among islands and there is no single "source."

For more than a century, biogeographers have accepted that the distinction between oceanic and continental islands is fundamental, although if the metapopulation model is correct and there is no mainland source, the distinction is irrelevant. In a metapopulation model, island endemics can be much older than their current island and usually are. The biology of real islands is no more unique than the biology of other ecological "islands," such as mountain peaks, landslides, or puddles. The individual "islands" may be more or less ephemeral, but ancient endemics can survive in these habitat types as dynamic metapopulations. For islands across the Pacific and elsewhere, molecular clock studies have indicated that clades are older than their islands and so the age of a real island, like that of any other stratum or ecological island, cannot be used to date taxa that are endemic to it. Many of the central Pacific endemics are only found on very young islands, but no one assumes that species endemic to the puddles or landslides of a region must be younger than each individual puddle or landslide. The faulty conclusion that taxon age must be less than or equal to island age has often been used as evidence for long-distance dispersal; dispersal theorists have argued that long-distance dispersal must have occurred on Hawaii, Tahiti, and the Galapagos, and so it can also be an important process elsewhere.

Instead of relying on long-distance dispersal with founder effect speciation, the model of Hawaiian biogeography suggested here proposes a major phase of mobilism and range expansion in the Cretaceous, followed by a phase of immobilism and vicariance of metapopulations by tectonic processes. Normal ecological dispersal occurred at all times within each metapopulation.

Widespread geological revolutions that may have contributed to the phase of mobilism and range expansion include the global sea level maximum in the Cretaceous (sea level never again reached these levels), the Jurassic–Cretaceous rifting of Gondwana, and the Cretaceous emplacement of large igneous provinces and volcanic island chains in the Pacific. Utsunomiya et al. (2008) emphasized that "In Earth history the global environment and life did not evolve gradually but rather by catastrophic changes. . . . The mid-Cretaceous was the last pulse period in Earth history . . . the paleotemperatures for the mid-Cretaceous reached its highest level in the last 150 million years. . . . Black shales were notably deposited worldwide and world oil resource production reached its peak during this period. . . . The oceanic plateaus in the western Pacific Ocean mostly formed in the mid-Cretaceous [and] the huge volcanic output contributed to global environmental changes" (p. 115).

The mid-Cretaceous plateaus and seamounts would have been colonized from earlier Cretaceous islands already in the region, for example, islands now represented by the guyots of the Mid-Pacific Mountains. During the subsequent phase of biotic immobilism, dismembering and dispersal of the plateaus and subsidence of the island chains would have led to vicariance. Finally, the subduction of large areas of crust around the margins of the Pacific would have led to the juxtaposition of regions and biotas that were formerly separate and the piling up of diversity in areas such as California and Melanesia.

9

Biogeography of Pantropical and Global Groups

Many groups show biogeographic connections between the central Pacific and tropical America. Examples include *Fitchia* and its allies (Fig. 6-15), the tribe Sicyeae in Cucurbitaceae (Fig. 6-16), and others (Figs. 7-3 to 7-6). These groups link the Pacific with western parts of the Americas and the Caribbean plate, terranes that formed in the Pacific and were later translated eastward. The Pacific–western American clades often have their sister groups in eastern parts of the Americas, and this break can be seen in Mexico and Colombia, as discussed next.

EAST/WEST DIFFERENTIATION IN MEXICO: THE "MEXICAN Y" AND THE GUERRERO TERRANE

In the ducks mapped in Fig. 1-9, western Mexico and other western parts of America connect with Old World areas, whereas eastern Mexico connects with Brazil. The phylogenetic and biogeographic break between western and eastern Mexico has intercontinental significance. This raises the usual questions: Is the break a common pattern? Where exactly is the break? And is it related spatially to any important tectonic features?

Morphological and distributional studies of butterflies concluded that the most important pattern in the country is one termed the "Mexican Y" (Luis-Martínez et al., 2006: Fig. 143A; Vargas-Fernández et al., 2006: Fig. 106). In this pattern there are two "arms," one in

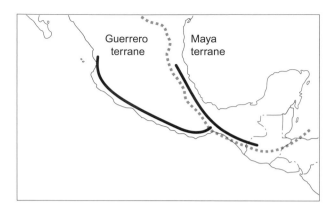

FIGURE 9-1. The "Mexican Y," the main track (set of nodes) in Mexican butterflies (Luis-Martínez et al., 2006; Vargas-Fernández et al., 2006). Dotted line = boundary between the Guerrero and Maya terranes (Umhoefer, 2003).

western Mexico (Guerrero, Oaxaca, etc.), and one in eastern Mexico north of the Tehuantepec isthmus, crossing to the Pacific side in southern Mexico (Fig. 9-1).

Primates occur along the eastern arm of the "Mexican Y," but none are known in the west. Climate alone cannot explain this, as primates are widespread in drier areas elsewhere. The absence of primates in the west is complemented by the endemism seen there in other mammals. All 12 mammal genera endemic to Mexico occur in western Mexico, on the Guerrero terrane, and nine are endemic there (Ceballos and Oliva, 2005). In the same way, the several Mexican groups of plants and animals that have global sister groups are all in the west, not the east, of the country (Heads, 2009a). *Romerolagus*, the volcano rabbit, belongs to both groups—it is endemic to Popocatépetl and three other volcanoes around Mexico City and is basal to a large group of Leporidae (*Lepus*, *Sylvilagus*, etc.) found throughout most of the world.

The traditional view of Mexican biogeography is that it represents a boundary zone between northern and southern elements or biotas. Of course, there are clear differences between the tropical biotas of the south and the temperate ones of the north, partly related to ecology. But Escalante et al. (2007) concluded that the underlying biogeographic pattern in Mexico was differentiation between east and west, and this is supported by the ducks, butterflies, and mammals cited above. The same break is now being revealed in many molecular studies, for example, in members of the passerine families Emberizidae (Navarro-Sigüenza et al., 2008; Bonaccorso et al., 2008) and Parulidae (Pérez-Emán et al., 2010).

Western Mexico, in particular the western arm of the "Mexican Y," is equivalent to the Guerrero terrane in the broad sense of Umhoefer (2003) (including the Cortes, Juarez, Mixteca, Santa Ana, and Xolapa terranes, along with the Guerrero terrane in the strict sense). The Guerrero terrane comprises submarine and locally subaerial volcanic rocks and sediments dated as Late Jurassic to middle Late Cretaceous (Cenomanian) (Centeno-García et al., 2008). The rocks have been correlated with similar arc rocks in the Greater Antilles and northern South America, and also with accreted arc terranes in the southwest Pacific. The Guerrero terrane includes mid-oceanic-ridge basalt, oceanic-island basalt, and island-arc basalt. Centeno-García et al. (2008) inferred a transtensional regime with considerable strike-slip faulting and intra-arc rifting from Middle Jurassic to mid-Cretaceous (Cenomanian). This produced a series of arc–backarc systems and complex paleogeography in what is now western Mexico and probably caused significant allopatric differentiation. Deformation of the arc assemblages and development of foreland and other basins took place in the Santonian to Maastrichtian (Late Cretaceous) and date the amalgamation of the Guerrero terrane with the continental margin. During the final accretion, a new continental arc developed and the transtensional regime was replaced with one of compression. A major Late Cretaceous–early Paleogene orogeny is recorded throughout Mexico, coeval with the Sevier and Laramide orogenies in western North America. This orogenic event is associated with the final amalgamation of the Guerrero terrane and also with the Mexican fold-and-thrust belt of the Sierra Madre Oriental. As with the earlier rifting, the change from a tectonic regime of transtension to one of compression would have had a profound impact on the regional biota.

THE ABSENCE OF PRIMATES FROM MANY PACIFIC MARGIN TERRANES

Primates and the Asian orders Dermoptera and Scandentia form the superorder Archonta. This group is widespread in the Old World and New World, and includes Arctic records in the fossil group Plesiadapiformes. Yet apart from humans and human introductions, Archonta have never been found, living or fossil, east of Sulawesi (in New Guinea and Australia) or in parts of western America (western Mexico and also Chile west of the Andes). In contrast, their sister group Glires (rodents and lagomorphs) has major diversity and endemism in both areas. This pattern does not seem to have been discussed, and no explanation has been offered.

One possibility is that the common ancestor of Archonta + Glires was globally distributed, with Glires developing around the Pacific, Archonta evolving elsewhere, and the overlap of the two groups developing later. The overlap would have involved major range expansion in Glires to cover most of the Earth's land surface. In Archonta, there would have been more modest expansion onto the terranes that became western Colombia and Central America. To summarize, the main nodes in the global Archonta + Glires seem to have been located around the southwestern Indian Ocean, the southwest Pacific/Southeast Asia region, and the eastern Pacific. Later, following the overlap of Archonta and Glires, mid-Cretaceous breaks developed in the Atlantic Ocean and in western Colombia.

OLD WORLD SUBOSCINES: PRESENT IN THE AMERICAS ONLY ON THE CHOCÓ TERRANES

The most conspicuous geographic breaks in pantropical groups are, of course, the Atlantic, Indian, and Pacific Oceans, but the phylogenetic breaks do not always coincide with these. In the group of passerines considered next, the phylogenetic breaks are at the Atlantic Ocean and at the Romeral fault in western Colombia. The latter break separates accreted terranes to the west from the craton to the east and is also an important biogeographic break in groups such as primates (Chapter 4).

The monotypic bird genus *Sapayoa* has a classic Chocó distribution, as it occurs in a narrow strip of rainforest from eastern Panama through western Colombia to northwestern Ecuador. The affinities of the species, aptly named *S. aenigma*, have puzzled ornithologists for years. It is now recognized as the only New World member of the otherwise Old World group, the Old World suboscines. This comprises one of the main clades of passerines (Moyle, 2006), as seen in the following phylogeny of the order.

Acanthisittidae: New Zealand

Suboscines (formerly Mesomyodi, now Tyranni)

New World suboscines (Tyrannides): South and Central America, a few extending to North America.

Old World suboscines (Eurylaimides): Africa and Asia to the Solomon Islands; disjunct in Chocó.

Oscines (Passeri): worldwide. Basal clades all in Australasia.

In the suboscines, the two clades are almost completely allopatric and overlap only in Chocó. This suggests simple vicariance of a pantropical ancestor, with local dispersal in and around Chocó.

Although the oldest passerine fossil is only Eocene in age, the main passerine groups show notable allopatry, as Ericson et al. (2003) emphasized in their Figure 3 (the interpretation of the allopatric sectors as dispersal routes is not accepted here). Ericson et al. (2003: 4) proposed a Gondwanan origin for the passerine order and suggested that "major groups of extant passerines derive from vicariant events following from the breakup of this former supercontinent." While this Gondwana center of origin may be too restrictive, Gondwana breakup was probably important for divergence of the main clades. As Ericson et al. (2003: 6) wrote, the new information "refutes the hypothesis that the diversification of all extant birds is the result of an explosive Tertiary radiation," and "the passerine radiation could be considerably older than the Early Tertiary." They wrote that this "may be surprising," and it would be if the fossil record is read literally, but the distribution patterns of the main clades are not surprising in the context of comparative biogeography.

Fjeldså et al. (2003) described the distribution of *Sapayoa* and its Old World allies as "peculiar," and Chesser (2004: 18) wrote that the discovery of the Old World affinities of *Sapayoa* is a "remarkable result [which] provides the only known instance of pantropical distribution among the large avian order Passeriformes." The Old World suboscines barely enter the American tropics, and the suboscines as a whole might provide a better example of pantropical distribution. Other pantropical clades in passerines include the "core martins" and the genus *Petrochelidon*, both in Hirundinidae (Sheldon et al., 2005). The trans-tropical Pacific disjunction in the Old World suboscines is a large one and perhaps remarkable in the context of the passerines, but it is common in other groups. Even in passerines, connections of Asia with the Americas (via Hawaii) are seen in *Corvus*. In the Old World family Sylviidae s.s., the only New World member is *Chamaea*. This is endemic to the west coast of North America (southern Oregon to Baja California) and is related to *Paradoxornis* of Southeast Asia (Nepal to Vietnam and Taiwan) (Gelang et al., 2009).

Fjeldså et al. (2003: S238) wrote that the distribution of *Sapayoa* and its Old World allies "may be best explained in terms of a Gondwanic and Late Cretaceous origin of the passerine birds, as this particular lineage dispersed from the Antarctic landmass, reaching the Old World tropics via the drifting Indian Plate, and South America via the West Antarctic Peninsula." This accepts an unnecessary theoretical framework of "southern origins," based on the assumption that the basal passerines (in New Zealand) represent a center of origin. Irestedt et al. (2006) suggested an alternative model of "northern origins," with a *Sapayoa* ancestor having reached America via Beringia or Greenland. Southern or northern origins both require ad hoc long-distance dispersal and extinction of the *Sapayoa* lineage throughout South America or North America, but this is not required either by the fossil record or the trans-tropical Pacific biogeography. The distinctive geochemical similarities between the Ontong Java Plateau and Nauru terranes (by the Solomon Islands, where the Old World suboscines break off range), and the terranes accreted to the Panama–Chocó area are relevant here. Both these groups of terranes have accreted to the Pacific margins, westward and eastward, respectively, from a central Pacific origin (Fig. 6-1).

Fjeldså et al. (2003) wrote that "The restricted distribution of *Sapayoa* at the northwestern corner of South America is most easily interpreted as being relictual: the Chocó area of endemism has many species representing deep phylogenetic branches (compared, for instance, with Amazonian forest birds)." The question is: Why? What is the reason for the phylogenetic break between the accreted Chocó terranes and the rest of South America? This is attributed here to the tectonics of the accreted terranes and the Romeral fault zone (Chapter 4).

The Pacific–America affinities complete the circuit of the tropics that began with the New World monkeys in Chapter 4, and so we are now in a position to consider pantropical groups and their evolution. The rest of this chapter examines patterns in the phylogeny and biogeography of some pantropical groups of angiosperms.

Authors in the 1960s proposed that angiosperms evolved as late as the Cretaceous. This has now been contradicted by the molecular work, which shows that angiosperms are sister to one or more of the main gymnosperm clades and are much older than was thought, either Triassic or Permian (Smith et al., 2010; Magallón, 2010). Some workers still accept a Cretaceous age for the large angiosperm clade, the eudi-cots, but this is doubtful, and even fossil-calibrated studies suggest they are Jurassic (Magallón, 2010).

THE BIOGEOGRAPHY OF RUTACEAE—ORANGES AND LEMONS

Trees in the orange, mahogany, and tree of heaven families (Rutaceae, Meliaceae, and Simaroubaceae) are important components of lowland for-est throughout the tropics; Rutaceae also occur in drier and cooler regions. Together the three families form a well-supported clade in Sapindales (Stevens, 2010). Rutaceae are the most diverse, with 160 genera and 1,800 species, and apart from trees also include many small-leaved, "ericoid" shrubs in South Africa and Australia. Groppo et al. (2008) emphasized that the molecular groupings of genera in Rutaceae are better correlated with the geographic distributions of the genera than with the traditional fruit characters.

Recent molecular clock studies of Rutaceae, Meliaceae, and Simaroubaceae have concluded that the main clades in the families origi-nated in the Cenozoic, too late to have been affected by the rifting of Gondwana, and so their distributions must all be explained by chance transoceanic dispersal. Nevertheless, the clock studies have all relied on treating fossil-calibrated clock dates, which are minimum ages, as maxi-mum ages. Thus, the clade ages proposed are probably too young, and the many disjunctions across ocean basins could be the result of plate tectonics.

Transmogrification of Fossil-based Dates into Maximum Dates

Many papers begin by accepting fossil-calibrated dates as minimum dates, but somewhere between the "Methods" and the "Conclusions" these dates are transformed by some mysterious process into maxi-mum dates and are used to indicate the age of clades (Chapter 2). Pfeil and Crisp (2008) used molecular clock dating to test between disper-sal and vicariance hypotheses for Rutaceae. In their section "The Age of Rutaceae" they treated the fossil-calibrated dates as minimum ages, which is logical. But in their section on "Biogeography," they drifted away from this position and instead accepted fossil-calibrated dates as maximum or absolute dates. Thus, while they wrote that "The mini-mum age of Rutaceae s.s. is between late Eocene and late Paleocene" (p. 1628), they concluded, "We have found no evidence that the crown of Rutaceae s.s. is as old as the Cretaceous, but instead it dates only to the late Paleocene" (p. 1630). In the abstract they claimed that Rutaceae s.s. originated "in"—not "in or before"—the Eocene. Likewise, they dated *Citrus* s.lat. "at a maximum of 11.8 Ma." In this way the authors were able to eliminate vicariance as a possible mechanism. This is a typical case involving the transmogrification of fossil-based minimum ages for clades into maximum ages. A key sentence near the end of Pfeil and Crisp (2008) reveals the contradiction: "Although all fossil calibrations *in principle imply* only minimum age, a lack of older fossils that we can confidently use to calibrate phylogenies in current knowledge *also sets a boundary* on the maximum ages of lineages for which we have evidence" (p. 1630; emphasis added). This is not logical—a lack of older fossils does not set anything except the minimum age. The authors' own claim (p. 1630) that they treated fossil-calibrated dates as minimum estimates would contradict this. But the claim itself is incorrect—Pfeil and Crisp (2008) were only able to "rule out" vicariance and conclude, for example, that "lineages must have arrived in New Caledonia by long distance dispersal" (p. 1629) because they treated the dates as *maximum* estimates.

It seems as if the authors wanted to accept fossil dates as both minimum ages (the logical approach, followed in their initial analysis) and maximum ages (the traditional approach, followed in their biogeographic discussion), but this is not possible. Nevertheless, Pfeil and Crisp's (2008) dates for Rutaceae have already been accepted, with Stevens (2010) agreeing that "the family is relatively young, and [so] distributions are unlikely to be much affected by continental drift events."

Pfeil and Crisp (2008: 1630) suggested that there is "no evidence" for Cretaceous Rutaceae, although what they meant is no *fossil* evidence. Their entire argument is based on the absence of fossils and does not account for the fact that some of the oldest Rutaceae fossils are already very "modern" in appearance (see below). There are two options: One is to use detailed studies of the record itself to estimate sampling error in the record and the likelihood that the clade existed earlier. This approach is fraught with problems, especially in groups such as Rutaceae that have a very sparse early fossil record. Another, much simpler, method avoids relying on the fossil record to give maximum clade ages and instead relates the geographic breaks in the molecular clades with regional tectonics. This method is applied below and indicates that the main clades in Rutaceae originated in the Early Cretaceous, with the family itself somewhat older.

Calibrating the Time Course of Phylogeny in Rutaceae Using Tectonics and Molecular Clade Distributions Instead of Oldest Fossil Ages

Many groups of plants and animals resemble Rutaceae in having a fossil record that extends only to the Paleogene (often in the northern hemisphere) but extant distributions that suggest Gondwana rifting. The orders of birds and mammals, including primates, provide many well-known cases. In another example, the cichlid fishes comprise one main clade in Madagascar and Sri Lanka, and the other in Madagascar, Africa, and South America. Instead of calibrating the cichlid phylogeny using the oldest fossils, ichthyologists are now using the biogeography of the clades together with tectonics (Chapter 2). This avoids using the fossil record for anything but minimum dates. The method is explicit and logical, and does not involve any twisting of the data. The results are very interesting and led the authors to reject "recent explosive radiation" as a model for cichlid evolution in Africa.

If the time course of the phylogeny is calibrated using the best of molecular biology (the biogeography of the molecular clades) and the best of hard rock geology (the tectonic dates) rather than the dangerously scanty fossil record, all support for chance dispersal in Rutaceae vanishes. Thus it is not necessary to propose an invasion of Australasian *Citrus* and other clades from Asia (Pfeil and Crisp, 2008); it is much more likely that *Citrus* and its clades evolved *in situ* and so their distributions could reflect tectonics. This can be tested further with detailed study of the phylogenetic/biogeographic breaks in the molecular clades and their relationship with regional geology.

Phylogenetic and Biogeographic Breaks in Rutaceae

Groppo et al. (2008) and Bayer et al. (2009) together sampled 83 of the 160 genera in Rutaceae, and strong geographic groupings are already evident. The four main clades, with a phylogeny: 1 (2 (3 + 4)), are as follows.

1. Spathelioideae are the basal clade in the Rutaceae and include a trans-Atlantic disjunction (Razafimandimbison et al., 2010; Appelhans et al., 2011). The two clades are vicariant, occurring in:

   • Amazon and Orinoco regions (*Sohnreyia*), Jamaica, Cuba, and the Bahamas (*Spathelia* and *Dictyloma*),

- Canary Islands, western Mediterranean (*Cneorum*) (former records of *Cneorum* from Cuba were discounted by Oviedo et al., 2009), Africa (*Bottegoa, Ptaeroxylon*), Madagascar (*Cedrelopsis*), Africa to tropical Asia and Australia (*Harrisonia*).

Assuming a simple vicariance process, the trans-Atlantic disjunction can be dated at ~120 Ma. A similar Cuba–Canary Islands connection is also seen in groups such as the gecko *Tarentola*, found in Cuba, Cape Verde and Canary Islands, North Africa, and Europe. In a molecular clock study of the genus, Carranza et al. (2000) dated the basal node, between the Cuban subgenus and the others, at 23 Ma, and so they concluded that *Tarentola* invaded Cuba "apparently via the North Equatorial current, a journey of at least 6000 km." The basis for the calibration was not made explicit, but was derived from unpublished study on the lacertid lizard *Gallotia* endemic to the Canary Islands. This probably involved the assumption that island endemics can be no older than their islands, but, as with fossils, this method will often give drastic underestimates for clade age. Fernández-Palacios et al. (2011) concluded that high islands may have been present in the Macaronesia region for much longer than is indicated by the age of the oldest current island (27 Ma).

2. The next basal clade in Rutaceae includes *Ruta* and allies (Rutoideae) plus *Citrus* and allies (Aurantioideae) (Fig. 9-2; Salvo et al., 2008). The Rutoideae–Aurantioideae clade includes the trans-Atlantic genus *Thamnosma* (in southern U.S./Mexico, Africa; Thiv et al., 2011)—a Cretaceous disjunction—and *Chloroxylon* in Madagascar, Sri Lanka, and southern India, another Cretaceous disjunction. The large degree of allopatry between Rutoideae and Aurantioideae is notable; for example, the former is in Madagascar and western South Africa, the latter in eastern South Africa.

The Mediterranean genus *Ruta* comprises nine species in two main clades, distributed as follows (Salvo et al., 2010):

- Canary Islands (three endemic species), western and northeastern Mediterranean (one species). Not in the Azores, Italy, Corsica, or Sardinia.
- The Azores, widespread around the Mediterranean including Italy (three species); also Corsica and Sardinia (two endemic species).

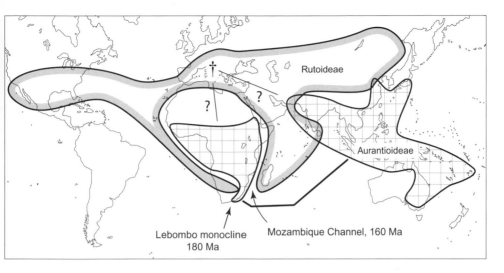

FIGURE 9-2. Distribution of the two sister groups Rutoideae and Aurantioideae (Rutaceae) (Salvo et al., 2008). Dagger symbol = fossil *"Citrus"* (which could be related to the Asian *Citrus* or the African *Citropsis*).

The key point, not mentioned by Salvo et al. (2010), is that the two clades are largely vicariant, with overlap only in the western and northeastern Mediterranean. The overlap could be explained by range expansion of a single species, *R. montana*, in the first clade.

Deducing the origin of *Ruta* means examining the distribution of its sister group, namely *Thamnosma* (southern U.S./Mexico and Africa) and *Boenninghausenia* (Himalayas to Japan). In this case, there is no overlap and the three genera can be interpreted as the result of simple vicariance. Salvo et al. (2010) did not refer to the precise allopatry of the three genera. Instead, they used fossil calibrations and a center of origin program to deduce a dispersal model for *Ruta* in which the genus originated outside the Mediterranean. This is not considered further here.

3. The smaller clade at the last main node in the family is widespread along and north of the Tethys belt: *Dictamnus/Skimmia* (Europe through central Asia to Japan and the Philippines), *Orixa* (Japan), and *Casimiroa* (Mexico to Costa Rica) (Poon et al., 2007).

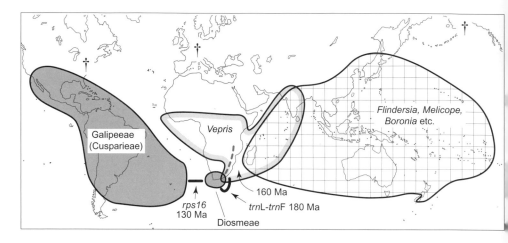

FIGURE 9-3. Distribution of a large clade in Rutaceae (Groppo et al., 2008). The third widespread clade *Zanthoxylum* + *Toddalia*, pantropical and also in temperate east Asia and North America, is not shown here. It may be sister to the *Vepris-Flindersia* group. Sequences from *rps16* link Diosmeae with Galipeeae. *trnL-trnF* data instead link Diosmeae with *Vepris* and *Flindersia*, etc. Broken line = *Calodendrum* (Diosmeae). Dagger symbols = fossil "*Evodia*" (*Flindersia/Melicope* group).

4. The remainder of the Rutaceae, the bulk of the family, comprises one large group (Fig. 9-3) with three sub-clades, each with a distinctive geographical distribution (Groppo et al., 2008).

The first subclade, a trans-Atlantic disjunct, comprises several tropical American groups (Galipeeae [= Cusparieae], *Balfourodendron*, *Adiscanthus*, etc.) and Diosmeae, with 250 species in *southern* Africa (mainly in the Cape region; Trinder-Smith et al., 2007). The trans-Atlantic disjunction again indicates divergence of the groups at 130 Ma.

The second large subclade, mainly allopatric with the first, has one branch in *tropical* Africa, Madagascar, and India (*Vepris*), the other in Madagascar, Asia, Australasia, and the Pacific (*Flindersia*, *Melicope*, etc.).

The third subclade (*Zanthoxylum*, *Toddalia*, etc.; cf. Poon et al., 2007) is not mapped here. It is a diverse, pantropical group (also present in north temperate Asia and America) with possible trans-Atlantic, trans-Indian Ocean, and trans-Pacific disjunctions. It may be sister to the second subclade, although as yet there is no significant statistical support. A well-sampled study of the ubiquitous *Zanthoxylum* would be of great interest.

Whether the diverse "ericoid shrub" clade of Rutaceae in South Africa, the Diosmeae, belongs with the non-ericoid American Galipeeae

(= Cusparieae) in a trans-Atlantic connection or with the Australian "ericoids" in a trans-Indian Ocean connection is an important question for taxonomy, but for biogeography and phylogeny, both connections are already indicated (Fig. 9-3). *rps16* sequences group Diosmeae with Galipeeae (Groppo et al., 2008), whereas *trnL-trnF* sequences group them with Old World groups (*Vepris*, *Boronia*, *Melicope*, etc.; Scott et al., 2000; Groppo et al., 2008). This is an interesting pattern, but the revealing ambivalence is lost in the combined tree of Groppo et al. (2008) as the trans-Indian Ocean affinity is swamped by the *rps16* data.

Ericoid shrubs can be described as "bract plants," equivalent to the (mainly sterilized) inflorescences of larger, more complex plants (Heads, 1994). The conversion of a normal tree with a large inflorescence into a small-leaved shrub is a simple morphogenetic process and requires only two steps:

- suppression of the entire plant shoot except for the inflorescence, and
- sterilization of most of the inflorescence bracts (turning them into "leaves").

If these reductions occur in a plant with a racemose (monopodial) inflorescence, the result is an ericoid shrub, as in the South African Diosmeae. If the two processes occur in a plant with a cymose (sympodial) inflorescence, the result is a "divaricating" shrub, as in *Melicope simplex* of New Zealand. The gene or genes coding for the two processes seem to have been widespread around the regions that became the Indian Ocean/Australia/New Zealand and less common around the Atlantic or Pacific (cf. Ericaceae, Plantaginaceae s.lat., Thymelaeaceae tribe Gnidieae, Rubiaceae tribe Anthospermeae, etc.). In some families, the Indian Ocean "bract plant" genes in a global ancestor were passed on in "monophyletic" Indian and Pacific Ocean clades (Rubiaceae tribe Anthospermeae, Thymelaeaceae tribe Gnidieae); in others they represent Indian Ocean "parallelism" (South African Ericoideae vs. Australian epacrids in Ericaceae s.lat.) (Stevens, 2010), but in Diosmeae both aspects are seen. *rps16* suggests that the ericoid habit of Indian Ocean Rutaceae is a parallelism; *trnL-trnF* indicates that it defines a monophyletic group. The two outcomes are perhaps not that different. The double affinity of the Cape plants across both the Atlantic and Indian Oceans could reflect polymorphism in a widespread ancestor (South America–South Africa–Australia) that existed prior to the modern clades. The order Bruniales is another example of ericoid Cape plants (Bruniaceae) related to non-ericoid South American plants (Columelliaceae).

In both main clades of Rutaceae (Figs. 9-2 and 9-3), important phylogenetic/geographic breaks occur in the southern Africa/Madagascar region, and these could represent vicariance events in the widespread, almost global ancestors of the two groups. In southern Africa the Aurantioideae (Fig. 9-2) are restricted to the southeast and are absent from western South Africa and Madagascar; Rutoideae are restricted to western South Africa and Madagascar. The last two areas have quite different climates, and the pattern is probably not due to ecology. Nevertheless, the pattern is a standard one, and the same two breaks are seen in primates. The first, between western and eastern South Africa, can be related to the main tectonic feature there, the Lebombo monocline (Chapters 3 and 5). This is a volcanic rifted margin that was active at ~180 Ma and was a precursor to the opening of the Mozambique Channel. The second break corresponds to the opening of the channel itself at 160 Ma. Madagascar has probably moved into the range of the aurantioids, perhaps dismembering them. The two massings of aurantioids may have been connected to the south of Madagascar. Thiv et al. (2011) discussed the disjunction in mainland Africa between southern Africa and Somalia in Rutoideae (represented here by *Thamnosma*) but did not refer to the Aurantioideae (with *Citropsis*, *Balsamocitrus*, etc.), and these fill the gap.

The same two breaks in Figure 9-2 occur in the taxa in Figure 9-3. Despite its high diversity in Madagascar and the Mascarenes, *Vepris* is absent from western South Africa. Likewise, although *Melicope* etc. range from Madagascar to Hawaii, they are not on the African mainland. The breaks are: the Atlantic (130 Ma) (between Cusparieae and Diosmeae—the *rps16* link), the Lebombo monocline (between Diosmeae and *Vepris/Boronia* etc.—the *trnL-trnF* link), and the Mozambique Channel (the limit of the *Melicope* group). Analysis of the overlap of *Vepris* and the *Melicope* group in Madagascar and India would require phylogenies and detailed distributional data for these groups.

Citrus

In their study of Aurantioideae, Bayer et al. (2009) included a detailed analysis of *Citrus*, and here they found two main groups, a "northern clade" and a "largely southern clade" (Fig. 9-4). The same pattern occurs in *Uvaria* (Annonaceae); a clade is disjunct between Australia/New Guinea (Clade 24) and India/Sri Lanka (Clade 23), and the large gap is filled by

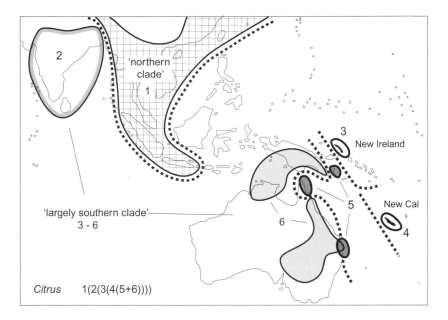

FIGURE 9-4. Distribution of the indigenous species of *Citrus* (Bayer et al., 2009). The "northern clade" extends to Japan and there are also possible leaf fossils in Europe. New Cal = New Caledonia. The phylogeny is indicated with breaks shown as dashed lines.

its sister group, endemic to Indochina and Sundaland (Clade 21)(Zhou et al., 2011). In *Citrus* the northern clade includes the oranges and lemons as well as wild species; the exact limits of the natural distribution are not known, although the general picture is clear enough. The phylogeny for the southern clade is: India (New Ireland (New Caledonia (Brisbane, Cape York, New Guinea islands + east Australia and New Guinea))). This could reflect a center of origin in India, with dispersal to New Ireland, then to New Caledonia, and so on. Or it could reflect a sequence of differentiation in an ancestor that was already widespread from India to New Caledonia. The locations of the breaks (dashed lines in Fig. 9-4) coincide with tectonic boundaries involving the accretion of India in the Eocene, the extrusion of Indochina from Asia, and rifting around New Caledonia and the Coral Sea (Heads, 2008a, 2008b). The Bismarck Archipelago/Solomon Islands region is often seen as a "sink" for dispersal, but this may not be correct. For example, the genus *Melonycteris* is endemic to this region and is basal or near-basal in the Old World fruit-bats, Pteropodidae, suggesting that old fractures took place in this region early in the history of widespread, diverse complexes (Heads, 2009a).

Rutaceae in the Pacific

The rifting of Gondwana in the Cretaceous to form the Atlantic and Indian Oceans is well known. Contemporary events in the Asia–Pacific region (including eastern Tethys) are less well understood but just as important for biogeography. In central Asia, the accretion of the Lhasa and West Burma terranes (calved off from northern Gondwana in the Late Jurassic, accreted to Asia by Late Cretaceous; see Chapter 5) has special relevance for Asia–Australasia distributions such as those in Aurantioideae. India was only the last of many Gondwana terranes to be accreted to Asia.

Tasman Sea–Coral Sea disjunctions and breaks are important in Aurantioideae and also in the *Melicope* group. The Australia–New Caledonia and Australia–New Zealand disjunctions could be the result of Mesozoic–Paleogene rifting. and this would fit with the Cretaceous chronology indicated above. In the *Melicope* group, five genera keyed together in Hartley's (2001a) morphological study range in New Caledonia (*Dutailliopsis*, *Dutaillyea*, *Comptonella*, *Picrella* ["*Zieridium*"]) and in east Australia, disjunct at Fraser Island and around Cairns (*Pitaviaster*). (Poon et al., 2007, did not sample the New Caledonian plants; *Melicope vitiflora* from Fiji is grouped with *Pitaviaster* but with little support.) Whatever the final arrangement, the Coral Sea disjunctions in this group and in *Citrus* could be the result of rifting and sea-floor spreading (cf. Heads, 2008a, 2008b).

As yet, only half of the Rutaceae genera have been sequenced, and the affinities of several groups are still unclear. *Plethadenia* of Cuba and Hispaniola may have trans-Pacific connections; based on morphology, Beurton (2000) interpreted it as an outlier of the Australasian/Pacific *Evodia/Boronia* group. *Pitavia* of southern Chile has been compared with Australian genera such as *Acradenia* and *Acronychia*, implying the usual South Pacific disjunction. Discussing the central Pacific Rutaceae, Hartley (2001b) emphasized that old taxa can survive as endemics on young volcanic islands around old centers of volcanism. Hartley accepted Pacific genera of Rutaceae as Cretaceous in origin on the basis of biogeographic disjunctions, among other evidence.

Rutaceae are diverse in the central Pacific. New Caledonia has 95 indigenous species in 20 genera (eight endemic) (Jaffré et al., 2001; Hartley, 2003; Bayer et al., 2009) and Hawaii has 52 species (48 in *Melicope* s.lat., four in *Zanthoxylum*) (Wagner et al., 2009). The following clade in

Melicope (including *Platydesma* and *Pelea*) was mentioned in Chapter 7 (from Harbaugh, Wagner, Allan, and Zimmer 2009).

> *Melicope simplex* etc.: Lord Howe Island, New Zealand, and Society Islands.

> > *Melicope rubra* etc.: Australia, New Guinea, Vietnam, Taiwan.

> > > *Melicope* sect. *Pelea*: Hawaiian Islands (48 spp.) and the Marquesas (4 spp.).

> > > *Platydesma*: Hawaiian Islands.

The three main subgroups divide up the Pacific with no overlap and with the usual breaks in the Tasman Sea (between Lord Howe Island and Australia) and between the Society and Marquesas Islands. The first break can be related to the Cretaceous rifting of the Lord Howe Rise from the mainland. The second break (between the Society and Marquesas Islands) is located at the center of the South Pacific super-swell (Chapter 6), a regional swell 4,000 km across and 680 m high that has been the site of extensive, multiscale volcanism since the Jurassic. The same break between the Society and Marquesas Islands occurs in many other taxa (for example, *Metrosideros* [Myrtaceae]; Chapter 6) and could have been caused by the same mid-Cretaceous events.

Fossil Rutaceae

The Rutaceae fossil record cannot give absolute dates for the taxa, but it does provide key evidence. Fossil seeds "closely resembling" those of modern *Euodia*, *Phellodendron*, and *Zanthoxylum* are known from the Oligocene (~25 Ma). Likewise, the fact that Eocene material of *Ailanthus* (Simaroubaceae) 50 m.y. old is "nearly identical" (Pfeil and Crisp, 2008) to extant plants shows the potential of taxa in this order to show long-term stasis. No precursors of these fossils are recognized in the record; they appear "out of nowhere," indicating prior gaps of an indefinite extent in the record.

Pan (2010) described excellent fossil material of *Vepris* (Fig. 9-3) and *Clausena* (Aurantioideae) from the Oligocene of Ethiopia, and the discovery that both fossil clades seem to be so close to modern species is of special interest. The *Vepris* is "very similar" to the extant *V. glomerata* and *V. sansibarensis*, while the *Clausena* "differs little" from the extant *C. anisata*, and the differences seemed so minor that Pan refrained from describing new species. The modern appearance of the Eocene–Oligocene fossils suggests that in Rutaceae the Cenozoic is only relevant for differentiation at or below species level.

MELIACEAE—THE MAHOGANY FAMILY

Rutaceae form a well-supported clade with Meliaceae (50 genera, 565 species) and Simaroubaceae (22 genera, 100 species). Fossil-calibrated divergence estimates "indicate that Meliaceae originated *after* the last known connection between Africa and South America" (Muellner et al., 2006; italics added). Fossils indicate that a clade evolved *before*, not after, a particular date and Muellner et al. eliminated trans-Atlantic vicariance only because the minimum (fossil) dates were transmogrified into maximum clade ages. Muellner et al. continued: "Although we are aware that our calculated values are minimum estimates our fossil calibration points would have to be at least 30–40 m.y. older to make the oldest Old–New World divergence events within the Meliaceae consistent with a continental break-up scenario" (p. 246). A difference between the age of a clade's origin and the age of its oldest fossil will always be expected. With the data of Muellner et al., the magnitude of this difference in Meliaceae can be estimated as at least 30–40 m.y., giving a useful estimate of the sampling error in the fossil record.

Muellner et al. (2006: 247) concluded: "Investigations employing new tools for biogeographic reconstruction of plants, most importantly DNA sequence data in conjunction with new fossil findings and a refined knowledge of geological history, have shown that long-distance, transoceanic dispersal may have played a major role in shaping the distribution of many taxa." What these studies have really shown is that center of origin/dispersal theory can be supported, but only by transmogrifying minimum dates into maximum dates.

Muellner et al. (2006) wrote that reliable age estimates depend on "completeness of the fossil record of a plant group . . . uncontroversial geology and age of the fossils . . . unambiguous assignment of the fossil taxa to extant taxa . . . and unambiguous, well-supported positions of the corresponding extant taxa in the phylogenetic trees." They proposed that two taxa fulfilled all these requirements: *Ailanthus* and *Toona*. But, as noted above, the oldest *Ailanthus* fossils are "nearly identical" (Pfeil and Crisp, 2008) to modern plants and appear in the fossil record without any obvious precursors. The simplest explanation is that their record is very incomplete.

Fossils from the Late Cretaceous of Wyoming have been attributed to an extant genus of Meliaceae, *Guarea*. Muellner et al. (2006) treated this as "controversial," although both the experts they consulted gave personal communications that the identification "may be correct." Muellner et al. did not use the fossil in the analysis as this calibration produced "unrealistically old ages" for clades. Instead, the fossil material provides a possible minimum age for Meliaceae genera and, given the distributions, the actual age of most of the genera is suggested to be mid-Cretaceous.

Apart from the issue of transmogrification, the analysis of meliaceae by Muellner et al. (2006) is outstanding, and 44 of the 50 genera were sampled. Many affinities are shown to straddle late Mesozoic ocean basins:

- Trans-Atlantic affinities include *Trichilia*: Central and South America, Africa and Madagascar, + *Guarea*: Central and South America, Africa.

- Trans-Indian Ocean disjunction is seen in *Ekebergia/ Quivisianthe*: Africa and Madagascar, + *Sandoricum*: Peninsular Malaysia to New Guinea.

- Trans-tropical Pacific Ocean affinities are seen in *Chukrasia*: India to Borneo, + *Schmardaea*: South America, in *Toona*: Pakistan to eastern Australia, + *Cedrela*: Central and South America (Muellner et al., 2010), and in a large clade of eight genera (*Aglaia, Dysoxylum*, etc.) that ranges between India and the southwest Pacific (to Samoa and Niue), + *Cabralea*: Central and South America.

Muellner et al. (2008) studied ITS sequences in the Aglaieae and again transmogrified their fossil-calibrated dates. This "provided evidence that the group is *not old enough* for major tectonic events to have influenced current distribution by vicariance. On the contrary, our data *clearly indicate* [following transmogrification] that long-distance dispersal has played a major role in the distribution and divergence of taxa within the Aglaieae" (p. 1785; italics added). The authors concluded that "Our fossil calibrations would have to be at least 40–50 Myr older to make divergence events consistent with continental break-up scenarios" (p. 1782). The authors see this 40–50 m.y. gap as militating against vicariance, but, again, it may be a more or less accurate estimate of the gap between age of origin and age of fossilization.

SIMAROUBACEAE—THE TREE OF HEAVEN FAMILY

The phylogeny of Simaroubaceae has been clarified by Clayton et al. (2007a), while Clayton et al. (2007b) calibrated the time course of the phylogeny. They used oldest fossils and again transmogrified the data; their results "add to a growing body of knowledge concerning the importance of the North Atlantic landbridge as a migration route for tropical groups, but also suggest long-distance dispersal has played a vital role in shaping current species distributions."

Instead, as in Rutaceae and Meliaceae, the molecular phylogeny of Simaroubaceae (Clayton et al., 2007a) indicates several key clades with disjunctions across ocean basins, and these can all be used for dating:

- Trans-tropical Pacific groups in the family include the basal clade. This comprises American genera plus *Picrasma*: India and southern China to Japan, New Guinea, and the Solomon Islands, also in Mexico, the Caribbean, and Central and South America.

- Trans-Atlantic groups include *Quassia* of West Africa (Nigeria to Angola) and tropical America (Brazil, Guianas, Central America, and Mexico), and *Pierreodendron* (= *Mannia*) of West Africa (Togo to Gabon) + *Simarouba* of eastern Brazil to Peru and Honduras.

- Trans-Indian Ocean groups include *Samadera* of Madagascar, southern India, Sri Lanka, and Burma to Australia and the Solomons, and also *Soulamea* of the Seychelles, Borneo to New Caledonia (most species), and Fiji.

SAPINDACEAE

The sister group of Meliaceae–Rutaceae–Simaroubaceae is the family Sapindaceae, widespread in warmer areas. Buerki et al. (2011) gave a phylogeny of ~60% of the genera and constrained the most recent common ancestor of Sapindaceae and Simaroubaceae to a maximum age of 125 Ma, although this was based on a fossil calibration.

EVOLUTION OF RUTACEAE AND ALLIES

Transmogrified dates are often used to rule out earlier events such as vicariance, although this is not logical. Instead, a fossil or fossil-calibrated date is valuable because, as a minimum age, it can be used to rule out *later* events as relevant.

The fossil-calibrated chronologies proposed for Rutaceae (Pfeil and Crisp, 2008), Meliaceae (Muellner et al., 2006, 2008), and Simaroubaceae (Clayton et al., 2007b) were based entirely on the transmogrification of fossil-based dates into clade ages. Because of this, the authors supported the traditional, dispersalist interpretation of biogeography in areas such as the Pacific. In this view, the region was originally devoid of Meliaceae, Rutaceae, Simaroubaceae, and their ancestors, and was only colonized later by immigrants from Asia and America. All, or nearly all, of the allopatric differentiation has been achieved by chance and founder dispersal, and the tectonic development of the region is seen as too old to have any relevance for the history of the modern taxa there.

Instead, the model proposed here suggests that the high diversity of groups such as Rutaceae in Brazil, South Africa, Western Australia, New Caledonia, and the Hawaiian Islands is the direct result of phylogeny and vicariance producing allopatric, regional blocks of taxa. This interpretation does not involve transmogrification but suggests that many groups have been affected by events in Early and Late Cretaceous time. Extensive overlap of clades in the families, including many genera, implies much secondary dispersal and range expansion, perhaps in the mid-Cretaceous, followed by a new phase of immobilism through the Cenozoic in which the modern subgenera and species differentiated. Long-distance dispersal across ocean basins and founder speciation are not required at any stage.

Many other pantropical groups have a biogeographic structure that shows parallels with the Rutaceae and their allies. Examples include the following groups of angiosperms.

CANELLALES

This pantropical order (Fig. 9-5) comprises Winteraceae and Canellaceae (Marquínez et al., 2009). Doyle (2000) cited possible fossil pollen of Winteraceae in Africa (*Walkeripollis*), although Van der Ham and van Heuven (2002) did not accept the identification. The two families in Canellales are almost completely allopatric, with primary breaks in northern Madagascar, the São Paulo/Rio de Janeiro region, and Central America/Venezuela. Breaks within the families at the Atlantic, Indian, Tasman, and Pacific basins occurred later, but perhaps not much later. The only dispersal required in the modern clades

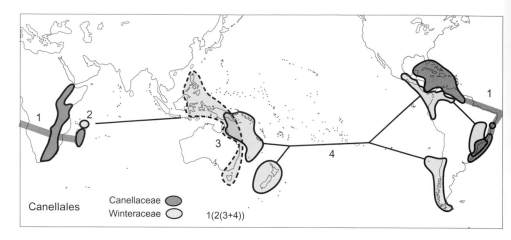

FIGURE 9-5. Distribution of the two families in Canellales. 1 = Canellaceae, 2–4 = Winteraceae (Marquínez et al., 2009; Stevens, 2010). The phylogeny is indicated.

shown in Fig. 9-5 is in Papua New Guinea/northeastern Queensland and in southeastern Brazil, perhaps in Costa Rica and Madagascar.

CUNONIACEAE TRIBE CUNONIEAE (OXALIDALES)

This is another widespread tropical clade (Fig. 9-6; data from Hopkins and Bradford, 1998; Hopkins et al., 1998; Rozefelds et al., 2001; Bradford, 2002). The trees in the group are often abundant in rainforest, especially in cooler, montane areas. The family Cunoniaceae as a whole is present in southwestern Australia, but not in tropical Africa. The higher-level clade Cunoniaceae + Elaeocarpaceae (including Cephalotaceae and Brunelliaceae) is absent in Africa outside the Cape area (Stevens, 2010); the remaining families in Oxalidales are either diverse in Africa (Oxalidaceae, Connaraceae) or endemic there (Huaceae), suggesting initial vicariance. Fossils of several genera (including "*Weinmannia*") are known from Australia, and there may have also been extinction in Africa, although these extinctions may not affect the overall pattern of allopatry in the extant groups.

The sequence of differentiation could be interpreted as dispersal from a center of origin near the basal group in eastern Australia, to New Caledonia (second basal group), from there to the Mascarenes, and across the Atlantic to South America. *Cunonia* would then have to disperse back again from South Africa to New Caledonia, and *Weinmannia*

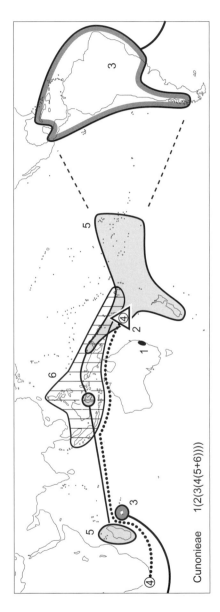

FIGURE 9-6. Distribution of Cunoniaceae tribe Cunonieae, showing the six main clades. The phylogeny is indicated. 1 = *Vesselowskya*, 2 = *Pancheria* (New Caledonia), 3 = *Weinmannia* s.str., 4 = *Cunonia* (dotted line), 5 = "*Weinmannia*" sect. *Leiospermum*, 6 = "*Weinmannia*" sect. *Fasciculatae* (Hopkins and Bradford, 1998; Hopkins et al., 1998; Rozefelds et al., 2001; Bradford, 2002). Fossils of "*Weinmannia*" are also recorded from Australia.

would have to disperse back again from the Tasman to Madagascar. This convoluted process requires a whole series of improbable, 10,000-km-long dispersal events; a sequence of differentiation in a group widespread through the southern hemisphere is much simpler.

In this model, the ancestral Cunonieae came into existence by vicariance with other tribes in Australia, where Cunonieae, apart from *Vesselowskya* and fossil "*Weinmannia*," are absent. This ancestral complex has broken down into subordinate components around two main nodes, one in the southwest Pacific and one in the southwest Indian Ocean. Repeated differentiation continued around these nodes over the same period, resulting in a phylogeny which gives the appearance of repeated long-distance dispersal jumps.

The sequence of differentiation proceeds in a centrifugal wave from the central Indian Ocean basin/Australia outward. The basal nodes that separate clade 1 (McPherson–Macleay overlap) from the rest, and then separate clade 2 (New Caledonia) from the rest, both occur in the Tasman Sea region. The rest of the pattern continues a parallel arc structure, 3 versus the rest, 4 versus the rest, and 5 versus 6. As noted, the two main zones of differentiation occur in the southwest Indian Ocean and the southwest Pacific Ocean. There are also breaks in the Pacific (between clade 3 and the rest), corresponding to the East Pacific Rise, and in the Atlantic (within clade 3).

Clades 5 and 6 show some secondary overlap with each other, but the main components of their distributions are allopatric. For example, clade 5 approaches mainland New Guinea (on Karkar Island, 15 km offshore) but is not on it, whereas clade 6 is well established there. Conversely, clade 5 is in New Caledonia while clade 6 is absent there, despite its presence in nearby Vanuatu. The comparative poverty of *Weinmannia* in New Caledonia, with only four species, is balanced by the high diversity of its vicariants there, including *Pancheria* (32 species, all endemic) and *Cunonia* (23 species, all endemic) (Jaffré et al., 2001). These figures represent high levels of diversity and endemism for trees, considering the size of the island.

The origin of the group and its subsequent evolution are compatible with a tectonic model in which terranes have rifted off the northern Gondwana margin at different times and then moved across prior Tethys basins (Paleo-, Meso-, and Ceno-Tethys) until they collided with Laurasia (Chapter 5). During the translation of each new set of terranes across the Tethys region, subduction has destroyed the earlier basin at the Laurasia margin, while any lighter crust, along with its biota, has been accreted to Laurasia. In the same way, the current diversity in New Caledonia probably results from the fusion of terranes derived from the Gondwana (Australia) side and others derived from the Pacific side (Heads, 2008b). The break between the Polynesian and American populations in the central Pacific may be related to the Cretaceous volcanism there. This was the greatest igneous event in Earth history and resulted in the emplacement of huge oceanic plateaus (Chapter 6). The collision of one of these, the Ontong Java Plateau, with the Solomon Islands has had profound effects on regional tectonics and biogeography. It may also explain the piling up of clades of Cunoniae in the triangle New Caledonia–northern New Guinea–Fiji.

MONIMIACEAE (LAURALES)

This family is widespread in tropical and south temperate rainforests (Fig. 9-7) and is one of the most abundant groups in the montane forest of New Guinea. Maximum diversity occurs in the Moluccas–New Guinea–northeastern Queensland region, where four main lineages overlap.

The distribution and phylogeny (Renner et al., 2010) show:

- trans-Indian Ocean (Gondwana) connections,
- trans-Atlantic connections between the Gulf of Guinea and Brazil,
- southern South Pacific (Gondwana) connections, and
- trans-tropical Pacific connections.

The initial split is between the southern clade 1 (Mascarenes, Australasia, Patagonia) and the others. Differentiation in the latter began in what is now the southwest Indian Ocean and Atlantic Ocean (between 2, 3, and 4 and their sister groups) and then occurred further east, with breaks (and subsequent overlap) around New Guinea and the Pacific basin. The breaks here are 5 versus the rest, 6 versus 7 and 8, and 7 versus 8. The main nodes are arranged in an arc: southwest Indian Ocean, Sri Lanka, southwest Pacific Ocean. There also breaks in the Atlantic (between 3 in Africa and its sister group 4–8, present in Brazil–Madagascar) and between Chile and Brazil/Peru. The genus *Monimia* is endemic on the Mascarenes and is dated as much older than the islands it currently occupies (see Chapter 2).

Renner et al. (2010) wrote that "Tree topology, fossils, inferred divergence times and ancestral area reconstruction fit with the break-up

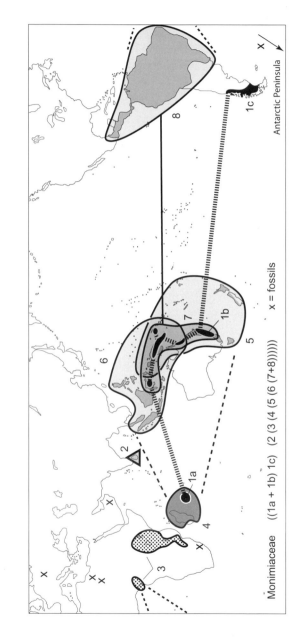

Monimiaceae   ((1a + 1b) 1c) (2 (3 (4 (5 (6 (7+8))))))   x = fossils

FIGURE 9-7. Distribution of Monimiaceae (based on Renner et al., 2010). The phylogeny is indicated. X = fossil localities.

of East Gondwana having left a still discernible signature consisting of sister clades in Chile and Australia." In contrast, they suggested that there is no support for the breakup of West Gondwana (Africa/South America) causing disjunctions in the Monimiaceae. But if a widespread ancestor is assumed, the break between clade 3 in Africa and clades 4–8 in Madagascar, Australasia, and America is consistent with separation by rifting in the Mozambique Channel and, later, the Atlantic. This suggests that a break between two clades can develop at one time in one part of the world and much later in another region.

Among the most interesting aspects of the beautiful pattern revealed by Renner et al. (2010) are the two trans-Pacific connections. The break in clade 1 between Australasia and Patagonia may run south of New Zealand (cf. more polar projections in Google Earth) and is easily explained by rifting in parts of Gondwana around West Antarctica. The other trans-Pacific connection is between the Australasian clade 7 and the American 8 (*Mollinedia*, etc.). Clade 7 has an identical distribution to that of 1b, but the distribution of 8 in America is quite different from that of 1c. The trans-Pacific connection between 7 and 8 is probably via the central Pacific, rather than the Antarctic/Pacific affinity in clade 1. Renner et al. (2010) approached the problem of the 7–8 connection using dates; they suggested young (Cenozoic) ages for many clades in the family, although the phylogeny was fossil-calibrated, and so these are minimum ages. Based on the conversion of these into maximum ages, they ruled out earlier vicariance and so inferred a dispersal model. They concluded: "The South American *Mollinedia* clade is only 28–16 Myr old and appears to have arrived via trans-Pacific dispersal from Australasia. . . . The crown ages of the other major clades range from 20 to 29 Ma, implying over-water dispersal between Australia, New Caledonia, New Zealand, and across the Indian Ocean to Madagascar and the Mascarenes."

Renner et al. (2010) admitted that the young age inferred for the *Mollinedia* clade (diverged at ~28 Ma, crown group age 16 Ma) is "most surprising," but these are minimum estimates and the actual age may be much older. They also commented that "Given the family's distinct pollen (in some genera), it is surprising that there are no pollen fossils." As the authors wrote, "The family is of disproportionate biogeographical interest because of its highly disjunct range and deep fossil record." While the fossil record is valuable, it is not perfect; the oldest fossil is only Late Cretaceous (83 Ma) and, as indicated, there is no known fossil pollen at all.

Renner et al. (2010) described the apparent youth of the *Mollinedia* clade as "[probably] the most unexpected biogeographical finding of this study. . . . The tree topology, chronogram and biogeographical reconstruction . . . , taken together, indicate that the ancestor of the *Mollinedia* clade arrived from Australasia by long-distance dispersal. The ability of Monimiaceae to disperse across salt water is attested by their occurrence on islands of the Bismarck Archipelago, all the way east to New Ireland Province." The presence of plants on islands is not necessarily evidence that they dispersed there from another region; they may have survived on other islands around the Bismarck/Solomons arc since the Mesozoic. The local dispersal this implies is quite different from the events required for trans-tropical Pacific dispersal. Renner et al. (2010) were intrigued by the trans-tropical Pacific affinity and referred to at least 20 other angiosperm genera or tribes in which it occurs. These groups show a wide range of ecology and means of dispersal, suggesting that these factors were not involved in producing the pattern.

Renner et al. (2010) and others have based their idea of dispersal across the tropical Pacific ($\sim$15,000 km) on a logical error, the transmogrification of fossil-calibrated dates, that is, minimum ages, into maximum ages that are then used to rule out earlier vicariance. In support of their result, if not their method, Renner et al. (2010: 1228) argued that "The resurrection of transoceanic dispersal as an explanation for range disjunctions has become so pervasive that long-distance dispersal (LDD) now seems a more plausible *a priori* explanation for most disjunctions than does continental drift (cf. de Queiroz, 2005)." But whether or not an explanation is pervasive in the scientific community is—or should be—irrelevant to whether or not it is scientifically plausible. Despite these problems with interpretation, the pantropical patterns now being described in work such as that of Renner et al. (2010) are among the most fascinating in the field and deserve much more study and discussion.

10

Evolution in Space, Time, and Form

Beyond Centers of Origin, Dispersal, and Adaptation

This book has dealt at some length with ideas on the center of origin and dispersal. The third component of the CODA model is adaptation by natural selection. Some current views of geneticists on this are discussed below, but first several conclusions reached in earlier chapters are summarized.

EVOLUTION IN SPACE AND TIME

Many of the studies cited in earlier chapters indicate that phylogenetic and geographic breaks in the distributions of plants and animals correspond with tectonic features. This spatial patterning is compatible with a history in which Earth and life have evolved together. This in turn suggests that standard models of the time scale of evolution may be misleading. Clades may be one, two, or even three orders of magnitude older than suggested in analyses that depend on clock models of evolution and fossil calibrations. If geological and biological evolution are linked, as suggested here, the ages of clades in the fossil record must be minimum ages only.

An alternative possibility is that the fossil record reflects the ages of clades more or less accurately. This would mean that any aspects of distribution not attributable to ecology are the result of chance dispersal in the Cenozoic. Any biogeographic similarities would be pseudopatterns caused by pseudocongruence. As molecular study continues, in more and more distribution patterns the component clades of a pattern are shown to have disparate branch lengths, suggesting (if clock-like evolution is assumed) that *all* patterns are due to chance dispersal at different times.

The choice involves accepting either that the fossils are younger to much younger than their clades, which seems likely, or that geographic distribution patterns, each repeated many times in different groups with different ecology and means of dispersal, are inherently meaningless. Accepting the first option opens up a vast field of enquiry, while accepting the second would mean there is little point in studying and comparing distributions, as they were generated stochastically. One implication of the first possibility is that the center of origin/dispersal component of the CODA model is flawed.

The debate has always involved the time factor. Dispersalists have long argued that while vicariance would be an acceptable model of evolution, the groups are not old enough for this, as shown by the fossil record and fossil-calibrated studies. But while fossils provide valid, useful evidence, fossil age must be always be distinguished from clade age. Broad trends are evident in the fossils; for example, in most groups the structure in Mesozoic members shows a different level of organization from the modern, Cenozoic forms. In practice, though, the fossil record is severely limited when it comes to dating individual lineages. In the molecular era, workers often claim to have found new evidence for dispersal, based on the age of the clade. But the "molecular" ages are based on fossil identifications and calibrations, along with a slightly relaxed version of the clock model of evolution that was introduced by an advocate of chance dispersal (Matthew, 1915).

EVOLUTION IN FORM: SELECTION, ADAPTATION, AND THE ORIGINS OF THE CODA MODEL

Modern ideas of morphological evolution began with Lamarck (1809). In his model, new circumstances (new habits, habitats, or both) produce new needs (*besoins*) in a group, and it is these needs that cause evolution. The *extrinsic need*, rather than any intrinsic propensity, is the key factor. This was the view that Darwin favored in his earlier writings, and later it formed the basis of the neo-Darwinian Modern Synthesis. This approach incorporates the concepts of center of origin, dispersal, and adaptation, and the evolutionary sequence is:

1. An organism develops new habits or disperses into a new habitat or locality.

2. The new habit or habitat means the organism has new needs.

3. These extrinsic needs cause the evolution of new, adaptive structure.

The question then is: Why did the organism move into the new habitat to begin with and how did it survive there before it adapted? The flatfishes (order Pleuronectiformes) provide a good example of the problem. The head twists and one eye appears to migrate to the other side of the body. In the orthodox neo-Darwinian explanation, this is "because their ancestors lay down on one side" (Dawkins, 2006: 121). Why the fishes lay down to start with is not explained; instead, this model focuses on the idea that the new sea-floor habitat and posture meant that the animal had new needs. In particular, the head needed to twist, and Dawkins emphasized that the "absurd distortion" in the flatfishes is carried "to the point of grotesqueness."

In an alternative, comparative approach to the problem, left–right asymmetry is seen as a trend in many animal phyla. Rather than simply being a "wonder of nature" or a one-off freak, the flatfish morphology is a point on a series, a long-term trajectory that can be analyzed with respect to phylogeny and morphogenesis. Left–right asymmetry occurs not just in Pleuronectiformes but, to a greater or lesser extent, in all the orders of chordates and echinoderms. It is present from the embryo stage onward and is evident in the most deep-seated organs, such as the brain and the heart (Boorman and Shimeld, 2002; Raya and Izpisúa Belmonte, 2006), as well as in superficial structures such as hair whorls in mammals and scale spirals in fishes. A purely bilateral form would not be viable. In an alternative to the Lamarckian/neo-Darwinian model, flatfishes lay down on one side because their head was twisted to start with and lying on the seafloor was the best option available. In the same way, left–right twisting in the jaw occurs in many vertebrates and has determined aspects of feeding ecology in some cichlids (Stewart and Albertson, 2010) and snakes (Hoso et al., 2007), for example.

Although Darwin adopted a pan-selectionist viewpoint in his earlier writing, such as the first edition of the *Origin of Species*, in his later work the emphasis shifted as he began to investigate "laws of growth" (Heads, 2009c). In the *Descent of Man* he wrote:

> I now admit that . . . in the earlier editions of my "Origin of Species" I perhaps attributed too much to the action of natural selection. . . . I was not . . . able to annul the influence of my former belief, then almost universal, that each species had been purposely created; and this led to my tacit assumption that every detail of structure, excepting rudiments, was of some special,

though unrecognised service. Any one with this assumption on his mind would naturally extend too far the action of natural selection. (Darwin, 1882: 61)

Inherent, long-term structural trends—Darwin's laws of growth—may be the primary factors determining morphological and molecular evolution, with selection pruning off forms that are not viable. In this view, structure determines function. Instead, neo-Darwinians follow the early Darwin in suggesting that function, like habitat, determines structure and somehow comes first. For example, Dawkins (2005: 497) argued that "Sponges live by passing a ceaseless current of water right through their body, from which they filter food particles. Consequently they are full of holes." Instead, the holes probably came first and the situation can be summed up: Sponges are full of holes, consequently a ceaseless current of water passes through their body.

The date palms and allied fan palms (Phoeniceae and Trachycarpeae) provide an example from plant life. With geological and climatic change since the Cretaceous, different members of this clade have found themselves stranded around the world in areas such as desert oases in Arabia, lowland and montane rainforest in Southeast Asia, and small atolls in the central Pacific. In the CODA model, the groups invaded these extreme habitats from a center of origin and then developed adaptations for living there. In a vicariance model, the direct ancestors of the Phoeniceae/Trachycarpeae already occupied Arabia, Asia, and the central Pacific before the divergence of the modern tribes or genera, and before subsidence, desertification, and orogeny had produced the modern atolls, deserts, and mountains. The palms had sufficient genetic flexibility to survive and evolve *in situ* during the processes that created the modern habitats. If a group has suitable pre-adaptations ("exaptations"), it will survive these sorts of changes. If not, it goes extinct. In this model there is no need for organisms to invade harsh new environments before they have developed the adaptations required to live there.

EXTRINSIC NEEDS VS. INTRINSIC GENETIC FACTORS AS CAUSES OF EVOLUTIONARY DIRECTION

The Modern Synthesis was based on Darwin's earlier, pan-selectionist concepts and it interpreted organic structure as the result of either natural selection or random drift. In contrast, Darwin's later writings and the work of the first geneticists suggested that selection is not the only factor causing non-random evolution or even the most important one. It has been suggested that the initial *divergence* of clades with different ecology is not the result of ecological adaptation (see Chapter 1 on ecological speciation). Likewise, the *diversity* of a clade may not be related to its ecology or its age but to "an intrinsic, clade-specific trait" (Lovette et al., 2002), a propensity for evolution inherent in the group's genome (Chapter 8). Other ideas on internal factors as a cause of evolution have appeared in recent work on genomics, and some of the historical background to this is considered first.

Selection, Mutation, and the Modern Synthesis

In the Modern Synthesis, selection is the only source of direction in evolution (Huxley, 1942: 56; Mayr, 1960: 355; Simpson, 1967: 159). Fisher (1930) and Haldane (1932) interpreted evolution as a "winner takes all" contest between the opposing pressures of mutation and selection. They reasoned that mutation rates are small and so selection beats mutation (Stoltzfus, 2006). Fisher (1930: 20) wrote: "The whole group of theories which ascribe to hypothetical physiological mechanisms, controlling the occurrence of mutations, a power of directing the course of evolution, must be set aside. . . . The sole surviving theory is that of Natural Selection." In the Fisher–Haldane model, the variation is random and does not favor one kind of outcome over another. Adaptation takes place at many loci simultaneously by means of infinitesimal variation so abundant that the process does not depend on the rate of new mutations. Thus mutation is not an effective evolutionary force and "evolution" can be reduced to shifting the frequencies of alleles already in the "gene pool" (Stoltzfus, 2006). When evolution is redefined in this way, the introduction process disappears and its effects become inaccessible.

Stoltzfus (2006) summarized the history as follows. In Darwin's early theory there was *variation on demand*: "Altered conditions of life" automatically turn on the flow of variation, producing abundant, infinitesimal, hereditary fluctuations leading to adaptation by selection, the engine of evolution. In neo-Darwinism the engine has a tank of fuel, the gene pool, that automatically keeps itself full. The gene pool theory of neo-Darwinism held that variation is "soaked up like a sponge" in a population and maintained, ensuring its constant availability. In Darwin's own model there is no storage tank, but fluctuating variation supplies fuel on demand, directly to the engine. Though not equivalent, both views represent variation as merely an abundant source of fuel, with no influence on where evolution goes. The abundance of raw materials "ensures that selection may spring into action to build anything, anywhere, anytime" (Stoltzfus, 2010). The parallels with chance dispersal are obvious. In the Modern Synthesis, evolutionary direction is determined by the needs imposed by the conditions of life and adaptation to these, not by intrinsic propensities.

In contrast, the Mendelians recognized mutation as a source of discontinuity, initiative, direction, and creativity in evolution. In the debate between the two schools of thought, the Modern Synthesis saw itself as rescuing evolutionary biology from the Mendelian heresy by showing that genetics is consistent with pan-selectionism—if mutation is denied any direct importance. In reality, the Mendelians had already synthesized genetics and selection, but they rejected Darwin's early views of heredity and pan-selectionism.

The source of evolutionary initiative in the Modern Synthesis is not the occurrence of *mutations*, which are individually insignificant (Dobzhansky et al., 1977: 72) and merely "replenish the supply of variability in the gene pool" (Stebbins, 1966: 29). The source of any initiative is the *change in conditions* that brings on selection of variation already present (e.g., Dobzhansky, 1955: 282; Mayr, 1963: 613; Dobzhansky et al., 1977: 6; Stebbins, 1982: 160). The following quotations illustrate this perspective:

> Evolution is not primarily a genetic event. Mutation merely supplies the gene pool with genetic variation; it is selection that induces evolutionary change. (Mayr, 1963: 613)

> Darwinism and the Modern Synthesis can be defined as: "the theory that selection is the only direction-giving factor in evolution." (Mayr, 1980: 3)

> Mutations are rarely if ever the direct source of variation upon which evolutionary change is based. Instead, they replenish the supply of variability in the gene pool which is constantly being reduced by selective elimination of unfavorable variants. . . . Consequently, we should not expect to find any relationship between rate of mutation and rate of evolution. (Stebbins, 1966: 29)

> Natural selection directs evolution not by accepting or rejecting mutations as they occur, but by sorting new adaptive combinations out of a gene pool of variability which has been built up through the combined action of mutation, gene recombination, and selection over many generations. (Stebbins, 1966: 31)

> If ever it could have been thought that mutation is important in the control of evolution, it is impossible to think so now. (Ford, 1971: 361)

As late as 1932 T.H. Morgan was asserting that "natural selection does not play the role of a creative principle in evolution," but ten years later all but a very few biologists were agreed on an evolutionary theory based firmly on Darwin's own ideas knitted with subsequent developments in genetics. (Berry, 1982: 14)

The Geneticists: Mendelian and Mutationist

In the Modern Synthesis account of history, the early Mendelians—the "geneticists"—are portrayed as naive, foolishly rejecting selection. This is not correct; what they did do was integrate mutation with selection and deny that selection was the only directional force in evolution. For T.H. Morgan (1916) and the other geneticists, mutation produced the variation while selection pruned off subviable forms, and evolution was a synthesis of the two forces. They saw "mutation as the basis of evolution, as the material upon which natural selection works (Punnett, 1911: 139). Morgan (1916) concluded:

> Such a view gives us a somewhat different picture of the process of evolution from the old idea of a ferocious struggle between the individuals of a species with the survival of the fittest and the annihilation of the less fit. . . . New and advantageous characters survive by incorporating themselves into the race, improving it and opening to it new opportunities. In other words, the emphasis may be placed less on the competition between the individuals of a species (because the destruction of the less fit does not *in itself* lead to anything that is new) than on the appearance of new characters and modifications of old characters that become incorporated in the species.

In Modern Synthesis historiography (Berry, 1982; Ridley, 1985; Dawkins, 1986: 305), the mutationists, led by such eminent geneticists as Morgan, de Vries, and Bateson, rejected selection because they did not understand it. In fact, these authors all had a clear understanding of selection and they did not reject it. The only problem for Modern Synthesis historiography is that they were not pan-selectionists.

Although Morgan was a pioneer geneticist, a Nobel laureate (1933), and a distinguished figure in American science, his concept of evolution was rejected by the Modern Synthesis in favor of random mutation and pan-selectionism. In the same way, the pioneer vicariance biogeography developed in the early 1900s by A.E. Ortmann (University of Pittsburgh), C.H. Eigenmann (Indiana University), and K. Andersen (British Museum) was rejected by the Modern Synthesis in favor of chance dispersal.

The Importance of Parallelism

Parallel evolution has always been recognized by morphologists; countless molecular phylogenies now show that it is ubiquitous and even more important than was thought. Its interpretation has been controversial, though. The views of the early geneticists—the mutationists—on directionality emerge clearly in their interpretation of parallel evolution, which they thought indicated non-random tendencies of variation. This hypothesis was rejected in the Modern Synthesis. Mayr and others argued that given the abundance of variation in the "gene pool" and the ability of natural selection to shape this gene pool to fit circumstances, it was not safe to assume that shared characters had a shared basis in homologous genes. Mayr wrote that' "In the early days of Mendelism there was much search for homologous genes that would account for such similarities. Much that has been learned about gene physiology makes it evident that the search for homologous genes is quite futile except in very close relative . . . . If there is only one efficient solution for a certain functional demand, very different gene complexes will come up with the same solution" (Mayr, 1963: 609). Despite the significance of this theory in the Modern Synthesis, it has been refuted by discoveries in modern genetics showing, for example, that the same genes control aspects of development in insects and in vertebrates.

Some of the most striking parallelisms include homeosis, a class of mutations first described in insects in which segment identity is changed. For example, a leg may develop in place of an antenna. The early geneticists saw these mutations as important for evolution, and in flies such as some Asilidae the "normal" antennae may be leg-like. Homeotic mutations in animals involve the same family of genes (the Hox genes); the equivalent in plants is the MADS-box family. Filler (2007) reviewed homeotic evolution in mammals and supported "an enlarged role for a mutational view (Stoltzfus, 2006) of evolutionary drive to update classic Darwinian and New Synthesis models."

NEW IDEAS FROM GENETICS ON INTERNAL CAUSES OF EVOLUTIONARY DIRECTION

The vast amounts of new genetic data have stimulated reexamination of basic evolutionary concepts, and important critiques are now appearing. Some of these are cited next.

The Significance of Non-random Mutation: Climbing Mount Probable

In the Modern Synthesis, the direction of evolution is determined solely by selection and, in particular, the Modern Synthesis is distinguished from the earlier work on mutation "by its utter denial of any internal causes of directionality" (Yampolsky and Stoltzfus, 2001: 73). As already mentioned, the Modern Synthesis writers argued that mutation rates were too low with respect to selection pressure for mutation to be relevant. In contrast, Yampolsky and Stoltzfus (2001: 73) showed that mutational bias in the introduction of variation "can strongly influence the course of evolution. . . . Bias in the introduction of variation may be expected to contribute to homoplasy, parallelism and directionality." Yampolsky and Stoltzfus (2001) discussed evidence showing "Non-randomness in mutation has a *predictable* effect on the outcome of an evolutionary process. This predictable effect may be said to represent an *orienting*, *directional* or *shaping* influence. . . . Bias in the introduction of novelty by mutation is a prior bias on the course of evolution, and may be said to be an 'internal' cause of orientation or directionality" (p. 81).

One example of bias cited above is the variation seen in the base composition of genomes. Gibson et al. (2005) analyzed the mitochondrial genomes of 69 mammals. They found that "Mammalian mitochondria exhibit striking variations in base composition both within genomes and across species. . . . It is now obvious that the proportions of C and T both change between different species and that changes are congruent at all sites examined within the genome, consistent with a directional mutation pressure" (Gibson et al., 2005: 262). In primates, Matsui et al. (2009) sequenced complete mitochondrial genomes for 26 species and found significant variations of C and T compositions across the species.

Hughes (2009: 332) drew the general conclusion that "'the maturation of modern evolutionary biology has been hindered by a conception of natural selection as an all-powerful force shaping every aspect of phenotypic evolution." Instead, Stoltzfus and Yampolsky (2009: 637) wrote that "biases in the introduction of variation, including mutational biases, may impose predictable biases on evolution, with no necessary dependence on neutrality." They presented "a new analysis partitioning the variance in mean rates of amino acid replacement during human–chimpanzee divergence to components of codon mutation and amino acid exchangeability. The results indicate that mutational effects are not merely important but account for most of the variance explained."

Stoltzfus and Yampolsky (2009: 643) concluded: "First, evolution has tendencies or propensities, an idea that has faced resistance from those committed to the idea that evolution is unpredictable (see Beatty, 2008, for an analysis of the resistance to the idea of trends or directions). . . . Second, these tendencies are predictable." Neo-Darwinian authors who accepted random mutation and selection as the basis for evolution have interpreted evolution as "climbing Mount Improbable" (Dawkins, 2006). In contrast, Stoltzfus and Yampolsky (2009) saw intrinsic bias in mutation as a cause of *non*-randomness and so interpreted evolution as "climbing Mount Probable."

Mayr (2001) suggested that while selectively neutral mutations occur at the molecular level, they do not affect phenotypic characters and so they are of little interest to evolutionists. Even Kimura (1983), in advocating the neutral theory of molecular evolution, accepted a neo-Darwinian interpretation of phenotypic evolution. In contrast, Nei (2007: 12235) concluded that "Phenotypic evolution occurs primarily by mutation of genes that interact with one another in the developmental process. . . . It appears that the driving force of phenotypic evolution is mutation, and natural selection is of secondary importance."

Cutter et al. (2009: 1199) discussed evolution in the model organism *Caenorhabditis*, a nematode, and called for researchers "to generate and critically assess nonadaptive hypotheses for genomic and developmental patterns, in addition to adaptive scenarios . . . natural selection is only one of several core evolutionary forces that can lead to nonrandom patterns in genomes. Equally fundamental, mutation, recombination, and genetic drift are agents of evolutionary change that can shape genomes in a nonadaptive, selectively neutral manner. Consequently, it is essential to generate both neutral, nonadaptive hypotheses as well as adaptive hypotheses as testable alternatives."

Genome Evolution and the Evolution of Complexity

Lynch and Conery (2003) showed that the huge differences in genome organization seen in the massive size range (average for bacteria 0.5–10 Mb, average for animals 100–100,000 Mb) may be explained without the need for selection and adaptation. In a subsequent review, Lynch (2007a) described the frailty of adaptive hypotheses for the origins of organismal complexity. He wrote that "The vast majority of biologists engaged in evolutionary studies interpret virtually every aspect of biodiversity in adaptive terms. This narrow view of evolution has become untenable in light of recent observations from genomic sequencing and population genetic theory. Numerous aspects of genomic architecture, gene structure, and developmental pathways are difficult to explain without invoking the nonadaptive forces of genetic drift and mutation. In addition, emergent biological features such as complexity, modularity, and evolvability . . . may be nothing more than indirect by-products of processes operating at lower levels of organization . . . the origins of many aspects of biological diversity, from gene-structural embellishments to novelties at the phenotypic level, have roots in nonadaptive processes, with the population genetic environment imposing strong directionality on the paths that are open to evolutionary exploitation" (p. 8597). Nonadaptive forces *may* result in adaptive fit to an environment, but selection was not the process that produced this. If carried far enough, nonadaptive trends can also lead to extinction. As Lynch (2007a: 8597) observed:

> Evolutionary biology is treated unlike any science by both academics and the general public. For the average person, evolution is equivalent to natural selection, and because the concept of selection is easy to grasp, a reasonable understanding of comparative biology is often taken to be a license for evolutionary speculation . . . the myth that all of evolution can be explained by adaptation continues to be perpetuated by our continued homage to Darwin's treatise (1859) in the popular literature. For example, Dawkins' (1976, 1986, 2006) agenda to spread the word on the awesome power of natural selection has been quite successful, but it has come at the expense of reference to any other mechanisms, a view that is in some ways profoundly misleading.

In a consideration of "internal versus external evolutionary forces," Lynch (2007a: 8598) wrote: "The literature is permeated with dogmatic statements that natural selection is the only guiding force of evolution, with mutation creating variation but never controlling the ultimate direction of evolutionary change." Instead of this "religious adherence to the adaptationist paradigm," Lynch (p. 8599) proposed "the passive emergence of genome complexity by nonadaptive processes."

Lynch (2007a) discussed the non-random effects of mutation and recombination: "[I]t is now known that genome composition is governed by biases in mutation and gene conversion [at recombination], some of which (e.g., mobile-element proliferation) operate via internal drive-like mechanisms." (p. 8599). "The hypothesis that expansions in the complexity of genomic architecture are largely driven by nonadaptive evolutionary forces is capable of explaining a wide range of previously disconnected observations" (p. 8600).

With respect to multicellularity, Lynch (2007a: 8600) quoted King (2004): "the historical predisposition of eukaryotes to the unicellular lifestyle begs the question of what selective advantages might have been conferred by the transition to multicellularity." Lynch also quoted Jacob (1977): "[I]t is natural selection that gives direction to changes, orients chance, and slowly, progressively produces more complex structures, new organs, and new species." As Lynch (2007a: 8600) pointed out:

> The vast majority of biologists almost certainly agree with such statements. But where is the direct supportive evidence for the assumption that complexity is rooted in adaptive processes? No existing observations support such a claim, and given the massive global dominance of unicellular species over multicellular eukaryotes, both in terms of species richness and numbers of individuals, if there is an advantage of organismal complexity, one can only marvel at the inability of natural selection to promote it. Multicellular species experience reduced population sizes, reduced recombination rates, and increased deleterious mutation rates, all of which diminish the efficiency of selection. . . . It may be no coincidence that such species also have substantially higher extinction rates than do unicellular taxa.

Lynch (2007a: 8603) concluded: "many aspects of biology that superficially appear to have adaptive roots almost certainly owe their existence in part to nonadaptive processes. Moreover, if the conclusion that nonadaptive processes have played a central role in driving evolutionary patterns is correct, the origins of biological complexity should no longer be viewed as extraordinarily low-probability outcomes of unobservable adaptive challenges, but expected derivatives of the special population-genetic features of DNA-based genomes."

Koonin (2009a: 1023) asked the question: Has there been a consistent trend toward increasing organizational and genomic complexity during the ~3.5 billion years of organic evolution? He found that "The most likely answer is, no. Even very conservative reconstructions of ancestral genomes of archaea and bacteria indicate that these genomes were comparable in size and complexity to those of relatively simple modern forms. . . . Furthermore, reconstructions for some individual groups, and not only parasites, point to gene loss and genome shrinking as the prevailing mode of evolution." Koonin (2009a) cited the discovery of large and complex genomes in animals such as Cnidaria that diverged from the main line of metazoan evolution prior to the origin of the Bilateria. This "suggests that there was little if any increase in genomic complexity during the evolution of the metazoan (although organizational complexity did increase); instead, recurrent gene loss in different lineages was the most prevalent evolutionary process" (p. 1023).

The Evolution of Gene Regulatory Networks

Gene expression is controlled by complex pathways and networks of interacting regulator genes. Lynch (2007b) wrote: "Many physicists, engineers and computer scientists, and some cell and developmental biologists, are convinced that biological networks exhibit properties that could only be products of natural selection . . . ; however, the matter has rarely been examined in the context of well-established evolutionary principles [and] there has been no formal demonstration of the adaptive origin of any genetic network." Lynch showed that many features of known transcriptional networks can arise readily through non-adaptive drift, mutation, and recombination, raising questions about whether natural selection is sufficient or even necessary for the origin of gene-network topologies. Lynch (2007b: 804) continued: "Three observations motivate the hypothesis that a considerable amount of regulatory-pathway evolution is driven by non-adaptive processes. First, one of the most puzzling aspects of many genetic pathways is their seemingly baroque structure. . . . Second, many lines of evidence support the idea that the regulatory machinery underlying complex adaptations is capable of undergoing frequent and dramatic shifts without altering the outwardly expressed phenotype."

Biased Gene Conversion

Biased gene conversion in mammalian genomes was reviewed by Duret and Galtier (2009). They wrote that:

> Recombination is typically thought of as a symmetrical process resulting in large-scale reciprocal genetic exchanges [during crossover] between homologous chromosomes. Recombination events, however, are also accompanied by . . . unidirectional exchanges known as gene conversion. . . . A large body of evidence suggests that gene conversion is GC-biased in many eukaryotes, including mammals and human. AT/GC heterozygotes produce more GC- than AT-gametes, thus conferring a population advantage to GC-alleles in high-recombining regions. This apparently unimportant feature of our molecular machinery has major evolutionary consequences. Structurally, GC-biased gene conversion explains the spatial distribution of GC-content

in mammalian genomes—the so-called isochore structure [an isochore is a region of the genome with relatively homogeneous base composition]. . . . Did GC-rich isochores evolve because they confer some selective advantage, or do they simply result from nonadaptive evolutionary processes? . . .

In humans, of the 36 crossover hotspots analyzed so far, 8 show distorted segregation ratios. . . . The base composition of homologous genomic regions is strongly correlated between amniote species (mammals, birds, and reptiles) . . . no such isochore organization is observed in amphibians or fishes. . . . This isochore organization is still evolving: in many, but not all, mammals, GC-rich isochores are eroding, i.e., their GC-content is decreasing . . . neither selectionist nor mutational models provide satisfactory explanations for the origin of GC-rich isochores, [and so] we focused on a third possible hypothesis: biased gene conversion . . . sequences subject to a high level of recombination should be GC-rich. And indeed, the analysis of human sequences demonstrated a genome-wide positive correlation between crossover rate and GC-content.

Duret and Galtier (2009) concluded that biased gene conversion mimics natural selection: "By definition, gBGC [GC-biased gene conversion] results in the non-random transmission of alleles to the next generation . . . the impact of gBGC on substitution patterns can be very strong, even in regions that are under selective pressure (coding sites or regulatory elements). Indeed, in some cases, gBGC overcomes purifying selection and leads to the fixation of deleterious AT → GC mutations. . . . Thus, recombination hotspots might constitute the Achilles' heels of our genome."

Genetic Conservation and Functional Relevance

If a sequence or property appears to be conserved across evolutionary time, it is often interpreted as being functionally important. Yet Koonin (2009a) concluded that even "this 'sacred,' central tenet of evolutionary biology . . . is not an absolute and the nonadaptive alternative is to be taken seriously." Simulations of regulatory element evolution show that apparently conserved patterns can be produced by neutral or indirect forces (Muers, 2010). Lusk and Eisen (2010: 6) wrote: "Lynch (2007a) has eloquently argued that biologists are often too quick to assume that organismal and genomic complexity must arise from selection for complex structures and too slow to adopt nonadaptive hypotheses. Our results lend additional support to this view, and extend it to show that indirect and non-adaptive forces can not only produce structure, but also create an illusion that this structure is being conserved . . . it is essential that we give careful consideration to the neutral and indirect forces that we now know can produce evolutionary mirages of structure and function."

Animals as different as sponges and humans show conservation of their overall genome architecture. But in the tunicate *Oikopleura*, a member of the sister group of vertebrates, Denoeud et al. (2010) found that multiple genomic features (including transposon diversity, developmental gene repertoire, physical gene order, and intron–exon organization) are "shattered." Despite this, the organism retains chordate morphology. The authors concluded that the "Ancestral architecture of animal genomes can be deeply modified and may therefore be largely nonadaptive."

Evolution of the Genetic Code

Massey (2010) cited increasing evidence for the emergence of beneficial traits in biological systems in the absence of direct selection, and he explored the case of the standard genetic code. The code is structured such that single nucleotide substitutions are more likely to result in changes between similar amino acids. This "error minimization" is often assumed to be an adaptation. Yet Massey (2010: 81) showed that "direct selection of the error minimization property is mechanistically difficult. In addition, it is apparent that error minimization may arise simply as a result of code expansion, this is termed the 'emergence' hypothesis. The emergence of error minimization in the genetic code is likened to other biological examples, where mutational robustness arises from the innate dynamics of complex systems; these include neutral networks and a variety of subcellular networks . . . the term 'pseudaptation' is used for such traits that are beneficial to fitness, but are not directly selected for."

Genome Reduction

Rho et al. (2009) wrote that "The evolutionary patterning of genome architecture by nonadaptive forces is supported by population genetic theory, estimates of the relative power of the major forces of evolution, and comparative analyses of whole-genome sequences. Nevertheless, some biologists still adhere to the idea that even the most arcane aspects of genome evolution, including expansions of genome size by mobile element proliferation, are direct products of natural selection (e.g., Gregory 2005; Kirschner and Gerhart 2005; Caporale 2006)." In contrast, Rho et al. (2009) observed that "widespread reductions in genome size have occurred in multiple lineages of mammals subsequent to the Cretaceous–Tertiary (KT) boundary. . . . Although the mechanisms driving such change remain unclear, these results provide a compelling example of a broad syndrome of genomic changes being driven by apparently nonadaptive events, while also demonstrating that mammalian genome architecture is currently in a nonequilibrium state . . . our results challenge the notion that genome size reflects a finely tuned structural determinant of the adaptive phenotypes of organisms." Genome reduction may be important in many groups. Janes et al. (2010) described "the events leading from an ancestral amniote genome—predicted to be large . . . to the small and highly streamlined genomes of birds."

Morphological Reduction

Trends in morphology include reduction, suppression, and fusion in complex, multipartite structures such as the angiosperm flower and the vertebrate skull. This is especially clear in the pharynx and the jaws. Trends lead to particular morphologies, and these determine the ecological space that is viable for the species. If suitable ecology is available in the biogeographic region, the organism will survive. For example, members of the genus *Dasypeltis* are among the few snakes that are specialist egg-eaters. During the course of evolution, they have lost most of their teeth. "Consequently," as Adriaens and Herrel (2009) reasoned, "they cannot capture and transport other prey types and are stuck in an ecological and evolutionary dead end in being obligate egg eaters."

In the model organism the zebrafish (*Danio rerio*), the restriction of teeth to a single pair of pharyngeal bones is a distinctive feature. Stock (2007) observed that "Such reduction of the dentition, characteristic of the order Cypriniformes, has never been reversed, despite subsequent and extensive diversification of the group in numbers of species and variety of feeding modes." Stock proposed studies (mutant screens and experimental alteration of gene expression) to "reveal the relative contribution to trends in dental evolution of biases in the generation of variation and sorting of this variation by selection or drift."

Non-adaptive Evolution in Primates

The authors cited above, and others such as Kurland and Berg (2010), have denied the suggestion that all genomic evolution is "adaptive" and have instead accepted Lynch's (2007a) argument for nonadaptive events. Harris (2010: 13) reviewed "evidence that suggests that primate and human evolution has been strongly influenced by nonadaptive processes. . . . Evolutionary biology has tended to focus on adaptive evolution by positive selection as the *primum mobile* of evolutionary trajectories in species while underestimating the importance of nonadaptive evolutionary processes . . . adaptive evolution is only one of the forces of evolutionary change, others being mutation, random genetic drift, and recombination. These forces are described as nonadaptive because the evolutionary change they produce is due to factors unrelated to individual differences in relative fitness. The importance of nonadaptive forces in influencing evolutionary change is becoming increasingly clear." Harris (2010: 25) wrote that non-adaptive processes (such as genetic "surfing") "can produce markedly high measures of genetic differentiation between different geographic populations, a pattern traditionally interpreted as evidence of local adaptation".

Harris (2010: 38) concluded: "[W]e might ask: to what extent have the unique features of human evolution (i.e., in brain size and organization, bipedal features, hand morphology, etc.) as well as differences in morphology among human population been influenced by nonadaptive processes?" Spocter (2009) has interpreted trends in hominid brain and body size in terms of non-adaptive mechanisms, and a revision of other non-adaptive trends in primate morphology would be of great interest, especially if these were integrated with corresponding trends in behavior and ecology. Long-term non-adaptive trends may be responsible for most aspects of morphology and ecology.

Toward a Postmodern Evolutionary Synthesis

Koonin (2009b) concluded that "The edifice of the modern synthesis has crumbled, apparently, beyond repair. The hallmark of the Darwinian discourse of 2009 is the plurality of evolutionary processes and patterns." In particular, the Modern Synthesis idea that all evolution is due to adaptation by selection is outdated; Koonin (2009a) wrote that "genomes show very little if any signs of optimal design, and random drift constrained by purifying in all likelihood contributes (much) more to genome evolution than Darwinian selection. . . . Moreover, with pan-adaptationism gone forever, so is the notion of evolutionary progress that is undoubtedly central to traditional evolutionary thinking, even if this is not always made explicit. . . . The summary of the state of affairs on the 150th anniversary of the Origin is somewhat shocking. In the postgenomic era, all major tenets of the modern synthesis have been, if not outright overturned, replaced by a new and incomparably more complex vision of the key aspects of evolution. . . . Although at present only isolated elements of a new, 'postmodern' synthesis of evolutionary biology are starting to be formulated, such a synthesis is indeed feasible."

NEW IDEAS ON COMMUNITY ECOLOGY

The new work in genetics and the biogeographic patterns presented in the earlier chapters all undermine the CODA model of evolution. While clades survive in their areas and habitats because they have suitable morphology and physiology, these attributes did not develop due to extrinsic needs after a clade occupied its habitat and do not explain why the group was present there to begin with. These ideas have implications for evolutionary ecology.

Exploring a forest or a reef may be our childhood introduction to real diversity, and it is not surprising that the local community has become a basic concept of ecology. Until now, ecologists have attempted to understand a community by examining it in itself, dissecting the processes that occur within it in ever greater detail. In the traditional adaptationist model, the community is structured by the physical environment, predation, competition, and other local interactions. The history over geological time is not considered. Ricklefs (2008) has taken an alternative position and described the "Disintegration of the Ecological Community." In this paper he criticized studies on communities carried out at smaller, ecological scales without consideration of the broader biogeographic context. He argued that "the seemingly indestructible concept of the community as a local, interacting assemblage of species has hindered progress toward understanding species richness at local to regional scales" (p. 741). Instead, Ricklefs suggested that the broader, regional distributions of species reveal more about the processes generating diversity patterns than does the co-occurrence of species at any given point. "*The local community is an epiphenomenon* that has relatively little explanatory power in ecology and evolutionary biology. Local coexistence cannot provide insight into the ecogeographic distributions of species within a region, from which local assemblages of species derive, nor can local communities be used to test hypotheses concerning the origin, maintenance, and regulation of species richness, either locally or regionally. Ecologists are moving toward a community concept based on interactions between populations over a continuum of spatial and temporal scales within entire regions, *including the population and evolutionary processes that produce new species*" (p. 741; italics added). A clade may be present in a habitat because it evolved in the region, not because it migrated into the area and then adapted. In the same way, a particular community may be present in a locality not because of current conditions but because its precursor communities were already in the region, if not the exact locality, tens of millions of years ago. Ancestral lemurs were in ancestral Madagascar before lemurs or Madagascar existed as such.

The approach favored by Ricklefs (2008) is also adopted in panbiogeography. An area of alpine vegetation may have been derived from an uplifted lowland biota, and the only members of this that have survived have been some of the smaller, hardier groups, such as grasses and subshrubs of open communities and prostrate plants of river gravels. The main factors determining the taxonomic structure of the alpine community are the lowland community from which it was derived and the uplift, which may have occurred millions of years ago.

Ricklefs (2008) emphasized that the patterns of differentiation at a small, ecological scale must be integrated with patterns at a larger scale in space and time, and that local communities must ultimately be interpreted in the broader context. He concluded (p. 746): "[T]he region is the appropriate scale for an ecological and evolutionary concept of community." This is a move in the right direction, toward a global understanding, but what exactly are the "regions" and the regional biotas from which the local communities have been derived? The traditional biogeographic regions for terrestrial life correspond to the continents, and although these regions were a key part of the CODA model and the Modern Synthesis, they are oversimplistic and misleading. Most terrestrial groups, if their phylogeny is traced far enough, prove instead to be based around the Atlantic, Indian, or Pacific Oceans (e.g. haplorhine and strepsirrhine primates, Fig. 3-1; the angiosperm order Canellales, Fig. 9-5), and so the main subglobal units correspond more closely to tectonic basins than to continents. In this way, the conceptual disintegration of the regions follows the disintegration of the community (Ricklefs, 2008) and the adapted genome (Lynch, 2007a; Stoltzfus, 2006). Many local, apparently purely ecological phenomena, such as anomalies in diversity and elevation, have long phylogenetic and biogeographic roots, and their resolution may require study on a regional or even global scale.

The Modern Synthesis was based on the concepts of chance dispersal and random mutation, with structure answering the needs imposed by the environment. All these concepts seem flawed, and the CODA program appears to have sent genetics and biogeography on a century-long detour. A group's ecology and habitat are determined by its biogeography and morphology, and a group's morphology is determined by long-term, inherent trends (Darwin's "laws of growth") pruned by selection. Adaptation by selection is a secondary factor in evolution and no longer the sole or even the primary factor.

The new molecular phylogenies that are being produced show a more and more detailed picture of global biogeographic structure, and the intriguing patterns that are emerging should lead to a renaissance in biogeography. Molecular biologists will soon realize that the same geographic patterns repeat endlessly in groups with very different ecology and means of dispersal, and the stultifying idea that the patterns are determined by chance can be replaced with the idea that the Earth and its living layer have always evolved as one.

Glossary of Geological Terms

ACCRETED TERRANE A *terrane* which has been thrust into or over another and sutured to it at a major fault.

ACTIVE MARGIN A continental margin that is an active plate boundary. (Contrast with *passive margin*.)

ARC (See *volcanic arc*.)

BACK-ARC REGION The opposite side of an arc from the trench and the subducting plate. (Contrast with *forearc*.)

BASIN A structural basin is formed by tectonic warping of horizontal strata, due to *extension* or crustal flexure. It may be older, younger, or the same as age as the sedimentary basin—the strata that fill it. After its formation a basin may later be uplifted or deformed so that it is no longer a topographic basin.

BASIN INVERSION The contraction of a previously formed basin, usually accompanied by *fault inversion*.

COMPRESSION The set of stresses directed toward the center of a rock mass. If maximum compressive stress is horizontal, *shortening* and *thickening* can occur with the development of *reverse* and *thrust faults*. If maximum compressive stress is vertical, *extension* and *thinning* can occur, with *normal faults*.

CRATON An old, stable part of continental crust. (Contrast with *orogen*.)

DEPOZONE A distinct depositional zone, as in a *foreland basin* system.

DEXTRAL MOVEMENT (RIGHT-LATERAL MOVEMENT) Relative movement in which the block on the other side of the fault from the observer moves to the right. (Contrast with *sinistral movement*.)

DEXTRAL (= RIGHT-LATERAL) FAULT A strike-slip fault with dextral (right-lateral) slip. (Contrast with *sinistral fault*.)

DIP The dip of a tilted bed, fault surface, or other planar feature is the steepest angle of descent of the feature relative to a horizontal plane. The dip direction is at 90° to the *strike*.

DIP-SLIP FAULT A fault where the relative movement (slip) on the fault plane parallels the *dip* of the fault. (Contrast with *strike-slip fault*.)

EPEIROGENY Regional-scale uplift of large continental areas without associated folding, metamorphism, or deformation. (Contrast with *orogeny*.)

EXTENSION The increase in surface area of a region of crust, caused by tensional stress. Extensional structures include divergent plate boundaries and normal faults. (Contrast with *shortening*.)

FAULT INVERSION The reversal of displacement on a fault; the result of a changing stress field.

FLOWER STRUCTURE (= "palm tree structure"). A faulted splay structure formed near the surface where a major *strike-slip fault* breaks up into many separate faults. In cross-section it resembles the head of a flower. In zones of *transpression*, a positive ("everted") flower structure develops; in zones of *transtension*, a negative (sunken) flower structure develops.

FOLD AND THRUST BELT A series of mountainous foothills, formed by folds and thrust faults in the foreland adjacent to an orogenic belt, as deformation propagates outward from the orogen.

FOOT WALL If the plane of a *dip-slip fault* is not exactly vertical, the lower block is the foot wall. (Contrast *hanging wall*.)

FOREARC The region in front of an arc, on the same side as the trench and the subducting plate. (Contrast *back-arc region*.)

FORELAND BASIN A *basin* that develops adajacent to an *orogen*, between it and a craton. Thickening in the orogen causes gravitational loading that deforms the crust, leading to basin formation.

HANGING WALL If a dip-slip fault plane is not exactly vertical, the upper block is the hanging wall. (Contrast *foot wall*.)

ISLAND ARC A *volcanic arc* formed on oceanic crust, producing islands.

LEFT-LATERAL (See *sinistral movement*.)

NORMAL FAULT A *dip-slip fault* in which the hanging wall moves downward relative to the foot wall. Normal faults develop when the crust is extended. (Contrast *reverse fault*.)

OROGEN A mountain belt produced by *orogeny*. (Contrast with *craton*.)

OROGENY (= OROGENESIS) Subduction-related mountain building at a fold belt. (Contrast with *epeirogeny*.)

PASSIVE MARGIN A continental margin that is not an active plate boundary. (Contrast with *active margin*.)

REVERSE FAULT A *dip-slip fault* in which the *hanging wall* moves up relative to the *foot wall*. Reverse faults develop where the crust is shortened or contracted. (Contrast with *normal fault*.)

RIGHT-LATERAL (See *dextral movement*.)

SHORTENING The decrease in surface area of a region of crust, caused by contractional stress. Shortening structures include convergent ridges and *thrust faults*. (Contrast with *extension*.)

SINISTRAL (= LEFT-LATERAL) FAULT A *strike-slip fault* with sinistral (left-lateral) slip. (Contrast with *dextral fault*.)

SINISTRAL (LEFT-LATERAL) MOVEMENT Relative movement in which the block on the other side of the fault from the observer moves to the left. (Contrast with *dextral movement*.)

SLAB ROLLBACK The sinking or rolling back of a descending slab and the trench, instead of the usual subduction downward and forward.

STRIKE The strike line of a bed, fault, or other planar feature is a line representing the intersection of the feature with a horizontal plane. The strike is at 90° to the *dip* of the planar feature.

STRIKE-SLIP FAULT (= "transcurrent fault"). A fault in which the slip is parallel to the *strike* of the fault. The fault surface is usually more or less vertical. (Contrast *dip-slip fault*.)

SUBDUCTION The sinking of one plate beneath another into the mantle, as seen at a convergent plate boundary.

TERRANE Fault-bounded block of crust with an independent tectonic and sedimentary history from other terranes. (See also *accreted terrane*.)

THICKENING Vertical thickening of the crust, mainly caused by *shortening* and terrane accretion. Emplacement of igneous rock and sedimentation also thickens the crust. (Contrast with *thinning*.)

THINNING Vertical thinning of the crust, mainly caused by *extension*. (Contrast with *thickening*.)

THRUST FAULT A fault with the same sense of motion as a *reverse fault*, but with the *dip* of the fault plane less than 45°.

TRANSCURRENT FAULT (See *strike-slip fault*.)

TRANSFORM FAULT A *strike-slip fault* that marks a plate boundary.

TRANSPRESSION Oblique shear involving strike-slip (transcurrent) movement and *shortening* (with compression). (Contrast with *transtension*.)

TRANSTENSION Oblique shear involving strike-slip (transcurrent) movement and *extension*. (Contrast with *transpression*.)

VOLCANIC ARC A subduction-related chain of volcanoes that forms at a plate margin behind the edge of the overriding plate. It may be built up on oceanic crust (producing an island arc) or continental crust.

WRENCH FAULT A regional-scale, continental *strike-slip fault*.

Bibliography

Ackermann, R.R., and J.M. Cheverud. 2002. Discerning evolutionary processes in patterns of tamarin (genus *Saguinus*) craniofacial variation. *American Journal of Physical Anthropology*, 117, 260–271.

Adam, C., and A. Bonneville. 2005. Extent of the South Pacific superswell. *Journal of Geophysical Research*, 110(B09408), 1–14. doi:10.1029/2004JB003465.

Adamowicz, S.J., S. Menu-Marque, S.A. Halse, J.C. Topan, T.S. Zemlak, P.D.N. Hebert, and J.D.S. Witt. 2010. The evolutionary diversification of the Centropagidae (Crustacea, Calanoida): A history of habitat shifts. *Molecular Phylogenetics and Evolution*, 55, 418–430.

Adamowicz, S.J., A. Petrusek, J.K. Colbourne, P.D.N. Hebert, and J.D.S. Witt. 2009. The scale of divergence: A phylogenetic appraisal of intercontinental allopatric speciation in a passively dispersed freshwater zooplankton genus. *Molecular Phylogenetics and Evolution*, 50, 423–436.

Adamson, A.M. 1939. Review of the fauna of the Marquesas Islands and discussion of its origin. *Bulletin of the Bernice P. Bishop Museum*, 159, 1–93.

Adriaens, D., and A. Herrel. 2009. Functional consequences of extreme morphologies in the craniate trophic system. *Physiological and Biochemical Zoology*, 82, 1–6.

Agassiz, L. 1850. Geographical distribution of animals. *Christian Examiner and Religious Miscellany*, 48, 181–204.

Aitchison, J.C., J.R Ali, and A.M. Davis. 2007. When and where did India and Asia collide? *Journal of Geophysical Research*, 112(B05423), 1–19.

Aitchison, J.C., and A.M. Davis. 2004. Evidence for the multiphase nature of the India–China collision from the Yarlung-Tsangpo suture zone, Tibet. In J. Malpas, C.J.N. Fletcher, J.R. Ali, and J.C. Aitchison (eds.), *Aspects of the tectonic evolution of China. Geological Society of London Special Publication*, 226, 217–233.

Ali, J.R., and M. Huber. 2010. Mammalian biodiversity on Madagascar controlled by ocean currents. *Nature*, 463, 653–656.

Allwood, J., D. Gleeson, G. Mayer, S. Daniels, J.R. Beggs, and T.R. Buckley. 2010. Support for vicariant origins of the New Zealand Onychophora. *Journal of Biogeography*, 37, 669–681.

Alvarez Alonso, J. 2002. *Characteristic avifauna of white-sand forests in northern Peruvian Amazonia.* M.Sc. thesis, Louisiana State University.

Amadon, D. 1943. Birds collected during the Whitney South Sea expedition. 52. Notes on some non-passerine genera. 3. *American Museum Novitates*, 1237, 1–21.

Ameghino, F. 1906. Les formations sédimentaires du Crétacé Supérieur et du Tertiaire de Patagonie. *Anales del Museo Nacional de Historia Natural de Buenos Aires*, ser. 3, 8, 1–568.

Anderson, D.L., 2005. Scoring hotspots: The plume and plate paradigms. In G.R. Foulger, J.H. Natland, D.C. Presnall, and D.L. Anderson (eds.), *Plates, plumes, and paradigms. Geological Society of America Special Paper*, 388, 31–54.

Anderson, D.L., and J.H. Natland. 2005. A brief history of the plume hypothesis and its competitors: Concept and controversy. In G.R. Foulger, J.H. Natland, D.C. Presnall, and D.L. Anderson (eds.), *Plates, plumes, and paradigms. Geological Society of America Special Paper*, 388, 119–145.

Anderson, R.S., and A.A. Lanteri. 2000. New genera and species of weevils from the Galapagos Islands, Ecuador, and Cocos Island, Costa Rica (Coleoptera; Curculionidae; Entiminae; Entimini). *American Museum Novitates*, 3299, 1–41.

Andrade, F.A.G., M.E.B. Fernandes, S.A. Marques-Aguiar, and G.B. Lima. 2008. Comparison between the Chiropteran fauna from *terra firme* and mangrove forests on the Bragança peninsula in Pará, Brazil. *Studies on Neotropical Fauna and Environment*, 43, 169–176.

Andrews, K.R., L. Karczmarski, W.W.L. Au, S.H. Rickards, C.A. Vanderlip, B.W. Bowen, and R.J. Toonen. 2010. Rolling stones and stable homes: Social structure, habitat diversity and population genetics of the Hawaiian spinner dolphin (*Stenella longirostris*). *Molecular Ecology*, 19, 732–748.

Antoine, P.-O., R. Salas-Gismondi, P. Baby, M. Benammi, S. Brusset, D. de Franceschi, N. Espurt, C. Goillot, F. Pujos, J. Tejada, and M. Urbina. 2007. The Middle Miocene (Laventan) Fitzcarrald fauna, Amazonian Peru. *Cuadernos del Museo Geominero* (Madrid), 8, 19–24.

Antonelli, A., A. Quijada-Mascareñas, A.J. Crawford, J.M. Bates, P.M. Velazco, and W. Wüster. 2010. Molecular studies and phylogeography of Amazonian tetrapods and their relation to geological and climatic models. In C. Hoorn and F.P. Wesselingh (eds.), *Amazonia, landscape and species evolution: A look into the past* (pp. 386–404). Wiley-Blackwell, Oxford.

Araripe, J., C.H. Tagliaro, P.S. Rêgo, I. Sampaio, S.F. Ferrari, and H. Schneider. 2008. Molecular phylogenetics of large-bodied tamarins, *Saguinus* spp. (Primates, Platyrrhini). *Zoologica Scripta*, 37, 461–467.

Arbogast, B.S., S.V. Edwards, J. Wakeley, P. Beeli, and J.B. Slowinski. 2002. Estimating divergence times from molecular data on phylogenetic and population genetic time scales. *Annual Review of Ecology and Systematics*, 33, 707–740.

Arboleya, M.-L., A. Teixel, M. Charroud, and M. Julivert. 2004. A structural transect through the High and Middle Atlas of Morocco. *Journal of African Earth Sciences*, 39, 319–327.

Arnason, U., J.A. Adegoke, A. Gullberg, E.H. Harley, A. Janke, and M. Kullberg. 2008. Mitogenomic relationships of placental mammals and molecular estimates of their divergence. *Gene*, 421, 37–51.

Arnason, U., A. Gullberg, A. Janke, and X. Xu. 1996. Pattern and timing of evolutionary divergences among hominoids based on analyses of complete mtDNAs. *Journal of Molecular Evolution*, 43, 650–661.

Arnedo, M.A., and R.G. Gillespie. 2006. Species diversification patterns in the Polynesian jumping spider genus *Havaika* Prószyński, 2001 (Araneae, Salticidae). *Molecular Phylogenetics and Evolution*, 41, 472–495.

Asquith, A. 1995. Evolution of *Sarona* (Heteroptera, Miridae): Speciation on geographic and ecological islands. In W.L. Wagner and V.A. Funk (eds.), *Hawaiian biogeography: Evolution on a hot spot archipelago* (pp. 90–120). Smithsonian Institution Press, Washington, DC.

Asquith, A. 1997. Hawaiian Miridae (Hemiptera: Heteroptera): The evolution of bugs and thought. *Pacific Science*, 51, 356–365.

Athens, J.S., J.V. Ward, and D.W. Blinn. 2007. Vegetation history of Laysan Island, Northwestern Hawaiian Islands. *Pacific Science*, 61, 17–37.

Avise, J.C. 1992. Molecular population structure and the biogeographic history of a regional fauna: A case history with lessons for conservation biology. *Oikos*, 63, 62–76.

Avise, J.C. 2000. *Phylogeography: The history and formation of species*. Harvard University Press, Cambridge, MA.

Avise, J.C. 2007. Twenty-five key evolutionary insights from the phylogeographic revolution in population genetics. In S. Weiss and N. Ferrand (eds.), *Phylogeography of southern European refugia* (pp. 7–21). Springer, Dordrecht.

Azuma, Y., Y. Kumazawa, M. Miya, K. Mabuchi, and M. Nishida. 2008. Mitogenomic evaluation of the historical biogeography of cichlids: Toward reliable dating of teleostean divergences. *BMC Evolutionary Biology*, 8(215), 1–13.

Baby, P., W. Gil Rodriguez, R. Barragán, C. Bernal, M. Rivadeneira, and C. Davila, C. 1999. Structural style and timing of hydrocarbon entrapments in the Oriente-Marañón foreland basin. Ingepet '99 seminar, Lima, October 26–29 1999. CD-ROM 1, pp. 1–6.

Baird, R.W., A.M. Gorgone, D.J. McSweeney, A.D. Ligon, M.H. Deakos, D.L. Webster, G.S. Schorr, K.K. Martien, D.R. Salden, and S.D. Mahaffy. 2009. Population structure of island-associated dolphins: Evidence from photo-identification of common bottlenose dolphins (*Tursiops truncatus*) in the main Hawaiian Islands. *Marine Mammal Science*, 25, 251–274.

Bajpai, S., R.F. Kay, B.A. Williams, D.P. Das, V.V. Kapur, and B.N. Tiwari. 2008. The oldest Asian record of Anthropoidea. *Proceedings of the National Academy of Sciences USA*, 105, 11093–11098.

Baldwin, B.G. 1996. Phylogenetics of the California tarweeds and the Hawaiian silversword alliance (Madiinae; Heliantheae *sensu lato*). In D.J.N. Hind and

H. Beentje (eds.), *Compositae: Systematics* (pp. 377–391). Royal Botanic Gardens, Kew.

Baldwin, B.G. 1997. Adaptive radiation of the Hawaiian silversword alliance: Congruence and conflict of phylogenetic evidence from molecular and non-molecular investigations. In T.J. Givnish and K.J. Sytsma (eds.), *Molecular evolution and adaptive radiation* (pp. 103–128). Cambridge University Press, New York.

Baldwin, B.G. 1998. Evolution in the endemic Hawaiian Compositae. In T.F. Stuessy and M. Ono (eds.), *Evolution and speciation in island plants* (pp. 49–74). Cambridge University Press, Cambridge.

Baldwin, B.G. 1999. New combinations and new genera in the North American tarweeds (Compositae—Madiinae). *Novon*, 9, 462–471.

Baldwin, B.G. 2007. Adaptive radiation of shrubby tarweeds (*Deinandra*) in the California islands parallels diversification of the Hawaiian silversword alliance (Compositae—Madiinae). *American Journal of Botany*, 94, 237–248.

Baldwin, B.G. 2009. Silverswords. In R.G. Gillespie and D.A. Clague (eds.), *Encyclopedia of islands* (pp. 835–839). University of California Press, Berkeley.

Baldwin, B.G., and M.J. Sanderson. 1998. Age and rate of diversification of the Hawaiian Silversword alliance (Compositae). *Proceedings of the National Academy of Sciences USA*, 95, 9402–9406.

Baldwin, B.G., and W.L. Wagner. 2010. Hawaiian angiosperm radiations of North American origin. *Annals of Botany*, 105, 849–879.

Baldwin, B.G., K.R. Wood, and T.J. Motley. 2008. Molecular evidence for an American origin of *Apostates* (Compositae), a monotypic endemic of Rapa (Austral Islands). Abstract, American Botanical Society Conference.

Ballard, H.E., and K.J. Sytsma. 2000. Evolution and biogeography of the woody Hawaiian violets (*Viola*, Violaceae): Arctic origins, herbaceous ancestry and bird dispersal. *Evolution*, 54, 1521–1532.

Banford, H.M., E. Bermingham, B.B. Collette, and S.S. McCafferty. 1999. Phylogenetic systematic of the *Scomberomorus regalis* (Teleostei: Scombridae) species group: Molecules, morphology and biogeography of Spanish mackerels. *Copeia*, 1999, 596–613.

Barber, A.J., and M.J. Crow. 2008a. The origin and emplacement of the West Burma-West Sumatra ribbon continent. Extended abstract. *Proceedings of the International Symposium on Geoscience Resources and Environments of Asian terranes* (4th IGCP and 5th APSEG). Bangkok. www.geo.sc.chula.ac.th/.

Barber, A.J., and M.J. Crow. 2008b. Structure of Sumatra and its implications for the tectonic assembly of Southeast Asia and the destruction of Paleotethys. *Island Arc*, 18, 3–20.

Barber, A.J., M.J. Crow, and J. Milsom (eds.). 2005. *Sumatra: Geology, resources and tectonic evolution*. Geological Society of London, London.

Barnett, A.A., and D. Brandon-Jones. 1997. The ecology, biogeography and conservation of the uakaris, *Cacajao* (Pitheciinae). *Folia Primatologica*, 68, 223–235.

Barraclough, T.G., and A.P. Vogler. 2000. Detecting the geographical pattern of speciation from species-level phylogenies. *American Naturalist*, 155, 419–434.

Barreda, V.D., L. Palazzesi, M.C. Telleria, L. Katinas, J.V. Crisci, K. Bremer, M.G. Passalia, R. Corsolini, R., Rodriguez Brizuela, and F. Bechis. 2010. Eocene Patagonia fossils of the daisy family. *Science*, 329, 1621.

Barrier, M., B.G. Baldwin, R.H. Robichaux, and M.D. Purugganan. 1999. Interspecific hybrid ancestry of a plant adaptive radiation: Allopolyploidy of the Hawaiian silversword alliance (Asteraceae) inferred from floral homeotic gene duplications. *Molecular Biology and Evolution*, 16, 1105–1113.

Basile, C., J. Mascle, and R. Guiraud. 2005. Phanerozoic geological evolution of the equatorial Atlantic domain. *Journal of African Earth Sciences*, 43, 275–282.

Baumgarten, A., and G.B. Williamson. 2007. The distributions of howling monkeys (*Alouatta pigra* and *A. palliata*) in southeastern Mexico and Central America. *Primates*, 48, 310–315.

Bayer, R.J., D.J. Mabberley, C. Morton, C.H. Miller, I.K. Sharma, B.E. Pfeil, S. Rich, R. Hitchcock, and S. Sykes. 2009. A molecular phylogeny of the orange subfamily (Rutaceae: Aurantioideae) using nine cpDNA sequences. *American Journal of Botany*, 96, 668–685.

Beard, K.C. 2004. *The hunt for the dawn monkey: Unearthing the origins of monkeys, apes and humans.* University of California Press, Berkeley.

Beard, K.C. 2006. Mammalian biogeography and anthropoid origins. In S. Lehman and J.G. Fleagle (eds.), *Primate biogeography* (pp. 439–467). Springer, New York.

Beard, K.C. 2008. The oldest North American primate and mammalian biogeography during the Paleocene-Eocene Thermal Maximum. *Proceedings of the National Academy of Sciences USA*, 105, 3815–3818.

Beard, K.C., L. Marivaux, Y. Chaimanee, J.-J. Jaeger, B. Marandat, P. Tafforeau, A.N. Soe, S.T. Tun, and A.A. Kyaw. 2009. A new primate from the Eocene Pondaung Formation of Myanmar and the monophyly of Burmese amphipithecids. *Proceedings of the Royal Society B*, 276, 3285–3294.

Beatty, J. 2008. Chance variation and evolutionary contingency: Darwin, Simpson, *The Simpsons*, and Gould. In M. Ruse (ed.), *Oxford handbook of the philosophy of biology* (pp. 189–210). Oxford University Press, Oxford.

Beaumont, M.A., and M. Panchal. 2008. On the validity of nested clade phylogeographical analysis. *Molecular Ecology*, 17, 2563–2565.

Bellemain, E., and R.E. Ricklefs. 2008. Are islands the end of the colonization road? *Trends in Ecology and Evolution*, 23, 461–468.

Berger, A.J. 1972. *Hawaiian birdlife*. University of Hawaii Press, Honolulu.

Bernard, H., I. Matsuda, G. Hanya, and A.H. Ahmad. 2011. Characteristics of night sleeping trees of Proboscis Monkeys (*Nasalis larvatus*) in Sabah, Malaysia. *International Journal of Primatology*. 32, 259–267.

Berry, R.J. 1982. *Neo-Darwinism*. Edward Arnold, London.

Beurton, C. 2000. The genus *Plethadenia* (Rutaceae). *Willdenowia*, 30, 115–123.

Bickel, D.J. 2009. *Amblypsilopus* (Diptera: Dolichopodidae: Sciapodinae) from the southwest Pacific, with a focus on the radiation in Fiji and Vanuatu. *Bishop Museum Occasional Papers*, 103, 3–61.

Bickel, D.J., and C.E. Dyte. 2007. Dolichopodidae. In N.L. Evenhuis (ed.), *Catalog of the Diptera of the Australasian and Oceanian regions*, Web version. Accessed 2009 at hbs.bishopmuseum.org/aocat/.

Biffin, E., E.J. Lucas, L.A. Craven, I. Ribeiro da Costa, M.G. Harrington, and M.D. Crisp. 2010. Evolution of exceptional species richness among lineages of fleshy-fruited Myrtaceae. *Annals of Botany*, 106, 79–93.

Biju, S.D., and F. Bossuyt. 2003. New frog family from India reveals an ancient biogeographical link with the Seychelles. *Nature*, 425, 711–714.

Bininda-Emonds, O.R.P., M. Cardillo, K.E. Jones, R.D.E. MacPhee, R.M.D. Beck, R. Grenyer, S.A. Price, R.A. Vos, J.L. Gittleman, and A. Purvis. 2007. The delayed rise of present-day mammals. *Nature*, 446, 507–512.

Bird, C.E., B.S. Holland, B.W. Bowen, and R.J. Toonen, R.J. 2007. Contrasting phylogeography in three endemic Hawaiian limpets (*Cellana* spp.) with similar life histories. *Molecular Ecology*, 16, 3173–3186.

Blackburn, D.C., D.P. Bickford, A.C. Diesmos, D.T. Iskandar, and R.M. Brown. 2010. An ancient origin for the enigmatic flat-headed frogs (Bombinatoridae: *Barbourula*) from the islands of Southeast Asia. *PLoS One*, 5(8), e12090.

Bloch, J.I., M.T. Silcox, D.M. Boyer, and E.J. Sargis, E.J. 2007. New Paleocene skeletons and the relationship of plesiadapiforms to crown-clade primates. *Proceedings of the National Academy of Sciences USA*, 104, 1159–1164.

Bolnick, D.I., and B.M. Fitzpatrick. 2007. Sympatric speciation: models and empirical evidence. *Annual Review of Ecology, Evolution and Systematics*, 38, 459–487.

Bonaccorso, E., A.G. Navarro-Sigüenza, L.A. Sánchez-González, A.T. Peterson, and J. García-Moreno. 2008. Genetic differentiation of the *Chlorospingus ophthalmicus* complex in Mexico and Central America. *Journal of Avian Biology*, 39, 311–321.

Bonacum, J., P.M. O'Grady, M. Kambysellis, and R. DeSalle. 2005. Phylogeny and age of diversification of the *planitibia* species group of the Hawaiian *Drosophila*. *Molecular Phylogenetics and Evolution*, 37, 73–82.

Bonneville, A. 2009. French Polynesia, geology. In R.G. Gillespie and D.A. Clague (eds.), *Encyclopedia of islands* (pp. 338–342). University of California Press, Berkeley.

Boorman, C.J., and S.M. Shimeld. 2002. The evolution of left–right asymmetry in chordates. *BioEssays*, 24, 1004–1011.

Borda, E., and M.E. Siddall. 2010. Insights into the evolutionary history of Indo-Pacific bloodfeeding terrestrial leeches (Hirudinida: Arhynchobdellida: Haemadipsidae). *Invertebrate Systematics*, 24, 456–472.

Bossuyt, F., and M.C. Milinkovitch. 2001. Amphibians as indicators of early Tertiary "out-of-India" dispersal of vertebrates. *Science*, 292, 93–95.

Bosworth, W. 1994. A model for the three-dimensional evolution of continental rift basins, north-east Africa. *International Journal of Earth Sciences*, 83, 671–688.

Boubli, J.P., M.N.F. da Silva, M.V. Amado, T. Hrbek, F. Boavista Pontual, and I.P. Farias. 2008. A taxonomic reassessment of *Cacajao melanocephalus* Humboldt (1811), with the description of two new species. *International Journal of Primatology*, 29, 723–741.

Boubli, J.P., and A.D. Ditchfield. 2000. The time of divergence between the two species of uakari monkeys: *Cacajao calvus* and *Cacajao melanocephalus*. *Folia Primatologica*, 71, 387–391.

Bouchenak-Khelladi, Y., G.A. Verboom, V. Savolainen, and T.R. Hodkinson. 2010. Biogeography of the grasses (Poaceae): a phylogenetic approach to reveal evolutionary history in geographical space and geological time. *Botanical Journal of the Linnean Society*, 162, 543–557.

Bouchet, P., J.-P. Rocroi, J. Fryda, B. Hausdorf, W. Ponder, A. Valdès, and A. Warén. 2005. Classification and nomenclator of gastropod families. *Malacologia*, 47, 1–397.

Bouetard, A., P. Lefeuvre, R. Gigant, S. Bory, M. Pignal, P. Besse, and M. Grisoni. 2010. Evidence of transoceanic dispersion of the genus *Vanilla* based on plastid DNA phylogenetic analysis. *Molecular Phylogenetics and Evolution*, 55, 621–630.

Bradford, J.C. 2002. Molecular phylogenetics and morphological evolution in Cunonieae (Cunoniaceae). *Annals of the Missouri Botanic Garden*, 89, 491–503.

Brandon-Jones, D. 1993. The taxonomic affinities of the Mentawai Island sureli, *Presbytis potenziani* (Bonaparte 1856) (Mammalia: Primata: Cercopithecidae). *Raffles Bulletin of Zoology*, 41, 33–357.

Brandon-Jones, D., A.A. Eudey, T. Geissmann, C.P. Groves, D.J. Melnick, J.C. Morales, M. Shekelle, and C.-B. Stewart. 2004. Asian primate classification. *International Journal of Primatology*, 25, 97–164.

Bremer, K. 1994. *Asteraceae: Cladistics and classification*. Timber Press, Portland, OR.

Bremer, K., E.-M. Friis, and B. Bremer. 2004. Molecular phylogenetic dating of asteroid flowering plants shows Early Cretaceous diversification. *Systematic Biology*, 53, 496–505.

Brenner, E.D., D.W. Stevenson, and R.W. Twigg. 2003. Cycads: Evolutionary innovations and the role of plant derived neurotoxins. *Trends in Plant Science*, 8, 446–452.

Brieger, F.G. 1981. Subtribus Dendrobiinae. In F.G. Brieger, R. Maatsch, and K. Senghas (eds.), *Rudolph Schlechter, die Orchideen: Ihre Beschreibung, Kultur und Züchtung*, 3rd ed., Band 1, Teil A, Lieferung 11–12 (pp. 636–752.) Paul Parey, Berlin.

Briggs, J.C. 2003. Marine centers of origin as evolutionary engines. *Journal of Biogeography*, 30, 1–18.

Brooke, M. de L. 2004. *Albatrosses and petrels across the world*. Oxford University Press, New York.

Brower, A.V.Z. 1994. Rapid morphological radiation and convergence among races of the butterfly *Heliconius erato* inferred from patterns of mitochondrial DNA evolution. *Proceedings of the National Academy of Sciences USA*, 91, 6491–95.

Brown, G.K., G. Nelson, and P.Y. Ladiges. 2006. Historical biogeography of *Rhododendron* section *Vireya* and the Malesian Archipelago. *Journal of Biogeography*, 33, 1929–1944.

Brown, J.H., and M.V. Lomolino. 2000. Concluding remarks: Historical perspective and the future of island biogeography. *Global Ecology and Biogeography*, 9, 87–92.

Browne, R.A., D.J. Anderson, J.N. Houser, F. Cruz, K.J. Glasgow, C. N. Hodges, and G. Massey. 1997. Genetic diversity and divergence of endangered Galápagos and Hawaiian petrel populations. *The Condor*, 99, 812–815.

Bryan, E.H., Jr. 1926. Insects of Hawaii, Johnston Island and Wake Island. *Bulletin of the Bernice P. Bishop Museum*, 31, 1–94.

Buerki, S., F. Forest, N. Alvarez, J.A.A. Nylander, N. Arrigo, and I. Sanmartín. 2011. An evaluation of new parsimony-based versus parametric inference methods in biogeography: A case study using the globally distributed plant family Sapindaceae. *Journal of Biogeography*, 38, 531–550.

Bumby, A.J., and R. Guiraud. 2005. The geodynamic setting of the Phanerozoic basins of Africa. *Journal of African Earth Sciences*, 43, 1–12.

Burgess, N.D., T.M. Butynski, N.J. Cordeiro, N.H. Doggart, J. Fjeldså, K.M. Howell, F.B. Kilahama, S.P. Loader, J.C. Lovett, B. Mbilinyi, M. Menegon, D.C. Moyer, E. Nashanda, A. Perkin, F. Rovero, W.T. Stanley, and S.N. Stuart, S.N. 2007. The biological importance of the eastern arc mountains of Tanzania and Kenya. *Biological Conservation*, 134, 209–231.

Burney, D.A., H.F. James, L.P. Burney, S.L. Olson, W. Kikuchi, W.L. Wagner, M. Burney, D. McCloskey, D. Kikuchi, F.V. Grady, R. Gage II, and R. Nishek. 2001. Fossil evidence for a diverse biota from Kaua'i and its transformation since human arrival. *Ecological Monographs*, 71, 615–641.

Burns, K.J., and R.A. Racicot. 2009. Molecular phylogenetics of a clade of lowland tanagers: Implications for avian participation in the Great American Interchange. *The Auk*, 126, 635–648.

Burrell, A.S., C.J. Jolly, A.J. Tosi, and T.R. Disotell. 2009. Mitochondrial evidence for the hybrid origin of the kipunji, *Rungwecebus kipunji* (Primates: Papionini). *Molecular Phylogenetics and Evolution*, 51, 340–348.

Burridge, C.P. 1997. Molecular phylogeny of *Nemadactylus* and *Acantholatris* (Perciformes: Cirrhitoidea: Cheilodactylidae), with implications for taxonomy and biogeography. *Molecular Phylogenetics and Evolution*, 13, 93–109.

Butynski, T.M. 2003. The robust chimpanzee *Pan troglodytes*: Taxonomy, distribution, abundance and conservation status. In R. Kormos, C. Boesch, M.I. Bakarr, and T.M. Butynski (eds.), *West African chimpanzees: Status survey and conservation action plan* (pp. 5–12). IUCN, Gland.

Caccone, A., G.D. Amato, and J.R. Powell. 1988. Rates and patterns of scnDNA and mtDNA divergence within the Drosophila melanogaster Subgroup. *Genetics*, 118, 671–683.

Cain, S.A. 1943. Criteria for the indication of center of origin in plant geographical studies. *Torreya*, 43, 132–154.

Candeiro, C.R.A., A.G. Martinelli, L.S. Avilla, and T.H. Rich. 2006. Tetrapods from the Upper Cretaceous (Turonian-Maastrichtian) Bauru Group of Brazil: A reappraisal. *Cretaceous Research*, 27, 923–946.

Canetti, E. [1935] 1962. *Auto da fé*. Cape, London.

Caporale, L.H. (ed.). 2006. *The implicit genome*. Oxford University Press, New York.

Carlquist, S. 1966a. The biota of long-distance dispersal II: Loss of dispersibility in Pacific Compositae. *Evolution*, 20, 30–48.

Carlquist, S. 1966b. The biota of long-distance dispersal III. Loss of dispersibility in the Hawaiian flora. *Brittonia*, 18, 310–335.

Carlquist, S. 1995. Introduction. In W.L. Wagner and V.A. Funk (eds.), *Hawaiian biogeography: Evolution on a hot spot archipelago* (pp. 1–13). Smithsonian Institution Press, Washington DC.

Carranza, S., E.N. Arnold, J.A. Mateo, and L.F. López-Jurado. 2000. Long-distance colonization and radiation in gekkonid lizards, *Tarentola* (Reptilia: Gekkonidae), revealed by mitochondrial DNA sequences. *Proceedings of the Royal Society, London B*, 267, 637–649.

Carson, H.L., and D.A. Clague. 1995. Geology and biogeography of the Hawaiian Islands. In W.L. Wagner and V.A. Funk (eds.), *Hawaiian biogeography: Evolution in a hot spot archipelago* (pp. 14–29). Smithsonian Institution Press, Washington, DC.

Carson, H.L., and A.R. Templeton. 1984. Genetic revolutions in relation to speciation phenomena: The founding of new populations. *Annual Review of Ecology and Systematics*, 15, 97–131.

Cartelle, C., and W.C. Hartwig. 1996. A new extinct primate among the Pleistocene megafauna of Bahia, Brazil. *Proceedings of the National Academy of Sciences USA*, 93, 6405–6409.

Carvajal, A., and G.H. Adler. 2005. Biogeography of mammals on tropical Pacific Islands. *Journal of Biogeography*, 32, 1561–1569.

Castoe, T.A., J.M. Daza, E.N. Smith, M.M. Sasa, U. Kuch, J.A. Campbell, P.T. Chippindale, and C.L. Parkinson. 2009. Comparative phylogeography of pitvipers suggests a consensus of ancient Middle American highland biogeography. *Journal of Biogeography*, 36, 88–103.

Caswell, J.L., S. Mallick, D.J. Richter, J. Neubauer, C. Schirmer, S. Gnerre, and D. Reich. 2008. Analysis of chimpanzee history based on genome sequence alignments. *PLoS Genetics*, 4(e1000057), 1–14.

Ceballos, G., and G. Oliva (eds.). 2005. *Los mamíferos silvestres de México*. CONABIO—Fondo la Cultura Económica, Mexico City.

Cecca, F. 2008. La dimension biogéographique de l'évolution de la Vie. *Comptes Rendus Palevol*, 8, 119–132.

Centeno-García, E., M. Guerrero-Suastegui, and O. Talavera-Mendoza. 2008. The Guerrero Composite Terrane of western Mexico: Collision and subsequent rifting in a supra-subduction zone. In A. Draut, P.D. Clift, and D.W. Scholl (eds.), *Formation and applications of the sedimentary record in arc collision zones. Geological Society of America Special Paper*, 436, 279–308.

Chakrabarty, P. 2004. Cichlid biogeography: Comment and review. *Fish and Fisheries*, 5, 97–119.

Chamberlin, T.C. [1890] 1965. The method of multiple working hypotheses. *Science*, 148, 754–759.

Chapman, C.A., A. Gautier-Hion, J.F. Oates, and D.A. Onderdonk. 1999. African primate communities: Determinants of structure and threats to survival. In J.G. Fleagle, C.H. Janson, and K.E. Reed (eds.), *Primate communities* (pp. 1–38). Cambridge University Press, New York.

Chapman, F.M. 1926. The distribution of bird-life in Ecuador. *Bulletin of the American Museum of Natural History*, 55, 1–784.

Chapman, F.M. 1931. The upper zonal bird-life of Mts. Roraima and Duida. *Bulletin of the American Museum of Natural History*, 63, 1–66.

Charles, C., and S. Sandin. 2009. Line Islands. In R.G. Gillespie and D.A. Clague (eds.), *Encyclopedia of islands* (pp. 553–557). University of California Press, Berkeley.

Chase, M.W. 2005. Classification of Orchidaceae in the age of DNA data. *Curtis's Botanical Magazine*, 22, 2–7.

Chatterjee, H.J. 2006. Phylogeny and biogeography of gibbons: A dispersal/vicariance analysis. *International Journal of Primatology*, 27, 699–712.

Chatterjee, H.J, S.Y.W. Ho, I. Barnes, and C. Groves. 2009. Estimating the phylogeny and divergence times of primates using a supermatrix approach. *BMC Evolutionary Biology*, 9, 259.

Che, J., W.-W. Zhou, J.-S. Hu, F. Yan, T.J. Papenfuss, D.B. Wake, and Y.-P. Zhang. 2010. Spiny frogs (Paini) illuminate the history of the Himalayan region and Southeast Asia. *Proceedings of the National Academy of Sciences USA*, 107, 13765–13770.

Chen, J.-H., D. Pan, C. Groves, Y.-X. Wang, E. Narushima, H. Fitch-Snyder, P. Crow, V. Ngoc Thanh, O. Ryder, H.-W. Zhang, Y. Fu, and Y. Zhang. 2006. Molecular phylogeny of *Nycticebus* inferred from mitochondrial genes. *International Journal of Primatology*, 27, 1187–1200.

Chesser, R.T. 2004. Molecular systematics of New World suboscine birds. *Molecular Phylogenetics and Evolution*, 32, 11–24.

Chi, C.T., and S.L. Dorobek. 2004. Cretaceous paleomagnetism of Indochina and surrounding regions: Cenozoic tectonic implications. In J. Malpas, C.J.N. Fletcher, J.R. Ali and J.C. Aitchison (eds.), *Aspects of the tectonic evolution of China. Geological Society of London Special Publication*, 226, 273–287.

Chicangana, G. 2005. The Romeral fault system: A shear and deformed extinct subduction zone between oceanic and continental lithospheres in northwestern South America. *Earth Science Research Journal*, 9, 51–66.

Ching, J.F.C. 1999. *Geological and geophysical study of Peru's Andean foothills and adjacent basins*. Unpublished M.Sc. thesis, University of South Carolina.

Chivers, D.J. 1986. Southeast Asian primates. In K. Benirschke (eds.), *Primates: The road to self-sustaining populations* (pp. 127–151). Springer, New York.

Cibois, A., J.-C. Thibault, and E. Pasquet. 2007. Uniform phenotype conceals double colonization by reed-warblers of a remote Pacific archipelago. *Journal of Biogeography*, 34, 1150–1166.

Clague, D.A. 1996. The growth and subsidence of the Hawaiian-Emperor volcanic chain. In A. Keast and S. Miller (eds.), *The origin and evolution of Pacific island biotas, New Guinea to eastern Polynesia* (pp. 35–50). SPB Academic Publishing, Amsterdam.

Clague, D.A., J.C. Braga, D. Bassi, P.D. Fullaar, W. Renema, and J.M. Webster. 2010. The maximum age of Hawaiian terrestrial lineages: geological constraints from Kōko Seamount. *Journal of Biogeography*, 37, 1022–1033.

Clague, D.A., and G.B. Dalrymple. 1989. Tectonics, geochronology and origin of the Hawaiian-Emperor volcanic chain. In E.L. Winterer, D.M. Hussong, and R.W. Decker (eds.), *The eastern Pacific Ocean and Hawaii. The geology of North America*, Vol. N (pp. 188–217). Geological Society of America, Boulder, CO.

Claridge, E. 2006. *The systematics and diversification of* Rhyncogonus *(Entiminae: Curculionidae: Coleoptera) in the Central Pacific*. Ph.D. dissertation, University of California, Berkeley.

Clark, J.R., R.H. Ree, M.G. King, W.L. Wagner, and E.H. Roalson. 2008. A comparative study in ancestral range reconstruction methods: Retracing the uncertain histories of insular lineages. *Systematic Biology*, 57, 693–707.

Clark, J.R., W.L. Wagner, and E.H. Roalson. 2009. Patterns of diversification and ancestral range construction in the southeast Asian angiosperm lineage *Cyrtandra* (Gesneriaceae). *Molecular Phylogenetics and Evolution*, 53, 982–994.

Clarke, J.A., C.P. Tambussi, J.I. Noriega, G.M. Erickson, and R.A. Ketcham. 2005. Definitive fossil evidence for the extant avian radiation in the Cretaceous. *Nature*, 433, 305–308.

Clayton, J.W., E.S. Fernando, P.S. Soltis, and D.E. Soltis. 2007a. Molecular phylogeny of the tree-of-heaven family (Simaroubaceae) based on chloroplast and nuclear markers. *International Journal of Plant Science*, 168, 1325–1339.

Clayton, J.W., E.S. Fernando, P.S. Soltis, and D.E. Soltis. 2007b. Historical biogeography and diversification in the tree-of-heaven family (Simaroubaceae, Sapindales). Abstract 1422 in Botanical Society of America, Botany and Plant Biology Conference. At www.2007.botanyconference.org.

Clegg, S.M., S.M. Degnan, J. Kikkawa, C. Moritz, A. Estoup, and I.P.F. Owens. 2002. Genetic consequences of sequential founder events by an island-colonizing bird. *Proceedings of the National Academy of Sciences USA*, 99, 8127–8132.

Clemens, W.A. 2004. *Purgatorius* (Plesiadapiformes, Primates?, Mammalia), a Paleocene immigrant into northeastern Montana: Stratigraphic occurrences and incisor proportions. *Bulletin of the Carnegie Museum of Natural History*, 36, 3–13.

Clement, W.L., M.C. Tebbitt, L.L. Forrest, J.E. Blair, L. Brouillet, T. Eriksson, and S.M. Swensen. 2004. Phylogenetic position and biogeography of *Hillebrandia sandwicensis* (Begoniaceae): A rare Hawaiian relict. *American Journal of Botany*, 91, 905–917.

Clements, F.E., and V.E. Shelford. 1939. *Bioecology.* Wiley, New York.

Clift, P.D., and G.M.H. Ruiz. 2007. How does the Nazca Ridge subduction influence the modern Amazonian foreland basin? Comment and reply. *Tectonics*. doi 10.1130/G24355C.1.

Clouard, V., and A. Bonneville. 2005. Ages of seamounts, islands, and plateaus on the Pacific plate. *Geological Society of America Special Paper*, 388, 71–90.

Cobbold, P.R., K.E. Meisling, and V.S. Mount. 2001. Reactivation of an obliquely rifted margin, Campos and Santos basins, southeastern Brazil. *Bulletin of the American Association of Petroleum Geologists*, 85, 1925–1944.

Cobbold, P.R., E.A. Rossello, P. Roperch, C. Arriagada, L.A. Gómez, and C. Lima. 2007. Distribution, timing, and causes of Andean deformation across South America. *Geological Society of London, Special Publication*, 272, 321–343.

Colgan, D.J., and S. Soheili. 2008. Evolutionary lineages in *Emballonura* and *Mosia* bats (Mammalia: Microchiroptera) from the southwestern Pacific. *Pacific Science*, 62, 219–232.

Collar, N. 2005. Family Turdidae (thrushes). In J. del Hoyo, A. Elliott, and D.A. Christie (eds.), *Handbook of birds of the world: Cuckoo-shrikes to Thrushes*, Vol. 10 (pp. 514–807). Lynx Edicions, Barcelona, Spain.

Collins, A.C. 2008. The taxonomic status of spider monkeys in the twenty-first century. In C.J. Campbell (ed.), *Spider monkeys: Behavior, ecology and evolution* (pp. 50–78). Cambridge University Press, Cambridge.

Collins, A.C., and J.M. Dubach. 2000. Biogeographic and ecological forces responsible for speciation in *Ateles*. *International Journal of Primatology*, 21, 421–444.

Comes, H. P., A. Tribsch, and C. Bittkau. 2008. Plant speciation in continental island floras as exemplified by *Nigella* in the Aegean archipelago. *Philosophical Transactions of the Royal Society B*, 363, 3083–3096.

Comin-Chiaramonti, P., C. de Barros Gomes, A. de Min, M. Ernesto, A. Marzolie, and C. Riccomini. 2007. Eastern Paraguay: An overview of the post-Paleozoic magmatism and geodynamic implications. *Rendiconti Lincei*, 18, 139–192.

Conran, J.G., J.M. Bannister, and D.E. Lee. 2009. Earliest orchid macrofossils: Early Miocene *Dendrobium* and *Earina* (Orchidaceae: Epidendroideae) from New Zealand. *American Journal of Botany*, 96, 464–474.

Coombs, M.L., D.A. Clague, G.F. Moorre, and B.L. Consens. 2004. Growth and collapse of Waianae Volcano, Hawaii, as revealed by exploration of its submarine flanks. *Geochemistry Geophysics Geosystems*, 5(8). doi:10.1029/2004GC000717.

Cornet, B. 1989. Late Triassic angiosperm-like pollen from the Richmond rift basin of Virginia, USA. *Palaeontographica. Abteilung B: Paläophytologie*, 213, 37–87.

Cortés-Ortiz, L. 2009. Molecular phylogenetics of the Callitrichidae with an emphasis on the marmosets and *Callimico*. In S.M. Ford, L.M. Porter, and L.C. Davis (eds.), *The smallest anthropoids: The marmoset/Callimico radiation* (pp. 3–24). Springer, New York.

Cortés-Ortiz, L., E. Bermingham, C. Rico, E. Rodríguez-Luna, I. Sampaio, and M. Ruiz-García. 2003. Molecular systematics and biogeography of the Neotropical monkey genus, *Alouatta*. *Molecular Phylogenetics and Evolution*, 26, 64–81.

Costa, J.B.S., R.L. Bemerguy, Y. Hasui, M.S. Borges, C.R.P. Ferreira, Jr., P.E.L. Bezerra, M.L. Costa, and J.M.G. Fernandes. 1996. Neotectônico da região Amazônica: Aspectos tectônicos, geomorphológicos e deposicionais. *Geonomos* (Belo Horizonte), 4, 23–44.

Costa, W.J.E.M. 2010. Historical biogeography of cynolebiasine annual killifishes inferred from dispersal–vicariance analysis. *Journal of Biogeography*, 37, 1995–2004.

Costello, A., and T.J. Motley. 2007. Phylogenetics of the *Tetraplasandra* group (Araliaceae) inferred from ITS, 5S-NTS, and morphology. *Systematic Botany*, 32, 464–477.

Courtillot, V.E., and P.R. Renne. 2003. On the ages of flood basalt events. *Comptes Rendus Geoscience*, 335, 113–140.

Covert, H.H. 2002. The earliest fossil primates. In W.C. Hartwig (ed.), *The primate fossil record* (pp. 13–20). Cambridge University Press, Cambridge.

Cowie, R.H., and B.S. Holland. 2006. Dispersal is fundamental to biogeography and the evolution of biodiversity on oceanic islands. *Journal of Biogeography*, 33, 193–198.

Cowie, R.H., and B.S. Holland. 2008. Molecular biogeography and diversification of the endemic terrestrial fauna of the Hawaiian Islands. *Philosophical Transactions of the Royal Society B*, 363, 3363–3376.

Cox, C.B., and P.D. Moore. 2010. *Biogeography: An ecological and evolutionary approach*. 8th ed. Blackwell, Oxford.

Coyne, J.A. 1994. Ernst Mayr and the origin of species. *Evolution*, 48, 19–30.

Coyne, J.A. 2009. *Why evolution is true*. Viking Penguin, New York.

Coyne, J.A., and H.A. Orr. 2004. *Speciation*. Sinauer Associates, Sunderland, MA.

Craddock, E.M. 2000. Speciation processes in the adaptive radiation of Hawaiian plants and animals. *Evolutionary Biology*, 31, 1–54.

Craw, R.C., J.R. Grehan, and M.J. Heads. 1999. *Panbiogeography: Tracking the history of life*. Oxford University Press, New York.

Crawford, A.J., S. Meffre, and P.A. Symonds. 2003. 120 to 0 Ma tectonic evolution of the southwest Pacific and analogous geological evolution of the 60 to 220 Ma Tasman Fold belt system. *Geological Society of Australia Special Publication*, 22, 377–397.

Crisp, M.D., and L.G. Cook. 2005. Do early branching lineages signify ancestral traits? *Trends in Ecology and Evolution*, 20, 122–128.

Crisp, M.D., and L.G. Cook. 2007. A congruent molecular signature of vicariance across multiple plant lineages. *Molecular Phylogenetics and Evolution*, 43, 1106–1117.

Croizat, L. 1964. *Space, time, form: The biological synthesis*. Published by the Author, Caracas, Venezuela.

Croizat, L. 1975. Biogeografía analítica y sintética ("Panbiogeografía") de las Américas. *Boletín de la Academia de Ciencias Físicas, Matemáticas y Naturales* (Caracas), 35, 103–106. (Later published as *Biblioteca de la Academia de Ciencias Físicas, Matemáticas y Naturales* [Caracas], 15 and 16, 1–890. 1976].)

Croizat, L. 1977. Carlos Darwin y sus teorías. *Boletín de la Academia de Ciencias Físicas, Matemáticas y Naturales* (Caracas), 37(113), 15–90.

Cronk, Q.C., M. Kiehn, W.L. Wagner, and J.F. Smith. 2005. Evolution of *Cyrtandra* (Gesneriaceae) in the Pacific Ocean: The origin of a supertramp clade. *American Journal of Botany*, 92, 1017–1024.

Cropp, S., and S. Boinski. 2000. The Central American squirrel monkey (*Saimiri oerstedii*): Introduced hybrid or endemic species? *Molecular Phylogenetics and Evolution*, 16, 350–365.

Crow, J.F. 2008. Commentary: Haldane and beanbag genetics. *International Journal of Epidemiology*, 37, 442–445.

Crow, J.F. 2009. Mayr, mathematics and the study of evolution. *Journal of Biology*, 8(13), 1–4.

Crow, K.D., H. Munehara, and G. Bernardi. 2010, Sympatric speciation in a genus of marine reef fishes. *Molecular Ecology*, 19, 2089–2105.

Cuénoud, P., M.A. del Pero Martinez, P.-A. Loizeau, R. Spichiger, S. Andrews, and J.-F. Manen. 2000. Molecular phylogeny and biogeography of the genus *Ilex* L. (Aquifoliaceae). *Annals of Botany*, 85, 111–122.

Cunningham, C.W., and T.M. Collins. 1994. Developing model systems for molecular biogeography: Vicariance and interchange in marine invertebrates.

In B.S. Schierwater, B.B. Streit, G.P. Wagner, and R. DeSalle (eds), *Molecular ecology and evolution: Approaches and applications* (pp. 405–433). Birkhäuser, Basel, Switzerland.

Cutter, A.D., A. Dey, and R.L. Murray. 2009. Evolution of the *Caenorhabditis elegans* genome. *Molecular Biology and Evolution*, 26, 1199–1234.

Darlington, P.D. [1957] 1966. *Zoogeography: The geographical distribution of animals*. Wiley, New York.

Darwin, C. 1859. *On the origin of species by means of natural selection or the preservation of favoured races in the struggle for life*. 1st ed. Murray, London.

Darwin, C. 1875. *The movements and habits of climbing plants*. 2nd ed. Murray, London.

Darwin, C. 1882. *The descent of man*. 2nd ed. Murray, London.

Darwin, F. 1887. *The life and letters of Charles Darwin*, Vol. 1. Murray, London.

Davenport, T.R.B., W.T. Stanley, E.J. Sargis, D.W. de Luca, N.E. Mpunga, S.J. Machaga, and L.E. Olson. 2006. A new genus of African monkey, *Rungwecebus*: Morphology, ecology and molecular systematics. *Science*, 312, 1378–1381.

Davis, A.S., L.B. Gray, D.A. Clague, and J.R. Hain. 2002. The Line Islands revisited: $^{40}Ar/^{39}Ar$ geochronologic evidence for episodes of volcanism due to lithospheric extension. *Geochemistry, Geophysics, Geosystems*, 3(3), 1018. doi:10.1029/2001GC000190.

Dawkins, R. 1976. *The selfish gene*. Oxford University Press, New York.

Dawkins, R. 1986. *The blind watchmaker*. W.W. Norton, New York.

Dawkins, R. [2004] 2005. *The ancestor's tale: A pilgrimage to the dawn of life*. Phoenix, London.

Dawkins, R. [1996] 2006. *Climbing Mount Improbable*. Penguin, London.

DeCelles, P.G., and K.A. Giles. 1996. Foreland basin systems. *Basin Research*, 8, 105–123.

DeCelles, P.G., P. Kapp, L. Ding, and G.E. Gehrels. 2007. Late Cretaceous to middle Tertiary basin evolution in the central Tibetan Plateau: Changing environments in response to tectonic partitioning, aridification, and regional elevation gain. *Bulletin of the Geological Society of America*, 119, 654–680.

de Granville, J.J. 1974. Aperçu sur la structure des pneumatophores de deux espèces des sols hydromorphes en Guyane. *Cahiers d'ORSTOM. Série Biologique*, 23, 3–22.

Deignan, H.G. 1963. Birds in the tropical Pacific: Pacific basin biogeography. In J.L. Gressitt (ed.), *Pacific Basin biogeography: A symposium* (pp. 263–269). Bishop Museum, Honolulu, HI.

de Jong, R. 2007. Estimating time and space in the evolution of the Lepidoptera. *Tijdschrift voor Entomologie*, 150, 319–346.

del Hoyo, J., A. Elliott, J. Sargatal, and D.A. Christie. 1992–2011. *Handbook of the birds of the world*. 16 vols. Lynx, Barcelona, Spain.

Delson, E. 1994. Evolutionary history of the colobine monleys in palaeoenvironmental perspective. In A.G. Davies and J.F. Oates (eds.), *Colobine monkeys: Their ecology, behavior and evolution* (pp. 11–44). Cambridge University Press, Cambridge, UK.

Delson, E., and I. Tattersall. 2002. Fossil primates. In *McGraw-Hill encyclopedia of science and technology*. 9th ed., Vol. 7 (pp. 472–479). McGraw-Hill, New York, NY.

Denoeud, F., and 54 others. 2010. Plasticity of animal genome architecture unmasked by rapid evolution of a pelagic tunicate. *Science*, 330, 1381–1385.

de Oliveira, A.A., and D.C. Daly. 1999. Geographic distribution of tree species occurring in the region of Manaus, Brazil: Implications for regional diversity and conservation. *Biodiversity and Conservation*, 8, 1245–1259.

de Queiroz, A. 2005. The resurrection of oceanic dispersal in historical biogeography. *Trends in Ecology and Evolution*, 20, 68–73.

DeSalle, R., and A.R. Templeton. 1992. The mtDNA genealogy of closely related *Drosophila silvestris*. *Journal of Heredity*, 83, 211–16.

Désamoré, A., A. Vanderpoorten, B. Laenen, S.R. Gradstein, and P.J.R. Kok. 2010. Biogeography of the Lost World (Pantepui region, northeastern South America): Insights from bryophytes. *Phytotaxa*, 9, 254–265.

de Weerdt, W.H. 1990. Discontinuous distribution of the tropical West Atlantic hydrocoral *Millepora squarrosa*. *Beaufortia*, 41, 195–203.

de Wit, M.J. 2003. Madagascar: Heads it's a continent, tails it's an island. *Annual Review of Earth and Planetary Science*, 31, 213–248.

Diamond, J.M. 1972. *Avifauna of the eastern highlands of New Guinea*. Nuttall Ornithological Club, Cambridge, MA.

Dickinson, E.C. 2003. *The Howard and Moore complete checklist of birds of the world*. Princeton University Press, Princeton, NJ.

Dickinson, W.R. 2004. Evolution of the North American cordillera. *Annual Review of Earth and Planetary Science*, 32, 13–45.

Dijkstra, L.-D.B. 2007. Gone with the wind: Westward dispersal across the Indian Ocean and island speciation in *Hemicordulia* dragonflies (Odonata: Corduliidae). *Zootaxa*, 1438, 27–48.

Doadrio, I., S. Perea, L. Alcaraz, and N. Hernandez. 2009. Molecular phylogeny and biogeography of the Cuban genus *Girardinus* Poey, 1854 and relationships within the tribe Girardini (Actinopterygii, Poeciliidae). *Molecular Phylogenetics and Evolution*, 50, 16–30.

do Amaral, I.L., J. Adis, and G.T. Prance. 1997. On the vegetation of a seasonal mixedwater inundation forest near Manaus, Brazilian Amazonia. *Amazoniana*, 14, 335–347.

Doan, T.M. 2003. A south-to-north biogeographic hypothesis for Andean speciation: Evidence from the lizard genus *Proctoporus* (Reptilia, Gymnophthalmidae). *Journal of Biogeography*, 30, 361–374.

Dobzhansky, T. 1955. *Genetics and the origin of species*. Wiley & Sons, New York.

Dobzhansky, T., F.J. Ayala, G.L. Stebbins, and J.W. Valentine. 1977. *Evolution*. W.H. Freeman, San Francisco, CA.

Domeier, M.L., and N. Nasby-Lucas. 2008. Migration patterns of white sharks *Carcharodon carcharias* tagged at Guadalupe Island, Mexico, and identification of an eastern Pacific shared offshore foraging area. *Marine Ecology Progress Series*, 370, 221–237.

Donoghue, M. J., and B.R. Moore. 2003. Toward an integrative historical biogeography. *Integrative and Comparative Biology*, 43, 261–270.

Doody, K., and O. Hamerlynck. 2003. Biodiversity of Rufiji District: A summary. *Rufiji Environment Management Project Technical Report* 44. Available at http://coastalforests.tfcg.org/pubs/.

Douady, C.J., F. Catzeflis, D.J. Kao, M.S. Springer, and M.J. Stanhope. 2002. Molecular evidence for the monophyly of Tenrecidae (Mammalia) and the timing of the colonization of Madagascar by Malagasy tenrecs. *Molecular Phylogenetics and Evolution*, 22, 357–363.

Doyle, J.A. 2000. Paleobotany, relationships, and geographic history of Winteraceae. *Annals of the Missouri Botanic Garden*, 87, 303–316.

Dragon, J.A., and D.S. Barrington. 2009. The systematics of the *Carex aquatilis* and *Carex lenticularis* lineages: Geographically and ecologically divergent sister clades of *Carex* section *Phacocystis* (Cyperaceae). *American Journal of Botany*, 96, 1896–1906.

Dransfield, J., M. Rakotoarinivo, W.J. Baker, R.P. Bayton, J.B. Fisher, J.W. Horn, B. Leroy, and X. Metz. 2008. A new coryphoid palm genus from Madagascar. *Botanical Journal of the Linnean Society*, 156, 79–91.

Drew, H., and P.H. Barber. 2009. Sequential cladogenesis of the reef fish *Pomacentrus moluccensis* (Pomacentridae) supports the peripheral origin of marine biodiversity in the Indo-Australian archipelago. *Molecular Phylogenetics and Evolution*, 53, 335–339.

Drew, J., G.R. Allen, L. Kaufman, and P.H. Barber. 2008. Endemism and regional colour and genetic differences in five putatively cosmopolitan reef fishes. *Conservation Biology*, 22, 965–975.

Duangjai, S., R. Samuel, J. Munzinger, F. Forest, B. Wallnöfer, M.H.J. Barfuss, G. Fischer, and M.W. Chase. 2009. A multi-locus plastid phylogenetic analysis of the pantropical genus *Diospyros* (Ebenaceae), with an emphasis on the radiation and biogeographic origins of the New Caledonian endemic species. *Molecular Phylogenetics and Evolution*, 52, 602–620.

Dumont, J.F., S. Lamotte, and F. Kahn. 1990. Wetland and upland ecosystems in Peruvian Amazonia: Plant species diversity in the light of some geological and botanical evidence. *Forest Ecology and Management*, 33/34, 125–139.

Dunbar-Co, S., A.M. Wieczorek, and C.W. Morden. 2008. Molecular phylogeny and adaptive radiation of the endemic Hawaiian *Plantago* species (Plantaginaceae). *American Journal of Botany*, 95, 1177–1188.

Durand, D., and R. Hoberman. 2006. Diagnosing duplications: Can it be done? *Trends in Genetics*, 22, 156–164.

Duret, L., and N. Galtier. 2009. Biased gene conversion and the evolution of mammalian genomic landscapes. *Annual Review of Genomics and Human Genetics*, 10, 285–311.

Dyer, B.S., and M.W. Westneat. 2010. Taxonomy and biogeography of the coastal fishes of Juan Fernández Archipelago and Desventuradas Islands, Chile. *Revista de Biología Marina y Oceanografía*, 45(S1), 589–617.

Eaton, M.J., A. Martin, J. Thorbjarnarson, and G. Amato. 2009. Species-level diversification of African dwarf crocodiles (genus *Osteolamus*): A geographic and phylogenetic perspective. *Molecular Phylogenetics and Evolution*, 50, 496–506.

Eggens, F., M. Popp, M. Nepokroeff, W.L. Wagner, and B. Oxelman. 2007. The origin and number of introductions of the Hawaiian endemic *Silene* species (Caryophyllaceae). *American Journal of Botany*, 94, 210–218.

Ehrendorfer, F., and R. Samuel. 2000. Comments on S.B. Hoot's interpretation of southern hemisphere relationships in *Anemone* (Ranunculaceae) based on molecular data [Am. J. Bot. 2000; 87(6, Suppl.), 154–155]. *Taxon*, 49, 781–784.

Eick, G.N., D.S. Jacobs, and C.A. Matthee. 2005. A nuclear DNA phylogenetic perspective on the evolution of echolocation and historical biogeography of extant bats (Chiroptera). *Molecular Biology and Evolution*, 22, 1869–1886.

Eigenmann, C.H. 1920. The Magdalena basin and the horizontal and vertical distribution of its fishes. *Indiana University Studies*, 7, 21–34.

Eigenmann, C.H. 1921. The nature and origin of the fishes of the Pacific slope of Ecuador, Peru and Chile. *Proceedings of the American Philosophical Society*, 60, 503–523.

Eiler, J.M., K.A. Farley, J.W. Valley, A.W. Hofmann, and E.M. Stolper. 1996. Oxygen isotope constraints on the sources of Hawaiian volcanism. *Earth and Planetary Science Letters*, 144, 453–468.

Eisenberg, J.F., K.H. Redford, and F.A. Reid. 2000. *Mammals of the Neotropics: Ecuador, Bolivia, Brazil*. University of Chicago Press, Chicago.

Eizirik, E., W.J. Murphy, and S.J. O'Brien. 2001. Molecular dating and biogeography of the early placental mammal radiation. *Journal of Heredity*, 92, 212–219.

Eizirik, E., W.J. Murphy, M.S. Springer, and S.J. O'Brien. 2004. Molecular phylogeny and dating of early primate divergences. In C.F. Ross and R.F. Kay (eds.), *Anthropoid origins: New visions* (pp. 45–63). Kluwer/Plenum, New York.

Ekman, S. 1953. *Zoogeography of the sea*. Sidgwick and Jackson, London.

Eldredge, N., J.N. Thompson, P.M. Brakefield, S. Gavrilets, D. Jablonski, J.B.C. Jackson, R.E. Lenski, B.S. Lieberman, M.A. McPeek, and W. Miller III. 2005. The dynamics of evolutionary stasis. *Paleobiology*, 31, 133–145.

Ellouz, N., M. Patriat, M., J.-M. Gaulier, R. Bouatmani, and S. Sabounji. 2003. From rifting to alpine inversion: Mesozoic and Cenozoic subsidence history of some Moroccan basins. *Sedimentary Geology*, 156, 185–212.

Endress, M.E., D.H. Lorence, and P.K. Endress. 1997. Structure and development of the gynoecium of *Lepinia marquisensis* and its systematic position in the Apocynaceae. *Allertonia*, 7, 267–272.

Engel, M.S. 2000. A new interpretation of the oldest fossil bee (Hymenoptera: Apidae). *American Museum Novitates*, 3296, 1–11.

Ereshefsky, M. 2010. Mystery of mysteries: Darwin and the species problem. *Cladistics*, 27, 67–79.

Ericson, P.G.P. 2008. Current perspectives on the evolution of birds. *Contributions to Zoology*, 77, 109–116.

Ericson, P.G.P., M. Irestedt, and U.S. Johansson. 2003. Evolution, biogeography, and patterns of diversification in passerine birds. *Journal of Avian Biology*, 34, 3–15.

Escalante, T., G. Rodríguez, N. Cao, M.C. Ebach, and J.J. Morrone. 2007. Cladistic biogeographic analysis suggests an early Caribbean diversification in Mexico. *Naturwissenschaften*, 94, 561–565.

Etnoyer, P.J., J. Wood, and T.C. Shirley. 2010. How large is the seamount biome? *Oceanography*, 23, 206–209.

Evans, B.J., J. Supriatna, N. Andayani, M.I. Setidai, D.C. Cannatella, and D.J. Melnick. 2003. Monkeys and toads define areas of endemism on Sulawesi. *Evolution*, 57, 1436–1443.

Evenhuis, N.L. 1999. Water-skating *Campsicnemus* of the Marquesas Islands, including two new species (Diptera: Dolichopodidae). *Journal of the New York Entomological Society*, 107, 289–296.

Fabre, P.-H., A. Rodrigues, and E.J.P. Douzery. 2009. Patterns of macroevolution among Primates inferred from a supermatrix of mitochondrial and nuclear DNA. *Molecular Phylogenetics and Evolution*, 53, 808–825.

Fairhead, J.D. 1986. Geophysical controls on sedimentation within the African Rift systems. *Geological Society of London Special Publication*, 25, 19–27.

Fairhead, J.D. 2003. Mesozoic plate tectonic controls on rift basin development in north central Africa: A major Cretaceous basin system. (Abstract). American Association of Petroleum Geologists Annual Meeting, 2003. http://aapg.confex.com/aapg.

Favela, J., and D.L. Anderson. 2000. Extensional tectonics and global volcanism. In E. Boschi, G. Ekström, and A. Morelli (eds.), *Problems in geophysics for the new millennium* (pp. 463–498). Compositori, Bologna, Italy.

Feder, J.L., S.H. Berlocher, J.B. Roethele, H. Dambroski, J.J. Smith, W.L. Perry, V. Gavrilovic, K.E. Filchak, J. Rull, and M. Aluja. 2003. Allopatric genetic origins for sympatric host-plant shifts and race formation in *Rhagoletis*. *Proceedings of the National Academy of Sciences USA*, 100, 10314–10319.

Fenchel, T. 2003. Biogeography for bacteria. *Science*, 301, 925–926.

Fernandes, M.E.B. 1991. Tool use and predation of oysters (*Crassostrea rhizophorae*) by the tufted capuchin, *Cebus apella apella*, in brackish water mangrove swamp. *Primates*, 32, 529–531.

Fernandes, M.E.B. 2000. Association of mammals with mangrove forests: A world wide review. *Boletim do Laboratorio de Hidrobiologia* (São Luís, MA), 13, 83–108.

Ferrari, S.F., L. Sena, M.P.C. Schneider, and J.S. Silva, Jr. 2010. Rondon's marmoset, *Mico rondoni* sp.n., from southwestern Brazilian Amazonia. *International Journal of Primatology*, 31, 693–714.

Filardi, C.E., and R.G. Moyle. 2005. Single origin of a pan-Pacific bird group and upstream colonization of Australasia. *Nature*, 438, 216–219.

Filler, A.G. 2007. Homeotic evolution in the Mammalia: Diversification of therian axial seriation and the morphogenetic basis of human origins. *PLoS One*, 2(10), e1019.

Fine, P.V.A., D.C. Daly, G.V. Muñoz, I. Mesones, and K.M. Cameron. 2005. The contribution of edaphic heterogeneity to the evolution and diversity of Burseraceae trees in the western Amazon. *Evolution* 59, 1464–1478.

Fiorillo, A.R. 2008. Dinosaurs of Alaska: Implications for the Cretaceous origin of Beringia. *Geological Society of America Special Paper*, 442, 313–326.

Fishbein, M., C. Hibsch-Jetter, D.E. Soltis, and L. Hufford., 2001. Phylogeny of Saxifragales (angiosperms, eudicots): Analysis of a rapid, ancient radiation. *Systematic Biology*, 50, 817–847.

Fisher, R.A. 1930. *The genetical theory of natural selection*. Oxford University Press, London, UK.

Fitton, J.G., J.J. Mahoney, P.J. Wallace, and A.D. Saunders. 2004. Origin and evolution of the Ontong Java Plateau: Introduction. In J.G. Fitton, J.J. Mahoney, P.J. Wallace, and A.D. Saunders (eds.), *Origin and evolution of the Ontong Java Plateau. Geological Society of London, Special Publication*, 229, 1–8.

Fiz, O., P. Vargas, M.L. Alarcón, and J.J. Aldasoro. 2006. Phylogenetic relationships and evolution in *Erodium* (Geraniaceae) based on *trnL-trnF* sequences. *Systematic Botany*, 31, 739–763.

Fjeldså, J., D. Zuccon, M. Irestedt, U.S. Johansson, and P.G.P. Ericson. 2003. *Sapayoa aenigma*: A New World representative of "Old World suboscines." *Proceedings of the Royal Society of London B*, 270, S238–S241.

Fleagle, J.G. 1999. *Primate adaptation and evolution*. Academic Press, San Diego.

Fleagle, J.G. 2002. The primate fossil record. *Evolutionary Anthropology*, 11(Suppl. 1), 20–23.

Fleagle, J.G., and C.C. Gilbert, C.C. 2006. The biogeography of primate evolution: The role of plate tectonics, climate and chance. In S.M. Lehman and J.G. Fleagle (eds.), *Primate biogeography* (pp. 375–418). Springer, New York.

Fleagle, J.G., and R.F. Kay. 1997. Platyrrhines, catarrhines and the fossil record. In W.G. Kinzey (ed.), *New World primates* (pp. 3–24). Aldine Transaction, Edison, NJ.

Fleischer, R.C. 2009. Honeycreepers, Hawaiian. In R.G. Gillespie and D.A. Clague (eds.), *Encyclopedia of islands* (pp. 410–413). University of California Press, Berkeley.

Fleischer, R.C., H.F. James, and S.L. Olson. 2008. Convergent evolution of Hawaiian and Australo-Pacific honeyeaters from distant songbird ancestors. *Current Biology*, 18, 1927–1931.

Fleischmann, A., B. Schäferhoff, G. Heubl, F. Rivadavia, W. Barthlott, and K.F. Müller. 2010. Phylogenetics and character evolution in the carnivorous plant genus *Genlisea* A. St.-Hil. (Lentibulariaceae). *Molecular Phylogenetics and Evolution*, 56, 768–783.

Florence, J. 2004. *Flore de la Polynésie française*, Vol. 2. IRD Éditions, Publications scientifiques du Muséum national d'Histoire naturelle, Paris.

Florin, A.-B. 2001. Bottlenecks and blowflies: Speciation, reproduction and morphological variation in *Lucilia. Acta Universitatis Upsaliensis: Comprehensive Summaries of Uppsala Dissertations from the Faculty of Science and Technology*, 660, 1–40.

Foissner, W. 2006. Biogeography and dispersal of micro-organisms: A review emphasizing protists. *Acta Protozoologica*, 45, 111–136.

Ford, E. B. 1971. *Ecological genetics*. 3rd ed. Chapman & Hall, London.

Fortey, R. 2004. *The Earth: An intimate history*. Harper, London.

Foulger, G.R. 2007. The "plate" model for the genesis of melting anomalies. *Geological Society of America Special Paper*, 430, 1–28.

Foulger, G.R., and D.M. Jurdy (eds.) 2007a. Plates, plumes, and planetary processes. *Geological Society of America Special Paper*, 430, 1–998.

Foulger, G.R., and D.M. Jurdy. 2007b. Preface: Plates, plumes, and planetary processes. *Geological Society of America Special Paper*, 430, vii–viii.

Foulger, G.R, J.H. Natland, D.C. Presnall, and D.L. Anderson (eds.). 2005. Plates, plumes, and paradigms. *Geological Society of America Special Paper*, 388, 1–881.

Fragaszy, D.M., E. Visalberghi, and L.M. Fedigan, L.M. 2004. *The complete capuchin: The biology of the genus Cebus*. Cambridge University Press, Cambridge.

Franzen, J.L., P.D. Gingerich, J. Habersetzer, J.H. Hurum, W. von Koenigswald, and B.H. Smith. 2009. Complete primate skeleton from the Middle Eocene of Messel in Germany: Morphology and paleobiology. *PLoS One*, 4(e5723), 1–26.

Frasier, C.L., V.A. Albert, and L. Struwe. 2008. Amazonian lowland, white sand areas as ancestral regions for South American biodiversity: Biogeographic and phylogenetic patterns in *Potalia* (Angiospermae). *Organisms, Diversity and Evolution*, 8, 44–57.

Frey, J.K. 1993. Modes of peripheral isolate formation and speciation. *Systematic Zoology*, 42, 373–381.

Frey, M.A. 2010. The relative importance of geography and ecology in species diversification: From a tropical marine intertidal snail (*Nerita*). *Journal of Biogeography*, 37, 1515–1528.

Friesen, V.L., A.L. Smith, E. Gómez-Díaz, M. Bolton, R.W. Furness, J. González-Solís, and L.R. Monteiro. 2007. Sympatric speciation by allochrony. *Proceedings of the National Academy of Sciences USA*, 104, 18589–18594.

Fuchs, J., E. Pasquet, A. Couloux, J. Fjeldså, and R.C.K. Bowie. 2009. A new Indo-Malayan member of Stenostiridae (Aves: Passeriformes) revealed by multilocus sequence data: Biogeographical implications for a morphologically diverse clade of flycatchers. *Molecular Phylogenetics and Evolution*, 53, 384–393.

Fuchs, J., J.-M. Pons, S.M. Goodman, V. Bretagnolle, M. Melo, R.C.K. Bowie, D. Currie, R. Safford, M.Z. Virani, S. Thomsett, A. Hija, C. Cruaud, and E. Pasquet. 2008. Tracing the colonization history of the Indian Ocean scops-owls (Strigiformes: *Otus*) with further insight into the spatio-temporal origin of the Malagasy avifauna. *BMC Evolutionary Biology*, 8, 197.

Fukuda, T., J. Yokoyama, and H. Ohashi. 2001. Phylogeny and biogeography of the genus *Lycium* (Solanaceae): Inferences from chloroplast DNA sequences. *Molecular Phylogenetics and Evolution*, 19, 246–258.

Funk, D.J., and P. Nosil. 2008. Comparative analyses of ecological speciation. In K.J. Tilmon (ed.), *Specialization, speciation, and radiation: The evolutionary biology of herbivorous insects* (pp. 117–135). University of California Press, Berkeley, CA.

Funk, V.A., R.J. Bayer, S. Keeley, R. Chan, L. Watson, B. Gemeinholzer, E. Schilling, J.L. Panero, B.G. Baldwin, N. Garcia-Jacas, A. Susanna, and R.K. Jansen. 2005. Everywhere but Antarctica: Using a tree to understand the diversity and distribution of the Compositae. *Biologiske Skrifter/Kongelige Danske Videnskabernes Selskab*, 55, 343–374.

Funk, V.A., and J.M. Bonifacino. 2009. Compositae classification: Re-visited, re-evaluated, re-everythinged, Part 1. Abstract. Botany Conference 2009 at http://2009.botanyconference.org.

Funk, V.A., R. Chan, and A. Holland. 2007. *Cymbonotus* (Compositae: Arctotideae: Arctotidinae): An endemic Australian genus embedded in a southern African clade. *Botanical Journal of the Linnean Society*, 153, 1–8.

Funk, V.A., and W.L. Wagner. 1995a. Biogeographic patterns in the Hawaiian Islands. In W.L. Wagner and V.A. Funk (eds.), *Hawaiian biogeography: Evolution on a hot spot archipelago* (pp. 379–419). Smithsonian Institution Press, Washington, DC.

Funk, V.A., and W.L. Wagner. 1995b. Biogeography of seven ancient Hawaiian plant lineages. In W.L. Wagner and V.A. Funk (eds.), *Hawaiian biogeography: Evolution on a hot spot archipelago* (pp. 160–194). Smithsonian Institution Press, Washington, DC.

Gaither, M.R., R.J. Toonen, D.R. Robertson, S. Planes, and B.W. Bowen. 2010. Genetic evaluation of marine biogeographic barriers: Perspectives from two widespread Indo-Pacific snappers (*Lutjanus kasmira* and *Lutjanus fulvus*). *Journal of Biogeography*, 37, 133–147.

Galley, C., and H.P. Linder. 2006. Geographical affinities of the Cape flora, South Africa. *Journal of Biogeography*, 33, 236–250.

Ganders, F. R., M. Berbee, and M. Pirseyedi. 2000. ITS base sequence phylogeny in *Bidens* (Asteraceae): Evidence for the continental relatives of Hawaiian and Marquesan *Bidens*. *Systematic Botany*, 25, 122–133.

Garb, J.E., and R.G. Gillespie. 2006. Island hopping across the Central Pacific: Mitochondrial DNA detects sequential colonization of the Austral Islands by crab spiders (Araneae: Thomisidae). *Journal of Biogeography*, 33, 201–228.

Garb, J.E., and R.G. Gillespie. 2009. Diversity despite dispersal: Colonization history and phylogeography of Hawaiian crab spiders inferred from multilocus genetic data. *Molecular Ecology*, 18, 1746–1764.

Gaston, K.J. 2003. *Structure and dynamics of geographic ranges*. Oxford University Press, Oxford.

Gavrilets, S., and J.B. Losos. 2009. Adaptive radiation: Contrasting theory with data. *Science*, 323, 732–737.

Gay, L., G. Neubauer, M. Zagalska-Neubauer, J.-M. Pons, D.A. Bell, and P.-A. Crochet. 2009. Speciation with gene flow in the large white-headed gulls: Does selection counterbalance introgression? *Heredity*, 102, 133–146.

Gebo, D.L. 2002. Adapiformes: Phylogeny and adaptation. In W.C. Hartwig (ed.), *The primate fossil record* (pp. 21–44). Cambridge University Press, Cambridge.

Geiger, J.M.O., and T.A. Ranker. 2005. Molecular phylogenetics and historical biogeography of Hawaiian *Dryopteris* (Dryopteridaceae). *Molecular Phylogenetics and Evolution*, 34, 392–407.

Geiger, J.M.O., T.A. Ranker, J.M.R. Neale, and S.T. Klimas. 2007. Molecular biogeography and origins of the Hawaiian fern flora. *Brittonia*, 59, 142–158.

Geist, D., and K. Harpp. 2009. Galápagos Islands, geology. In R.G. Gillespie and D.A. Clague (eds.), *Encyclopedia of islands* (pp. 367–372). University of California Press, Berkeley.

Gelang, M., A. Cibois, E. Pasquet, U. Olsson, P. Alström, and P.G.P. Ericson. 2009. Phylogeny of babblers (Aves, Passeriformes): Major lineages, family limits and classification. *Zoologica Scripta*, 38, 225–236.

Gemmill, C.E.C., G.J. Allan, W.L. Wagner, and E.A. Zimmer. 2002. Evolution of insular Pacific *Pittosporum* (Pittosporaceae): Origin of the Hawaiian radiation. *Molecular Phylogenetics and Evolution*, 22, 31–42.

Genner, M.J., O. Seehausen, D.H. Lunt, D.A. Joyce, P.W. Shaw, G.R. Carvalho, and G.F. Turner. 2007. Age of cichlids: New dates for ancient fish radiation. *Molecular Biology and Evolution*, 24, 1269–1282.

Gerlach, J. 2005. The impact of rodent eradication on the larger invertebrates of Fregate Island, Seychelles. *Phelsuma*, 13, 44–54.

Gheerbrant, E., and J.-C. Rage. 2006. Paleobiogeography of Africa: How distinct from Gondwana and Laurasia? *Palaeogeography, Palaeoclimatology, Palaeoecology*, 241, 224–246.

Gibb, G., and D. Penny. 2010. Two aspects along the continuum of pigeon evolution: A South-Pacific radiation and the relationship of pigeons within Neoaves. *Molecular Phylogenetics and Evolution*, 56, 698–706.

Gibson, A., V. Gowri-Shankar, P.G. Higgs, and M. Rattray. 2005. A comprehensive analysis of mammalian mitochondrial genome base composition and improved phylogenetic methods. *Molecular Biology and Evolution*, 22, 251–264.

Gibson, S.A., R.N. Thompson, O.H. Leonardos, A.P. Dickin, and J.G. Mitchell. 1999. The limited extent of plume-lithosphere interactions during continental flood-basalt genesis: Geochemical evidence from Cretaceous magmatism in southern Brazil. *Contributions to Mineralogy and Petrology*, 137, 147–169.

Gillespie, R.G. 2009. Adaptive radiation. In R.G. Gillespie and D.A. Clague (eds.), *Encyclopedia of islands* (pp. 1–7). University of California Press, Berkeley.

Gillespie, R.G, E.M. Claridge, and S.J. Goodacre. 2008. Biogeography of the fauna of French Polynesia. *Philosophical Transactions of the Royal Society B*, 363, 3335–3346.

Gillespie, R.G., and G.K. Roderick. 2002. Arthropods on islands: Colonization, speciation, and conservation. *Annual Review of Entomology*, 47, 595–632.

Gillespie, R.G., A.J. Rivera, and J.E. Garb. 1998. Sun, surf and spiders: Taxonomy and phylogeography of Hawaiian Araneae. In P.A. Selden (ed.), *Proceedings of the 17th European Colloquium of Arachnology, Edinburgh 1997* (pp. 41–52). British Arachnological Society, Burnham Beeches, Bucks.

Gingerich, P.D. 1976. Cranial anatomy and evolution of early Tertiary Plesiadapidae (Mammalia, Primates). *University of Michigan Papers in Paleontology*, 15, 1–141.

Gingras, M.K., M.E. Räsänen, S.G. Pemberton, and L.P. Romero. 2002. Ichnology and sedimentology reveal depositional characteristic of bay-margin parasequences in the Miocene Amazon foreland basin. *Journal of Sedimentary Research*, 72, 871–883.

Giresse, P. 2005. Mesozoic-Cenozoic history of the Congo Basin. *Journal of African Earth Sciences*, 43, 301–315.

Givnish, T.J. 2003. How a better understanding of adaptations can yield better use of morphology in plant systematics: toward Eco-Evo-Devo. In T.F. Stuessy, V. Mayer, and E. Hörandl (eds.), *Deep morphology: Toward a renaissance of morphology in plant systematics* (pp. 273–295). A.R.G. Gantner, Ruggell.

Givnish, T.J., T.M. Evans, J.C. Pires, and K.J. Sytsma. 1999. Polyphyly and convergent morphological evolution in Commelinales and Commelinidae: evidence from *rbcL* sequence data. *Molecular Phylogenetics and Evolution*, 12, 360–385.

Givnish, T.J., K.C. Millam, A.R. Mast, T.B. Paterson, T.J. Theim, A.L. Hipp, J.M. Henss, J.F. Smith, K.R. Wood, and K.J. Sytsma. 2009. Origin, adaptive radiation and diversification of the Hawaiian lobeliads (Asterales: Campanulaceae). *Proceedings of the Royal Society B*, 276, 407–416.

Givnish, T.J., K.J. Sytsma, J.F. Smith, and W.S. Hahn. 1995. Molecular evolution, adaptive radiation, and geographic speciation in *Cyanea* (Campanulaceae, Lobelioideae). In W.L. Wagner and V. Funk (eds.), *Hawaiian biogeography: Evolution on a hotspot archipelago* (pp. 288–337). Smithsonian Institution Press, Washington DC.

Glazko, G.V., E.V. Koonin, and I.B. Rogozin. 2005. Molecular dating: Ape bones agree with chicken entrails. *Trends in Genetics*, 21, 89–92.

Godinot, M. 2006a. Primate origins: A reappraisal of historical data favouring tupaiid affinities. In M.J. Ravosa and M. Dagosto (eds.), *Primate origins: Adaptations and evolution* (pp. 83–142). Springer, New York.

Godinot, M. 2006b. Lemuriform origins as viewed from the fossil record. *Folia Primatologica*, 77, 446–464.

Godinot, M., and F. de Lapparent de Broin. 2003. Arguments for a mammalian and reptilian dispersal from Asia to Europe during the Paleocene-Eocene boundary interval. *Deinsea*, 10, 255–275.

Goldani, A., G.S. Carvalho, and J.C. Bicca-Marques. 2006. Distribution patterns of neotropical primates (Platyrrhini) using parsimony analysis of endemicity. *Brazilian Journal of Biology*, 66, 61–74.

Gonder, M.K., T.R. Disotell, and J.F. Oates. 2006. New genetic evidence on the evolution of chimpanzee populations and implications for taxonomy. *International Journal of Primatology*, 27, 1103–1127.

Gonedelé Bi, S., I. Koné, J.-C.K. Béné, A.E. Bitty, B.K. Akpatou, Z. Goné Bi, K. Ouattara, and D.A. Koffi. 2008. Tanoé forest, south-eastern Côte-d'Ivoire identified as a high priority site for the conservation of critically endangered primates in West Africa. *Tropical Conservation Science*, 1, 265–278.

Goodacre, S.L., and C.M. Wade. 2001. Molecular evolutionary relationships between partulid land snails of the Pacific. *Proceedings of the Royal Society of London B*, 268, 1–7.

Goodall-Copestake, W.P., D.J. Harris, and P.M. Hollingsworth. 2009. The origin of a mega-diverse genus: Dating *Begonia* (Begoniaceae) using alternative datasets, calibrations and relaxed clock methods. *Botanical Journal of the Linnean Society*, 159, 363–380.

Goodall-Copestake, W.P., S. Pérez-Espona, D.J. Harris, and P.M. Hollingsworth. 2010. The early evolution of the mega-diverse genus *Begonia* (Begoniaceae) inferred from organelle DNA phylogenies. *Biological Journal of the Linnean Society*, 101, 243–250.

Goodman, M., L.I. Grossman, and D.E. Wildman. 2005. Moving primate genomics beyond the chimpanzee genome. *Trends in Genetics*, 21, 511–517.

Goosens, B., L. Chikhi, M.F. Jalil, M. Ancrenaz, I. Lackman-Acrenaz, M. Mohamed, P. Andau, and M.W. Bruford. 2005. Patterns of genetic diversity and migration in increasingly fragmented and declining orang-utan (*Pongo pygmaeus*) populations from Sabah, Malaysia. *Molecular Ecology*, 14, 441–456.

Goswami, A., and P. Upchurch. 2010. The dating game: A reply to Heads (2010). *Zoologica Scripta*, 39, 406–409.

Gould, S.J. 2002. *The structure of evolutionary theory*. Belknap Press, Harvard University, Cambridge, MA.

Grandcolas, P., J. Murienne, T. Robillard, L. Desutter-Grandcolas, H. Jourdan, E. Guilbert, and L. Deharveng. 2008. New Caledonia: A very old Darwinian island? *Philosophical Transactions of the Royal Society B*, 363, 3309–3317.

Grant, P.R. 2001. Reconstructing the evolution of birds on islands: 100 years of research. *Oikos*, 92, 385–403.

Gregory, T.R. (ed.). 2005. *The evolution of the genome*. Elsevier Academic Press, Boston.

Grehan, J.R., and J.H. Schwarz. 2009. Evolution of the third orangutan: Phylogeny and biogeography of hominid origins. *Journal of Biogeography*, 36, 1823–1844.

Gressitt, J.L. 1978. Evolution of the endemic Hawaiian cerambycid beetles. *Pacific Insects*, 18, 137–167.

Griffiths, C.J. 1993. The geological evolution of East Africa. In J.C. Lovett and S.K. Wasser (eds.), *Biogeography and ecology of the rain forests of eastern Africa* (pp. 9–21). Cambridge University Press, Cambridge, UK.

Groppo, M., J.R. Pirani, M.L.F. Salatino, S.R. Blanco, and J.A. Kallunki. 2008. Phylogeny of Rutaceae based on two noncoding regions from cpDNA. *American Journal of Botany*, 95, 985–1005.

Groves, C.P. 1972. Systematics and phylogeny of gibbons. In D.M. Rumbaugh (ed.), *Gibbon and Siamang*, Vol. 1 (pp. 1–89). Karger, Basel.

Groves, C.P. 2001. *Primate taxonomy*. Smithsonian Institution Press, Washington, DC.

Groves, C.P. 2005. Primates. In D.E. Wilson and D.M. Reeder (eds.), *Mammal species of the world: A taxonomic and geographic reference*, 3rd ed. (pp. 111–184). Johns Hopkins University Press, Baltimore, MD.

Groves, C. 2006. Taxonomy and biogeography of the primates of western Uganda. In M.E. Newton-Fisher, H. Notman, J.D. Paterson, and V. Reynolds (eds.), *Primates of western Uganda* (pp. 3–20). Springer, New York.

Groves, C., and M. Shekelle. 2010. The genera and species of Tarsiidae. *International Journal Primatology*, 31, 1071–1082.

Guiraud, M., and J.-C. Plaziat. 1993. Seismites in the fluviatile Bima sandstones: Identification of paleoseisms and discussion of their magnitudes in a Cretaceous synsedimentary strike-slip basin (Upper Benue, Nigeria). *Tectonophysics*, 225, 493–522.

Gunnell, G.F., and R.L. Ciochon. 2008. Revisiting primate postcrania from the Pondaung Formation of Myanmar: The purported anthropoid astragalus.

In J.G. Fleagle and C.C. Gilbert (eds.), *Elwyn Simons: A search for origins* (pp. 211–228). Springer, New York.

Gunnell, G.F., R.L. Ciochon, P.D. Gingerich, and P.A. Holroyd. 2002. New assessment of *Pondaungia* and *Amphipithecus* (Primates) from the late Middle Eocene of Myanmar, with a comment on "Amphipithecidae." *Contributions from the Museum of Paleontology, University of Michigan*, 30, 337–372.

Gunnell, G.F., and K.D. Rose. 2002. Tarsiiformes: Evolutionary history and adaptation. In W.C. Hartwig (ed.), *The primate fossil record* (pp. 45–82). Cambridge University Press, Cambridge.

Gutscher, M.-A., J.-L. Olivet, D. Aslanian, J.-P. Eissen, and R. Maury. 1999. The "lost Inca Plateau": Cause of flat subduction beneath Peru? *Earth and Planetary Science Letters*, 171, 335–341.

Haig, D., and S. Henikoff. 2004. Genomes and evolution: Deciphering the genomic palimpsest. *Current Opinion in Genetics and Development*, 14, 559–602.

Haldane, J.B.S. 1932. *The causes of evolution*. Longmans, Green and Co., New York.

Hall, R., B. Clements, and H.R. Smyth. 2009. Sundaland: Basement character, structure and plate tectonic development. Proceedings, Indonesian Petroleum Association Thirty-Third Annual Convention and Exhibition, May 2009.

Hall, R., and M.E.J. Wilson. 2000. Neogene sutures in eastern Indonesia. *Journal of Asian Earth Sciences*, 18, 781–808.

Hallé, F. 1978. Arbres et forêts des Iles Marquises. *Cahiers du Pacifique*, 21, 315–357.

Hames, W.E., P.R. Renne, and C. Ruppel. 2000. New evidence for geologically instantaneous emplacement of earliest Jurassic Central Atlantic magmatic province basalts on the North American margin. *Geology*, 28, 859–862.

Hamilton, A.M., J.H. Hartman, and C.C. Austin. 2009. Island area and species diversity in the southwest Pacific Ocean: Is the lizard fauna of Vanuatu depauperate? *Ecography*, 32, 247–258.

Hamilton, A.M., G.R. Zug, and C.C. Austin. 2010. Biogeographic anomaly or human introduction: A cryptogenic population of tree skink (Reptilia: Squamata) from the Cook Islands, Oceania. *Biological Journal of the Linnean Society*, 100, 318–328.

Hamilton, W.B. 2007. Driving mechanisms and 3-D circulation of plate tectonics. *Geological Society of America Special Paper*, 433, 1–25.

Han, K.L., M.B. Robbins, and M.J. Braun. 2010. A multi-gene estimate of phylogeny in the nightjars and nighthawks (Caprimulgidae). *Molecular Phylogenetics and Evolution*, 55, 443–453.

Harbaugh, D.T., and B.G. Baldwin. 2007. Phylogeny and biogeography of the sandalwoods (*Santalum*, Santalaceae): Repeated dispersals throughout the Pacific. *American Journal of Botany*, 94, 1028–1040.

Harbaugh, D.T., M. Nepokroeff, R.K. Rabeler, J. McNeill, E.A. Zimmer, and W.L. Wagner. 2010. A new lineage-based tribal classification of the family Caryophyllaceae. *International Journal of Plant Science*, 171, 185–198.

Harbaugh, D.T., H.L. Oppenheimer, K.R. Wood, and W.L. Wagner. 2010. Taxonomic revision of the endangered Hawaiian red-flowered sandalwoods

(*Santalum*) and discovery of an ancient hybrid species. *Systematic Botany*, 35, 827–838.

Harbaugh, D.T., W.L. Wagner, G.J. Allan, and E.A. Zimmer. 2009. The Hawaiian Archipelago is a stepping stone for dispersal in the Pacific: An example from the plant genus *Melicope* (Rutaceae). *Journal of Biogeography*, 36, 230–241.

Harbaugh, D.T., W.L. Wagner, D.M. Percy, H.F. James, and R.C. Fleischer. 2009. Genetic structure of the polymorphic *Metrosideros* (Myrtaceae) complex in the Hawaiian Islands using nuclear staellite data. *PloS One*, 4(3), e4698, 1–7.

Harrington, M.G., and P.A. Gadek, P.A. 2009. A species well-travelled: The *Dodonaea viscosa* (Sapindaceae) complex based on phylogenetic analyses of nuclear ribosomal ITS and ETSf sequences. *Journal of Biogeography*, 36, 2313–2323.

Harris, E.E. 2010. Nonadaptive processes in primate and human evolution. *Yearbook of Physical Anthropology*, 53, 13–45.

Harrison, T. 2001. Archaeological and ecological implications of the primate fauna from prehistoric sites in Borneo. *Bulletin of the Indo-Pacific Association*, 20, 133–146.

Hartley, R.W., and P.A. Allen. 1994. Interior cratonic basins of Africa: Relation to continental break-up and role of mantle convection. *Basin Research*, 6, 95–113.

Hartley, T.G. 2001a. Morphology and biogeography in Australasian-Malesian Rutaceae. *Malayan Nature Journal*, 55, 197–219.

Hartley, T.G. 2001b. On the taxonomy and biogeography of *Euodia* and *Melicope* (Rutaceae). *Allertonia*, 8, 1–328.

Hartley, T.G. 2003. *Neoschmidia*, a new genus of Rutaceae from New Caledonia. *Adansonia*, sér. 3, 25, 7–12.

Hartwig, W.C., A.L. Rosenberger, P.A. Garber, and M.A. Norconk. 1996. On atelines. In M.A. Norconk, A.L. Rosenberger, and P.A. Garber (eds.), *Adaptive radiations of neotropical primates* (pp. 427–432). Springer, New York.

Haugaasen, T., and C. Peres. 2005. Primate assemblage structure in Amazonian flooded and unflooded forests. *American Journal of Primatology*, 67, 243–258.

Haugaasen, T., and C. Peres. 2006. Floristic, edaphic and structural characteristics of flooded and unflooded forests in the lower Rio Purús region of central Amazonia, Brazil. *Acta Amazonica*, 36, 25–36.

Haugaasen, T., and C. Peres. 2008. Population abundance and biomass of large-bodied birds in Amazonian flooded and unflooded forests. *Bird Conservation International*, 18, 87–101.

Havran, J.C., K.J. Sytsma, and H.E. Ballard, Jr. 2009. Evolutionary relationships, interisland biogeography, and molecular evolution in the Hawaiian violets (*Viola*: Violaceae). *American Journal of Botany*, 96, 2087–2099.

Heads, M. 1994. Morphology, architecture and taxonomy in the *Hebe* complex (Scrophulariaceae). *Bulletin du Muséum national d'Histoire naturelle* Paris, sér. 4, 16, sect. B., *Adansonia*, 163–191.

Heads, M. 2001. Birds of paradise, biogeography and ecology in New Guinea: A review. *Journal of Biogeography*, 28, 893–925.

Heads, M. 2002. Birds of paradise, vicariance biogeography and terrane tectonics in New Guinea. *Journal of Biogeography*, 29, 261–284.

Heads, M. 2003. Ericaceae in Malesia: Vicariance biogeography, terrane tectonics and ecology. *Telopea*, 10, 311–449.

Heads, M. 2005a. Dating nodes on molecular phylogenies: A critique of molecular biogeography. *Cladistics*, 21, 62–78.

Heads, M. 2005b. Towards a panbiogeography of the seas. *Biological Journal of the Linnean Society*, 84, 675–723.

Heads, M. 2005c. The history and philosophy of panbiogeography. In J. Llorente and J.J. Morrone (eds.), *Regionalización biogeográfica en Iberoamérica y tópicos afines* (pp. 67–123). Universidad Nacional Autónoma de México, Mexico City.

Heads, M. 2006a. Panbiogeography of *Nothofagus* (Nothofagaceae): Analysis of the main species massings. *Journal of Biogeography*, 33, 1066–1075.

Heads, M. 2006b. Seed plants of Fiji: An ecological analysis. *Biological Journal of the Linnean Society*, 89, 407–431.

Heads, M. 2008a. Biological disjunction along the West Caledonian fault, New Caledonia: A synthesis of molecular phylogenetics and panbiogeography. *Botanical Journal of the Linnean Society*, 158, 470–488.

Heads, M. 2008b. Panbiogeography of New Caledonia, south-west Pacific: Basal angiosperms on basement terranes, ultramafic endemics inherited from volcanic island arcs, and old taxa endemic to young islands. *Journal of Biogeography*, 35, 2153–2175.

Heads, M. 2009a. Globally basal centres of endemism: The Tasman-Coral Sea region (south-west Pacific), Latin America and Madagascar/South Africa. *Biological Journal of the Linnean Society*, 96, 222–245.

Heads, M. 2009b. Inferring biogeographic history from molecular phylogenies. *Biological Journal of the Linnean Society*, 98, 757–774.

Heads, M. 2009c. Darwin's changing ideas on evolution: From centres of origin and teleology to vicariance and incomplete lineage sorting. *Journal of Biogeography*, 36, 1018–1026.

Heads, M. 2009d. Vicariance. In R.G. Gillespie and D.A. Clague (eds.), *Encyclopedia of islands* (pp. 947–950). University of California Press, Berkeley.

Heads, M. 2010a. Evolution and biogeography of primates: A new model based on plate tectonics, molecular phylogenetics and vicariance. *Zoologica Scripta*, 39, 107–127.

Heads, M. 2010b. The biogeographical affinities of the New Caledonian biota: A puzzle with 24 pieces. *Journal of Biogeography*, 37, 1179–1201.

Heads, M. 2011. Old taxa on young islands: a critique of the use of island age to date island-endemic clades and calibrate phylogenies. *Systematic Biology*, 60, 204–218.

Heads, M., and R.C. Craw. 2004. The Alpine fault biogeographic hypothesis revisited. *Cladistics*, 20, 184–190.

Heckman, K.L., C.L. Mariani, R. Rasoloarison, and A.D. Yoder. 2007. Multiple nuclear loci reveal patterns of incomplete lineage sorting and complex species history within western mouse lemurs (*Microcebus*). *Molecular Phylogenetics and Evolution*, 43, 353–367.

Hedges, S.B. 2010. Molecular clocks, flotsam, and Caribbean islands. In C.B. Cox and P.D. Moore (eds.), *Biogeography: An ecological and evolutionary approach*, 8th ed. (pp. 353–354). Blackwell, Oxford.

Hedges, S.B., A. Couloux, and N. Vidal. 2009. Molecular phylogeny, classification, and biogeography of West Indian racer snakes of the Tribe Alsophiini (Squamata, Dipsadidae, Xenodontinae). *Zootaxa*, 2067, 1–28.

Heesy, C.P., N.J. Stevens, and K.E. Samonds. 2006. Biogeographic origins of primate higher taxa. In S.M. Lehman and J.G. Fleagle (eds.), *Primate biogeography* (pp. 419–437). Springer, New York.

Heine, C., and R.D. Müller. 2005. Late Jurassic rifting along the Australian north west shelf: Margin geometry and spreading ridge configuration. *Australian Journal of Earth Sciences*, 52, 27–39.

Hendrian and K. Kondo. 2007. Molecular phylogeny of *Ochrosia sensu lato* (Apocynaceae) based on rps16 intron and ITS sequence data: Supporting the inclusion of *Neisosperma*. *Chromosome Botany*, 2, 133–140.

Hennig, W., 1966. *Phylogenetic systematics*. University of Illinois Press, Urbana, IL.

Hermoza, W., S. Brusset, P. Baby, W. Gil, M. Roddaz, N. Guerrero, and M. Bolaños. 2005. The Huallaga foreland basin evolution: Thrust propagation in a deltaic environment, northern Peruvian Andes. *Journal of South American Earth Sciences*, 19, 21–34.

Hershkovitz, P. 1977. *Living New World monkeys (Platyrrhini): With an introduction to Primates*, Vol. 1. University of Chicago Press, Chicago.

Hershkovitz, P. 1984. Two new species of night monkeys, genus *Aotus* (Cebidae, Platyrrhini): A preliminary report on *Aotus* taxonomy. *American Journal of Primatology*, 4, 209–243.

Hesselbo, S.P., S.A. Robinson, F. Surlyk, and S. Piasecki. 2002. Terrestrial and marine extinction at the Triassic-Jurassic boundary synchronized with major carbon-cycle perturbation: A link to initiation of massive volcanism? *Geology*, 30, 251–254.

Heymann, E.W., and R. Aquino. 2010. Peruvian Red Uakaris (*Cacajao calvus ucayalii*) are not flooded-forest specialists. *International Journal of Primatology*, 31, 751–758.

Hickerson, M.J., B.C. Carstens, J. Cavender-Bares, K.A. Crandall, C.H. Graham, J.B. Johnson, L. Rissler, P.F. Victoriano, and A.D. Yoder. 2010. Phylogeography's past, present, and future: 10 years after Avise, 2000. *Molecular Phylogenetics and Evolution*, 54, 291–301.

Higley, D.K. 2001. *The Putumayo-Oriente-Maranon Province of Colombia, Ecuador and Peru: Mesozoic-Cenozoic and Paleozoic petroleum systems*. U.S. Geological Survey Digital Data Series 63.

Hillier, J.K. 2006. Pacific seamount volcanism in space and time. *Geophysical Journal International*. doi:10.1111/j.1365-246X.2006.03250.x.

Hjerman, D.Ø. 2009. Metapopulations. In R.G Gillespie and D.A. Clague (eds.), *Encyclopedia of islands* (pp. 629–631). University of California Press, Berkeley.

Hobbs, C., and B.G. Baldwin. 2008. Origin of Hawaiian *Artemisia* (Compositae—Anthemideae). Abstract. Botany Conference 2008. Available at http://2008.botanyconference.org.

Hoch, H. 2006. Systematics and evolution of *Iolania* (Hemiptera: Fulgoromorpha: Cixiidae) from Hawai'i. *Systematic Entomology*, 31, 302–320.

Hoffstetter, R. 1974. Phylogeny and geographical deployment of the primates. *Journal of Human Evolution*, 3, 327–350.

Hofman, A., L.R. Maxson, and J.W. Arntzen. 1991. Biochemical evidence pertaining to the taxonomic relationships within the family Chameleonidae. *Amphibia-Reptilia*, 12, 245–265.

Hohenegger, J., and H. Zapfe. 1990. Craniometric investigation on *Mesopithecus* in comparison with two recent colobines. *Beiträge zur Paläontologie von Österreich*, 16, 111–144.

Holland, B.S. 2009. Land snails. In R.G. Gillespie and D.A. Clague (eds.), *Encyclopedia of islands* (pp. 537–542). University of California Press, Berkeley.

Holland, B.S., and R.H. Cowie. 2006. Dispersal and vicariance in Hawaii: Submarine slumping does not create deep interisland channels. *Journal of Biogeography*, 33, 2154–2157.

Holland, B.S., and R.H. Cowie. 2007. A geographic mosaic of passive dispersal: Population structure of the endemic Hawaiian amber snail *Succinea caduca* (Mighels, 1845). *Molecular Ecology*, 16, 2422–2435.

Holland, B.S., and R.H. Cowie. 2009. Land snail models in island biogeography: A tale of two snails. *American Malacological Bulletin*, 27, 59–68.

Holloway, J.D., and R. Hall. 1998. SE Asian geology and biogeography. In R. Hall and J.D. Holloway (eds.), *Biogeography and geological evolution of SE Asia* (pp. 1–23). Backhuys, Amsterdam, Netherlands.

Hoorn, C. 2006. Mangrove forests and marine incursions in Neogene Amazonia (Lower Apaporis River, Colombia). *Palaios*, 21, 197–209.

Hoorn, C., F.P. Wesselingh, H. ter Steege, M.A. Bermudez, A. Mora, J. Sevink, I. Sanmartín, A. Sanchez-Meseguer, C.L. Anderson, J.P. Figueiredo, C. Jaramillo, D. Riff, F.R. Negri, H. Hooghiemstra, J. Lundberg, T. Stadler, T. Särkinen, and A. Antonelli. 2010. Amazonia through time: Andean uplift, climate change, landscape evolution, and biodiversity. *Science*, 330, 927–931.

Hopkins, H.C.F., and J.C. Bradford. 1998. A revision of *Weinmannia* (Cunoniaceae) in Malesia and the Pacific. 1: Introduction and an account of the species of western Malesia, the Lesser Sunda Islands and the Moluccas. *Adansonia*, sér. 3, 20, 5–41.

Hopkins, H.C.F., R.D. Hoogland, and J.C. Bradford. 1998. A revision of *Weinmannia* (Cunoniaceae) in Malesia and the Pacific. 3: New Guinea, Solomon Islands, Vanuatu and Fiji, with notes on the species of Samoa, Rarotonga, New Caledonia and New Zealand. *Adansonia*, sér. 3, 20, 67–106.

Hormiga, G., M. Arnedo, and R.G. Gillespie. 2003. Speciation on a conveyor belt: Sequential colonization of the Hawaiian Islands by *Orsonwelles* spiders (Araneae, Linyphiidae). *Systematic Biology*, 52, 70–88.

Horner, D.S., K. Lefkimmiatis, A. Reyes, C. Gissi, C. Saccone, and G. Pesole. 2007. Phylogenetic analyses of complete mitochondrial genome sequences suggest a basal divergence of the enigmatic rodent *Anomalurus*. *BMC Evolutionary Biology*, 7(16), 1–12.

Hoso, M., Asami, T., and M. Hori. 2007. Right-handed snakes: Convergent evolution of asymmetry for functional specialization. *Biology Letters*, 3, 169–172.

Hovikoski, J., M. Räsänen, M. Gingras, M. Roddaz, S. Brusset, W. Hermoza, and L.R. Pittman. 2005. Miocene semidiurnal tidal rhythmites in Madre de Dios, Peru. *Geology*, 33, 177–180.

Howarth, D.G., and D.E. Gardner. 1997. Phylogeny of *Rubus* subgenus *Idaeobatus* (Rosaceae) and its implications toward colonization of the Hawaiian Islands. *Systematic Botany*, 22, 433–441.

Howarth, F.G. 1990. Hawaiian terrestrial arthropods: An overview. *Bishop Museum Occasional Papers*, 30, 4–26.

Hu, Y., J. Meng, Y. Wang, and C. Li. 2005. Large Mesozoic mammals fed on young dinosaurs. *Nature*, 433, 149–152.

Hubbell, S.P. 2001. *The unified neutral theory of biodiversity and biogeography.* Princeton University Press, Princeton, NJ.

Hubert, N., F. Duponchelle, J. Nuñez, C. Garcia-Davila, D. Paugy, and J.F. Renno. 2007. Phylogeography of the piranha genera *Serraselmus* and *Pygocentrus*: Implications for the diversification of the neotropical ichthyofauna. *Molecular Ecology*, 16, 2115–2136.

Hubert, N., and J.F. Renno. 2006. Historical biogeography of South American freshwater fishes. *Journal of Biogeography*, 33, 1414–1436.

Hufford, L. 2004. Hydrangeaceae. In K. Kubitzki (ed.), *The families and genera of vascular plants*, Vol. 6: *Flowering plants dicotyledons: Celastrales, Oxalidales, Rosales, Cornales, Ericales* (pp. 202–215). Springer, Berlin.

Hufford, L., M.L. Moody, and D.E. Soltis. 2001. A phylogenetic analysis of Hydrangeaceae based on sequences of the plastid gene *matK* and their combination with *rbcL* and morphological data. *International Journal of Plant Science*, 162, 835–846.

Hughes, A.L. 2009. Evolution in the post-genome era. *Perspectives in Biology and Medicine*, 52, 332–337.

Humboldt, A. von. [1814–1825] 1995. *Personal narrative of a journey to the equinoctial regions of the New Continent.* Penguin, London.

Humphries, E.M., and K. Winker. 2011. Discord reigns among nuclear, mitochondrial and phenotypic estimates of divergence in nine lineages of trans-Beringian birds. *Molecular Ecology*, 20, 573–583.

Hundsdoerfer, A.K., D. Rubinoff, M. Attié, M. Wink, and I.J. Kitching. 2009. A revised molecular phylogeny of the globally distributed hawkmoth genus *Hyles* (Lepidoptera: Sphingidae), based on mitochondrial and nuclear DNA sequences. *Molecular Phylogenetics and Evolution*, 52, 852–865.

Husen, S., E. Kissling, and R. Quintero. 2002. Tomographic evidence for a subducted seamount beneath the Gulf of Nicorya, Costa Rica: The cause of the 1990 *Mw* = 7.0 Gulf of Nicoya earthquake. *Geophysical Research Letters*, 29(8), 1328. doi:10.1029/2001GL014045.

Hutton, F.J. 1872. On the geographic relations of the New Zealand fauna. *Transactions of the New Zealand Institute*, 5, 227–256.

Huxley, J.S. 1942. *Evolution: The modern synthesis.* Allen & Unwin, London.

Huxley, T.H. [1870] 1896. *Discourses: Biological and geological essays.* D. Appleton, New York.

Ingle, S., J.J. Mahoney, H. Sato, M.F. Coffin, J.-I. Kimura, N. Hiano, and M. Nakanishi. 2007. Depleted mantle wedge and sediment fingerprint from unusual basalts from the Manihiki Plateau, central Pacific Ocean. *Geology*, 35, 595–598.

Irestedt, M., J.I. Ohlsson, D. Zuccon, M. Källersjö, and P. Ericson. 2006. Nuclear DNA from old collections of avian study skins reveals the evolutionary history of the Old World suboscines (Aves, Passeriformes). *Zoologica Scripta*, 35, 567–580.

Isler, M.L., J. Alvarez Alonso, P.R. Isler, and B.M. Whitney. 2001. A new species of *Percnostola* antbird (Passeriformes: Thamnophilidae) from Amazonian Peru, and an analysis of species limits within *Percnostola rufifrons. Wilson Bulletin*, 113, 164–176.

Israfil, H., S.M. Zehr, A.R. Mootnick, M. Ruvolo, and M.E. Steiper. 2011. Unresolved molecular phylogenies of gibbons and siamangs (Family: Hylobatidae) based on mitochondrial, Y-linked, and X-linked loci indicate a rapid Miocene radiation or sudden vicariance event. *Molecular Phylogenetics and Evolution*, 58, 447–455.

Ito, G., and P.E. van Keken. 2007. Hotspots and melting anomalies. In D. Bercovici (ed.), *Treatise on geophysics*, Vol. 7: *Mantle dynamics* (pp. 371–435). Elsevier, Amsterdam.

IUCN. 2009. *The IUCN redlist of threatened species.* Available at www.iucnredlist.org/.

Iwamoto, M., Y. Hasegawa, and A. Koizumi. 2005. A Pliocene colobine from the Nakatsu group, Kanagawa, Japan. *Anthropological Science*, 113, 123–127.

Iwaniuk, A.N., S.L. Olson, and H.F. James. 2009. Extraordinary cranial specialization in a new genus of extinct duck (Aves: Anseriformes) from Kauai, Hawaiian Islands. *Zootaxa*, 2296, 47–67.

Jablonski, N.G. (ed.). 1993. *Theropithecus: The rise and fall of a primate genus.* Cambridge University Press, Cambridge.

Jablonski, N.G. 1998. The evolution of the doucs and snub-nosed monkeys and the question of phyletic unity of the odd-nosed colobines. In N.G. Jablonski (ed.), *The natural history of the doucs and snub-nosed monkeys* (pp. 13–52). World Scientific, Singapore.

Jackson, J.B.C., P. Jung, A.G. Coates, and L.S. Collins. 1993. Diversity and extinction of tropical American mollusks and closure of the Isthmus of Panama. *Science*, 260, 1624–1626.

Jacob, F. 1977. Evolution and tinkering. *Science*, 196, 1161–1166.

Jacobs Cropp, S., A. Larson, A., and J.M. Cheverud. 1999. Historical biogeography of tamarins, genus *Saguinus:* The molecular phylogenetic evidence. *American Journal of Physical Anthopology*, 108, 65–89.

Jaeger, J.J., K.C. Beard, Y. Chaimanee, M. Salem, M. Benammi, O. Hlal, P. Coster, A.A. Bilal, P. Duringer, M. Schuster, X. Valentin, B. Marandat, L. Marivaux, E. Métais, O. Hammuda, and M. Brunet. 2010. Late middle Eocene epoch of Libya yields earliest known radiation of African anthropoids. *Nature*, 467, 1095–1099.

Jaffré, T., P. Morat, J.-M. Veillon, F. Rigault, and G. Dagostini. 2001. *Composition et caractérisation de la flore indigène de Nouvelle-Calédonie.* Institut de Recherche pour le Développement, Nouméa.

Jaillard, E. 1994. Kimmeridgian to Paleocene tectonic and geodynamic evolution of the Peruvian (and Ecuadorian) margin. In J.A. Salfity (ed.), *Cretaceous tectonics in the Andes* (pp. 101–167). Vieweg und Sohn, Braunschweig.

Jaillard, E., P. Bengtson, and A.V. Dhondt. 2005. Late Cretaceous marine transgressions in Ecuador and northern Peru: A refined stratigraphic framework. *Journal of South American Earth Sciences*, 19, 307–323.

Jaillard, E., P. Bengtson, M. Ordoñez, W. Vaca, A. Dhondt, J. Suárez, and J. Toro. 2008. Sedimentary record of terminal Cretaceous accretions in Ecuador: The Yunguilla Group in the Cuenca area. *Journal of South American Earth Sciences*, 25, 133–144.

James, H.F. 2004. The osteology and phylogeny of the Hawaiian finch radiation (Fringillidae: Drepanidini), including extinct taxa. *Zoological Journal of the Linnean Society*, 41, 207–255.

James, H.F., and S.L. Olson. 2003. A giant new species of Nukupuu (Fringillidae: Drepanidini: *Hemignathus*) from the island of Hawaii. *The Auk*, 120, 970–981.

Janečka, J.E., W. Miller, T.H. Pringle, F. Wiens, A. Zitzmann, K. Helgen, M.S. Springer, and W.J. Murphy. 2007. Molecular and genomic data identify closest living relative of primates. *Science*, 318, 792–794.

Janes, D.E., C.L. Organ, M.K. Fujita, A.M. Shedlock, and S.V. Edwards. 2010. Genome evolution in Reptilia, the sister group of mammals. *Annual Review of Genomics and Human Genetics*, 11, 239–264.

Ji, Q., Z.-X. Luo, C.-X. Yuan, and A.R. Tabrum, A.R. 2006. A swimming mammaliaform from the Middle Jurassic and ecomorphological diversification of early mammals. *Science*, 311, 1123–1127.

Jia, D., G. Wei, Z. Chen, B. Li, Q. Zeng, and G. Yang. 2006. Longmen Shan fold-thrust belt and its relation to the western Sichuan Basin in central China: New insights from hydrocarbon exploration. *Bulletin of the American Association of Petroleum Geologists*, 90, 1425–1447.

Jockusch, E.L., and D.B. Wake. 2002. Falling apart and merging: Diversification of slender salamanders (Plethodontidae: *Batrachoseps*) n the American West. *Biological Journal of the Linnean Society*, 76, 361–391.

Jolivet, L., H. Maluski, O. Beyssac, B. Goffé, C. Lepvrier, P.T. Thi, and N.V. Vuong. 1999. Oligocene-Miocene Bu Khang extensional gneiss dome in Vietnam: Geodynamic implications. *Geology*, 27, 67–70.

Jolivet, P. 2008. La faune entomologique en Nouvelle-Calédonie. *Le Coléoptériste*, 11, 35–47.

Jolivet, P., and K.K. Verma. 2008a. Eumolpinae: A widely distributed and much diversified subfamily of leaf beetles (Coleoptera, Chrysomelidae). *Terrestrial Arthropod Review*, 1, 3–37.

Jolivet, P., and K.K. Verma. 2008b. On the origin of the chrysomelid fauna of New Caledonia. In P. Jolivet, J.A. Santiago-Blay, and M. Schmitt (eds.), *Research on Chrysomelidae*, Vol. 1 (pp. 309–319). Brill, Leiden.

Jones, A. W., and R. S. Kennedy. 2008. Evolution in a tropical archipelago: Comparative phylogeography of Philippine fauna and flora reveals complex patterns of colonization and diversification. *Zoological Journal of the Linnean Society*, 95, 620–639.

Jønsson, K.A., R.C.K. Bowie, J.A.A. Nylander, L. Christidis, J.A. Norman, and J. Fjeldså. 2010. Biogeographical history of cuckoo-shrikes (Aves: Passeriformes): Transoceanic colonization of Africa from Australo-Papua. *Journal of Biogeography*, 37, 1767–1781.

Jønsson, K.A., M. Irestedt, R.C.K. Bowie, L. Christidis, and J. Fjeldså. 2011. Systematics and biogeography of Indo-Pacific ground-doves. *Molecular Phylogenetics and Evolution*, 59, 538–543.

Jordaens, K., P. van Riel, A.M. Frias Martins, and T. Backeljau. 2009. Speciation on the Azores islands: Congruent patterns in shell morphology, genital anatomy, and molecular markers in endemic land snails (Gastropoda, Leptaxinae). *Biological Journal of the Linnean Society*, 97, 166–176.

Jordan, S., C. Simon, and D. Polhemus. 2003. Molecular systematics and adaptive radiation of Hawaii's endemic damselfly genus *Megalagrion* (Odonata: Coenagrionidae). *Systematic Biology*, 52, 89–109.

Jordan, S., C. Simon, D. Foote, and R.A. Englund. 2005. Phylogeographic patterns of Hawaiian *Megalagrion* damselflies (Odonata: Coenagrionidae) correlate with Pleistocene island boundaries. *Molecular Ecology*, 14, 3457–3470.

Jorgensen, S.J., C.A. Reeb, T.K. Chapple, S. Anderson, C. Perle, S.R. Van Sommeran, C. Fritz-Cope, A.C. Brown, P. Klimley, and B.A. Block. 2010. Philopatry and migration of Pacific white sharks. *Proceedings of the Royal Society B*, 277, 679–688.

Jourdan, F., H. Bertrand, U. Schärer, J. Blichert-Toft, G. Féraud, and A.B. Kampunzu. 2007. Major and trace element and Sr, Nd, Hf, and Pb isotope compositions of the Karoo Large Igneous province, Botswana-Zimbabwe: Lithosphere vs. mantle plume contribution. *Journal of Petrology*, 48, 1043–1077.

Kadarusman, P. 2000. Petrology and *P-T* evolution of garnet peridotites from central Sulawesi, Indonesia. *Journal of Metamorphic Geology*, 18, 193–209.

Kalkman, C. 1979. Dispersal and distribution of Malesian angiosperms. In K. Larsen and L. Holm-Nielsen (eds.), *Tropical botany* (pp. 135–141). Academic Press, London.

Kamilar, J., S. Martin, and A. Tosi. 2009. Combining biogeographic and phylogenetic data to examine primate speciation: an example using cercopithecin monkeys. *Biotropica*, 41, 514–519.

Kaneko, S., Y. Isag, and F. Nobushima. 2008. Genetic differentiation among populations of an oceanic island: The case of *Metrosideros boninensis*, an endangered endemic tree species in the Bonin Islands. *Plant Species Biology*, 23, 119–128.

Kaneshiro, K.Y., R.G. Gillespie, and H.L. Carson. 1995. Chromosomes and male genitalia of Hawaiian *Drosophila*: Tools for interpreting phylogeny and geography. In W.L. Wagner and V.A. Funk (eds.), *Hawaiian biogeography: Evolution on a hot spot archipelago* (pp. 57–71). Smithsonian Institution Press, Washington, DC.

Kappeler, P.M. 2000. Lemur origins: Rafting by groups of hibernators? *Folia Primatologica*, 71, 422–425.

Karanth, K.P., L. Singh, R.V. Collura, and C.-B. Stewart. 2008. Molecular phylogeny and biogeography of langurs and leaf monkeys of South Asia (Primates: Colobinae). *Molecular Phylogenetics and Evolution*, 46, 683–694.

Kårehed, J., I. Groeninckx, S. Dessein, T.J. Motley, and B. Bremer. 2008. The phylogenetic utility of chloroplast and nuclear DNA markers and the phylogeny of the Rubiaceae tribe Spermacoceae. *Molecular Phylogenetics and Evolution*, 49, 843–866.

Kay, R.F., J.F. Fleagle, T.R.T. Mitchell, M. Colbert, T. Bown, and D.W. Powers. 2008. The anatomy of *Dolichocebus gaumanensis*, a stem platyrrhine monkey from Argentina. *Journal of Human Evolution*, 54, 323–382.

Kay, R.F., B.A. Williams, C.F. Ross, M. Takai, and N. Shigehara. 2004. Anthropoid origins: A phylogenetic analysis. In C.F. Ross and R.F. Kay (eds.), *Anthropoid origins: New visions* (pp. 91–136). Kluwer/Plenum, New York.

Kear, J. (ed.) 2005. *Ducks, geese and swans*. Oxford University Press, New York.

Keller, C., C. Roos, L.F. Groeneveld, J. Fischer, and D. Zinner. 2009. Introgressive hybridization in southern African baboons shapes patterns of mtDNA variation. *American Journal of Physical Anthropology*, 142, 125–136.

Keller, G.R., R.F. Wendlandt, and M.H.P. Bott. 1995. West and Central African Rift system. In K.H. Olsen (ed.), *Continental rifts: Evolution, structure, tectonics* (pp. 437–460). Elsevier, Amsterdam.

Keller, R.A., M.R. Fisk, and W.M. White. 2000. Isotopic evidence for Late Cretaceous plume-ridge interaction at the Hawaiian hotspot. *Nature*, 405, 673–675.

Kelley, S. 2007. The geochronology of large igneous provinces, terrestrial impact craters, and their relationship to mass extinctions on Earth. *Journal of the Geological Society*, 164, 923–936.

Kennedy, M., and R.D.M. Page. 2002. Seabird supertrees: Combining partial estimates of procellariiform phylogeny. *The Auk*, 119, 88–108.

Kenny, A. 2006. *A new history of Western philosophy*, Vol. 3: *The rise of modern philosophy*. Clarendon Press, Oxford, UK.

Keppel, G., P.D. Hodgkiss, and G.M. Plunkett. 2008. Cycads in the insular Southwest Pacific: Dispersal or vicariance? *Journal of Biogeography*, 35, 1004–1015.

Keppel, G., A.J. Lowe, and H.P. Possingham. 2009. Changing perspectives on the biogeography of the tropical South Pacific: Influences of dispersal, vicariance and extinction. *Journal of Biogeography*, 36, 1035–1054.

Kerr, A.C., G.F. Marriner, J. Tarney, A. Nivia, A.D. Saunders, M.F. Thirlwall, and C.W. Sinton. 1997. Cretaceous basaltic terranes in western Colombia: Elemental, chronological and Sr-Nd isotopic constraints on petrogenesis. *Journal of Petrology*, 38, 677–702.

Kerr, A.C., and J. Tarney. 2005. Tectonic evolution of the Caribbean and northwestern South America: The case for accretion of two Late Cretaceous oceanic plateaus. *Geology*, 33, 269–272.

Kidd, S.E. 2005. *Molecular phylogenetics of the Hawaiian geraniums*. Unpublished M.Sc thesis, Bowling Green State University, Ohio.

Kim, H.G., S.C. Keeley, P.S. Vroom, and R.K. Jansen. 1998. Molecular evidence for an African origin of the Hawaiian endemic *Hesperomannia* (Asteraceae). *Proceedings of the National Academy of Sciences of the USA*, 95, 15440–15445.

Kim, S.-C., L. Chunghee, and J.A.Mejías. 2007. Analysis of chloroplast DNA *mat*K gene and ITS of nrDNA sequences reveals polyphyly of the genus

Sonchus and new relationships among the subtribe Sonchinae (Asteraceae: Cichorieae). *Molecular Phylogenetics and Evolution*, 44, 578–597.

Kim, S.-C., D.J. Crawford, M. Tadesse, M. Berbee, F.R. Ganders, M. Pirseyedi, and E.J. Esselman. 1999. ITS sequences and phylogenetic relationships in *Bidens* and *Coreopsis* (Asteraceae). *Systematic Botany*, 24, 480–493.

Kimball, R.T., and D.J. Crawford. 2004. Phylogeny of Coreopsideae (Asteraceae) using ITS sequences suggests lability in reproductive characters. *Molecular Phylogenetics and Evolution*, 33, 127–139.

Kimura, M. 1983. *The neutral theory of molecular evolution*. Cambridge University Press, Cambridge.

King, N. 2004. The unicellular ancestry of animal development. *Developmental Cell*, 7, 313–325.

Kingdon, J. 1974. *East African mammals*, Vol. 2A. Academic Press, New York.

Kirschner, M., and J. Gerhart. 2005. *The plausibility of life*. Yale University Press, New Haven, CT.

Klaver, C.J.J., and W. Böhme. 1986. Phylogeny and classification of the Chameleonidae (Sauria) with special reference to hemipenis morphology. *Bonner Zoologiche Monographien*, 22, 1–64.

Klaver, C.J.J., and W. Böhme. 1997. *Chamaeleonidae (Animal Kingdom, 112)*. Walter de Gruyter, Berlin.

Kley, J., C.R. Monaldi, and J.A. Salfity. 1999. Along-strike segmentation of the Andean foreland: Causes and consequences. *Tectonophysics*, 301, 75–94.

Knowles, L.L. 2008. Why does a method that fails continue to be used? *Evolution*, 61, 2713–2717.

Knowles, L.L., and W.P. Maddison. 2002. Statistical phylogeography. *Molecular Ecology*, 11, 2623–2635.

Knowlton, N., and L.A. Weigt. 1998. New dates and new rates for divergence across the Isthmus of Panama. *Proceedings of the Royal Society of London B*, 265, 2257–2263.

Knowlton, N., L.A. Weigt, L. A. Solórzano, D.K. Mills, and E. Bermingham. 1993. Divergence in proteins, mitochondrial DNA, and reproductive capability across the Isthmus of Panama. *Science* 260, 1629–1632.

Kodandaramaiah, U. 2010. Use of dispersal–vicariance analysis in biogeography: A critique. *Journal of Biogeography*, 37, 3–11.

Köhler, F., and M. Glaubrecht. 2007. Out of Asia and into India: On the molecular phylogeny and biogeography of the endemic freshwater gastropod *Paracrostoma* Cossmann, 1900 (Caenogastropoda: Paschychilidae). *Biological Journal of the Linnean Society*, 91, 627–651.

Köhler, F., and C. Dames. 2009. Phylogeny and systematrics of the Pachychilidae of mainland south-east Asia: Novel insights from morphology and mitochondrial DNA (Mollusca, Caenogastropoda, Cerithioidea). *Zoological Journal of the Linnean Society*, 157, 679–699.

Koonin, E.V. 2009a. Darwinian evolution in the light of genomics. *Nucleic Acids Research*, 37, 1011–1034.

Koonin, E.V. 2009b. The Origin at 150: Is a new evolutionary synthesis in sight? *Trends in Genetics*, 25, 473–475.

Koopman, M.M., and D.A. Baum. 2008. Phylogeny and biogeography of tribe Hibisceae (Malvaceae) on Madagascar. *Systematic Botany*, 33, 364–374.

Kopp, H., C. Kopp, J. Phipps Morgan, E.R. Flueh, W. Weinrebe, and W.J. Morgan. 2003. Fossil hotspot-ridge interaction in the Musicians Seamount province: Geophysical investigations of hot spot volcanism at volcanic elongated ridges. *Journal of Geophysical Research*, 108 (B3), 2160. doi:10.1029/2002JB002015.

Koppers, A.A.P. 2009. Pacific region. In R.G. Gillespie and D.A. Clague (eds.), *Encyclopedia of islands* (pp. 702–715). University of California Press, Berkeley.

Koppers, A.A.P., H. Staudigel, J. Phipps Morgan, and R.A. Duncan. 2007. Nonlinear ^{40}Ar/^{39}Ar age systematics along the Gilbert Ridge and Tokelau Seamount Trail and the timing of the Hawaii-Emperor Bend. *Geochemistry Geophysics Geosystems*, 8, Q06L13. doi:10.1029/2006GC001489.

Kosaki, R.K., R.L. Pyle, J.E. Randall, and D.K. Irons. 1991. New records of fishes from Johnston Atoll, with notes on biogeography. *Pacific Science*, 45, 186–203.

Kreft, H. and W. Jetz. 2010. A framework for delineating biogeographical regions based on species distributions. *Journal of Biogeography*, 37, 2029–2053.

Krell, F.-T., and P.S. Cranston. 2004. Which side of the tree is more basal? *Systematic Entomology*, 29, 279–281.

Kubitzki, K. 1989. The ecogeographical differentiation of Amazonian inundation forests. *Plant Systematics and Evolution*, 162, 285–304.

Kumar, S., and S.B. Hedges. 1998. A molecular time scale for vertebrate evolution. *Nature*, 392, 917–920.

Kumazawa, Y., and M. Nishida. 2000. Molecular phylogeny of osteoglossoids: A new model for Gondwanan origin and plate tectonic transportation of the Asian arowana. *Molecular Biology and Evolution*, 17, 1869–1878.

Kurland, C.G., and O.G. Berg. 2010. A hitchhiker's guide to evolving networks. In G. Caetano-Anollés (ed.), *Evolutionary genomics and systems biology* (pp. 363–383). Wiley, Hoboken, NJ.

Kuschel, G. 2003. Nemonychidae, Belidae, Brentidae (Insecta: Coleoptera: Curculionoidea). *Fauna of New Zealand*, 45, 1–91.

Kuschel, G. 2008. Curculionoidea (weevils) of New Caledonia and Vanuatu: Basal families and some Curculionidae. In P. Grandcolas (ed.), Zoologica Neocaledonica 6. *Mémoires du Muséum national d'Histoire naturelle*, 197, 99–249.

Lacoste, V., and B. de Thoisy. 2010. Phylogeny and phylogeography of squirrel monkeys (genus *Saimiri*) based on cytochrome *b* genetic analysis. *American Journal of Primatology*, 72, 242–253.

Lamarck, J.-B. 1809. *Philosophie zoologique*. Dentu, Paris.

Large, M.F., and J.E. Braggins. 2004. *Tree ferns*. CSIRO, Melbourne, Australia.

Lavergne, A., M. Ruiz-García, F. Catzeflis, S. Lacote, H. Contamin, O. Mercereau-Puijalon, V. Lacoste, and B. de Thoisy, B. 2010. Phylogeny and phylogeography of squirrel monkeys (genus *Saimiri*) based on cytochrome *b* genetic analysis. *American Journal of Primatology*, 72, 242–253.

Lee, G.H., M.A. Eissa, C.L. Decker, J.P. Castagna, D.J. O'Meara, and H.D. Marín. 2004. Aspects of the petroleum geology of the Bermejo field, northwestern Oriente basin, Ecuador. *Journal of Petroleum Geology*, 27, 1–22.

Le Goff, J. [1964] 2001. *Medieval civilization*. Blackwell, Oxford, UK.

Lehman, S.M., and J.G. Fleagle. 2006. Biogeography and primates: A review. In S.M. Lehman and J.G. Fleagle (eds.), *Primate biogeography* (pp. 1–36). Springer, New York, NY.

Lehman, S.M., R.W. Sussman, J. Philips-Conroy, and W. Prince. 2006. Ecological biogeography of primates in Guyana. In S.M. Lehman and J.G. Fleagle (eds.), *Primate biogeography* (pp. 105–126). Springer, New York.

Leloup, P.H., R. Lacassin, P. Tapponnier, U. Schärer, Z. Dalai, L. Xiaohan, Z. Liangshang, J. Shaocheng, and P.T. Trinh. 1995. The Ailao Shan-Red River shear zone (Yunnan, China), Tertiary transform boundary of Indochina. *Tectonophysics*, 251, 3–10.

Lemeunier, J.R., L. David, M. Tsacas, and M. Ashburner. 1986. The *melanogaster* species group. In M. Ashburner, H.L. Carson, and J. N. Thompson (eds.), *The genetics and biology of Drosophila*, Vol. 3e (pp. 147–256). Academic Press, New York.

Leray, M., R. Beldade, S.J. Holbrook, R.J. Schmitt, S. Planes, and G. Bernardi. 2010. Allopatric divergence and speciation in coral reef fish: The three-spot *Dascyllus trimaculatus* species complex. *Evolution*, 64, 1218–1230.

Lessios, H.A. 2008. The Great American Schism: Divergence of marine organisms after the rise of the Central American Isthmus. *Annual Reviews of Ecology, Evolution and Systematics*, 39, 63–91.

Lessios, H.A., and D.R. Robertson. 2006. Crossing the impassable: Genetic connections in 20 reef fishes across the eastern Pacific barrier. *Proceedings of the Royal Society B*, 273, 2201–2208.

Levin, D.A. 2000. *The origin, expansion, and demise of plant species*. Oxford University Press, New York.

Levin, R.A., and J.S. Miller. 2005. Relationships within the tribe Lycieae (Solanaceae): Paraphyly of *Lycium* and multiple origins of gender dimorphism. *American Journal of Botany*, 92, 2044–2053.

Levin, R.A., N.R. Myers, and L. Bohs. 2006. Phylogenetic relationships among the "spiny solanums" (*Solanum* subgenus *Leptostemonum*, Solanaceae). *American Journal of Botany*, 93, 157–169.

Lewis, G., B. Schrire, B. Mackinder, and M. Lock. 2005. *Legumes of the world*. Royal Botanic Gardens, Kew.

Li, C., R.D. van der Hilste, E.R. Engdahl, and S. Burdick. 2008. A new global model for P wave speed variations in Earth's mantle. *Geochemistry Geophysics Geosystems*, 9, Q05018, 1–21.

Li, J., D. Zhang, and M.J. Donoghue. 2003. Phylogeny and biogeography of *Chamaecyparis* (Cupressaceae) inferred from DNA sequences of the nuclear ribosomal ITS region. *Rhodora*, 105, 106–117.

Li, X., P. Musikasinthorn, and Y. Kumazawa. 2006. Molecular phylogenetic analyses of snakeheads (Perciformes: Channidae) using mitochondrial DNA sequences. *Ichthyological Research*, 53, 148–159.

Liebherr, J.K. 2006. Recognition and description of *Blackburnia kavanaughi*, new species (Coleoptera: Carabidae, Platynini) from Kauai, Hawaii. *Journal of the New York Entomological Society*, 114, 17–27.

Liebherr, J.K., and A.E.Z. Short. 2006. *Blackburnia riparia*, new species (Coleoptera: Carabidae, Platynini): A novel element in the Hawaiian riparian insect fauna. *Journal of the New York Entomological Society*, 114, 1–16.

Lim, H.C., F. Zou, S.S. Taylor, B.D. Marks, R.G. Moyle, G. Voelker, and F. H. Sheldon. 2010. Phylogeny of magpie-robins and shamas (Aves: Turdidae: *Copsychus* and *Trichixos*): Implications for island biogeography in Southeast Asia. *Journal of Biogeography*, 37, 1894–1906.

Linder, H.P. 2008. Plant species radiations: Where, when, why? *Philosophical Transactions of the Royal Society B*, 363, 3097–3105.

Lindgren, A.R. 2010. Systematics and distribution of the squid genus *Pterygioteuthis* (Cephalopoda: Oegopsida) in the eastern tropical Pacific Ocean. *Journal of Molluscan Studies*, 76, 389–398.

Lindqvist, C., and V.A. Albert. 2002. Origin of the Hawaiian endemic mints within North American *Stachys* (Lamiaceae). *American Journal of Botany*, 89, 1709–1724.

Lindqvist, C., T.J. Motley, J.J. Jeffrey, and V.A. Albert. 2003. Cladogenesis and reticulation in the Hawaiian endemic mints. *Cladistics*, 19, 480–495.

Lobel, P.S., and L.K. Lobel. 2004. Annotated checklist of the fishes of Wake Atoll. *Pacific Science*, 58, 65–90.

Lomolino, M.V., and J.H. Brown. 2009. The reticulating phylogeny of island biogeography theory. *Quarterly Review of Biology*, 84, 357–390.

López-Martínez, N. 2003. La búsqueda del centro de origen en biogeografía histórica. *Graellsia*, 59, 503–522.

López-Martínez, N. 2009. Time asymmetry in the palaeobiogeographic history of species. *Bulletin de la Société Géologique de France*, 180, 45–55.

Lorence, D.H., and W.L. Wagner. 1997. A revision of *Lepinia* (Apocynaceae), with description of a new species from the Marquesas Islands. *Allertonia*, 7, 254–266.

Losos, J.B. 2010. Adaptive radiation, ecological opportunity, and evolutionary determinism. *American Naturalist*, 175, 623–639.

Losos, J.B., and D.L. Mahler. 2010. Adaptive radiation: The interaction of ecological opportunity, adaptation, and speciation. In M.A. Bell, D.J. Futuyma, W.F. Eanes, and J.S. Levinton (eds.), *Evolution after Darwin: The first 150 years*. Sinauer, Sunderland, MA.

Lovette, I.J., E. Bermingham, and R.E. Ricklefs. 2002. Clade-specific morphological diversification and adaptive radiation in Hawaiian songbirds. *Proceedings of the Royal Society of London B*, 269, 37–42.

Lovette, I.J., and D.R. Rubenstein. 2007. A comprehensive molecular phylogeny of the starlings (Aves: Sturnidae) and mockingbirds (Aves: Mimidae): Congruent mtDNA and nuclear trees for a cosmopolitan avian radiation. *Molecular Phylogenetics and Evolution*, 44, 1031–1056.

Lowell, J.D. 1995. Mechanics of basin inversion from worldwide examples. *Geological Society of London, Special Publication*, 88, 39–57.

Lowrey, T.K. 1995. Phylogeny, adaptive radiation, and biogeography of Hawaiian *Tetramolopium* (Asteraceae, Astereae). In W.L. Wagner and V.A. Funk (eds.), *Hawaiian biogeography: Evolution on a hot spot archipelago* (pp. 195–220). Smithsonian Institution Press, Washington, DC.

Lowrey, T.K., C.J. Quinn, R.K. Taylor, R. Chan, R.T. Kimball, and J.C. DeNardi. 2001. Molecular and morphological reassessment of relationships within the *Vittadinia* group of Astereae (Asteraceae). *American Journal of Botany*, 88, 1279–1289.

Lowrey, T.K., R. Whitkus, and W.R. Sykes. 2005. A new species of *Tetramolopium* (Asteraceae) from Mitiaro, Cook Islands: Biogeography, phylogenetic relationships, and dispersal. *Systematic Botany*, 30, 448–455.

Lowry, P.P., II. 1988. Notes on the Fijian endemic *Meryta tenuifolia* (Araliaceae). *Annals of the Missouri Botanical Garden*, 75, 389–391.

Lowry, P.P., II. 1998. Diversity, endemism, and extinction in the flora of New Caledonia: A review. In C.I. Peng and P.P. Lowry II (eds.), *Rare, threatened, and endangered floras of Asia and the Pacific rim* (pp. 181–206). Academica Sinica Monograph 16. Institute of Botany, Taipei.

Lowry, P.P., II, and G.M. Plunkett. 2010. Recircumscription of *Polyscias* (Araliaceae) to include six related genera, with a new infrageneric classification and a synopsis of species. *Plant Diversity and Evolution*, 128, 55–84.

Luis-Martínez, A., M. Trujano, J. Llorente-Bousquets, and I. Vargas-Fernández. 2006. Patrones de distribución de las subfamilias Danainae, Apaturinae, Biblidinae y Heliconiinae (Lepidoptera: Nymphalidae). In J.J. Morrone and J. Llorente-Bousquets (eds.), *Componentes bióticos principales de la entomofauna Mexicana*, Vol. 2 (pp. 771–865). Universidad Nacional Autónoma de México, Mexico City.

Lundberg, J.G. 1993. African–South American freshwater fish clades and continental drift: Problems with a paradigm. In P. Goldblatt (ed.), *Biological relationships between Africa and South America* (pp. 156–199). Yale University Press, New Haven.

Lundberg, J.G., L.G. Marshall, J. Guerrero, B. Horton, M.C.S.L. Malabarba, and F. Wesselingh. 1998. The stage for neotropical fish diversification: A history of tropical South American rivers. In L.R. Malabarba, R.E. Reis, R.P. Vari, Z.M.S. Lucena, and C.A.S. Lucena (eds.), *Phylogeny and classification of Neotropical fishes* (pp. 13–48). Edipucrs, Porto Alegre, RS.

Luo, Z.-X., and J.R. Wible. 2005. A Late Jurassic digging mammal and early mammalian diversification. *Science*, 308, 103–107.

Lusk, R.W., and M.B. Eisen. 2010. Evolutionary mirages: Selection on binding site composition creates the illusion of conserved grammars in *Drosophila* enhancers. *PLoS Genetics*, 6(1), e1000829.

Luzieux, L.D.A., F. Heller, R. Spikings, C.F. Vallejo, and W. Winkler. 2006. Origin and Cretaceous tectonic history of the coastal Ecuadorian forearc between 1°N and 3°S: Paleomagnetic, radiometric and fossil evidence. *Earth and Planetary Science Letters*, 249, 400–414.

Lynch, J.D. 1986. Origins of the high Andean herpetological fauna. In F. Vuilleumier and M. Monasterio (eds.), *High altitude tropical biology* (pp. 478–499). Oxford University Press, New York.

Lynch, M. 2007a. The frailty of adaptive hypotheses for the origins of organismal complexity. *Proceedings of the National Academy of Sciences USA*, 104, S8597–S8604.

Lynch, M. 2007b. The evolution of genetic networks by non-adaptive processes. *Nature Reviews Genetics*, 8, 803–813.

Lynch, M., and J.S. Conery. 2003. The origins of genome complexity. *Science*, 302, 1401–1404.

Mabberley, D.J., C.M. Pannell, and A.M. Sing. 1995. Meliaceae. *Flora Malesiana*, ser. 1(12), 1–407.

MacArthur, R.H., and E.O. Wilson. 1967. *The theory of island biogeography.* Princeton University Press, Princeton, NJ.

MacPhee, D. 2006. *Exhumation, rift-flank uplift, and the thermal evolution of the Rwenzori Mountains determined by combined (U-Th)/He and U-Pb thermochronometry.* Unpublished M.Sc. thesis, Massachusetts Institute of Technology.

MacPhee, R.D.E. 1994. Morphology, adaptations, and relationships of *Plesiorycteropus*, and a diagnosis of a new order of eutherian mammals. *Bulletin of the American Museum of Natural History*, 220, 1–214.

MacPhee, R.D.E. 2005. "First" appearances in the Cenozoic land-mammal record of the Greater Antilles: Significance and comparison with South American and Antarctic records. *Journal of Biogeography*, 32, 551–564.

MacPhee, R.D.E., and I. Horovitz. 2002. Extinct Quarternary platyrrhines of the Greater Antilles and Brazil. In W.C. Hartwig (ed.), *The primate fossil record* (pp. 189–200). Cambridge University Press, Cambridge.

MacPhee, R.D.E., I. Horovitz, O. Arredondo, and O.J. Vasquez. 1995. A new genus for the extinct Hispaniolan monkey *Saimiri bernensis* Rímoli, 1977: With notes on its systematic position. *American Museum Novitates*, 134, 1–21.

Magnacca, K.N., and B.N. Danforth. 2006. Evolution and biogeography of native Hawaiian *Hylaeus* bees (Hymenoptera: Colletidae). *Cladistics*, 22, 393–411.

Magnacca, K.N., and B.N. Danforth. 2007. Low nuclear DNA variation supports a recent origin of Hawaiian *Hylaeus* bees (Hymenoptera: Colletidae). *Molecular Phylogenetics and Evolution*, 43, 908–915.

Magnacca, K.N., and P.M. O'Grady. 2008. Revision of the "nudidrosophila" and "ateledrosophila" species groups of Hawaiian *Drosophila* (Diptera: Drosophilidae), with descriptions of twenty-two new species. *Systematic Entomology*, 33, 395–428.

Magnacca, K.N., and P.M. O'Grady. 2009. Revision of the Modified Mouthparts species group of Hawaiian *Drosophila* (Diptera: Drosophilidae): The *ceratostoma, freycinetiae, semifuscata,* and *setiger* subgroups, and unplaced species. *University of California Publications, Entomology*, 130, 1–93.

Maisels, F., S. Blake, M. Fay, G. Mobolambi, and V. Yako. 2006. A note on the distribution of Allen's swamp monkey, *Allenopithecus nigroviridis*, in northwestern Congo. *Primate Conservation*, 21, 93–95.

Malaquias, M.A.E., and D.G. Reid. 2009. Tethyan vicariance, relictualism and speciation: Evidence from a global molecular phylogeny of the opisthobranch genus *Bulla. Journal of Biogeography*, 36, 1760–1777.

Mann, P., L. Gahagan, and M.B. Gordon. 2003. Tectonic setting of the world's giant oil and gas fields. In M.T. Halbouty (ed.), *Giant oil and gas fields of the decade,* 1990–1999. *Memoirs of the American Association of Petroleum Geologists*, 78, 15–106.

Mann, P., R.D. Rogers, and L. Gahagan. 2007. Overview of late tectonic history and its unresolved tectonics problems. In J. Bundschuh and G.E. Alvarado (eds.), *Central America: Geology, resources and hazards* (pp. 205–241). Taylor and Francis, Leiden.

Mapes, R.W., A.C.R. Nogueira, D.S. Coleman, and A.M. Leguizamon Vega. 2006. Evidence for a continent scale drainage inversion in the Amazon basin since the Late Cretaceous. Paper 214-3. Abstracts of the Geological Society of America Annual Meeting, Philadelphia 2006.

Marivaux, L., P.-O. Antoine, S.R.H., Baqri, M. Benammi, Y. Chaimanee, J.-Y. Crochet, D. de Franceschi, N. Iqbal, J.-J. Jaeger, G. Métais, G. Roohi, and J.-L. Welcomme. 2005. Anthropoid primates from the Oligocene of Pakistan (Bugti Hills): Data on early anthropoid evolution and biogeography. *Proceedings of the National Academy of Sciences USA*, 102, 8436–8441.

Marivaux, L., L. Bocat, Y. Chaimanee, J.J. Jaeger, B. Marandat, P. Srisuk, P. Tafforeau, C. Yamee, and J.-L. Welcomme. 2006. Cynocephalid dermopterans from the Paleogene of South Asia (Thailand, Myanmar and Pakistan): Systematic, evolutionary and palaeobiogeographic implications. *Zoologica Scripta*, 35, 395–420.

Marivaux, L., Y. Chaimanee, P. Tafforeau, and J.-J. Jaeger. 2006. New strepsirrhine primate from the Late Eocene of Peninsular Thailand (Krabi Basin). *American Journal of Physical Anthropology*, 130, 425–434.

Marivaux, L., J.-L. Welcomme, P.-O. Antoine, G. Métais, I.M. Baloch, M. Benammi, Y. Chaimanee, S. Ducrocq, and J.-J. Jaeger. 2001. A fossil lemur from the Oligocene of Pakistan. *Science*, 294, 587–591.

Marko, P.B. 2002. Fossil calibration of molecular clocks and the divergence times of geminate species pairs separated by the Isthmus of Panama. *Molecular Biology and Evolution*, 19, 2005–2021.

Marks, K.M., and A.A. Tikku. 2001. Cretaceous reconstructions of East Antarctica, Africa and Madagascar. *Earth and Planetary Science Letters*, 186, 479–495.

Marquínez, X., L.G. Lohmann, M.L. Faria Salatino, A. Salatino, and F. González. 2009. Generic relationships and dating of lineages in Winteraceae based on nuclear (ITS) and plastid (*rpS16* and *psbA-trnH*) sequence data. *Molecular Phylogenetics and Evolution*, 53, 435–449.

Martin, A., and C. Simon. 1990. Differing levels of among-population divergence in the mitochondrial DNA of periodical cicadas related to historical biogeography. *Evolution*, 44, 1066–1080.

Martin, H. 1994. Australian Tertiary phytogeography: Evidence from palynology. In R.S. Hill (ed.), *History of the Australian vegetation: Cretaceous to Recent* (pp. 104–142). Cambridge University Press, Cambridge.

Martin, R.D. 1990. *Primate origins and evolution: A phylogenetic reconstruction.* Princeton University Press, Princeton, NJ.

Martin, R.D. 2008. Evolution of placentation in primates: Implications of mammalian phylogeny. *Evolutionary Biology*, 35, 125–145.

Martin, R.D., C. Soligo, and S. Tavaré. 2007. Primate origins: Implications of a Cretaceous ancestry. *Folia Primatologica*, 78, 277–296.

Marvaldi, A.E., R.G. Oberprieler, C.H.C. Lyal, T. Bradbury, and R.S. Anderson. 2006. Phylogeny of the Oxycoryninae *sensu lato* (Coleoptera: Belidae) and evolution of host-plant associations. *Invertebrate Systematics*, 20, 447–476.

Marx, H.E., N. O'Leary, Y.-W. Yuan, P. Lu-Irving, D.C. Tank, M.E. Múlgura, and R.G. Olmstead. 2010. A molecular phylogeny and classification of Verbenaceae. *American Journal of Botany*, 97, 1647–1663.

Marzoli, A., E.M. Piccirillo, P.R. Renne, G. Bellieni, M. Iacumin, J.B. Nyobe, and A.T. Tongwa. 2000. The Cameroon Volcanic Line revisited: Petrogenesis of continental basaltic magmas from lithospheric and asthenospheric mantle sources. *Journal of Petrology*, 41, 87–109.

Massey, S.E. 2010. Pseudaptations and the emergence of beneficial traits. In P. Pontarotti (ed.), *Evolutionary biology: Concepts, molecular and morphological evolution* (pp. 81–98). Springer, Berlin.

Masters, J.C. 2006. When, where and how? Reconstructing a timeline for primate evolution using fossil and molecular data. In L. Sineo and R. Stanyon (eds.), *Primate cytogenetics* (pp. 105–121). Florence University Press, Florence.

Masters, J.C., N.M. Anthony, and A. Mitchell. 2005. Reconstructing the evolutionary history of the Lorisidae using morphological, molecular, and geological data. *American Journal of Physical Anthropology*, 127, 465–480.

Masters, J.C., M. Boniotto, S. Crovella, C. Roos, L. Pozzi, and M. Delpero. 2007. Phylogenetic relationships among the Lorisoidea as indicated by craniodental morphology and mitochondrial sequence data. *American Journal of Primatology*, 69, 6–15.

Masters, J.C., M.J. de Wit, and R.J. Asher. 2006. Reconciling the origins of Africa, India and Madagascar with vertebrate dispersal scenarios. *Folia Primatologica*, 77, 399–418.

Masters, J.C., B.G. Lovegrove, and M.J. de Wit. 2007. Eyes wide shut: Can hypometabolism really explain the primate colonization of Madagascar? *Journal of Biogeography*, 34, 21–37.

Mathiasen, P., and A.C. Premoli. 2010. Out in the cold: Genetic variation of *Nothofagus pumilio* (Nothofagaceae) provides evidence for latitudinally distinct evolutionary histories in austral South America. *Molecular Ecology*, 19, 371–385.

Mathys, B.A., and J.L. Lockwood. 2011. Contemporary morphological diversification of passerine birds introduced to the Hawaiian archipelago. *Proceedings of the Royal Society B*, 278, 2392–2400.

Matsuda, I., A. Tuuga, and S. Higashi. 2009. Ranging behavior of proboscis monkeys in a riverine forest with special reference to ranging in inland forest. *International Journal of Primatology*, 30, 313–325.

Matsudaira, K., and T. Ishida. 2010. Phylogenetic relationships and divergence dates of the whole mitochondrial genome sequences among three gibbon genera. *Molecular Phylogenetics and Evolution*, 55, 454–459.

Matsui, A., F. Rakotondraparany, I. Munechika, M. Hasegawa, and S. Horai 2009. Molecular phylogeny and evolution of prosimians based on complete sequences of mitochondrial DNAs. *Gene*, 441, 53–66.

Matthee, C.A., C.R. Tilbury, and T. Townsend. 2004. A phylogenetic review of the African leaf chameleons: Genus *Rhampholeon* (Chamaeleonidae): The role of vicariance and climate change in speciation. *Proceedings of the Royal Society of London B*, 271, 1967–1975.

Matthew, W.D. 1915. Climate and evolution. *Annals of the New York Academy of Science*, 24, 171–318.

Maurin, O., M.W. Chase, M. Jordaan, and M. van der Bank. 2010. Phylogenetic relationships of Combretaceae inferred from nuclear and plastid DNA seqnence data: Implications for generic classification. *Botanical Journal of the Linnean Society*, 162, 453–476.

Mayr, E. 1931. Birds collected during the Whitney South Sea Expedition. XIV. *American Museum Novitates*, 488, 1–11.

Mayr, E. 1940. The origin and history of the bird fauna of Polynesia. *Proceedings of the Sixth Pacific Science Congress* (California), 4, 197–216.

Mayr, E. 1942. *Systematics and the origin of species*. Columbia University Press, New York.

Mayr, E. 1944. The birds of Timor and Sumba. *Bulletin of the American Museum of Natural History*, 83, 127–194.

Mayr, E. 1954. Change of genetic environment and evolution. In J. Huxley, A.C. Hardy, and E.B. Ford (eds.), *Evolution as a process* (pp. 157–180). Allen and Unwin, London.

Mayr, E. 1960. The emergence of evolutionary novelties. In S. Tax and C. Callender (eds.), *Evolution after Darwin: The University of Chicago centennial* (pp. 349–380). University of Chicago Press, Chicago.

Mayr, E. 1963. *Animal species and evolution*. Harvard University Press, Cambridge, MA.

Mayr, E. 1980. Some thoughts on the history of the evolutionary synthesis. In E. Mayr and W. Provine (eds.), *The evolutionary synthesis* (pp. 1–48). Harvard University Press, Cambridge, MA.

Mayr, E. 1982. Processes of speciation in animals. In C. Barigozzi (ed.), *Mechanisms of speciation* (pp. 1–19). Liss, New York.

Mayr, E. 1992. Controversies in retrospect. *Oxford Surveys in Evolutionary Biology*, 8, 1–34.

Mayr, E. 1997. *This is biology: The science of the living world*. Harvard University Press, Cambridge MA.

Mayr, E. 1999. Introduction. In E. Mayr (ed.), *Systematics and the origin of species* (pp. xiii–xvi). Harvard University Press, Cambridge MA.

Mayr, E. 2001. *What evolution is*. Basic Books, New York.

Mayr, E., and J. Diamond. 2001. *The birds of northern Melanesia: Speciation, ecology and biogeography*. Oxford University Press, New York.

Mayr, E., and W.H. Phelps, Jr. 1967. The origin of the bird fauna of the South Venezuelan highlands. *Bulletin of the American Museum of Natural History*, 136, 274–327.

McBride, C.S., R. van Velzen, and T.B. Larsen. 2009. Allopatric origin of cryptic butterfly species that were discovered feeding on distinct host plants in sympatry. *Molecular Ecology*, 18, 3639–3651.

McCall, R.A., 1997. Implications of recent geological investigations of the Mozambique Channel for the mammalian colonization of Madagascar. *Proceedings of the Royal Society of London B*, 264, 663–665.

McCracken, K.G., and M.D. Sorenson. 2005. Is homoplasy or lineage sorting the source of incongruent mtDNA and nuclear gene trees in the stiff-tailed ducks (*Nomonyx-Oxyura*)? *Systematic Biology*, 54, 35–55.

McDowall, R.M. 2007. Process and pattern in the biogeography of New Zealand: A global microcosm? *Journal of Biogeography*, 35, 197–212.

McGlaughlin, R.J., M.C. Blake, Jr., W.V. Sliter, C.M. Wentworth, and R.W. Graymer. 2009. The Wheatfield Fork terrane: A remnant of Siletzia (?) in Franciscan Complex Coastal belt of northern California. 2009 Portland Geological Society of America Annual Meeting (18–21 October 2009), Paper No. 201-10.

McGoogan, K., T. Kivell, M. Hutchison, H. Young, S. Blanchard, M. Keeth, and S.M. Lehman. 2007. Phylogenetic diversity and the conservation biogeography of African primates. *Journal of Biogeography*, 34, 1962–1974.

McHone, J.G. 2000. On-plume magmatism and rifting during the opening of the central Atlantic Ocean. *Tectonophysics*, 316, 287–296.

McHone, J.G. 2003. Volatile emissions from central Atlantic magmatic province basalts: Mass assumptions and environmental consequences. *Geophysical Monograph—American Geophysical Union*, 136, 241–254.

McKenna, M.C. 1980. Early history and biogeography of South America's extinct land mammals. In R.L. Ciochon and A.B. Chiarelli (eds.), *Evolutionary biology of the New World monkeys and continental drift* (pp. 43–77). Plenum, New York.

McKinnon, J.S., S. Mori, B.K. Blackman, L. David, D.M. Kingsley, L. Jamieson, L. Chou, and D. Schluter. 2004. Evidence for ecology's role in speciation. *Nature*, 429, 294–298.

McKinnon, J.S., and H.D. Rundle. 2002. Speciation in nature: The threespine stickleback model systems. *Trends in Ecology and Evolution*, 17, 480–488.

Medeiros, M.J. 2009. A revision of the endemic Hawaiian genus *Thyrocopa* (Lepidoptera: Xyloryctidae: Xyloryctinae). *Zootaxa*, 2202, 1–47.

Medeiros, M.J., D. Davis, F.G. Howarth, and R. Gillespie. 2009. Evolution of cave living in Hawaiian *Schrankia* (Lepidoptera: Noctuidae) with description of a remarkable new cave species. *Zoological Journal of the Linnean Society*, 156, 114–139.

Medeiros, M.J., and R.G. Gillespie. 2011. Biogeography and the evolution of flightlessness in a radiation of Hawaiian moths (Xyloryctidae: *Thyrocopa*). *Journal of Biogeography*, 38, 101–111.

Meijaard, E., and C.P. Groves. 2006. The geography of mammals and rivers in mainland Southeast Asia. In S.M. Lehman and J.G. Fleagle (eds.), *Primate biogeography* (pp. 305–329). Springer, New York.

Meijaard, E., and V. Nijman. 2000. Distribution and conservation of the proboscis monkey (*Nasalis larvatus*) in Kalimantan, Indonesia. *Biological Conservation*, 92, 15–24.

Meijaard, E., and V. Nijman. 2003. Primate hotspots on Borneo: Predictive value for general biodiversity and the effects of taxonomy. *Conservation Biology*, 17, 725–732.

Meimberg, H., A. Wistuba, P. Dittrich, and G. Heubl. 2001. Molecular phylogeny of Nepenthaceae based on cladistic analysis of plastid *trn*K intron sequence data. *Plant Biology*, 3, 164–175.

Menard, H.W. 1986. *Islands*. Scientific American Library, New York.

Mendelson, T.C., and K.L. Shaw. 2005. Rapid speciation in an arthropod. *Nature*, 433, 375–376.

Mendes, S.L., F.R. de Melo, J.P. Boubli, L.G. Dias, K.B. Strier, L.P.S. Pinto, V. Fagundes, B. Cosenza, and P. de Marco, Jr. 2005. Directives for the conservation of the northern muriqui, *Brachyteles hypoxanthus* (Primates, Atelidae). *Neotropical Primates*, 13(Suppl.), 5–18.

Menezes, N.L. de. 2006. Rhizophores in *Rhizophora mangle* L.: An alternative explanation of so-called "aerial roots." *Anais da Academia Brasileira de Ciências*, 78, 213–226.

Menzies, M.A., S.L. Klemperer, C.J. Ebinger, and J. Baker. 2002. Characteristics of volcanic rifted margins. In M.A. Menzies, S.L. Klemperer, C.J. Ebinger and J. Baker (eds.), *Volcanic rifted margins. Geological Society of America Special Paper*, 362, 1–14.

Merker, S., C. Driller, Dahruddin, Wirdateti, W. Sinaga, D. Perwitasari-Farajallah, and M. Shekelle. 2010. *Tarsius wallacei*: A new tarsier species from Central Sulawesi occupies a discontinuous range. *International Journal of Primatology*, 31, 1107–1122.

Merker, S., C. Driller, D. Perwitasari-Farajallah, J. Pamungkas, and H. Zischler. 2009. Elucidating geological and biological processes underlying the diversification of Sulawesi tarsiers. *Proceedings of the National Academy of Sciences USA*, 106, 8459–8464.

Meschede, M., and W. Frisch. 2002. The evolution of the Caribbean plate and its relation to global plate motion vectors: Geometric constraints for an inter-American origin. In T.A. Jackson (ed.), *Caribbean geology into the third millennium* (pp. 1–14). University of the West Indies Press, Kingston.

Metcalfe, I. 1998. Paleozoic and Mesozoic geological evolution of the SE Asian region: Multidisciplinary constraints and implications for biogeography. In R. Hall and J.D. Holloway (eds.), *Biogeography and geological evolution of SE Asia* (pp. 25–41). Backhuys, Leiden.

Metcalfe, I. 2006. Palaeozoic and Mesozoic tectonic evolution and palaeogeography of east Asia crustal fragments: The Korean peninsula in context. *Gondwana Research*, 9, 24–46.

Metcalfe, I. 2008. Gondwana dispersion and Asian accretion: An update. Extended abstract. *Proceedings of the International Symposium on Geoscience Resources and Environments of Asian terranes* (4th IGCP and 5th APSEG). Bangkok. www.geo.sc.chula.ac.th/.

Metcalfe, I. 2010. Tectonic framework and Phanerozoic evolution of Sundaland. *Gondwana Research*, 19, 3–21.

Meyer, D., D. Rinaldi, H. Ramlee, D. Perwitasari-Farajallah, J.K. Hodges, and C. Roos. 2011. Mitochondrial phylogeny of leaf monkeys (genus *Presbytis*, Eschscholtz, 1821) with implications for taxonomy and conservation. *Molecular Phylogenetics and Evolution*, 59, 311–319.

Miller, E.R., G.F. Gunnell, and R.D. Martin. 2005. Deep time and the search for anthropoid origins. *Yearbook of Physical Anthropology*, 48, 60–95.

Miller, J.A., and C. Harris. 2007. Petrogenesis of the Swaziland and northern Natal rhyolites of the Lebombo rifted volcanic margin, south east Africa. *Journal of Petrology*, 48, 185–218.

Miller, J.S., A. Kamath, J. Damashek, and R.A. Levin. 2011. Out of America to Africa or Asia: Inference of dispersal histories using nuclear and plastid DNA and the *S-RNase* self-incompatibility locus. *Molecular Biology and Evolution*, 28, 793–801.

Miller, M.J., E. Bermingham, J. Klicka, P. Escalante, F.S. Raposo do Amaral, J.T. Weir, and K. Winker. 2008. Out of Amazonia again and again: Episodic crossing of the Andes promotes diversification in a lowland forest flycatcher. *Proceedings of the Royal Society B*, 275, 1133–1142.

Miller, M.J., E. Bermingham, and R.E. Ricklefs. 2007. Historical biogeography of the New World solitaires (*Myadestes* spp.). *The Auk*, 124, 868–885.

Miranda Ribeiro, A. 1940. Commentaries on South American primates. *Memórias do Instituto Oswaldo Cruz* (Rio de Janeiro), 35, 779–851.

Mitchell, A.D., P.B. Heenan, B.G. Murray, B.P.J. Molloy, and P.J. de Lange. 2009. Evolution of the south-western Pacific genus *Melicytus* (Violaceae): Evidence from DNA sequence data, cytology and sex expression. *Australian Systematic Botany*, 22, 143–157.

Mittermeier, R.A., W.R. Konstant, F. Hawkins, E.E. Louis, O. Langrand, J. Ratsimbazafy, R. Rasoloarison, J.U. Ganzhorn, S. Rajaobelina, I. Tattersall, and D.M. Meyers. 2006. *Lemurs of Madagascar*. 2nd ed. Conservation International, Washington DC.

Mohriak, W., M. Nemčok, and G. Enciso. 2008. South Atlantic divergent margin evolution: Rift-border uplift and salt tectonics in the basins of SE Brazil. In R.J. Pankhurst, R.A.J. Trouw, B.B. de Brito Neves, and M.J. de Wit (eds.), *West Gondwana: Pre-Cenozoic correlations across the South Atlantic region. Geological Society of London Special Publication*, 294, 365–398.

Molnar, P. 1990. The structure of mountain ranges. In E. Moores (ed.), *Shaping the Earth: Tectonics of continents and oceans* (pp. 125–138). W.H. Freeman, New York.

Molvray, M., P.J. Kores, and M.W. Chase. 1999. Phylogenetic relationships within *Korthalsella* (Viscaceae) based on nuclear ITS and plastid *trnL-F* sequence data. *American Journal of Botany*, 86, 249–260.

Mooers, A.Ø., H.D. Rundle, and M.C. Whitlock. 1999. The effects of selection and bottlenecks on male mating success in peripheral isolates. *American Naturalist*, 153, 437–444.

Moore, J.G., W.R. Normark, and R.T. Holcomb. 1994a. Giant Hawaiian underwater landslides. *Science*, 264, 46–47.

Moore, J.G., W.R. Normark, and R.T. Holcomb. 1994b. Giant Hawaiian landslides. *Annual Review of Earth and Planetary Sciences*, 22, 119–144.

Mootnick, A., and C. Groves. 2005. A new generic name for the Hoolock gibbon (Hylobatidae). *International Journal of Primatology*, 26, 971–976.

Moreira-Muñoz, A. 2007. *Plant geography of Chile: An essay on postmodern biogeography*. Unpublished Ph.D. dissertation, Friedrich-Alexander University, Erlangen-Nürnberg.

Morgan, M.J., J.D. Roberts, and J.S. Keogh. 2007. Molecular phylogenetic dating supports an ancient endemic speciation model in Australia's biodiversity hotspot. *Molecular Phylogenetics and Evolution*, 44, 371–385.

Morgan, T.H. 1916. *A critique of the theory of evolution*. Princeton University Press, Princeton, NJ.

Morley, C.K. 2001. Combined escape tectonics and subduction rollback-backarc extension: A model for the evolution of Tertiary rift basins in Thailand, Malaysia and Laos. *Journal of the Geological Society*, 158, 461–474.

Morley, C.K. 2002. A tectonic model for the Tertiary evolution of strike-slip faults and rift basins in SE Asia. *Tectonophysics*, 347, 189–215.

Motley, T.J. In press. Phylogeny of *Kadua* (= *Hedyotis*): Origins, biogeography, and subgeneric classification. *Annals of the Missouri Botanical Garden.*

Motley, T.J., H. Dempewolf, D.H. Lorence, and W. Wagner. 2008. Biogeographic patterns and affinities of the Pacific genera *Oparanthus* and *Fitchia* (Coreopsideae: Asteraceae). Abstract. 2008 Botany Conference, Vancouver, BC. http://2008.botanyconference.org/.

Mouly, A., S.G. Razafimandimbison, A. Khodbandeh, and B. Bremer. 2009. Phylogeny and classification of the species-rich pantropical showy genus *Ixora* (Rubiaceae-Ixoreae) with indications of geographical monophyletic units and hybrids. *American Journal of Botany*, 96, 686–706.

Moya, A., A. Galiana, and F.J. Ayala. 1995. Founder-effect speciation theory: Failure of experimental corroboration. *Proceedings of the National Academy of Sciences USA*, 92, 3983–3986.

Moyle, R.G. 2006. A molecular phylogeny of kingfishers (Alcedinidae) with insights into early biogeographic history. *The Auk*, 123, 487–499.

Moyle, R.G., R.T. Chesser, R.O. Prum, P. Schikler, and J. Cracraft. 2006. Phylogeny and evolutionary history of Old World Suboscine birds (Aves: Eurylaimides). *American Museum Novitates*, 3544, 1–22.

Moyle, R.G., C.E. Filardi, C.E. Smith, and J. Diamond. 2009. Explosive Pleistocene diversification and hemispheric expansion of a "great speciator." *Proceedings of the National Academy of Sciences USA*, 106, 1863–1868.

Muchmore, W.B. 2000. The Pseudoscorpionida of Hawaii, Part I: Introduction and Chthonioidea. *Proceedings of the Hawaiian Entomological Society*, 34, 127–142.

Mueller, K. 1892. Remarks on Dr H. von Jhering's paper "On the ancient connections between New Zealand and South America." *Transactions of the New Zealand Institute*, 25, 428–434.

Muellner, A.N., C.M. Pannell, A. Coleman, and M.W. Chase. 2008. The origin and evolution of Indomalesian, Australasian and Pacific island biotas: Insights from Aglaieae (Meliaceae, Sapindales). *Journal of Biogeography*, 35, 1769–1789.

Muellner, A.N., T.D. Pennington, A.V. Koecke, and S.S. Renner. 2010. Biogeography of *Cedrela* (Meliaceae, Sapidales) in Central and South America. *American Journal of Botany*, 97, 511–518.

Muellner, A.N., V. Savolainen, R. Samuel, and M.W. Chase. 2006. The mahogany family "out-of-Africa": Divergence time estimation, global biogeographic patterns inferred from *rbcL* DNA sequences, extant, and fossil distribution of diversity. *Molecular Phylogenetics and Evolution*, 40, 236–250.

Muers, M. 2010. Evolution: Illusions of conservation. *Nature Reviews Genetics*, 11, 169.

Müller, C.J., and L.B. Beheregaray. 2010. Palaeo island-affinities revisited: Biogeography and systematics of the Indo-Pacific genus *Cethosia* Fabricius (Lepidoptera: Nymphalidae). *Molecular Phylogenetics and Evolution*, 57, 314–326.

Müller, K., and T. Borsch. 2005. Phylogenetics of Amaranthaceae based on *matK/trnK* sequence data: Evidence from parsimony, likelihood, and Bayesian analyses. *Annals of the Missouri Botanic Garden*, 92, 96–102.

Müller, R.D., M. Sdrolias, G. Gaina, B. Steinberger, and C. Heine. 2008. Long-term sea-level fluctuations driven by ocean basin dynamics. *Science*, 319, 1357–1362.

Mummenhoff, K., H. Brüggemann, and J.L. Bowman. 2001. Chloroplast DNA phylogeny and biogeography of *Lepidium* (Brassicaceae). *American Journal of Botany*, 88, 2051–2063.

Murienne, J. 2009a. Testing biodiversity hypotheses in New Caledonia using phylogenetics. *Journal of Biogeography*, 36, 1433–1434.

Murienne, J., 2009b. New Caledonia, biology. In R.G. Gillespie and D.A. Clague (eds.), *Encyclopedia of islands* (pp. 643–645). University of California Press, Berkeley.

Murphy, W.J., and G.E. Collier. 1997. A molecular phylogeny for aplocheiloid fishes (Atherinomorpha, Cyprinodontiformes): The role of vicariance and the origins of annualism. *Molecular Biology and Evolution*, 14, 790–799.

Muss, A., D.R. Robertson, C.A. Stepien, P. Wirtz, and B.W. Bowen. 2001. Phylogeography of *Ophioblennius*: The role of ocean currents and geography in reef fish evolution. *Evolution*, 55, 561–572.

Musser, G.G., K.M. Helgen, and D.P. Lunde. 2008. Systematic review of New Guinea Leptomys (Muridae, Murinae) with descriptions of two new species. *American Museum Novitates*, 3624, 1–60.

Myers, A.A., and J.K. Lowry. 2009 The biogeography of Indo-West Pacific tropical amphipods with particular reference to Australia. *Zootaxa*, 2260, 109–127.

Nagy, Z.T., U. Joger, M. Wink, F. Glaw, and M. Vences, M. 2003. Multiple colonization of Madagascar and Socotra by colubrid snakes: Evidence from nuclear and mitochondrial gene phylogenies. *Proceedings of the Royal Society of London B*, 270, 2613–2621.

Namoff, S., Q. Luke, F. Jiménez, A. Veloz, C.E. Lewis, V. Sosa, M. Maunder, and J. Francisco-Ortega. 2010. Phylogenetic analyses of nucleotide sequences confirm a unique plant intercontinental disjunction between tropical Africa, the Caribbean, and the Hawaiian Islands. *Journal of Plant Research*, 123, 57–65.

Natland, J.H., and E.L. Winterer. 2005. Fissure control on volcanic action in the Pacific. *Geological Society of America Special Paper*, 388, 687–710.

Navarro-Sigüenza, A.G., A.T. Peterson, A. Nyari, G.M. García-Deras, and J. García-Moreno. 2008. Phylogeography of the *Buarremon* brush-finch complex (Aves, Emberizidae) in Mesoamerica. *Molecular Phylogenetics and Evolution*, 47, 21–35.

Neall, V.E., and S.A. Trewick. 2008. The age and origin of the Pacific islands: A geological overview. *Philosophical Transactions of the Royal Society B*, 363, 3293–3308.

Near, T.J., and M.J. Sanderson. 2004. Assessing the quality of molecular divergence time estimates by fossil calibrations and fossil-based model selection. *Philosophical Transactions of the Royal Society of London B*, 359, 1477–1483.

Nei, M. 2002. Review of "Where do we come from? The molecular evidence for human descent" by J. Klein and N. Takahata. *Nature*, 417, 899–900.

Nei, M. 2007. The new mutation theory of phenotypic evolution. *Proceedings of the National Academy of Sciences USA*, 104, 12235–12242.

Neigel, J.E. 2002. Is F_{ST} obsolete? *Conservation Genetics*, 3, 167–173.

Nelson, G. 2006. Hawaiian vicariance. *Journal of Biogeography*, 33, 2154–2157.

Nesom, G.L. 2001. Taxonomic notes on *Keysseria* and *Ptynicarpa* (Asteraceae: Astereae, Lageniferinae). *Sida*, 19, 513–518.

Newman, W., and E. Gomez. 2002. On the status of giant clams, relics of Tethys (Mollusca: Bivalvia: Tridacnidae). *Proceedings of the 9th International Coral Reef Symposium* (Bali), 2, 927–936.

Nijman, V., and K.A.I. Nekaris. 2010. Checkerboard patterns, interspecific competition, and extinction: Lessons from distribution patterns of tarsiers (*Tarsius*) and slow lorises (*Nycticebus*) in insular Southeast Asia. *International Journal of Primatology*, 31, 1147–1160.

Nitta, J.H., and P.M. O'Grady. 2008. Mitochondrial phylogeny of the endemic Hawaiian craneflies (Diptera, Limoniidae, *Dicranomyia*): Implications for biogeography and species formation. *Molecular Phylogenetics and Evolution*, 46, 1182–1190.

Njome, M.S., and C.E. Suh. 2005. Tectonic evolution of the Tombel graben basement, southwestern Cameroon. *Episodes*, 28, 37–41.

Nnange, J.M., Y.H.P. Djomani, J.D. Fairhead, and C. Ebinger. 2001. Determination of the isostatic compensation mechanism of the region of the Adamawa Dome, West Central Africa, using the admittance technique of gravity data. *African Journal of Science and Technology*, 1, 29–35.

Nomade, S., K.B. Knight, E. Beutel, P.R. Renne, C. Verati, A. Marzoli, N. Youbi, and H. Bertrand. 2006. Chronology of the Central Atlantic province: Implications for the central Atlantic rifting processes and the Triassic-Jurassic biotic crisis. *Palaeogeography, Palaeoclimatology, Palaeoecology*, 244, 326–344.

Nonnotte, P., H. Guillou, B. Le Gall, M. Benoit, J. Cotton, and S. Scaillet. 2008. New K-Ar age determinations of Kilimanjaro volcano in the North Tanzanian diverging rift, East Africa. *Journal of Volcanology and Geothermal Research*, 173, 99–112.

Norman, J.E., and M.F. Ashley. 2000. Phylogenetics of Perissodactyla and tests of the molecular clock. *Journal of Molecular Evolution*, 50, 11–21.

Norton, I.O. 2007. Speculations on Cretaceous tectonic history of the northwest Pacific and a tectonic origin for the Hawaii hotspot. *Geological Society of America Special Paper*, 430, 451–470.

Nowak, K. 2008. Frequent water drinking by Zanzibar red colobus (*Procolobus kirkii*) in a mangrove forest refuge. *American Journal of Primatology*, 70, 1081–1092.

Nowak, K., and P.C. Lee. 2011. Demographic structure of Zanzibar red colobus in unprotected coral rag and mangrove habitats. *International Journal of Primatology*, 32, 24–45.

Nyblade, A.A. 2002. Crust and upper mantle structure in East Africa: Implications for the origin of Cenozoic rifting and volcanism and the form of magmatic rifted margins. *Geological Society of America Special Paper*, 362, 15–26.

O'Brien, T.G., M.F. Kinnaird, A. Nurcahyo, M. Iqbal, and M. Rusmanto. 2004. Abundance and distribution of sympatric gibbons in a threatened Sumatran rain forest. *International Journal of Primatology*, 25, 267–284.

O'Connor, J.M., P. Stoffers, J.R. Wijbrans, and T.J. Worthington. 2007. Migration of widespread long-lived volcanism across the Galápagos Volcanic Province: Evidence for a broad hotspot melting anomaly? *Earth and Planetary Science Letters*, 263, 339–354.

O'Grady, P., and R. DeSalle. 2008. Out of Hawaii: The origin and biogeography of the genus *Scaptomyza* (Diptera: Drosophilidae). *Biology Letters*, 4, 195–199.

O'Grady, P.M., Magnacca, K., and R.T. Lapoint. 2009. *Drosophila*. In R.G. Gillespie and D.A. Clague (eds.), *Encyclopedia of islands* (pp. 232–235). University of California Press, Berkeley.

O'Grady, P.M., and M. Zilversmit. 2004. Phylogenetic relationships within the *Drosophila haleakalae* species group inferred by molecular and morphological characters (Diptera: Drosophilidae). *Bishop Museum Bulletin in Entomology*, 12, 117–134.

Oates, J.F., R.A. Bergl, and J.M. Linder. 2004. Africa's Gulf of Guinea forests: Biodiversity patterns and conservation priorities. *Advances in Applied Biodiversity Science*, 6, 1–95.

Olmstead, R.G., L. Bohs, H.A. Migid, E. Santiago-Valentin, V.F. Garcia, and S.M. Collier. 2008. A molecular phylogeny of the Solanaceae. *Taxon*, 57, 1159–1181.

Olmstead, R., and D.C. Tank. 2008. Over-precision in molecular dating. Abstract 433 in Botanical Society of America Conference Abstracts, 2008. At: http://www.2008.botanyconference.org.

Olson, L.E., E.J. Sargis, and R.D. Martin. 2005. Intraordinal phylogenetics of treeshrews (Mammalia: Scandentia) based on evidence from the mitochondrial *12S* rRNA gene. *Molecular Phylogenetics and Evolution*, 35, 656–673.

Olson, L.E., E.J. Sargis, W.T. Stanley, K.B. Hildebrandt, and T.R. Davenport. 2008. Additional molecular evidence strongly supports the distinction between the recently described African primate *Rungwecebus kipunji* (Cercopithecidae, Papionini) and *Lophocebus*. *Molecular Phylogenetics and Evolution*, 48, 789–794.

Olson, S.L., and H.F. James. 1982. Prodromus of the fossil avifauna of the Hawaiian Islands. *Smithsonian Contributions to Zoology*, 365, 1–59.

Olson, S.L., and H.F. James. 1991. Descriptions of thirty-two new species of birds from the Hawaiian Islands, Part II : Passeriformes. *Ornithological Monographs*, 46, 1–88.

Omland, K.E., L.G. Cook, and M.D. Crisp. 2008. Tree thinking for all biology: The problem with reading phylogenies as ladders of progress. *BioEssays*, 30, 854–867.

Onuma, M. 2002. Daily ranging patterns of the proboscis monkey, *Nasalis larvatus*, in coastal areas of Sarawak, Malaysia. *Mammal Study*, 27, 141–144.

Opazo, J.C., D.E. Wildman, T. Prychitko, R.M. Johnson, and M. Goodman. 2006. Phylogenetic relationships and divergence times among New World monkeys (Platyrrhini, Primates). *Molecular Phylogenetics and Evolution*, 40, 274–280.

Orr, H.A. 2005. The genetic basis of reproductive isolation: insights from *Drosophila. Proceedings of the National Academy of Sciences USA*, 102 (Suppl. 1), 6522–6526.

Osterholz, M., L. Walter, L., and C. Roos. 2008. Phylogenetic position of the langur genera *Semnopithecus and Trachypithecus* among Asian colobines, and genus affiliiations of their species groups. *BMC Evolutionary Biology*, 8(58), 1–12.

Osterholz, M., L. Walter, and C. Roos. 2009. Retropositional events consolidate the branching order among New World monkey genera. *Molecular Phylogenetics and Evolution*, 50, 507–513.

Oviedo, R., A. Traveset, A. Valido, and G. Brull. 2009. Sobre la presencia de *Cneorum* (Cneoraceae) en Cuba: ¿Ejemplo de disyunción biogeográfica Mediterráneo-Caribe? *Annales del Jardin Botanico de Madrid*, 66, 25–33.

Padilha, A.L. N.B. Trivedi, I. Vitorello, and J.M da Costa. 1991. Geophysical constraints on tectonic models of the Taubaté Basin, southeastern Brazil. *Tectonophysics*, 196, 157–172.

Pagès, M., S. Calvignac, C. Klein, M. Paris, S. Hughes, and C. Hänni. 2008. Combined analysis of fourteen nuclear genes refines the Ursidae phylogeny. *Molecular Phylogenetics and Evolution*, 47, 73–83.

Pálfy, J. 2003. Volcanism of the central Atlantic magmatic province as a potential driving force in the end-Triassic mass extinction. *Geophysical Monograph—American Geophysical Union*, 136, 255–267.

Pan, R., C. Groves, and C. Oxnard. 2004. Relationships between the fossil colobine *Mesopithecus pentelicus* and extant cercopithecoids, based on dental metrics. *American Journal of Primatology*, 62, 287–299.

Panero, J.L., and V.A. Funk. 2008. The value of sampling anomalous taxa in phylogenetic studies: Major clades of the Asteraceae revealed. *Molecular Phylogenetics and Evolution*, 47, 757–782.

Parent, C.E., A. Caccone, and K. Petren. 2008. Colonization and diversification of Galápagos terrestrial fauna: A phylogenetic and biogeographical synthesis. *Philosophical Transactions of the Royal Society B*, 363, 3347–3361.

Parenti, L.R. 1981. A phylogenetic and biogeographic analysis of cyprinodontiform fishes (Teleostei, Atherinomorpha). *Bulletin of the American Museum of Natural History*, 168, 335–557.

Parkinson, C. 1998. An outline of the petrology, structure and age of the Pompangeo Schist Complex of central Sulawesi, Indonesia. *Island Arc*, 7, 231–245.

Parkinson, C., K. Miyazaki, K. Wakita, and D.A. Carswell. 1998. An overview and tectonic synthesis of the pre-Tertiary very-high-pressure metamorphic and associated rocks of Java, Sulawesi and Kalimantan, Indonesia. *Island Arc*, 7, 184–200.

Paulay, G. 1984. Adaptive radiation on an isolated oceanic island: The Cryptorhynchinae (Curculionidae) of Rapa revisited. *Biological Journal of the Linnean Society*, 26, 95–187.

Pax, D.L., R.A. Price, and H.J. Michaels. 1997. Phylogenetic position of the Hawaiian geraniums based on *rbc*L sequences. *American Journal of Botany*, 84, 72–78.

Paxinos, E.E., H.F. James, S.L. Olson, M.D. Sorenson, J. Jackson, and R.C. Fleischer. 2002. mtDNA from fossils reveals a radiation of Hawaiian geese recently derived from the Canada goose (*Branta canadensis*). *Proceedings of the National Academy of Sciences USA*, 99, 1399–1404.

Pennington, T.D., C. Reynel, and A. Daza. 2004. *Illustrated guide to the trees of Peru*. David Hunt, Sherborne.

Pennisi, E. 2009. On the origin of flowering plants. *Science*, 324, 28–31.

Percy, D.M., A.M. Garver, W.L. Wagner, H.F. James, C.W. Cunningham, S.E. Miller, and R.C. Fleischer. 2008. Progressive island colonization and ancient origin of Hawaiian *Metrosideros* (Myrtaceae). *Proceedings of the Royal Society B*, 275, 1479–1490.

Peres, C. 1999. Effects of subsistence hunting and forest types on the structure of Amazonian primate communities. In J.G. Fleagle, C.H. Janson, and K.E. Reed (eds.), *Primate communities* (pp. 268–283). Cambridge University Press, Cambridge.

Peres, C., and C.H. Janson. 1999. Species coexistence, distribution and environmental determinants of neotropical primate richness: A community zoogeographic analysis. In J.G. Fleagle, C.H. Janson, and K.E. Reed (eds.), *Primate communities* (pp. 55–74). Cambridge University Press, Cambridge.

Pérez-Emán, J.L., R.L. Mumme, and P.G. Jabłoński. 2010. Phylogeography and adaptive plumage evolution in Central American subspecies of the slate-thoated redstart (*Myioborus miniatus*). *Ornithological Monographs*, 67, 90–102.

Perkins, R.C.L. 1901. An introduction to the study of the Drepanididae, a family of birds peculiar to the Hawaiian Islands. *Ibis*, 8th ser., 1, 562–585.

Petit, R.J. 2008. The coup de grâce for the nested clade phylogeographic analysis? *Molecular Ecology*, 17, 516–518.

Pettigrew, J.D., B.C. Maseko, and P.R. Manger. 2008. Primate-like retinotectal decussation in an echolocating megabat, *Rousettus aegyptiacus*. *Neuroscience*. doi:10.1016/j.neuroscience.2008.02.019.x.

Pfeil, B.E., and M.D. Crisp. 2008. The age and biogeography of *Citrus* and the orange subfamily (Rutaceae: Aurantioideae) in Australasia and New Caledonia. *American Journal of Botany*, 95, 1621–1631.

Philipson, W.R. 1970. Floristics of Rarotonga. In R. Fraser (ed.), *The Cook Bicentenary Expedition in the southwestern Pacific* (pp. 49–54). Bulletin No. 8. Royal Society of New Zealand, Wellington.

Pienkowski, M.W., A.R. Watkinson, G. Kerby, L. Naughton-Treves, A. Treves, C. Chapman, and R. Wrangham. 1998. Temporal patterns of crop-raiding by primates: Linking food availability in croplands and adjacent forest. *Journal of Applied Ecology*, 35, 596–606.

Pilsbry, H. 1901. The genesis of mid-Pacific faunas. *Proceedings of the Academy of Natural Sciences of Philadelphia*, 52, 568–581.

Pindell, J., and L. Kennan. 2009. Tectonic evolution of the Gulf of Mexico, Caribbean and northern South America in the mantle reference frame: An update. In K. James, M.A. Lorente, and J. Pindell (eds.), *The geology and evolution of the region between North and South America. Geological Society of London, Special Publication*, 328, 1–55.

Pindell, J., L. Kennan, K.P. Stanek, W.V. Maresch, and G. Draper, G. 2006. Foundations of Gulf of Mexico and Caribbean evolution: Eight controversies resolved. *Geologica Acta*, 4, 303–341.

Pindell, J.L., and K.D. Tabbutt. 1995. Mesozoic-Cenozoic Andean paleogeography and regional controls on hydrocarbon systems. In A.J. Tankard, R. Suárez Soruco, and H.J. Welsink (eds.), *Petroleum basins of South America* (pp. 63–78). *Memoir of the American Association of Petroleum Geologists*, 62.

Pirie, M.D., L.W. Chatrou, J.B. Mols, R.H.J. Erkens, and J. Oosterhof. 2006. "Andean- centred" genera in the short-branch clade of Annonaceae: Testing biogeographical hypotheses using phylogeny reconstruction and molecular dating. *Journal of Biogeography*, 33, 31–46.

Planes, S., and C. Fauvelot. 2002. Isolation by distance and vicariance drive genetic structure of a coral reef fish in the Pacific Ocean. *Evolution*, 56, 378–399.

Plautz, H.L., E.C. Gonçalves, S.F. Ferrari, M.P.C. Schneider, and A. Silva. 2009. Evolutionary inferences on the diversity of the genus *Aotus* (Platyrrhini, Cebidae) from mitochondrial cytochrome *c* oxidase subunit II gene sequences. *Molecular Phylogenetics and Evolution*, 51, 382–387.

Plunkett, G.M., and P.P. Lowry II. 2007. Evolution and biogeography in Melanesian *Schefflera* (Araliaceae): A preliminary assessment based on ITS and ETS sequence data. Abstract. Botany and Plant Biology Conference 2007.

Plunkett, G.M., and P.P. Lowry II. 2010. Paraphyly and polyphyly in *Polyscias* sensu lato: Molecular evidence and the case for recircumscribing the "pinnate genera" of Araliaceae. *Plant Diversity and Evolution*, 128, 23–54.

Plunkett, G.M., P.P. Lowry II, D.G. Frodin, and J. Wen. 2005. Phylogeny and geography of *Schefflera*: Pervasive polyphyly in the largest genus of Araliaceae. *Annals of the Missouri Botanical Garden*, 92, 202–224.

Poon, W.-S., P.-C. Shaw, M.P. Simmons, and P.P.-H. But. 2007. Congruence of molecular, morphological, and biochemical profiles in Rutaceae: A cladistic analysis of the subfamilies Rutoideae and Toddalioideae. *Systematic Botany*, 32, 837–846.

Porter, L.M., and P.A. Garber. 2004. Goeldi's monkey: A primate paradox? *Evolutionary Anthropology*, 13, 104–115.

Porter, L.M., S.M. Sterr, and P.A. Garber. 2007. Habitat use and ranging behavior of *Callimico goeldii*. *International Journal of Primatology*, 28, 1035–1058.

Poux, C., P. Chevret, D. Huchon, W.W. de Jong, and E.J. Douzery. 2006. Arrival and diversification of caviomorph rodents and platyrrhine primates in South America. *Systematic Biology*, 55, 228–244.

Poux, C., O. Madsen, E. Marquard, D.R. Vieites, W.W. de Jong, and M. Vences. 2005. Asynchronous colonization of Madagascar by the four endemic clades of primates, tenrecs, carnivores, and rodents, as inferred from nuclear genes. *Systematic Biology*, 54, 719–730.

Prance, G.T. 1979. Notes on the vegetation of Amazonia III. The terminology of Amazonian forest types subject to inundation. *Brittonia*, 31, 26–38.

Pratt, H.D. 2001. Why the Hawaii creeper is an *Oreomystis*: What phenotypic characters reveal about the phylogeny of Hawaiian honeycreepers. *Studies in Avian Biology*, 22, 81–97.

Prenner, G., and P. Rudall. 2007. Comparative ontogeny of the cyathium in *Euphorbia* (Euphorbiaceae) and its allies: Exploring the organ-flower-inflorescence boundary. *American Journal of Botany*, 94, 1612–1629.

Price, J.J., K.P. Johnson, S.E. Bush, and D.H. Clayton. 2005. Phylogenetic relationships of the Papuan Swiftlet *Aerodramus papuensis* and implications for the evolution of avian echolocation. *Ibis*, 147, 79–796.

Price, J.P. 2009. Hawaiian Islands, biology. In R.G. Gillespie and D.A. Clague (eds.), *Encyclopedia of islands* (pp. 397–404). University of California Press, Berkeley.

Price, J.P., and D.A. Clague. 2002. How old is the Hawaiian biota? Geology and phylogeny suggest recent divergence. *Proceedings of the Royal Society of London B*, 269, 2429–2435.

Price, J.P., and D. Elliot-Fisk. 2004. Topographic history of the Maui Nui complex, Hawaiʻi. *Pacific Science*, 58, 27–45.

Price, J.P., and W.L. Wagner. 2004. Speciation in Hawaiian angiosperm lineages: Cause, consequence and mode. *Evolution*, 58, 2185–2200.

Pubellier, M., J. Ali, and C. Monnier. 2003. Cenozoic plate interaction of the Australia and Philippine Sea plates: "Hit-and-run" tectonics. *Tectonophysics*, 363, 181–199.

Pubellier, M., C. Monnier, R. Maury, and R. Tamayo. 2004. Plate kinematics, origin and tectonic emplacement of supra-obduction ophiolites in SE Asia. *Tectonophysics*, 392, 9–36.

Puigbó, P., Y.I. Wolf, and V. Koonin. 2009. Search for a "tree of life" in the thicket of the phylogenetic forest. *Journal of Biology*, 8, 59.

Punnett, R.C. 1911. *Mendelism*. 3rd ed. MacMillan, New York.

Puspita, S.D., R. Hall, and C.F. Elders. 2005. Structural styles of the offshore West Sulawesi fold belt, North Makassar Straits, Indonesia. *Proceedings of the 30th Annual Convention of the Indonesian Petroleum Association* (pp. 519–528).

Pyron, R.A., and F.T. Burbrink. 2010. Hard and soft allopatry: Physically and ecologically mediated modes of geographic speciation. *Journal of Biogeography*, 37, 2005–2015.

Raaum, R.L., K.N. Sterner, C.M. Noviello, C.-B. Stewart, and T.R. Disotell. 2005. Catarrhine primate divergence dates estimated from complete mitochondrial genomes: Concordance with fossil and nuclear DNA evidence. *Journal of Human Evolution*, 48, 237–257.

Rabinowitz, P.D., and S. Woods. 2006. The Africa-Madagascar connection and mammal migrations. *Journal of African Earth Sciences*, 44, 270–276.

Randall, J.K, P.S. Lobel, and E.H. Chave. 1985. Annotated checklist of the fishes of Johnston Island. *Pacific Science*, 39, 24–80.

Ranker, T.A., J.M.O. Geiger, S.C. Kennedy, A.R. Smith, C.H. Haufler, and B.S. Parris. 2003. Molecular phylogenetics and evolution of the endemic Hawaiian genus *Adenophorus* (Grammitidacae). *Molecular Phylogenetics and Evolution*, 26, 337–347.

Ranker, T.A., C.E.C. Gemmill, and P.G. Trapp. 2000. Microevolutionary patterns and processes of the native Hawaiian colonizing fern *Odontosoria chinensis* (Lindsaeaceae). *Evolution*, 54, 828–839.

Rasmussen, D.T. 1994. The different meanings of a tarsioid-anthropoid clade and a new model of anthropoid origins. In J. G. Fleagle and R. F. Kay (eds.), *Anthropoid origins* (pp. 335–360). Plenum, New York.

Rasmussen, D.T. 2002. The origin of primates. In W.C. Hartwig (ed.), *The primate fossil record* (pp. 5–11). Cambridge University Press, Cambridge.

Rasmussen, D.T., and K.A. Nekaris. 1998. Evolutionary history of lorisiform primates. *Folia Primatologica*, 69(Suppl. 1), 250–285.

Rassman, K. 1997. Evolutionary age of the Galápagos iguanas predates the age of the present Galápagos Islands. *Molecular Phylogenetics and Evolution*, 7, 158–172.

Raxworthy, C.J., M.R.J. Forstner, and R.A. Nussbaum. 2002. Chameleon radiation by oceanic dispersal. *Nature*, 415, 784–787.

Raya, A., and J.C. Izpisúa Belmonte. 2006. Left–right asymmetry in the vertebrate embryo: From early information to higher-level integration. *Nature Reviews Genetics*, 7, 283–293.

Razafimandimbison, S.G., M.S. Appelhans, H. Rabarison, T. Haevermans, A. Rakotondrafara, S.R. Rakotonandrasana, M. Ratsimbason, J.-N. Labat, P.J.A. Keßler, E. Smets, C. Cruaud, A. Couloux, and M. Randrianarivelojosia. 2010. Implications of a molecular phylogenetic study of the Malagasy genus *Cedrelopsis* and its relatives (Ptaeroxylaceae). *Molecular Phylogenetics and Evolution*, 57, 258–265.

Reding, D.M., J.T. Foster, H.F. James, H.D. Pratt, and R.C. Fleischer. 2009. Convergent evolution of "creepers" in the Hawaiian honeycreeper radiation. *Biology Letters*, 5, 221–224.

Ree, R.H., and S. Smith. 2008. LAGRANGE: Likelihood analysis of geographic range evolution, version 2.0. Software available from http://www.code.google.com/p/lagrange/.

Reece, J.S., D.G. Smith, and E. Holm. 2010. The moray eels of the *Anarchias cantonensis* Group (Anguilliformes: Muraenidae), with description of two new species. *Copeia*, 2010, 421–430.

Renema, W., D.R. Bellwood, J.C. Braga, K. Bromfield, R. Hall, K.G. Johnson, P. Lunt, C.P. Meyer, L.B. McMonagle, R.J. Morley, A. O'Dea, J.A. Todd, F.P. Wesselingh, M.E.J. Wilson, and J.M. Pandolfi. 2008. Hopping hotspots: Global shifts in marine biodiversity. *Science*, 321, 654–657.

Renner, S.S. 2005. Relaxed molecular clocks for dating historical plant dispersal events. *Trends in Ecology and Evolution*, 10, 550–558.

Renner, S.S., J.S. Strijk, D. Strasberg, and C. Thébaud, C. 2010. Biogeography of the Monimiaceae (Laurales): A role for East Gondwana and long-distance dispersal, but not West Gondwana. *Journal of Biogeography*, 37, 1227–1238.

Rheindt, F.E., and J.J. Austin. 2005. Major analytical and conceptual shortcomings in a recent taxonomic revision of the Procellariiformes: A reply to Penhallurick and Wink (2004). *Emu*, 105, 181–186.

Rho, M., M. Zhou, X. Gao, S. Kim, H. Tang, and M. Lynch. 2009. Independent mammalian genome contractions following the KT boundary. *Genome Biology and Evolution*, 1, 2–12.

Ribas, C.C., R.G. Moyle, C.Y. Miyitaki, and J. Cracraft. 2007. The assembly of montane biotas: Linking Andean tectonics and climatic oscillations to independent regimes of diversification in *Pionus* parrots. *Proceedings of the Royal Society B*, 274, 2399–2408.

Ribeiro, A.C. 2006. Tectonic history and the biogeography of the freshwater fishes from the coastal drainages of eastern Brazil: An example of faunal evolution associated with a divergent continental margin. *Neotropical Ichthyology*, 4, 225–246.

Rich, T.H., T.F. Flannery, P. Trusler, and P. Vickers-Rich. 2001. Corroboration of the Garden of Eden hypothesis. In I. Metcalfe (ed.), *Faunal and floral migrations and evolution in SE Asia-Australasia* (pp. 323–332). Balkema, Lisse.

Richard, A.F., S.J. Goldstein, and R.E. Dewar. 1989. Weed macaques: The evolutionary implications of macaque feeding ecology. *International Journal of Primatology*, 10, 569–594.

Richards, C.M., and W.S. Moore. 1996. A phylogeny for the African tree frog family Hyperoliidae based on mitochondrial rDNA. *Molecular Phylogenetics and Evolution*, 5, 522–532.

Ricklefs, R.E. 2008. Disintegration of the ecological community. *American Naturalist*, 172, 741–750.

Ridley, M. 1985. *The problems of evolution*. Oxford University Press, Oxford.

Rieppel, O., 2002. A case of dispersing chameleons. *Nature*, 415, 744–745.

Riley, J., and J.C. Wardill. 2001. The rediscovery of Cerulean paradise-flycatcher *Eutrichomyias rowleyi* on Sangihe, Indonesia. *Forktail*, 17, 45–55.

Ripley, S.D., and H. Birckhead. 1942. Birds collected during the Whitney South Sea expedition. 51. On the fruit pigeons of the *Ptilinopus purpuratus* group. *American Museum Novitates*, 1192, 1–14.

Rivera, M.A.J., F.G. Howarth, S. Taiti, and G.K. Roderick. 2002. Evolution in Hawaiian cave-adapted isopods (Oniscidea: Philosciidae): Vicariant speciation or adaptive shifts? *Molecular Phylogenetics and Evolution*, 25, 1–9.

Roberts, D.L., and R.M. Bateman. 2009. Orchids. In R.G. Gillespie and D.A. Clague (eds.), *Encyclopedia of islands* (pp. 696–700). University of California Press, Berkeley.

Roddaz, M., P. Baby, S. Brusset, W. Hermoza, and J.M. Darrozes. 2005. Forebulge dynamics and environmental control in Western Amazonia: The case study of the Arch of Iquitos (Peru). *Tectonophysics*, 399, 87–108.

Rodríguez-Trelles, F., R. Tarrio, and F.J. Ayala. 2003. Molecular clocks: Whence and whither? In P.C.J. Donoghue and M.P. Smith (eds.), *Telling the evolutionary time: Molecular clocks and the fossil record* (pp. 5–26). CRC Press, Boca Raton, FL.

Romer, A. 1966. *Vertebrate paleontology*. University of Chicago Press, Chicago.

Ronquist, F. 1997. Dispersal-vicariance analysis: A new approach to the quantification of historical biogeography. *Systematic Biology*, 46, 195–203.

Roos, C., J. Schmitz, and H. Zischler. 2004. Primate jumping genes elucidate strepsirrhine phylogeny. *Proceedings of the National Academy of Sciences USA*, 101, 10650–10654.

Roos, C., T. Ziegler, J.K. Hodges, H. Zischler, and C. Abegg, C. 2003. Molecular phylogeny of Mentawai macaques: Taxonomic and biogeographic implications. *Molecular Phylogenetics and Evolution*, 29, 139–150.

Roque-Albelo, L., and B. Landry. 2002. The Sphingidae (Lepidoptera) of the Galápagos Islands: Their identification, distribution, and host plants, with new records. *Bulletin de la Société Entomologique Suisse* 74, 217–226.

Rose, K.D. 1995. The earliest primates. *Evolutionary Anthropology*, 3, 159–173.

Rose, K.D., R.S. Rana, A. Sahni, K. Kumar, P. Missiaen, L. Singh, and T. Smith. 2009. Early Eocene primates from Gujarat, India. *Journal of Human Evolution*, 56, 366–404.

Rosen, B., 1988. Progress, problems and patterns in the biogeography of reef corals and other tropical marine organisms. *Helgoländer Meeresuntersuchungen*, 40, 269–301.

Rosenbaum, G., D. Giles, M. Saxon, P.G. Betts, R.F. Weinberg, and C. Duboz C. 2005. Subduction of the Nazca Ridge and the Inca Plateau: Insights into the formation of ore deposits in Peru. *Earth and Planetary Science Letters*, 239, 18–32.

Rosenberger, A.L. 1984. Fossil New World monkeys dispute the molecular clock. *Journal of Human Evolution*, 13, 737–742.

Rosenberger, A.L. 2002. Platyrrhine paleontology and systematics: The paradigm shifts. In W.C. Hartwig (ed.), *The primate fossil record* (pp. 151–160). Cambridge University Press, Cambridge.

Rosenberger, A.L. 2006. Protoanthropoidea (Primates, Simiiformes): A new primate higher taxon and a solution to the *Rooneyia* problem. *Journal of Mammalian Evolution*, 13, 139–146.

Rosenberger, A.L. 2010. The skull of *Tarsius*: Functional morphology, eyeballs, and the nonpursuit predatory lifestyle. *International Journal of Primatology*, 31, 1032–1054.

Rosenberger, A.L., S.B. Cooke, R. Rímoli, X. Ni, and L. Cardoso. 2011. First skull of *Antillothrix bernensis*, an extinct relict monkey from the Dominican Republic. *Proceedings of the Royal Society B*, 278, 67–74.

Rosenberger, A.L., and L.J. Matthews. 2008. *Oreonax*: Not a genus. *Neotropical Primates*, 15, 8–12.

Rosenberger, A.L., M.F. Tejedor, S.B. Cooke, and S. Pekar. 2008. Platyrrhine ecophylogenetics in space and time. In P.A. Garber, A. Estrada, J.C. Bicca-Marques, E.W. Heymann, and K.B. Strier (eds.), *South American primates: Comparative perspectives in the study of behavior, ecology, and conservation* (pp. 69–113). Springer, New York.

Ross, C.F. 2003. Review of "The primate fossil record," ed. by W.C. Hartwig. *Journal of Human Evolution*, 45, 195–201.

Ross, C.F., and R.F. Kay. 2004. Anthropoid origins: Retrospective and prospective. In C.F. Ross and R.F. Kay (eds.), *Anthropoid origins: New visions* (pp. 701–731). Kluwer/Plenum, New York.

Rossetti, D.F., and R.G. Netto. 2006. First evidence of marine influence in the Cretaceous of the Amazonas Basin, Brazil. *Cretaceous Research*, 27, 513–528.

Rossetti, D.F., and P.M. Toledo. 2007. Environmental changes in Amazonia as evidenced by geological and paleontological data. *Revista Brasileira de Ornitologia*, 15, 251–264.

Rossie, J.B., X. Ni, and K.C. Beard. 2006. Cranial remains of an Eocene tarsier. *Proceedings of the National Academy of Sciences USA*, 103, 4381–4385.

Rowan, R.G., and J.A. Hunt. 1991. Rates of DNA change and phylogeny from the DNA sequences of the alcohol dehydrogenase gene from five closely related species of Hawaiian *Drosophila*. *Molecular Biology and Evolution*, 8, 49–70.

Rowe, D.L., K.A. Dunn, R.M. Adkins, and R.L. Honeycutt. 2010. Molecular clocks keep dispersal hypotheses afloat: Evidence for trans-Atlantic rafting by rodents. *Journal of Biogeography*, 37, 305–324.

Rowe, K.C., M.L. Reno, D.M. Richmond, R.M. Adkins, and S.J. Steppan. 2008. Pliocene colonization and adaptive radiations in Australia and New Guinea (Sahul): Multilocus systematic of the old endemic rodents (Muroidea: Murinae). *Molecular Phylogenetics and Evolution*, 47, 84–101.

Rowe, T., T.H. Rich, P. Vickers-Rich, M. Springer, and M.O. Woodburne. 2008. The oldest platypus and its bearing on divergence timing of the platypus and echidna clades. *Proceedings of the National Academy of Sciences USA*, 105, 1238–1242.

Rozefelds, A.C., R.W. Barnes, and B. Pellow. 2001. A new species and comparative morphology of *Vesselowskya* (Cunoniaceae). *Australian Systematic Botany*, 14, 175–192.

Rozendaal, F.G., and F.R. Lambert. 1999. The taxonomic and conservation status of *Pinarolestes sanghirensis* Oustalet 1881. *Forktail*, 15, 1–13.

Rubinoff, D. 2008. Phylogeography and ecology of an endemic radiation of Hawaiian case-bearing moths (*Hyposmocoma*: Cosmopterigidae). *Philosophical Transactions of the Royal Society B*, 363, 3459–3465.

Rubinoff, D., and P. Schmitz. 2010. Multiple aquatic invasions by an endemic, terrestrial Hawaiian moth radiation. *Proceedings of the National Academy of Sciences USA*, 107, 5903–5906.

Ruiz, G.M.H., D. Seward, and W. Winkler. 2004. Detrital thermochonology: A new perspective on hinterland tectonics, an example from the Andean Amazon basin, Ecuador. *Basin Research*, 16, 413–430.

Ruiz-García, M., M.I. Castillo, C. Vásquez, K. Rodriguez, M. Pinedo-Castro, J. Shostell, and N. Leguizamon, N. 2010. Molecular phylogenetics and phylogeography of the white-fronted capuchin (*Cebus albifrons*; Cebidae, Primates) by means of mtCOII gene sequences. *Molecular Phylogenetics and Evolution*, 57, 1049–1061.

Rundell, R.J., and T.D. Price. 2009. Adaptive radiation, nonadaptive radiation, ecological speciation and nonecological speciation. *Trends in Ecology and Evolution*, 24, 394–399.

Rundle, H.D. 2003. Divergent environments and population bottlenecks fail to generate premating isolation in *Drosophila pseudoobscura*. *Evolution*, 57, 2557–2565.

Rundle, H.D., A.Ø. Mooers, and M.C. Whitlock. 1998. Single founder-flush events and the evolution of reproductive isolation. *Evolution*, 52, 1850–1855.

Rundle, H.D., and P. Nosil. 2005. Ecological speciation. *Ecology Letters*, 8, 336–352.

Rundle, H.D., and D. Schluter. 2004. Natural selection and ecological speciation in sticklebacks. In U. Dieckmann, M. Doebeli, J.A.J. Metz, and D. Tautz (eds.), *Adaptive speciation* (pp. 192–209). Cambridge University Press, Cambridge.

Ruse, M., and J. Travis. 2009. Introduction. In M. Ruse and J. Travis (eds.), *Evolution: The first four billion years* (pp. ix–xi). Harvard University Press, Cambridge, MA.

Ruthsatz, B. 1978. Las plantas en cojín de los semi-desiertos andinos del Noroeste Argentina. *Darwiniana*, 21, 494–539.

Rylands, A.B., A.F. Coimbra-Filho, and R.A. Mittermeir. 2009. The systematics and distributions of the marmosets (*Callithrix, Callibella, Cebuella,* and *Mico*) and callimico (*Callimico*). In S.M. Ford, L.M. Porter, and L.C. Davis (eds.), *The smallest anthropoids: The marmoset/Callimico radiation* (pp. 25–62). Springer, New York.

Rylands, A.B., G.A.B. da Fonseca, Y.L.R. Leite, and R.A. Mittermeier. 1996. Primates of the Atlantic forest: Origin, distributions, endemism, and communities. In M.A. Norconk, P.A. Garber, and A.L. Rosenberger (eds.), *Adaptive radiations in neotropical primates* (pp. 21–52). Springer, New York.

Saadi, A., M.N. Machette, K.M. Haller, R.L. Dart, L.-A. Bradley, and A.M.P.D. de Souza. 2002. *Map and database of Quaternary faults and lineaments in Brazil.* U.S. Geological Survey Open-File Report 02-230. Available at http://pubs.usgs.gov/of/2002/ofr-02-230/.

Sager, W.W. 2007. Divergence between paleomagnetic and hotspot-model-predicted polar wander for the Pacific plate with implications for hotspot fixity. *Geological Society of America Special Paper*, 430, 335–357.

Saha, A., A.R. Basu, J. Wakabayashi, and G.L. Wortman. 2005. Geochemical evidence for a subducted infant arc in Franciscan high-grade metamorphic tectonic blocks. *Geological Society of America Bulletin*, 117, 1318–1335.

Salem, A.H., D.A. Ray, J. Xing, P.A. Callinan, J.S. Myers, D.J. Hedges, R.K. Garber, D.J. Witherspoon, L.B. Jorde, and M.A. Batzer. 2003. Alu elements and hominid phylogenetics. *Proceedings of the National Academy of Sciences USA*, 100, 12787–12791.

Salter, R.E., N.A. MacKenzie, N. Nightingale, K.M. Aken, and P. Chai, P. 1985. Habitat use, ranging behaviour, and food habits of the proboscis monkey, *Nasalis larvatus* (van Wurmb), in Sarawak. *Primates*, 26, 436–451.

Salvo, G., G. Bacchetta, F. Ghahremaninejad, and E. Conti. 2008. Phylogenetic relationships of Ruteae (Rutaceae): New evidence from the chloroplast genome and comparisons with non-molecular data. *Molecular Phylogenetics and Evolution*, 49, 736–748.

Salvo, G., S.Y.W. Ho, G. Rosenbaum, R. Ree, and E. Conti. 2010. Tracing the temporal and spatial origins of island endemics in the Mediterranean region: A case study from the Citrus family (*Ruta* L., Rutaceae). *Systematic Biology*, 59, 705–722.

Sanchez Alvarez, J.O. 2007. *Structural and stratigraphic evolution of Shira Mountains, central Ucayali Basin, Perú.* Unpublished M.Sc thesis, Texas A&M University.

Sang, T., D.J. Crawford, S. Kim, and T.F. Stuessy. 1994. Radiation of the endemic genus *Dendroseris* (Asteraceae) on the Juan Fernandez Islands: Evidence from sequences of the ITS regions of nuclear ribosomal DNA. *American Journal of Botany*, 81, 1494–1501.

Sanmartín, I., and F. Ronquist. 2004. Southern hemisphere biogeography inferred by event-based models: Plant versus animal patterns. *Systematic Biology*, 53, 216–243.

Santos, C.M.D. 2007. On basal clades and ancestral areas. *Journal of Biogeography*, 34, 1470–1471.

Savin, S.M., and R.G. Douglas. 1985. Sea level, climate, and the Central American land bridge. In F.G. Stehli and S.D. Webb (eds.), *The great American biotic interchange* (pp. 303–324). Plenum, New York.

Schaefer, H., A. Kocyan, and S.S. Renner. 2008. *Linnaeosicyos* (Cucurbitaceae): A new genus for *Trichosanthes amara*, the Caribbean sister species of all Sicyeae. *Systematic Botany*, 33, 349–355.

Schaik, C.P. van, A. van Amerongen, and M.A. van Noordwijk. 1996. Riverine refuging by wild Sumatran long-tailed macaques (*Macaca fascicularis*). In J.E. Fa and D.G. Lindburgh (eds.), *Evolution and ecology of Macaque societies* (pp. 160–181). Cambridge University Press, Cambridge.

Schawaller, W, W.A. Shear, and P.M. Bonamo. 1991. The first Paleozoic pseudoscorpions (Arachnida, Pseudoscorpionida). *American Museum Novitates*, 3009, 1–17.

Scheiner, S.M. 1999. Towards a more synthetic view of evolution. *American Journal of Botany*, 86, 145–148.

Schellart, W.P., and G.S. Lister. 2005. The role of the East Asian active margin in widespread strike-slip deformation in East Asia. *Journal of the Geological Society*, 162, 959–972.

Schellart, W.P., G.S. Lister, and V.G. Toy. 2006. A Late Cretaceous and Cenozoic reconstruction of the southwest Pacific region: Tectonics controlled by subduction and slab rollback processes. *Earth-Science Reviews*, 76, 191–233.

Schenk, C.J., R.J. Viger, and C.P. Anderson. 1997. *Maps showing geology, oil and gas fields and geologic provinces of the South America region*. U.S. Geological Survey Open-File Report 97-470D.

Schluter, D. 2001. Ecology and the origin of species. *Trends in Ecology and Evolution*, 16, 372–380.

Schluter, D. 2009. Evidence for ecological speciation and its alternative. *Science*, 323, 737–741.

Schmalfuss, H. 2009. *World catalog of terrestrial isopods (Isopoda: Oniscidea)*. http://www.naturkundemuseum-bw.de/stuttgart/projekte/oniscidea-catalog/Cat_terr_isop.pdf.

Schmitz, J., C. Roos, and H. Zischler. 2005. Primate phylogeny: Molecular evidence from retroposons. *Cytogenetics and Genome Research*, 108, 26–37.

Schneider, H., T.A.S. Ranker, S.J. Russell, R. Cranfill, J.M.O. Geiger, R. Aguraiuja, K.R. Wood, M. Grundmann, K. Kloberdanz, and J.C. Vogel. 2005. Origin of the endemic fern genus *Diellia* coincides with the renewal of Hawaiian terrestrial life in the Miocene. *Proceedings of the Royal Society B*, 272, 455–460.

Schulte, J.A., II, J. Melville, and A. Larson. 2003. Molecular phylogenetic evidence for ancient divergence of lizard taxa on either side of Wallace's Line. *Proceedings of the Royal Society of London B*, 270, 597–603.

Schwarzbach, A.E., and J.W. Kadereit. 1999. Phylogeny of prickly poppies, *Argemone* (Papaveraceae), and the evolution of morphological and alkaloid characters based on ITS nrDNA sequence variation. *Plant Systematics and Evolution*, 218, 257–279.

Sclater, P.L. 1864. The mammals of Madagascar. *Quarterly Journal of Science*, 1, 213–219.

Searle, M.P., S.R. Noble, J.M. Cottle, D.J. Waters, A.H.G. Mitchell, T. Hlaing, and M.S.A. Horstwood. 2007. Tectonic evolution of the Mogok metamorphic belt, Burma (Myanmar) constrained by U-Th-Pb dating of metamorphic and magmatic rocks. *Tectonics*, 26, TC3014. doi:10.1029/2006TC002083.

Sebastian, P.M., R. Lira, H.B. Cross, I.R.H. Telford, T.J. Motley, and S.S. Renner. 2009. Biogeography of the *Sicyos* clade (Cucurbitaceae): from the New World to Australia, New Zealand, and Hawaii. Botany and Mycology 2009 (Snowbird, Utah, July 25–29). Abstract Book, p. 221 (abstract 792; also see http://2009.botanyconference.org/).

Seiffert, E.R. 2006. Early evolution and biogeography of lorisiform strepsirrhines. *American Journal of Primatology*, 69, 27–35.

Seiffert, E.R., E.L. Simons, W.C. Clyde, J.B. Rossie, Y. Attia, T.M. Bown, P. Chatrath, and M.E. Mathison. 2005. Basal anthropoids from Egypt and the antiquity of Africa's higher primate radiation. *Science*, 310, 300–304.

Seiffert, E.R., E.L. Simons, T.M. Ryan, and Y. Attia. 2005. Additional remains of *Wadilemur elegans*, a primitive stem galagid from the late Eocene of Egypt. *Proceedings of the National Academy of Sciences USA*, 102, 11396–11401.

Setoguchi, T., and A.L. Rosenberger. 1987. A fossil owl monkey from La Venta, Colombia. *Nature*, 326, 692–694.

Shapiro, L.H., J.S. Strazanac, and G.K. Roderick. 2006. Molecular phylogeny of *Banza* (Orthoptera: Tettigoniidae), the endemic katydids of the Hawaiian archipelago. *Molecular Phylogenetics and Evolution*, 41, 53–63.

Sharma, P., and G. Giribet. 2009. A relict in New Caledonia: phylogenetic relationships of the family Troglosironidae (Opiliones: Cyphophthalmi). *Cladistics*, 25, 279–294.

Shekelle, M., C. Groves, S. Merker, and J. Supriatna. 2008. *Tarsius tumpara*: A new tarsier species from Siau Island, North Sulawesi. *Primate Conservation*, 23, 55–64.

Shekelle, M., R. Meier, I. Wahyu, Wirdateti, and N. Ting. 2010. Molecular phylogenetics and chronometrics of Tarsiidae based on 12S mtDNA haplotypes: Evidence for Miocene origins of crown tarsiers and numerous species within the Sulawesian clade. *International Journal of Primatology*, 31, 1083–1106.

Sheldon, F.H., L.A. Whittingham, R.G. Moyle, B. Slikas, and D.W. Winkler. 2005. Phylogeny of swallows (Aves: Hirundinidae) estimated from nuclear and mitochondrial DNA sequences. *Molecular Phylogenetics and Evolution*, 35, 254–270.

Sherrod, D.R. 2009. Hawaiian islands: Geology. In R.G. Gillespie and D.A. Clague (eds.), *Encyclopedia of islands* (pp. 404–410). University of California Press, Berkeley.

Shih, H.-T., D.C.J. Yeo, and P.K.L. Ng. 2009. The collision of the Indian plate with Asia: Molecular evidence for its impact on the phylogeny of freshwater crabs (Brachyura: Potamidae). *Journal of Biogeography*, 36, 703–719.

Short, A.E.Z., and J.K. Liebherr. 2007. Systematics and biology of the endemic water scavenger beetles of Hawaii (Coleootera: Hydrophiliae, Hydrophilini). *Systematic Entomology*, 32, 601–624.

Silcox, M.T. 2008. The biogeographic origins of Primates and Euprimates: East, west, north, or south of Eden? In E.J. Sargis and M. Dagosto (eds.), *Mammalian evolutionary morphology: A tribute to Frederick S. Szalay* (pp. 199–232). Springer, Dordrecht.

Siler, C.D., J.R. Oaks, J.A. Esselstyn, A.C. Diesmos, and R.M. Brown. 2010. Phylogeny and biogeography of Philippine bent-toed geckos (Gekkonidae: *Cyrtodactylus*) contradict a prevailing model of Pleistocene diversification. *Molecular Phylogenetics and Evolution*, 55, 699–710.

Simberloff, D., and M.D. Collins. 2010. Birds of the Solomon Islands: the domain of the dynamic equilibrium theory and assembly rules, with comments on the taxon cycle. In J.B. Losos and R.E. Ricklefs (eds.), *The theory of island biogeography revisited* (pp. 237–263). Princeton University Press, Princeton, NJ.

Simmons, M.P., C.C. Clevinger, V. Savolainen, R.H. Archer, S. Mathews, and J.J. Doyle. 2001. Phylogeny of the Celastraceae inferred from phytochrome B gene sequence and morphology. *American Journal of Botany*, 88, 313–325.

Simons, E.L. 1976. The fossil record of primate phylogeny. In M. Goodman and R.E. Tashian (eds.), *Molecular anthropology* (pp. 35–61). Plenum, New York.

Simons, E.L. 1997. Preliminary description of the cranium of *Proteopithecus sylviae*, an Egyptian late Eocene anthropoidean primate. *Proceedings of the National Academy of Sciences USA*, 94, 14970–14975.

Simons, E.L. 2001. The cranium of *Parapithecus grangeri*, an Egyptian Oligocene anthropoidean primate. *Proceedings of the National Academy of Sciences USA*, 98, 7892–7897.

Simons, E.L. 2003. The fossil record of tarsier evolution. In P.C. Wright, E.L. Simons, and S. Gursky (eds.), *Tarsiers: Past, present and future* (pp. 9–31). Rutgers University Press New Brunswick, NJ.

Simons, E.L., and E.R. Seiffert. 1999. A partial skeleton of *Proteopithecus sylviae* (Primates, Anthropoidea): First associated dental and postcranial remains of an Eocene anthropoidean. *Comptes Rendus de l'Académie des Sciences*, IIA, 329, 921–927.

Simpson, G.G. 1937. The beginning of the Age of Mammals. *Biological Reviews*, 12, 1–46.

Simpson, G.G. 1940. Mammals and land bridges. *Journal of the Washington Academy of Sciences*, 30, 137–163.

Simpson, G.G. 1945. The principles of classification and a classification of mammals. *Bulletin of the American Museum of Natural History*, 85, 1–350.

Simpson, G.G. 1967. *The meaning of evolution*. 2nd ed. Yale University Press, New Haven, CT.

Smedmark, J.E.E., and A. Anderberg. 2007. Boreotropical migration explains hybridization between geographically distant lineages in the pantropical clade Sideroxyleae (Sapotaceae). *American Journal of Botany*, 94, 1491–1505.

Smith, A.B. 2007. Marine diversity through the Phanerozoic: Problems and prospects. *Journal of the Geological Society*, 164, 731–745.

Smith, A.D. 2007. A plate model for Jurassic to Recent intraplate volcanism in the Pacific Ocean basin. *Geological Society of America Special Paper*, 430, 471–495.

Smith, H.G., and D.M. Wilkinson. 2007. Not all free-living microorganisms have cosmopolitan distributions: the case of *Nebela* (*Apodera*) *vas* Certes (Protozoa: Amoebozoa: Arcellinida). *Journal of Biogeography*, 34, 1822–1831.

Smith, S.A., J.M. Beaulieu, and M.J. Donoghue. 2010. An uncorrelated relaxed-clock analysis suggests an earlier origin for flowering plants. *Proceedings of the National Academy of Sciences USA*, 107, 5897–5902.

Smith, T., K.D. Rose, and P.D. Gingerich. 2006. Rapid Asia-Europe-North America geographic dispersal of earliest Eocene primate *Teilhardina* during the Paleocene-Eocene thermal maximum. *Proceedings of the National Academy of Sciences USA*, 103, 11223–11227.

Smithsonian Institution. 2009. *South America centers of plant diversity and endemism. III. Amazonia.* Available at: http://botany.si.edu/projects/cpd/sa/sa5.htm.

Smyth, H.R., P.J. Hamilton, R. Hall, and P.D. Kinny. 2007. The deep crust beneath island arcs: Inherited zircons reveal continental fragment beneath East Java. *Earth and Planetary Science Letters*, 258, 269–282.

Sobolev, S.V., and A.Y. Babeyko. 2005. What drives orogeny in the Andes? *Geology*, 33, 617–620.

Solem, A. 1983. *Endodontoid land snails from Pacific islands* (*Mollusca: Pulmonata: Sigmurethra*), Part II : *Families Punctidae and Charopidae, zoogeography*. Field Museum of Natural History, Chicago, IL.

Soligo, C., and R.D. Martin. 2006. Adaptive origins of primates revisited. *Journal of Human Evolution*, 50, 414–430.

Soligo, C., O. Will, S. Tavaré, C.R. Marshall, and R.D. Martin. 2007. New light on the dates of primate origins and divergence. In M.J. Ravosa and M. Dagosto (eds.), *Primate origins: Adaptations and evolution* (pp. 29–49). Springer, New York.

Sorenson, M.D., A. Cooper, E.E. Paxinos, T.W. Quinn, H.F. James, S.L. Olson, and R.C. Fleischer. 1999. Relationships of the extinct moa-nalos, flightless Hawaiian waterfowl, based in ancient DNA. *Proceedings of the Royal Society of London B*, 266, 2187–2193.

Sparks, J.S., 2004. Molecular phylogeny and biogeography of the Malagasy and South Asian cichlids (Teleostei: Perciformes: Cichlidae). *Molecular Phylogenetics and Evolution*, 30, 599–614.

Sparks, J.S., and W.L. Smith. 2004. Phylogeny and biogeography of cichlid fishes (Teleostei: Perciformes: Cichlidae): A multilocus approach to recovering deep intrafamilial divergences and the cichlid sister group. *Cladistics*, 20, 1–17.

Sparks, J.S., and W.L. Smith. 2005. Freshwater fishes, dispersal ability, and nonevidence: "Gondwana life rafts" to the rescue. *Systematic Biology*, 54, 158–165.

Spocter, M.A. 2009. *The Panglossian paradigm revisited: The role of non adaptive mechanisms in hominid brain and body size evolution.* Ph.D. dissertation, University of the Witwatersrand, Johannesburg.

Sporck, M.J., and L. Sack. 2008. Exceptional diversification of leaf surfaces in the Hawaiian *Chamaesyce*. Abstract. Botanical Society of America Conference. http://2008.botanyconference.org.

Springer, M.S., W.J. Murphy, E. Eizirik, and S.J. O'Brien. 2003. Placental mammal diversification and the Cretaceous-Tertiary boundary. *Proceedings of the National Academy of Sciences USA*, 100, 1056–1061.

Springer, V.G. 1982. Pacific plate biogeography, with special reference to shore-fishes. *Smithsonian Contributions to Zoology*, 367, 1–182.

Springer, V.G., and J.T. Williams. 1994. The Indo-west Pacific blenniid fish genus *Istiblennius* reappraised: A revision of *Istiblennius*, *Blenniella*, and *Paralticus*, new genus. *Smithsonian Contributions to Zoology*, 565, 1–193.

St. George, I. 1993. The Pacific genus *Earina*. *The Orchadian*, 11, 56–65.

Stacy, E., N. DeBoer, J. Johansen, and T. Sakishima. 2010. Analysis of population structure reveals dispersal limitation and significant differentiation of extreme-habitat varieties of the dominant *Metrosideros polymorpha* on east Hawaii Island. Abstract, Botany Conference 2010. http://2010. botanyconference.org/

Stanford, C.B., and R.C. O'Malley. 2008. Sleeping tree choice by Bwindi chimpanzees. *American Journal of Primatology*, 70, 642–649.

Stankiewicz, J., and M. de Wit. 2006. A proposed drainage evolution model for central Africa: Did the Congo flow east? *Journal of African Earth Sciences*, 44, 75–84.

Stankiewicz, J., C. Thiart, J. Masters, and M.J. de Wit. 2006. Did lemurs have sweepstake tickets? An exploration of Simpson's model for the colonization of Madagascar by mammals. *Journal of Biogeography*, 33, 221–235.

Steadman, D.W. 2006. *Extinction and biogeography of tropical Pacific birds*. University of Chicago Press, Chicago, IL.

Stebbins, G.L. 1966. *Processes of organic evolution*. Prentice Hall, Englewood Cliffs, NJ.

Stebbins, G.L. 1982. *Darwin to DNA, molecules to humanity*. W.H. Freeman & Co., San Francisco.

Steckler, M.S., H. Akhter, and L. Seeber. 2008. Collision of the Ganges-Brahmaputra delta with the Burma Arc: Implications for earthquake hazard. *Earth and Planetary Science Letters*, 273, 367–378.

Stefanović, S., and M. Costea. 2008. Reticulate evolution in the parasitic genus *Cuscuta* (Convolvulaceae). *Botany*, 86, 791–808.

Stehli, F.G., and S.D. Webb. 1985. *The great American biotic interchange*. Plenum, New York.

Steiner, C., A. Hobson, P. Favre, G.M. Stampfli, and J. Hernandez. 1998. Mesozoic sequence of Fuerteventura (Canary Islands): Witness of Early Jurassic sea-floor spreading in the central Atlantic. *Bulletin of the Geological Society of America*, 110, 1304–1317.

Steiper, M.E., and N.M. Young. 2008. Timing primate evolution: lessons from the discordance between molecular and paleontological estimates. *Evolutionary Anthropology*, 17, 179–188.

Steppan, S.J., R.M. Adkins, P.Q. Spinks, and C. Hale. 2005. Multigene phylogeny of the Old World mice, Murinae, reveals distinct geographic lineages and the declining utility of mitochondrial genes compared to nuclear genes. *Molecular Phylogenetics and Evolution*, 37, 370–388.

Sterner, K.N., R.L. Raaum, Y.-P. Zhang, C.-B. Stewart, and T.R. Disotell. 2006. Mitochondrial data support an odd-nosed colobine clade. *Molecular Phylogenetics and Evolution*, 40, 1–7.

Stevens, P.F. 2010. Angiosperm phylogeny website. www.mobot.org/MOBOT/Research/APweb/ (January 2010).

Stewart, T.A., and R.C. Albertson. 2010. Evolution of a unique predatory feeding apparatus: Functional anatomy, development and a genetic locus for jaw laterality in Lake Tanganyika scale-eating cichlids. *BMC Biology*, 8:8, 1–11.

Stock, D.W. 2007. Zebrafish dentition in comparative context. *Journal of Experimental Zoology (Molecular Development and Evolution)*, 308B, 523–549.

Stoltzfus, A. 2006. Mutationism and the dual causation of evolutionary change. *Evolution and Development*, 8, 304–317.

Stoltzfus, A. 2010. *The curious disconnect.* http://www.molevol.org/cdblog/intro.

Stoltzfus, A., and L.Y. Yampolsky. 2009. Climbing Mount Probable: Mutation as a cause of nonrandomness in evolution. *Journal of Heredity*, 100, 637–647.

Stork, A.L., N.D. Selby, R. Heyburn, and M.P. Searle. 2008. Accurate relative earthquake hypocenters reveal structure of the Burma subduction zone. *Bulletin of the Seismological Society of America*, 98, 2815–2827.

Storti, F., R.E. Holdsworth, and F. Salvini. 2003. Intraplate strike-slip deformation belts. In F. Storti, R.E. Holdsworth and F. Salvini (eds.), *Intraplate strike-slip deformation belts. Geological Society of London Special Publication*, 210, 1–14.

Stuart, W.D., G.R. Foulger, and M. Barall. 2007. Propagation of the Hawaiian-Emperor volcano chain by Pacific plate cooling stress. *Geological Society of America Special Paper*, 430, 497–506.

Stuessy, T.F., T. Sang, and M. De Vore. 1996. Phylogeny and biogeography of the subfamily Barnadesioideae with implications for early evolution of the Compositae. In D.J.N. Hind and H.J. Beentje (eds.), *Compositae: Systematics.* Proceedings of the International Compositae Conference. Kew, 1994, Vol. 1 (pp. 463–490). Royal Botanic Gardens, Kew.

Sturge, R.J., F. Jacobsen, B.B. Rosensteel, R.J. Neale, and K.E. Omland. 2009. Colonization of South America from Caribbean Islands confirmed by molecular phylogeny with increased taxon sampling. *The Condor*, 111, 575–579.

Sun, W., X. Ding, Y.-H. Hu, and X.-H. Li. 2007. The golden transformation of the Cretaceous plate subduction in the west Pacific. *Earth and Planetary Science Letters*, 262, 533–542.

Swenson, U., J. Munzinger, and I.V. Bartish. 2007. Molecular phylogeny of *Planchonella* and eight new species from New Caledonia. *Taxon*, 56, 329–354.

Sykes, W.R. 1998. *Scaevola gracilis* (Goodeniaceae) in the Kermadec Islands and Tonga. *New Zealand Journal of Botany*, 36, 671–674.

Szalay, F.S., and E. Delson. 1979. *Evolutionary history of the Primates.* Academic Press, New York.

Tabuce, R., M. Mahboubi, P. Tafforeau, and J. Sudre. 2004. Discovery of a highly-specialized plesiadapiform primate in the early-middle Eocene of northwestern Africa. *Journal of Human Evolution*, 47, 305–321.

Tabuce, R., L. Marivaux, R. Lebrun, M. Adaci, M. Bensalah, P.-H. Fabre, E. Fara, H. Gomes Rodrigues, L. Hautier, J.-J. Jaeger, V. Lazzari, F. Mebrouk, S. Peigné, J. Sudre, P. Tafforeau, X. Valentin, and M. Mahboubi. 2009. Anthropoid *versus* strepsirhine status of the African Eocene primates *Algeripithecus* and *Azibius*: Craniodental evidence. *Proceedings of the Royal Society B*, 276, 4087–4094.

Taiti, S., M. Arnedo, S.E. Lew, and G.K. Roderick. 2003. Evolution of terrestriality in Hawaiian species of the genus *Ligia* (Isopoda, Oniscidea). *Crustaceana Monographs*, 2, 85–102.

Takacs, Z., J.C. Morales, T. Geissmann, and D.J. Melnick. 2005. A complete species-level phylogeny of the Hylobatidae based on mitochondrial $ND3$-$ND4$ gene sequences. *Molecular Phylogenetics and Evolution*, 36, 456–467.

Tanaka, R., A. Makishima, and E. Nakamura. 2008. Hawaiian double volcanic chain triggered by an episodic involvement of recycled material: constraints from temporal Sr-Nd-Hf-Pb isotopic trend of the Loa-type volcanoes. *Earth and Planetary Science Letters*, 365, 450–465.

Tanaka-Ueno, T., M. Matsui, S.L. Chen, O. Takenaka, and H. Ota. 1998. Phylogenetic relationships of brown frogs from Taiwan and Japan assessed by mitochondrial cytochrome *b* gene sequences (*Rana*: Ranidae). *Zoological Science*, 15, 283–288.

Tanner, L.H., S.G. Lucas, and M.G. Chapman. 2004. Assessing the record and causes of Late Triassic extinctions. *Earth-Science Reviews*, 65, 103–139.

Tattersall, I. 2007. Madagascar's lemurs: cryptic diversity or taxonomic inflation? *Evolutionary Anthropology*, 16, 12–23.

Tattersall, I. 2008. Vicariance vs. dispersal in the origin of the Malagasy mammal fauna. In J.G. Fleagle and C.C. Gilbert (eds.), *Elwyn Simons: A search for origins* (pp. 397–408). Springer, New York.

Taylor, B. 2006. The single largest oceanic plateau: Ontong Java–Manihiki–Hikurangi. *Earth and Plantary Science Letters*, 241, 372–380.

Taylor, E.B., J.W. Boughman, M. Groenenboom, M. Sniatynski, D. Schluter, and J.L. Gow. 2006. Speciation in reverse: morphological and genetic evidence of the collapse of a three-spined stickleback (*Gasterosteus aculeatus*) species pair. *Molecular Ecology*, 15, 343–355.

Taylor, E.B., and J.D. McPhail. 2000. Historical contingency and ecological determinism interact to prime speciation in sticklebacks, *Gasterosteus*. *Proceedings of the Royal Society of London B*, 267, 2375–2384.

Taylor, G.K., J. Gascoyne, and H. Colley. 2000. Rapid rotation of Fiji: Paleomagnetic evidence and tectonic implications. *Journal of Geophysical Research*, 105 B3, 5771–5781.

Teixell, A., M.-L. Arboleya, and M. Julivert. 2003. Tectonic shortening and topography in the central High Atlas. *Tectonics*, 22(5), 1051. doi:10.1029/2002TC001460.

Tello Saenz, C.A., P.C. Hackspacher, J.C. Hadler Neto, P.J. Iunes, S. Guedes, L.F.B. Ribeiro, and S.R. Paulo. 2003. Recognition of Cretaceous, Paleocene, and Neogene tectonic reactivation through apatite fission-track analysis in Precambrian areas of southeast Brazil: Association with the opening of the South Atlantic Ocean. *Journal of South American Earth Sciences*, 15, 765–774.

Templeton, A.R. 1980. The theory of speciation *via* the founder principle. *Genetics*, 94, 1011–1038.

Templeton, A.R. 1998. Nested clade analyses of phylogeographic data: Testing hypotheses about gene flow and population history. *Molecular Ecology*, 7, 381–397.

Templeton, A.R. 2008. The reality and importance of founder speciation in evolution. *BioEssays*, 30, 470–479.

Terrell, E.E., H.E. Robinson, W.L. Wagner, and D.H. Lorence. 2005. Resurrection of genus *Kadua* for Hawaiian Hedyotinae (Rubiaceae), with emphasis on seed and fruit characters and notes on South Pacific species. *Systematic Botany*, 30, 818–833.

Ter Steege, H., D. Sabatier, H. Castellanos, T. van Andel, J. Duivenvoorden, A. Adalardo de Oliveira, R. Ek, R. Lilwah, P. Maas, and S. Mori. 2000. An analysis of the floristic composition and diversity of Amazonian forests including those of the Guiana Shield. *Journal of Tropical Ecology*, 16, 801–828.

Thacker, C.E., and M.A. Hardman. 2005. Molecular phylogeny of basal gobioid fishes: Rhyacichthyidae, Odontobutidae, Xenisthmidae, Eleotridae (Teleostei: Perciformes: Gobioidei). *Molecular Phylogenetics and Evolution*, 37, 858–871.

Thinh, V.N., A.R. Mootnick, T. Geissmann, M. Li, T. Ziegler, and M. Agil, P. Moisson, T. Nadler, L. Walter, and C. Roos. 2010. Mitochondrial evidence for multiple radiations in the evolutionary history of small apes. *BMC Evolutionary Biology*, 10, 74.

Thiv, M., T. van der Niet, F. Rutschmann, M. Thulin, T. Brune, and H.P. Linder. 2011. Old–New World and trans-African disjunctions of *Thamnosma* (Rutaceae): Intercontinental long-distance dispersal and local differentiation in the succulent biome. *American Journal of Botany*, 98, 76–87.

Thomas, G.H., C.D.L. Orme, R.G. Davies, V.A. Olson, P.M. Bennett, K.J. Gaston, I.P.F. Owens, and T.M. Blackburn. 2008. Regional variation in the historical components of global avian species richness. *Global Ecology and Biogeography*, 17, 340–351.

Thomas, J.D. 2006. Marine hotspots revisited. *EOS, Transactions, American Geophysical Union* 87(36, Suppl.).

Thomas, R.H., and J.A. Hunt. 1991. The molecular evolution of the alcohol dehydrogenase locus and the phylogeny of Hawaiian *Drosophila*. *Molecular Biology and Evolution*, 8, 687–702.

Thomasson, M., and G. Thomasson. 1991. Essai sur la flore du Sud-Ouest malgache: Originalité, affinités et origines. *Bulletin du Muséum National d'Histoire Naturelle, Paris*, ser. 4, 13, 71–89.

Thompson, R.N., S.A. Gibson, J.G. Mitchell, A.P. Dickin, O.H. Leonardos, J.A. Brod, and J.C. Greenwood. 1998. Migrating Cretaceous-Eocene magmatism in the Serra do Mar alkaline province, SE Brazil: Melts from the deflected Trindade mantle plume? *Journal of Petrology*, 39, 1493–1526.

Thornton, I.W.B., and T.R. New. 2007. *Island colonization: The origin and development of island communities*. Cambridge University Press, Cambridge.

Ting, N. 2008. Mitochondrial relationships and divergence dates of the African colobines: Evidence of Miocene origins for the living colobus monkeys. *Journal of Human Evolution*, 55, 312–325.

Ting, N., A.J. Tosi, Y. Li, Y.-P. Zhang, and T.R. Disotell. 2008. Phylogenetic incongruence between nuclear and mitochondrial markers in the Asian colobines and the evolution of the langur and leaf monkeys. *Molecular Phylogenetics and Evolution*, 46, 466–474.

Tokeshi, M. 1999. *Species coexistence: Ecological and evolutionary perspectives.* Blackwell, Oxford.

Tosi, A.J. 2008. Forest monkeys and Pleistocene refugia: A phylogeographic window onto the disjunct distribution of the *Chlorocebus lhoesti* species group. *Zoological Journal of the Linnean Society*, 154, 408–418.

Tosi, A.J., D.J. Melnick, and T.R. Disotell. 2004. Sex chromososme phylogenetics indicate a single transition to terrestriality in the guenons (tribe Cercopithecini). *Journal of Human Evolution*, 46, 223–237.

Tosi, A.J., J.C. Morales, and D.J. Melnick. 2003. Paternal, maternal, and biparental molecular markers provide unique windows onto the evolutionary history of macaque monkeys. *Evolution*, 57, 1419–1435.

Townsend, T.M., K.A. Tolley, F. Glaw, W. Böhme, and M. Vences. 2011. Eastward from Africa: Palaeocurrent-mediated chameleon dispersal to the Seychelles islands. *Biology Letters*, 7, 225–228.

Travis, J., and D.N. Reznick. 2009. Adaptation. In M. Ruse and J. Travis (eds.), *Evolution: The first four billion years* (pp. 105–131). Harvard University Press, Cambridge, MA.

Trewick, S.A., and G.P. Wallis. 2001. Bridging the "beech-gap": New Zealand invertebrate phylogeography implicates Pleistocene glaciations and Pliocene isolation. *Evolution*, 55, 2170–2180.

Trinder-Smith, T., H.P. Linder, T. van der Niet, A.G. Verboom, and T.L. Nowell. 2007. Plastid DNA sequences reveal generic paraphyly within Diosmeae (Rutoideae, Rutaceae). *Systematic Botany*, 32, 847–855.

Tronchet, F., and G.M. Plunkett, J. Jérémie, and P.P. Lowry II. 2005. Monophyly and major clades of *Meryta* (Araliaceae). *Systematic Botany*, 30, 657–670.

Trusty, J.L., R.G. Olmstead, D.J. Bogler, A. Santos-Guerra, and J. Francisco-Ortega. 2004. Using molecular data to test a biogeographic connection of the Macaronesian genus *Bystropogon* (Lamiaceae) to the New World: A case of conflicting phylogenies. *Systematic Botany*, 29, 702–715.

Tsang, L.M., T.-Y. Chan, M.K. Cheung, and K.H. Chu. 2009. Molecular evidence for the southern hemisphere origin and deep sea diversification of spiny lobsters (Crustacea: Decapoda: Palinuridae). *Molecular Phylogenetics and Evolution*, 51, 304–311.

Turelli, M., N.H. Barton, and J.A. Coyne. 2001. Theory and speciation. *Trends in Ecology and Evolution*, 16, 330–42.

Turk, F.A. 1964. Form, size, macromutation and orthogenesis in the Arachnida: An essay. *Annals of the Natal Museum*, 16, 236–255.

Umhoefer, P.J. 2003. A model for the North America Cordillera in the early Cretaceous: Tectonic escape related to arc collision of the Guerrero terrane and a change in North America plate motion. In S.E. Johnson, S.R. Paterson,

J.M. Fletcher, G.H. Girty, D.L. Kimbrough, and A. Martín-Barajas (eds.), *Tectonic evolution of northwestern México and the southwestern USA. Geological Society of America Special Paper*, 374, 117–134.

Utsunomiya, A., N. Suzuki, and T. Ota. 2008. Preserved paleo-oceanic plateaus in accretionary complexes: implications for the contributions of the Pacific superplume to global environmental change. *Gondwana Research*, 14, 115–125.

Valente, L.M., G. Reeves, J. Schnitzler, I. Pizer Mason, M.F. Fay, T.G. Rebelo, M.W. Chase, and T.G. Barraclough. 2010. Diversification of the African genus *Protea* (Proteaceae) in the Cape biodiversity hotspot and beyond: Equal rates in different biomes. *Evolution*, 64, 745–760.

Vallejo, C., R.A. Spikings, L. Luzieux, W. Winkler, D. Chew, and L. Page. 2006. The early interaction between the Caribbean Plateau and the NW South American Plate. *Terra Nova*, 18, 264–269.

van Balgooy, M.M.J. 1966. Osteomeles. *Pacific Plant Areas*, 5, 294–295.

van Balgooy, M.M.J., and Ding Hou. 1966. Perrottetia. *Pacific Plant Areas*, 5, 94–95.

van Balgooy, M.M.J., P.H. Hovenkamp, and P.C. van Welzen. 1996. Phytogeography of the Pacific: Floristic and historical distribution patterns in plants. In A. Keast and S.E. Miller (eds.), *The origin and evolution of Pacific Island biotas, New Guinea to Polynesia: Patterns and processes* (pp. 191–213). SPB Academic, Amsterdam.

Van der Ham, R., and B.J. van Heuven. 2002. Evolutionary trends in Winteraceae pollen. *Grana*, 41, 4–9.

Vander Kloet, S.P. 1996. Taxonomy of *Vaccinium* sect. *Macropelma*. *Systematic Botany*, 21, 355–364.

van der Meijden, R. 1975. Gunnera L. *Pacific Plant Areas*, 3, 258.

van der Pluijm, B.A., and S. Marshak, 2004. *Earth structure: An introduction to structural geology and tectonics*. 2nd ed. Norton, New York.

vander Velde, N. 2009. Marshall Islands. In R.G. Gillespie and D.A. Clague (eds.), *Encyclopedia of islands* (pp. 610–612). University of California Press, Berkeley.

VanderWerf, E.A. 2007. Biogeography of 'elepaio: Evidence from inter-island song playbacks. *Wilson Journal of Ornithology*, 119, 325–333.

VanderWerf, E.A., L.C. Young, N.W. Yeung, and D.B. Carlon. 2010. Stepping stone speciation in Hawaii's flycatchers: Molecular divergence supports new island endemics within the elepaio. *Conservation Genetics*, 11, 1283–1298.

Vane-Wright, R.I., and R. de Jong. 2003. The butterflies of Sulawesi: annotated checklist for a critical island fauna. *Zoologische Verhandelingen* (Leiden), 343, 3–267.

van Roosmalen, M.G.M., and T. van Roosmalen. 2003. The description of a new marmoset genus, *Callibella* (Callitrichinae, Primates), including its molecular phylogenetic status. *Neotropical Primates*, 11, 1–10.

van Roosmalen, M.G.M., T. van Roosmalen, and R.A. Mittermeier. 2002. A taxonomic review of the Titi monkeys, genus *Callicebus* Thomas, 1903, with the description of two new species, *Callicebus bernhardi* and *Callicebus stephennashi*, from Brazilian Amazonia. *Neotropical Primates*, 10(Suppl.), 1–11.

van Steenis, C.G.G.J.. 1936. On the origin of the Malaysian mountain flora, Part 3: Analysis of floristic relationships (1st installment). *Bulletin du Jardin Botanique de Buitenzorg*, III, 14, 56–72.

Vargas, P., B.G. Baldwin, and L. Constance. 1999. A phylogenetic study of *Sanicula* sect. *Sanicoria* and *S.* sect. *Sandwicenses* (Apiaceae) based on nuclear rDNA and morphological data. *Systematic Botany*, 24, 228–248.

Vargas-Fernández, I., M. Trujano, J. Llorente-Bousquets, and A. Luis-Martínez. 2006. Patrones de distribución de las subfamilias Ithomiinae, Morphinae y Charaxinae (Lepidopotera: Nymphalidae). In J.J. Morrone and J. Llorente-Bousquets (eds.), *Componentes bióticos principales de la entomofauna Mexicana*, Vol. 2 (pp. 867–943). Universidad Nacional Autónoma de México, Mexico City.

Veevers, J.J. 2004. Gondwanaland from 650–500 Ma assembly through 320 Ma merger in Pangea to 185–100 Ma breakup: Supercontinental tectonics via stratigraphy and radiometric dating. *Earth-Science Reviews*, 68, 1–132.

Velazco, P.M., and B.D. Patterson. 2008. Phylogenetics and biogeography of the broad-nosed bats, genus *Platyrrhinus* (Chiroptera: Phyllostomidae). *Molecular Phylogenetics and Evolution*, 49, 749–759.

Vences, M., J. Kosuch, F. Glaw, and M. Veith. 2003. Molecular phylogeny of hyperoliid tree frogs: Biogeographic origin of Malagasy and Seychellean taxa and reanalysis of familial paraphyly. *Journal of Zoological Systematics and Evolutionary Research*, 41, 205–215.

Venditti, C., A. Meade, and M. Pagel. 2010. Phylogenies reveal new interpretation of speciation and the Red Queen. *Nature*, 463, 349–352.

Voje, K.L., C. Hemp, O. Flagstad, G.-P. Sætre, and N.C. Stenseth. 2009. Climatic change as an engine for speciation in flightless Orthoptera species inhabiting African mountains. *Molecular Ecology*, 18, 93–108.

Vonhof, H.B., F.P. Wesselingh, R.J.G. Kaandorp, G.R. Davies, J.E. van Hinte, J. Guerrero, M. Räsänen, L. Romero-Pittman, and A. Ranzi. 2003. Paleogeography of Miocene Western Amazonia: Isotopic composition of molluscan shells constrains the influence of marine incursions. *Bulletin of the Geological Society of America*, 115, 983–993.

Voss, R.S., and L.H. Emmons. 1996. Mammalian diversity in neotropical lowland rainforests: A preliminary assessment. *Bulletin of the American Museum of Natural History*, 230, 1–115.

Wade, C.M., P.B. Mordan, and F. Naggs. 2006. Evolutionary relationships among the pulmonate land snails and slugs (Pulmonata, Stylommatophora). *Biological Journal of the Linnean Society*, 87, 593–610.

Wagner W.L., and V.A. Funk. (eds.). 1995. *Hawaiian biogeography: Evolution on a hot spot archipelago*. Smithsonian Institution Press, Washington, DC.

Wagner, W.L., D.R. Herbst, and D.H. Lorence. 2009. *Flora of the Hawaiian Islands*. Available at http://botany.si.edu/Pacificislandbiodiversity/hawaiianflora/index.htm.

Wagner, W.L., D.R. Herbst, and S.H. Sohmer. 1990. *Manual of the flowering plants of Hawaii*. University of Hawaii Press, Honolulu.

Wagner, W.L., and H. Robinson. 2001. *Lipochaeta* and *Melanthera* (Asteraceae: Heliantheae subtribe Ecliptinae): Establishing their natural limits and a synopsis. *Brittonia*, 53, 539–561.

Wagstaff, S.J., I. Breitwieser, and U. Swenson. 2006. Origin and relationships of the austral genus *Abrotanella* (Asteraceae) inferred from DNA sequences. *Taxon*, 55, 95–106.

Wahlberg, N., and A.V.L. Freitas. 2007. Colonization and radiation in South America by butterflies in the subtribe Phyciodina (Lepidoptera: Nymphalidae). *Molecular Phylogenetics and Evolution*, 44, 1257–1272.

Wall, W., N. Douglas, Q.-Y. Xiang, W. Hoffmann, T. Wentworth, and M. Hohmann, M. 2010. Do endemic plants migrate? The phylogeography of *Pyxidanthera barbulata* (Diapensiaceae) as a test of the Pleistocene refugial paradigm. Abstract, Botany Conference 2010. http://2010.botanyconference.org/

Wallace, A.R. 1860. On the zoological geography of the Malay Archipelago. *Proceedings of the Linnean Society, Zoology, London*, 4, 173–184.

Wallace, A.R. [1869] 1962. *The Malay archipelago: The land of the orang-utan and the bird of paradise*. Dover, New York.

Wallace, A.R. 1876. *The geographical distribution of animals*. Macmillan, London.

Wallace, A.R. [1881] 1998. *Island life*. Prometheus, New York.

Wallander, E., and V.A. Albert. 2000. Phylogeny and classification of Oleaceae, based on *rps16* and *trn*L-F sequence data. *American Journal of Botany*, 87, 1827–1841.

Wallis, G.P., and S.A. Trewick. 2001. Finding fault with vicariance: A critique of Heads (1998). *Systematic Biology*, 50, 602–609.

Walsh, H.E., I.L. Jones, and V.L. Friesen. 2005. A test of founder effect speciation using multiple loci in the Auklets (*Aethia* spp.). *Genetics*, 171, 1–10.

Wang, C., X. Zhao, Z. Liu, P.C. Lippert, S.A. Graham, R.S. Coe, H. Yi, L. Zhu, S. Liu, and Y. Li. 2008. Constraints on the early uplift history of the Tibet Plateau. *Proceedings of the National Academy of Sciences USA*, 105, 4987–4992.

Wangchuk, T., D.W. Inouye, and M.P. Hare. 2008. The emergence of an endangered species: Evolution and phylogeny of the *Trachypithecus geei* of Bhutan. *International Journal of Primatology*, 29, 565–582.

Wanntorp, L., H.E. Wanntorp, and M. Källersjö. 2002. Phylogenetic relationships of *Gunnera* based on nuclear ribosomal DNA ITS Region, *rbcL* and *rps16* intron sequences. *Systematic Botany*, 27, 512–521.

Warren, B.H., E. Bermingham, R.C.K. Bowie, R.P. Prys-Jones, and C. Thébaud. 2003. Molecular phylogeography reveals island colonization history and diversification of western Indian Ocean sunbirds (*Nectarinia*: Nectariniidae). *Molecular Phylogenetics and Evolution*, 29, 67–85.

Warren, B.H., E. Bermingham, R.P. Prys-Jones, and C. Thébaud. 2006. Immigration, species radiation and extinction in a highly diverse songbird lineage: White-eyes on Indian Ocean islands. *Molecular Ecology*, 15, 3769–3786.

Warren, B.H., D. Strasberg, J.H. Bruggemann, R.P. Prys-Jones, and C. Thébaud. 2010. Why does the biota of the Madagascar region have such a strong Asiatic flavour? *Cladistics*, 26, 526–538.

Watkeys, M.K. 2002. Development of the Lebombo rifted volcanic margin of southeast Africa. *Geological Society of America Special Paper*, 362, 27–46.

Watts, A.B., D.T. Sandwell, W.H.F. Smith, and P. Wessel. 2006. Global gravity, bathymetry, and the distribution of submarine volcanism through space and time. *Journal of Geophysical Research*, 111(B08408). doi:10.1029/2005JB004083.

Weber, A., and L.E. Skog. 2007. *The genera of Gesneriaceae. Basic information with illustration of selected species.* 2nd ed. www.genera-gesneriaceae.at.

Wegener, A. [1912] 2002. The origins of continents. Geol. Rundsch. 3:276–292. Translation. *International Journal of Earth Sciences*, 91, S4–S17.

Wegener, A. 1915. *Die Entstehung der Kontinente und Ozeane.* Vieweg, Braunschweig.

Weir, J.T., and Schluter, D. 2008. Calibrating the avian molecular clock. *Molecular Ecology*, 17, 2321–2328.

Wells, R.E. 2007. Reconsidering the origin and emplacement of Siletzia. 103rd Annual Meeting of the Geological Society of America (4–6 May 2007). Paper no. 10-7.

Wen, J., and S.M. Ickert-Bond. 2009. Evolution of the Madrean-Tethyan disjunctions and the North and South American amphitropical disjunctions in plants. *Journal of Systematics and Evolution*, 47, 331–348.

Werner, R., and K. Hoernle. 2003. New volcanological and volatile data provide strong evidence for the continuous existence of Galápagos Islands over the past 17 million years. *International Journal of Earth Science*, 92, 904–911.

Wesener, T., M.J. Raupach, and P. Sierwald, P. 2010. The origins of the giant pill-millipedes from Madagascar (Diplopoda: Sphaerotheriida: Arthrosphaeridae). *Molecular Phylogenetics and Evolution*, 57, 1184–1193.

Wesener, T., and D. VandenSpiegel. 2009. A first phylogenetic analysis of giant pill-millipedes (Diplopoda: Sphaerotheriida), a new model Gondwanan taxon, with special emphasis on island gigantism. *Cladistics*, 25, 545–573.

Wessell, P. 2009. Seamounts, geology. In R.G. Gillespie and D.A. Clague (eds.), *Encyclopedia of islands* (pp. 821–825). University of California Press, Berkeley.

Westgate, J.W. 2008. Vertebrates from a middle Eocene estuarine mangrove community in the Rio Grande embayment. *Abstracts of the Geological Society of America 2008 Conference.* Available at http://gsa.confex.com/gsa/2008SC/finalprogram/abstract_136190.htm.

Westgate, J.W., and A. Salazar. 1999. *After the dinosaurs: A Texas tropical paradise recovered at Lake Casa Blanca.* University of Texas Press, Austin.

White, B.N. 1986. The isthmian link, antitropicality and American biogeography: Distributional history of the Atherinopsinae (Pisces: Atherinidae). *Systematic Zoology*, 35, 176–194.

White, G. [1789] 1977. *The natural history of Selborne.* Penguin, Harmondsworth.

White, R.V., and A.D. Saunders. 2004. Volcanism, impact and mass extinctions: Incredible or credible coincidences? *Lithos*, 70, 299–316.

Whitmore, T.C. 1973. Plate tectonics and some aspects of Pacific plant geography. *New Phytologist*, 72, 1185–1190.

Whitney, B.M., D.C. Oren, and R.T. Brumfield. 2004. A new species of *Thamnophilus* antshrike (Aves: Thamnophilidae) from the Serra do Divisor, Acre, Brazil. *The Auk*, 121, 1031–1039.

Whittaker, D.J. 2006. A conservation action plan for the Mentawai primates. *Primate Conservation*, 20, 95–105.

Whittaker, D.J., J.C. Morales, and D.J. Melnick. 2007. Resolution of the *Hylobates* phylogeny: Congruence of mitochondrial D-loop sequences with molecular, behavioral, and morphological data sets. *Molecular Phylogenetics and Evolution*, 45, 620–628.

Whittaker, D.J., N. Ting, and D.J. Melnick. 2006. Molecular phylogenetic affinities of the simakobu monkey (*Simias concolor*). *Molecular Phylogenetics and Evolution*, 39, 887–982.

Whittaker, R.J. 1998. *Island biogeography: Ecology, evolution, and conservation.* Oxford University Press, Oxford.

Whittaker, R.J., K.A. Triantis, and R.J. Ladle. 2010. A general dynamic theory of island biogeography: Extending the MacArthur-Wilson theory to accommodate the rise and fall of volcanic islands. In J.B. Losos and R.E. Ricklefs (eds.), *The theory of island biogeography revisited* (pp. 88–115). Princeton University Press, Princeton, NJ.

Wignall, P. 2005. The link between large igneous province eruptions and mass extinctions. *Elements*, 1, 293–297.

Wikström, N., M. Avino, S.G. Razafimandimbison, and B. Bremer. 2010. Historical biogeography of the Coffee family (Rubiaceae, Gentianales) in Madagascar: Case studies from the tribes Knoxieae, Naucleeae, Paederieae and Vanguerieae. *Journal of Biogeography*, 37, 1094–1113.

Wikström, N., V. Savolainen, and M.W. Chase. 2001. Evolution of the angiosperms: Calibrating the family tree. *Proceedings of the Royal Society of London B*, 268, 2211–2220.

Wildman, D.E., N.M. Jameson, J.C. Opazo, and S.V. Yi. 2009. A fully resolved genus level phylogeny of neotropical primates (Platyrrhini). *Molecular Phylogenetics and Evolution*, 53, 694–702.

Wilkinson, J.A., R.C. Drewes, and O.L. Tatum. 2002. A molecular phylogenetic analysis of the family Rhacophoridae with an emphasis on the Asian and African genera. *Molecular Phylogenetics and Evolution*, 24, 265–273.

Wilkinson, R.D., M.E. Steiper, C. Soligo, R.D. Martin, Z. Yang, and S. Tavaré. 2011. Dating primate divergences through an integrated analysis of palaeontological and molecular data. *Systematic Biology*, 60, 16–31.

Wilson, E.O. 1959. Adaptive shift and dispersal in a tropical ant fauna. *Evolution*, 13, 122–144.

Wilson, E.O. 2001a. Preface. In R.H. MacArthur and E.O. Wilson, *The theory of island biogeography* (pp. vii–ix). Princeton University Press, Princeton, NJ.

Wilson, E.O. 2001b. *The diversity of life.* Penguin, London.

Wilson, E.O. 2009. Foreword. In M. Ruse and J. Travis (eds.), *Evolution: The first four billion years* (pp. vii–viii). Harvard University Press, Cambridge, MA.

Wilson, E.O. 2010. Island biogeography in the 1960s: Theory and experiment. In J.B. Losos and R.E. Ricklefs (eds.), *The theory of island biogeography revisited* (pp. 1–13). Princeton University Press, Princeton, NJ.

Winkler, W., D. Seward, G.M.H. Ruiz, and N. Martin-Gombojav. 2005. The Andean Cordillera of Ecuador: Timing and mode of orogenic growth as revealed by sediments in the Amazon basin (heavy minerals and detrital zircon fission-tracks). Abstracts, 3rd Swiss Geoscience Meeting, Zürich (pp. 59–60).

Winkworth, R.C. 2009. Darwin and DNA: Explaining the New Zealand biota. *New Zealand Science Review*, 66, 93–94.

Wolfe, C.J., S.C. Solomon, G. Laske, J.A. Collins, R.S. Detrick, J.A. Orcutt, D. Bercovici, and E.H. Hauri. 2009. Mantle shear-wave velocity structure beneath the Hawaiian hot spot. *Science*, 326, 1388–1390.

Woodall, P.F. 2001. Family Alcedinidae (kingfishers). In J. del Hoyo, A. Elliott, and J. Sargatal (eds.), *Handbook of the birds of the worl* , Vol. 6 (pp. 130–249). Lynx Edicions, Barcelona.

World Wide Fund for Nature. 2009. Terrestrial ecoregions: Indochina mangroves. www.worldwildlife.org/wildworld/profiles.

Wörner, G., R.S. Harmon, and W. Wegner. 2009. Geochemical evolution of igneous rocks and changing magma sources during the formation and closure of the Central American land bridge of Panama. In S.M. Kay, V.A. Ramos, and W.R. Dickinson (eds.), *Backbone of the Americas: Shallow subduction, plateau uplift, and ridge and terrane collision. Geological Society of America Memoir*, 204, 183–196.

Worth, J.R.P., G.J. Jordan, J.R. Marthick, G.E. McKinnon, and R.E. Vaillancourt. 2010. Chloroplast evidence for geographic stasis of the Australian bird-dispersed shrub *Tasmannia lanceolata* (Winteraceae). *Molecular Ecology*, 19, 2949–2963.

Worthy, T.H. 2005. A new species of *Oxyura* (Aves: Anatidae) from the New Zealand Holocene. *Memoirs of the Queensland Museum*, 51, 259–275.

Wright, P.C. 1997. Behavioral and ecological comparisons of Neotropical and Malagasy Primates. In W.G. Kinzey (ed.), *New World primates* (pp. 127–142). Aldine Transaction, Edison, NJ.

Wright, P.C., E.L. Simons, and S. Gursky, S. 2006. Introduction. In P.C. Wright, E.L. Simons, and S. Gursky (eds.), *Tarsiers: Past, present and future* (pp. 1–6). Rutgers University Press, New Brunswick, NJ.

Wright, S.D., C.G. Yong, J.W. Dawson, D.J. Whittaker, and R.C. Gardner. 2000. Riding the ice age El Niño? Pacific biogeography and evolution of *Metrosideros* subg. *Metrosideros* (Myrtaceae) inferred from nuclear ribosomal DNA. *Proceedings of the National Academy of Sciences USA*, 97, 4118–4123.

Wright, S.D, C.G. Yong, S.R. Wichman, J.W. Dawson, and R.C. Gardner. 2001. Stepping stones to Hawaii: A trans-equatorial dispersal pathway for *Metrosideros* (Myrtaceae) inferred from nrDNA (ITS+ETS). *Journal of Biogeography*, 28, 769–774.

Wright, T.F., E.E. Schirtzinger, T. Matsumoto, J.R. Eberhard, G.R. Graves, J.J. Sanchez, S. Capelli, H. Müller, J. Scharpegge, G.K. Chambers, and R.C. Fleischer. 2008. A multilocus molecular phylogeny of the parrots (Psittaciformes): Support for a Gondwana origin during the Cretaceous. *Molecular Biology and Evolution*, 25, 2141–2156.

Xing, J., H. Wang, Y. Zhang, D.A. Ray, A.J. Tosi, T.R. Disotell, and M.A. Batzer. 2007. A mobile element-based evolutionary history of the guenons (tribe Cercopithecini). *BMC Biology*, 5(5), 1–10.

Yampolsky, L.Y., and A. Stoltzfus. 2001. Bias in the introduction of variation as an orienting factor in evolution. *Evolution and Development*, 3, 73–83.

Yang, S., J.G. Bishop, and M.S. Webster, M.S. 2008. Colonization genetics of an animal-dispersed plant (*Vaccinium membranaceum*) at Mount St. Helens, Washington. *Molecular Ecology*, 17, 731–740.

Yang, Y., C.W. Morden, L. Sack, M.J. Sporck, and P.E. Berry. 2009. Phylogeny and adaptive radiation of woody Hawaiian *Chamaesyce* from herbaceous and annual ancestors in subtropical North America (*Euphorbia*-Euphorbiaceae). Botany and Mycology 2009 (Snowbird, Utah, July 25–29) Abstract Book, p. 182 (abstract 649; also see http://2009.botany-conference.org/).

Yélamos, T. 1998. The *Aeletes* of the Hawaiian Islands (Coleoptera: Histeridae). *Bishop Museum Occasional Papers*, 54, 1–61.

Yeung, C.K.L., P.-W. Tsai, R.T. Chesser, R.C. Lin, C.-T. Yao, X.-H. Tian, and S.-H. Li. 2011. Testing founder effect speciation: divergence population genetics of the spoonbills *Platalea regia* and *Pl. minor* (Threskiornithidae, Aves). *Molecular Biology and Evolution*, 28, 473–482.

Yoder, A.D. 1997. Back to the future: A synthesis of strepsirrhine systematics. *Evolutionary Anthropology*, 6, 11–22.

Yoder, A.D., M.M. Burns, S. Zehr, T. Delefosse, G. Veron, S.M. Goodman, and J.J. Flynn. 2003. Single origin of Malagasy Carnivora from an African ancestor. *Nature*, 421, 734–737.

Yoder, A.D., and M.D. Nowak. 2006. Has vicariance or dispersal been the predominant biogeographic force in Madagascar? Only time will tell. *Annual Review of Ecology, Evolution and Systematics*, 37, 405–431.

Yoder, A.D., and Z. Yang. 2004. Divergence dates for Malagasy lemurs estimated from multiple gene loci: Geological and evolutionary context. *Molecular Ecology*, 13, 757–773.

Yuan, Y.-M., S. Wohlhauser, M. Möller, J. Klackenberg, M.W. Callmander, and P.K. Küpfer. 2005. Phylogeny and biogeography of *Exacum* (Gentianaceae): A disjunctive distribution in the Indian Ocean Basin resulted from long distance dispersal and extensive radiation. *Systematic Biology*, 54, 21–34.

Yumul, G.P., Jr., C.B. Dimalanta, R.A. Tamayo, Jr., R.C. Maury, H. Bellon, M. Polvé, V.B. Maglambayan, C.L. Querubin, and J. Cotton. 2004. Geology of the Zamboanga Peninsula, Mindanao, Philippines: An enigmatic South China continental fragment? In J. Malpas, C.J.N. Fletcher, J.R. Ali, and J.C. Aitchison (eds.), *Aspects of the tectonic evolution of China. Geological Society of London Special Publication*, 226, 289–312.

Zhang, K.J. 2000. Cretaceous paleogeography of Tibet and adjacent areas (China): Tectonic implications. *Cretaceous Research*, 21, 23–33.

Zhou, D., Y. Sun, H.-Z. Chen, H.-H. Xu, W.-Y. Wang, X. Pang, D.S. Cai, and D.-K. Hu. 2008. Mesozoic paleogeography and tectonic evolution of South China Sea and adjacent areas in the context of Tethyan and paleo-Pacific interconnections. *Island Arc*, 17, 186–207.

Zimmerman, E.C. 1940. Synopsis of the genera of Hawaiian Cossoninae with notes on their origin and distribution. *Occasional Papers of the Bernice P. Bishop Museum*, 15, 1.

Zimmerman, E.C. 1948. *Insects of Hawaii*. University of Hawaii Press, Honolulu.

Zinner, D., M.L. Arnold, and C. Roos. 2009. Is the new primate genus *Rungwecebus* a baboon? *PLoS One*, 4(3), e4859.

Zinner, D., L.F. Groeneveld, C. Keller, and C. Roos. 2009. Mitochondrial phylogeography of baboons (*Papio* spp.): Indication for introgressive hybridization? *BMC Evolutionary Biology*, 9, 83.

Zuccon, D., A. Cibois, E. Pasquet, and P.G.P. Ericson. 2006. Nuclear and mitochondrial sequence data reveal three major lineages of starlings, mynas and related taxa. *Molecular Phylogenetics and Evolution*, 41, 333–344.

Index

About the Author

Michael Heads was educated in New Zealand and has taught ecology and systematics at universities in Papua New Guinea, Fiji, Zimbabwe, and Ghana. He has also carried out field studies in the American tropics, mainly in Jamaica and Venezuela. His research interests are in tree architecture, biogeography, and the evolution of rainforest plants and animals.

Species and Systematics

COMPOSITION: MPS Limited, a Macmillan Company

TEXT: 10/13 Sabon

DISPLAY: Din

PRINTER AND BINDER: Sheridan Books